D1415547

BIOMONITORING
of
COASTAL WATERS
and
ESTUARIES

BIOMONITORING *of* COASTAL WATERS *and* ESTUARIES

Edited by
Kees J.M. Kramer
Senior Research Scientist
Laboratory for Applied Marine Research
Institute of Environmental Sciences - TNO
Den Helder, The Netherlands

CRC

CRC Press
Boca Raton Ann Arbor London Tokyo

Library of Congress Cataloging-in-Publication Data

Biomonitoring of coastal waters and estuaries / editor, Kees J. M. Kramer.
p. cm.
Includes bibliographical references and index.
ISBN 0-8493-4895-1
1. Marine organisms — Effect of water pollution on. 2. Environmental monitoring. 3. Indicators (biology) 4. Coastal ecology. 5. Kramer, Kees J. M.
QH545.W3B567 1994
574.5'2636—dc20 94-10065
for Library of Congress CIP

International Standard Book Number 0-8493-4895-1
Library of Congress Card Number 94-10065

Printed in the United States of America 4 5 6 7 8 9 0
Printed on acid-free paper

Preface

There is an increasing effort to protect the natural environment, including the marine waters. Estuaries and coastal waters are located in the territorial waters of the bordering countries and consequently under direct legal control and management. It is therefore not surprising that monitoring activities are carried out in these areas in order to support national and regional environmental policies.

Traditionally, in fact from the start of this century, monitoring was aiming at various physical parameters like temperature, currents and tides, and chemical constituents such as the salinity and nutrients. When it was realized in the 1960s that pollutants could affect mankind (and the ecosystem), further effort was made to detect inorganic and organic micropollutants in order to obtain information on their geographical distribution, transport mechanisms and fate. The first true monitoring for pollution started in the early 1970s, hampered (as we know now) by analytical methods that were still to be perfected. The compartments of interest were water, sediments and biota.

Since then, the monitoring of biota, or biomonitoring, has developed substantially. Not only were methods to detect pollutants in the organisms optimized (essentially the chemical monitoring of biota) and strategies developed, but biological effect parameters were defined and new methods continue to emerge.

The broad field of biological monitoring includes a wide variety of techniques. Passive and active biomonitoring, in which the tissues of suitable indicator species are analyzed for specific contaminants, provide information on the presence of contaminants over sometimes well defined periods of time and may identify contaminants present sporadically or at trace concentrations. Monitoring of physiological and/or detection of behavioral changes of organisms may be made automatically by on-line systems thus providing a valuable approach to monitoring of *e.g.* complex effluents. Laboratory or field based bioassays are used to determine the acute or chronic toxicity of a sample or a water body. Immunoassays can provide a rapid means of measuring chemical species, individually or as groups. Biomarkers, proteins and enzymes whose activity in living organisms is induced by exposure of the organism to toxic contaminants, can also be utilised for their sensitivity and ability to identify the occurrence of (broad) categories of contaminants. Biological or ecological surveys, which assess the presence or abundance of a particular indicator species or use some form of diversity index, can provide a valuable indicator of changing conditions. Where there is a discrepancy between the observed biological data and that which might be expected from knowledge of the physico-chemical conditions, a warning of environmental degradation resulting from sporadic contamination can be made.

The lack of a text that comprehensively reviews these new developments in chemical and biological monitoring of biota applied to the coastal and estuarine environments led to the start of this volume.

The scope for this book only includes adverse (man-induced) effects upon the ecosystem. Those techniques where field trials have proven their practical applicability to the monitoring of aquatic systems are included. The book should be of value to both the experienced

scientist and the graduate student alike, both in western societies as well as developing countries.
The author of each chapter was asked to include a short review of their field, some theoretical background information of the method(s) and to illustrate their topic with clear examples of applications. It was intended that these 'case studies' were to cover examples from all over the world, from both temperate and tropical regions in order to illustrate that most techniques have a wide applicability. Due to lack of space, detailed technical information is usually not provided in publications in scientific journals. For this reason I urged each author to provide a technical appendix describing essential information about the practical aspects (chemicals, logistics, instrumentation, statistics, etc.).
I am extremely happy that so many authors, all of them acknowledged specialists in their respective fields, have agreed to prepare a chapter, despite these boundary conditions, and I'm grateful for their willing acceptance and cooperation.
Shortly before this volume was sent to the publisher, it became known that one respected and productive colleague, dr. Geoffrey W. Bryan of Plymouth Marine Laboratory, had suddenly passed away, just before his retirement. It is with grief that we realize that he did not live long enough to see the realization of this volume.
For me it was a pleasure to edit this volume, but without the support of many, colleagues and relatives, it would have been a much more difficult task. In particular I must thank my colleagues Tim Bowmer, Pim de Kock, Bas Looyse, Martin Scholten and Nicoline Wrisberg for fruitful discussions and comments on manuscripts. I am especially grateful for the support of the staff of the TNO Electronic Publishing Department, in particular Arie Boer for his art-work and above all Jeanette van der Kruk whose endless DTP sessions led to the present lay out. The TNO Institute of Environmental Sciences is acknowledged for allowing me to use their services. Many thanks are due to Máira Maguire who corrected the english of many manuscripts.
Certainly, last but not least, I must particularly thank my partner Dien Geertsma, who endured for many endless evenings and week-ends over the past year my 'virtual' absence while working on this volume.

Kees Kramer

Bergen
January, 1994

to Dien

The Editor

Kees J.M. Kramer, Ph.D. (1950) graduated from the University of Amsterdam, The Netherlands (*cum laude*) in inorganic and environmental chemistry and chemical oceanography in 1975. In the same year he joined the Netherlands Institute for Sea Research at Texel where he worked on the chemical speciation of trace metals in the marine environment. In 1985 he received his Ph.D. from the University of Groningen, The Netherlands on a chemical speciation subject. From 1986 until the present he has been employed by the Institute for Environmental Sciences of the Netherlands Organisation for Applied Scientific Research (IMW-TNO), at present as a senior scientist at the Departments of Biology and Analytical Chemistry.
Dr. Kramer is a member of the Royal Netherlands Chemical Society, the International Commission on Bioindicators of the International Union of Biological Sciences and the Society for the History of Natural History. He is an adviser to the Netherlands Scientific Committee on Chemical Oceanography and secretary of the Dutch Working Group on Marine Monitoring.
Kees Kramer has written over 100 scientific papers, book chapters and reports; he has edited three books. He has served in the editorial boards of *Marine Chemistry*, the *Netherlands Journal for Sea Research* and *Chemical Speciation and Bioavailability*.
During the last decade he has been active in various training as well as educational programmes and demonstration projects on (bio)monitoring in various European countries. He has worked for shorter periods in Indonesia, India and Hong Kong on specialist missions.
His present research interests relate to the behaviour, transport and fate of trace metals in the marine environment through studying the compartments water, sediment and biota. Chemical speciation and the effects of different chemical species distributions upon biota has always been his special interest. Over the years he has concentrated on monitoring strategies for the marine environment, in particular working on biological early warning systems and the applications of chemical monitoring of biota. The quality control of environmental analyses, including aspects such as sampling, sample treatment and instrumental analysis, has received special attention recently.
Kees Kramer is an active collector of antiquarian books on oceanography, specialising in the earlier oceanographic expeditions.

List of Contributors

Tore Aunaas
SINTEF Applied Chemistry
Environmental Technology Group
Gryta 2
Brattøra
7034 Trondheim
Norway

Ian G. Baldwin
Water Research Centre
Henley Road
Medmenham
Marlow, Bucks. SL7 2HD
England

Geoffrey W. Bryan†
Plymouth Marine Laboratory
Citadel Hill
Plymouth, Devon PL1 2PB
England

Adriana Y. Cantillo
Coastal Monitoring Branch
National Oceanic and Atmospheric Administration
N/ORCA21
1305 East West Highway
Silver Spring, MD 20910
USA

Kun-Hsiung Chang
Institute of Zoology
Academia Sinica
Taipei
Taiwan, Republic of China

Chang-Feng Dai
Institute of Oceanography
National Taiwan University
Taipei
Taiwan, Republic of China

Willem C. de Kock
Laboratory for Applied Marine Research,
IMW-TNO
P.O. Box 57
1780 AB Den Helder
The Netherlands

Ab de Zwaan
Netherlands Institute of Ecology
Centre for Estuarine and Coastal Ecology
Vierstraat 28
4401 EA Yerseke
The Netherlands

Richard H.M. Eertman
Netherlands Institute of Ecology
Centre for Estuarine and Coastal Ecology
Vierstraat 28
4401 EA Yerseke
The Netherlands

Lars Förlin
Department of Zoophysiology
University of Göteborg
Medicinaregatan 18
413 90 Göteborg
Sweden

Stephen G. George
NERC Unit of Aquatic Biochemistry
University of Stirling
Stirling FK9 4LA
Scotland

Peter E. Gibbs
Plymouth Marine Laboratory
Citadel Hill
Plymouth, Devon PL1 2PB
England

Vadim M. Glaser
Department of Biology
Moscow State University
119899 Moscow
Russia

Anders Goksøyr
Laboratory of Marine Molecular Biology
University of Bergen, HIB
5020 Bergen
Norway

Astrid-Mette Husøy
Laboratory of Marine Molecular Biology
University of Bergen, HIB
5020 Bergen
Norway

Rong-Quen Jan
Institute of Zoology
Academia Sinica
Taipei
Taiwan, Republic of China

David W. Jeffrey
School of Botany
Trinity College
Dublin 2
Ireland

Sergey V. Kotelevtsev
Department of Biology
Moscow State University
119899 Moscow
Russia

Kees J.M. Kramer
Laboratory for Applied Marine Research
IMW-TNO
P.O. Box 57
1780 AB Den Helder
The Netherlands

Gunnar G. Lauenstein
Coastal Monitoring Branch
National Oceanic and Atmospheric
Administration
N/ORCA21
1305 East West Highway
Silver Spring, MD 20910
USA

Thomas P. O'Connor
Coastal Monitoring Branch
National Oceanic and Atmospheric
Administration
N/ORCA21
1305 East West Highway
Silver Spring, MD 20910
USA

Per-Erik Olsson
Department of Zoophysiology
University of Göteborg
Medicinaregatan 18
413 90 Göteborg
Sweden

David J.H. Phillips
Acer Environmental
Units 1-8, Howard Court
Manor Park
Daresbury, Cheshire WA7 1SJ
England

Aad C. Smaal
National Institute for Coastal and
Marine Management, RIKZ
P.O. Box 8039
4330 AK Middelburg
The Netherlands

Ludmila I. Stepanova
Department of Biology
Moscow State University
119899 Moscow
Russia

John Widdows
Plymouth Marine Laboratory
Prospect Place
Plymouth, Devon PL1 3DH
England

James G. Wilson
Environmental Science Unit
Trinity College
Dublin 2
Ireland

Paul P. Yevich and **Carolin A. Yevich**
retired: Environmental Research Laboratory
US Environmental Protection Agency
Narragansett, RI 02882
USA

Karl Erik Zachariassen
ALLFORSK Avh
Department of Ecotoxicology
University of Trondheim
7055 Dragvoll
Norway

Table of Contents

Chapter 1

Biological Early Warning Systems (BEWS)

Ian G. Baldwin[1] and **Kees J.M. Kramer**[2]

[1] Water Research Centre, Henley Road, Medmenham, Marlow, Bucks., SL7 2HD, England
[2] Laboratory for Applied Marine Research, IMW-TNO, P.O. Box 57, 1780 AB Den Helder, The Netherlands

ABSTRACT

In contrast to all other methods of biomonitoring presented in this volume, Biological Early Warning Systems, BEWS, aim to give a rapid response (alarm) to the occurrence of toxic pollutants within a period of minutes to one hour. The applications therefore offer typical examples that are the result of this time-scale. The examples provided in the chapter are, for the most part, capable of responding within this time-scale.
The basic principle of all is that of monitoring some function of physiology or behaviour in a test organism which is changed when the organism is exposed to a toxic substance at a sufficient concentration.
Various approaches have been adopted for monitoring biological functions, and several early warning systems have been designed and are applied. Although most systems are designed for use in the freshwater environments, a number have been used, or are of potential use, under marine conditions. Notably monitoring systems incorporating fish, bivalve molluscs, algae or bacteria seem promising in the marine environment.
Development of BEWS has, to date, largely focused on freshwater applications. At least 2 systems could, however, readily be used in coastal/estuarine conditions and others would be adapted for these conditions.
This chapter summarizes the various BEWS systems and their approaches, and gives examples of use, or potential use, in coastal waters and estuaries.

1. BACKGROUND

The traditional mainstay of monitoring natural waters and effluents is discrete sampling followed by chemical analysis in the laboratory. This provides information on the constituents of the sample in a precise, clearly defined manner, but only for a subset of chemical species, *i.e.* those specifically chosen for analysis, and at a particular point in time.

Biomonitoring of Coastal Waters and Estuaries. Edited by Kees J.M. Kramer.

This approach, exclusive of other methods of monitoring, is inadequate for three main reasons:

- it will fail to identify contaminants which are not included in the routine analysis but which may nonetheless occur, albeit rarely;
- it does not necessarily relate directly to the impact that the measured chemical species have on the biota;
- it may fail to identify sporadic contamination of the environment.

These inadequacies have always been recognised, however, and those involved in monitoring have sought to obtain supplementary information which, whilst less explicit in its identification of chemical species present, provides a more meaningful indication of the environmental effects of any contamination resulting from Man's activities. Broadly, the two approaches that are adopted to provide such supplementary (though not subsidiary) information are:

- continuous monitoring,
- biological monitoring.

Continuous monitoring, *i.e.* continuous, or very frequent, automated measurement of individual chemical species or general physico-chemical properties, can provide a very rapid warning of changed conditions or the occurrence of contaminants, but is restricted by the availability of suitable sensors and monitors, particularly if low power, battery operation of equipment is a prerequisite. Also, most of the physico-chemical sensors which are widely applicable (*e.g.* pH, dissolved oxygen, temperature, conductivity), whilst of value in showing long-term trends in environmental conditions, are poor indicators of sporadic events involving low concentrations of contaminants.

All methods in the other chapters of this volume are historic - they provide information about the past occurrence of contaminants that have in some way impacted upon the biota for the last few weeks or more. If it is necessary to determine and give warning of an event involving the discharge of a hazardous contaminant into the environment at the time of its occurrence, we must look to improved methods of continuous monitoring.

In contrast to the other biomonitoring techniques Biological Early Warning Systems (BEWS) have explicitly been developed to provide a rapid warning of the occurrence of contaminants at concentrations which could be of immediate threat to living organisms. In effect they are an integration of the two approaches defined earlier, *i.e.* they are automated continuous monitors which employ a biological organism or material as a primary sensing element. The secondary sensing element, that from which the final measurement is obtained or derived, may be variously electrical, optical or electro-/opto-chemical. It typically measures some physiological or behavioral function of the organism.

Historically, the main thrust of development of BEWS has been within the last twenty years and has been made effective by the introduction of the micro-computer. This has made possible the rapid and immediate analysis of large data sets in order to evaluate significant changes in the physiology or behaviour of the test organism(s). Prior to the introduction of the micro-computer methods of determining a response were relatively crude being confined either to gross fatal or near fatal responses in the organism which could be assessed visually or, at best, by the application of analogue electronics, or resulted in the painstakingly checking metres of historic strip chart recordings.

The main applications for which BEWS have been developed have been freshwater monitoring, usually to provide early warning of possible contamination of raw water at potable water intakes, and effluent monitoring (Cairns & van der Schalie, 1980; Gruber & Diamond, 1988). This chapter will review the developments that have taken place and the experience gained from the application of BEWS to date and assess their potential for application to the monitoring of coastal waters and estuaries, both now and in the future.

2. GENERAL PRINCIPLES AND PERFORMANCE CONSIDERATIONS

The basic principle of all BEWS is that of monitoring some function of physiology or behaviour in the test organism which is impaired or modified when the organism is exposed to a toxic substance at a sufficient concentration. Only these functions are of interest to BEWS, as only they allow a potential fast response time. It should be noted that the fast response (minutes) excludes a mechanism that is driven by an increased internal body concentration of the toxicant, except probably when dealing with algae or bacteria. The direct action will in most cases be the result of the response of sensory tissue. BEWS are, as such, broad-band monitors providing a warning of possible toxic conditions without identifying the chemical species involved, although some, by being restricted in the range of toxins to which they respond, can give some indication of the type of toxin causing the response, *e.g.* the response of algae to a herbicide.

A wide variety of test organisms have been used in the development of BEWS including fish species, daphnids and midge larvae, bivalve molluscs, algae and bacteria. In addition, isolated biological materials, cell organelles, enzymes and antigens, have been investigated for use in biosensors.

The test organism is usually contained in a specially designed unit with the sample water supplied by pumps. Attempts have been made to design BEWS that will operate *in situ* but, with the notable exception of one system (the Musselmonitor® described later), these have not yielded systems of general practical use as yet.

The secondary sensing element, *i.e.* the mechanism of measuring the behavioral or physiological response of the organisms, is generally specific to the species, or at least group, concerned and these will be presented in greater detail in the following sections. They generally fall into three categories:

- electrical, in which non-invasive electrodes measure the small potential difference generated by muscular activity in the water surrounding an organism;
- electro-magnetic, in which differences in a HF induced magnetic field (or a stationary magnet - reed relay combination) are an indication of movement of (parts of) the organism;
- optical, in which photosensitive cells or video tracking systems are used to determine location or activity;
- electro- or opto-chemical, in which a chemical sensor is used to measure a metabolite (*e.g.* oxygen), metabolic product (*e.g.* carbon dioxide) or cellular constituent (*e.g.* chlorophyll).

However, there are many variations on these general methods as will be seen. In BEWS it is usual for the secondary sensing element to be non-invasive, *i.e.* not inserted into the organism itself, to avoid stress to the organism which could interfere with its ability to show a measurable response to the additional stress imposed by the toxic substances that it is intended to detect. Developments of systems using unicellular algae and bacteria, however, may involve immobilisation of the organism on the surface of the electrode. Here the dividing line between a BEWS and a true biosensor becomes somewhat vague.

The raw signal obtained from a BEWS frequently bears no direct relationship to the function being monitored. A good example of this is an electrical signal generated by muscular activity in fish from which the gill beat frequency (ventilation frequency) may be derived. The signal processing necessary to derive the function may be complex and will certainly require the service of a microprocessor. Moreover, even having established a measurement it may not, in itself, be directly related to a condition imposed by exposure to a toxic substance. A fish ventilation frequency, to carry the example further, cannot at any point in time relate anything about the condition of the fish (not even if it is zero since many fish have a propensity to arrest ventilation for short periods).

Most (though not all) BEWS operate by determining a change from an established 'standard' condition in the test organism. The standard condition may be defined by a control group, but is more frequently derived by reference to historic data from the test organisms themselves. The method of data analysis is therefore an integral part of the BEWS and is as important to its overall performance as is the test species itself and the factor monitored. The data analysis must be capable of determining a significant change in the physiological or behavioral function being monitored without being unduly affected by inherent biological variation. With higher organisms, this usually involves the use of several test organisms in the BEWS and the data analysis technique must therefore preferably assess both the individual response and the group as a whole, in order to determine the significance of a change in the monitored function. When possible, individuals should be measured and compared with themselves; averaging the output from the start should in these cases be avoided. A further discussion on these statistical aspects is given in the Appendix.
As can be seen, a BEWS is a complex integration of three major components:

- a suitable test organism;
- a method of monitoring a suitable physiological or behavioral function;
- a means of processing the signals and data obtained, to yield a result on which alarm limits may be set to warn of a change which may be the result of exposure to a toxic substance.

Each of these system components will determine the range of substances to which it will respond, its sensitivity to each of those substances and its disposition to false responses.
The range of substances to which the system will respond is ultimately determined by the species employed in the monitor. Some species, even within a particular taxonomic family, may exhibit differing tolerance and susceptibility to particular toxic substances. If a species is particularly tolerant to a given toxic substance, or if symptoms of toxicity are chronic rather than acute, no monitoring method or data analysis can be expected to account for this deficiency. A good example of this is pesticides which function as acetyl-cholinesterase inhibitors thereby affecting the transmission of impulses across nerve synapses. A simple unicellular organism possessing no nervous system could not be expected to be sensitive to such a substance at any level of interest. Tolerance and susceptibility will also vary within a given population of the same species, a consideration which is to some extent accounted for by using multiple test organisms. The method of monitoring the behavioral or physiological response of an organism will also impact upon the range of substances to which it may respond. Fish, for instance, exhibit an avoidance reaction to heavy metals at very low concentrations, but the concentration must be considerably higher before a significant change in ventilation frequency occurs.
The sensitivity of the monitor to specific substances will, also depend on the method of data analysis. At its simplest, the data analysis will determine a high or low alarm value in the measured, or transformed, signal from the BEWS. The sensitivity will depend directly on how well this measured value relates to the concentrations of particular toxic substances in the sample and there is little scope for improving the sensitivity within the data analysis. However, as already stated, many BEWS operate by determining a significant change in the measured or transformed signal over time. There are, in essence, two ways of achieving this. One is to determine a change of sufficient magnitude to be significant (*i.e.* to have a low probability of occurrence given the underlying variation) and the other is to determine a change which, whilst it may not be of sufficient magnitude to be highly significant, is apparent in a sufficient number of organisms and observations to be so. The second method will obviously be more sensitive, or at least potentially so, since it does not require that changes in individual organisms be so great. As an example, the valve closure response of mussels is interpreted in two ways: full closure of the shells for a certain number of minutes ('close alarm'), or the detection of a certain decrease in the average valve opening over a given pe-

riod (decreased average alarm). However, these distinctive methods are only applicable where each organism is being monitored individually, which is not always the case, particularly where the organism is unicellular and maintained as a culture in a reaction vessel. The simple choice of method is not the only way in which the sensitivity may be affected by the data analysis. What, for instance, is the optimal period over which changes should be assessed? How long a period should be used to establish the 'standard' condition of the organisms? Is it reasonable to assume a normal distribution in the data being used to establish the standard condition? What, therefore, are the most appropriate statistical analyses to apply? Each of these considerations will be crucial in determining the sensitivity of the BEWS and also its disposition to false responses, which will be discussed next.
However sensitive the system may be and to whatever range of substances, a disposition towards false response, *i.e.* response in the absence of a contaminant of interest, may render it impractical. Just what is regarded as an unacceptable rate of false responses is highly dependent on the application and also the ease with which a false response may be identified and accounted for. In general, for environmental and effluent monitoring applications a false response rate of no greater than one per year is used as a target. All three components of the BEWS, mentioned above, will affect the disposition to false response. The organism must, of course, be capable of survival without being stressed in the full range of environmental conditions that will be encountered (unless, that is, the sample is conditioned prior to presentation to the organism). Whatever the secondary sensing element and data analysis employed, if environmental conditions are outside the survival conditions of the organism a 'false' response will ensue. The secondary sensing element will determine the disposition to false response in two ways: directly, by being itself affected by prevailing environmental conditions, and indirectly, if the physiological or behavioral function being monitored is affected by factors other than the presence of a toxic contaminant. Such factors include environmental conditions (*e.g.* pH, temperature, dissolved oxygen), disturbance by noise, vibration or physical proximity and 'natural' variation such as diurnal variation. Of course, as far as the organism is concerned, a response to these factors is not false but a true response to changing conditions, but such factors represent interference in an operational system. It will always be the aim when defining the data analysis method to minimise the rate of false responses within the constraints of the organism used and the physiological or behavioral function monitored. It has been suggested to express the typical response of a given organism to these natural factors in an algorithm, and to 'treat' the data according to the environmental conditions to obtain a 'normalized' data set (Borcherding, 1992). We believe that this method is not very reliable, as 'the' typical response will not exist. A carefully devised data analysis method will be capable of compensating for some factors contributing to the disposition to false response, particularly those caused by natural variation. In compensating for variation caused by other factors, the sensitivity of the system will inevitably suffer. A BEWS almost invariably becomes a compromise between an acceptably low rate of false response and a sensitivity which is adequate to meet the needs of the application. It will be obvious that when analyzing data, the situation where, *e.g.*, only few organisms of a group respond to a low set threshold, this will result in a very sensitive monitor, but one which is prone to false responses. The opposite is also true, a balance exists between sensitivity and reliability which, whilst it can be reduced by appropriate data analysis, can never be eliminated. As indicated earlier, the acceptable balance is usually determined by the application. For example, if in environmental monitoring a high sensitivity is useful to estimate the extent of episodic pollution based on subsequent analysis of samples automatically collected, a false alarm is not that critical. If on the other hand the result of an alarm generated by a BEWS results in the shut-down of an industrial production process, the validity of the alarm should be very high because of the apparent economic consequences.

With each of the three main components of a BEWS determining, both individually and as an integral whole, the performance of a BEWS it is evident that the performance cannot be readily predicted. Nor can the performance of one monitor be inferred from another using a similar method of monitoring. An important requirement in the development of any BEWS is therefore the careful and appropriate assessment of its performance. Without the availability of suitable evaluatory information, the use of a BEWS in the field becomes thwart with difficulties, particularly with respect to interpretation of, and confidence in, responses.

3. REVIEW OF BIOLOGICAL EARLY WARNING SYSTEMS

Various approaches for BEWS design have been tested in laboratory situations and in the field; some are only proposed in the literature, including a number of patents (Kramer & Botterweg, 1991). The following sections provide an overview of BEWS categorised according to the species, or group of species, used. An indication of the state of development of the systems and applications in which they are employed is provided together with an assessment of their (potential) use in estuarine and coastal water monitoring.

3.1. Fish monitors

With a primary stimulus for developing BEWS being the protection of potable water intakes, it is not surprising that considerable effort has been expended on developing systems which employ fish. Fish are the highest form of truly aquatic organism and as such may be expected to be closest to man in their range of, and sensitivity to, chemical substances which have a toxic effect. They are also widely used in bioassays to assess the toxicity of chemicals.
Fish monitors have been developed which use both behavioral and physiological measurements.

3.1.1. Detection of behavioral changes

Behavioral functions that may be measured are:

- chemotaxis (avoidance);
- (positive) rheotaxis (the characteristic of swimming into the water flow);
- motility;
- positional and swimming (locomotory) behaviour, and
- death detection.

These monitors variously use optical (light beams or video), electrical (electrodes measuring voltage from muscular activity), contact sensors or ultrasound methods to determine the location of the fish in the monitor.
Avoidance monitors are designed such that there is a dual stream of water in a so-called 'fluviarium', one of known good quality and one of the test water (Cherry & Cairns, 1982). A wide range of metals and some poly-electrolytes have been shown to induce highly significant avoidance reactions at concentrations as low as 0.03% of the 96 hour LC_{50} (Sprague & Drury, 1969; Hadjinicolaou & Spraggs, 1988; Hartwell *et al.*, 1989) but avoidance of organic substances is less pronounced, usually being identifiable only after long exposure (48 hours) or at concentrations too high to be of practical value except, possibly, for effluent monitoring (Sprague & Drury, 1969; Folmar, 1976; De Graeve, 1982). Although at least one avoidance monitor is manufactured commercially, there is no published information on the operational performance characteristics of these monitors with respect to rate of false responses. Their main routine application is in monitoring potable water supplies but examples also exist of their use in effluent monitoring, an application to which they are probably better suited. They are generally simple in construction, can be used with a wide variety of fish species and could be readily developed for use in salt-water conditions.

Rheotaxis monitors, which measure impairment of the ability of fish to swim against the water current, have been developed by Poels (1977) and separately by Juhnke & Besch (1971) and Besch *et al.* (1977). Both of these monitors are in commercial production but the latter has received the most widespread use with over 100 monitors sold throughout Europe, though mostly in Germany and The Netherlands. In these systems, like the Aqua-Tox-Control®, the (freshwater) gold ide *(Leuciscus idus)* has been used successfully. It is marketed as an effluent monitoring device but has also been applied to general freshwater monitoring and potable water supplies. In the present set-up of the instrument the fish are not monitored separately. In recent versions the fish are periodically forced to swim upstream a current generated by a pumping system. When they fail to do so they touch a screen at the back side of the tank. Each touching of the screen is counted, and a threshold is set to a given number of (total) pulses before the alarm is activated. In some designs the screen generates electrical pulses to reactivate the fish and minimize false alarms.

Little information exists on its sensitivity but evaluatory tests on the River Rhine have shown a good relationship between responses and increased concentrations of various compounds (Botterweg *et al.,* 1989), although response times, at up to four hours (Besch *et al.,* 1977), can be too slow for some applications. The system is very tolerant of interference, but no formal data on the rate of false responses exists. Any fish species of suitable size (6 - 12 cm), which exhibits a (positive) rheotaxis response may be used in the monitor and it could be readily used for monitoring in salt water conditions.

Motility and *swimming behaviour monitors* have been developed with apparatus of some sophistication, including multiple (2000) photocells (Kleerkoper *et al.,* 1970; Morgan, 1979), scanning light beam (Lemly & Smith, 1986), ultrasonic echoes (Morgan *et al.,* 1982), implanted magnets (Petry, 1982) and video coupled to on-line computerised image analysis (Smith & Bailey, 1988; Kress & Nachtigall, 1989) to monitor a wide variety of factors associated with motility and swimming and positional behaviour. In general the systems detect the location of the fish, either continuously or semi-continuously, and compare the present activity with previously recorded values under non-polluted conditions. Both multi-chamber and single chamber devices have been reported.

These approaches have largely been confined to laboratory studies and the small amount of data published suggests that, as with avoidance monitors, sensitivity to metals is far greater than sensitivity to organics. Interest in these systems arises not only from the use of fish, however. They are to a large extent species independent and can readily accommodate a mixture of species in the same monitor. They could as easily be developed for use in salt as in fresh water. The BahavioQuant® which uses a video tracking system for the behavioural detection of organisms has been developed recently.

Death detection monitors are monitoring systems that are based on the detection of dead fish, either floating on the surface blocking *e.g.* a light beam or obstructing a water outlet, are not to be considered as true early warning sensors, as they usually require ample reaction time and fail to be a sensitive device. Nevertheless, several patents have been applied for, that propose these devices (Kramer & Botterweg, 1991).

3.1.2. Detection of physiological variables

The most common physiological functions that are used for fish monitors are those that may be derived from monitoring gill movement, *viz:*

- ventilation frequency;
- ventilation volume or amplitude;
- cough frequency;
- electrical pulses.

In *ventilation monitors* the gill movement is monitored by immersing electrodes in close proximity to the fish to measure the small voltage changes associated with muscular activity, a method developed by Spoor *et al.* (1971). The electrical signal also gives information on the strength of the movement (ventilatory depth) and the coughing rate (Drummond & Carlson, 1977). The method has also been demonstrated to be capable of determining heart beat, but no practical monitor has been developed using this measurement. Each fish is kept in its own container, the signals recorded and processed separately. Fourier transformation techniques may assist in the data analysis of the complex ventilation patterns. Of the gill movement related detection, ventilation frequency has proved the most reliable in practice and commercial systems have been developed using this method including the WRc Fish Monitor® (Mk 3) and the Bio-Sensor® (Gruber & Cairns, 1981; Gruber *et al.*, 1991; Evans & Johnson, 1984; Evans & Walwork, 1988).

There is a large body of literature on the ventilatory response of fish to exposure to toxic substances which dates back as far as 1929 (Belding, 1929). Rainbow trout *(Oncorhynchus mykiss)* and bluegills *(Lepomis macrochirus)* are the most commonly studied and used fish, largely because of their ease of availability and use in standard toxicological studies. In theory, any fish species providing a sufficiently large signal from gill movement could be used, but in practice the characteristics of some fish (carp, *Cyprinus carpio* for instance, which have a tendency to arrest ventilation for extended periods) render them unsuitable. The response of rainbow trout to organic contaminants has been shown to fall within the range of 10% and 200% of the 96 hour LC_{50} value with a response time typically within 30 minutes and a false alarm rate of 4 per year (Baldwin *et al.*, 1993a,b). They are generally tolerant of changes in environmental factors, the exception being when such changes occur rapidly. Tidal influence was identified as being an interfering factor in one monitor operated in the field. They are less tolerant of operational disturbance (excessive vibration, proximity, water supply problems) and must be operated with some care. They are widely used in Europe, primarily the UK, and the USA for the purposes of potable water intake monitoring for which they are well suited. Their use for effluent monitoring has also been investigated (Gruber & Cairns, 1981; Morgan *et al.*, 1982) but their use in this application has been very limited, possibly because interference from rapid changes in factors other than toxicity is more likely to occur in effluents than in natural waters.

There are no examples of their use in salt waters and they would certainly be unsuitable for use in estuarine conditions.

The *electric pulses monitor* is another, very specific, example of physiological monitoring of fish that is used for the elephant-nosed mormyrid *(Gnathonemus petersi)*. This is a warm water fish from the African tropics which discharges short electrical pulses, originally evolved for navigation and food location in highly turbid waters but an extended use for communication has apparently evolved. The rate of discharge alters significantly if the fish is exposed to toxic substances (Geller, 1983). A commercial system has been developed and this has recently been refined further (Lewis *et al.*, 1990; Kay & Lewis, 1993). The specific nature of this monitor does not permit its use for salt waters, which would be outside the survival conditions of the fish.

3.2. Mussel monitors

Bivalve molluscs are common both in fresh water and the marine environment. Several physiological response techniques have been tested under laboratory conditions for their suitability to detect environmental changes (Akberali & Trueman, 1985). Examples that offer possible use in a BEWS are the heart beat (Sabourin & Tullis, 1981; Grace & Gainey, 1987), the respiration and filtration rate (Abel, 1976; Manley & Davenport, 1979; Manley, 1983). Because of the time-scale of the effect, the byssus-thread formation (*e.g.* Martin *et al.*, 1975; Roberts, 1975) falls beyond the scope of BEWS. The pumping rate detection (Famme *et al.*,

1986; Redpath & Davenport, 1988; Salánki *et al.* 1991) seems promising for incorporation in a BEWS system. As yet none of these techniques have lead to a practical BEWS.
The closure of the valves is a typical example of an escape behaviour response. Mussels have under normal conditions their shells open for respiration and feeding. It has been shown that they close their shells under (natural or anthropogenic) stress for an extended period. This response was used to study a number of natural and anthropogenic effects, including a series of toxicants, like trace metals (Davenport, 1977; Manley & Davenport, 1979), pesticides (Salánki & Varanka, 1978) and other trace organics (Sabourin & Tullis, 1981; Slooff *et al.*, 1983). Interestingly the closure of the shells as escape mechanism is not the only behavioral effect. Certain pollutants cause a dramatic increase in the valve movement activity. An increase in close/open frequency from 2 to 100 h^{-1} has been recorded *e.g.* as the result of the addition of free chlorine or organic solvents (Kramer *et al.*, 1989).
Warning systems have been developed using this valve movement response, the Musselmonitor® (de Zwart & Slooff, 1987; Kramer *et al.*, 1989; de Zwart *et al.* 1993), and the Dreissena Monitor® (Borcherding, 1992).
The Musselmonitor® is designed to operate *in situ* (although a flow-through laboratory version exists), and all parts are contained in a waterproof housing. Detection of the valve movement response is based on a high frequency (HF) electromagnetic induction system. The electronic sensor consists of two small coils, glued to opposite shell halves of the mussel. One coil acts as transmitter, generating a magnetic field, the other coil as receiver. Within the span of bivalves, a linear response is obtained. The Musselmonitor® is depicted in Figure 1.

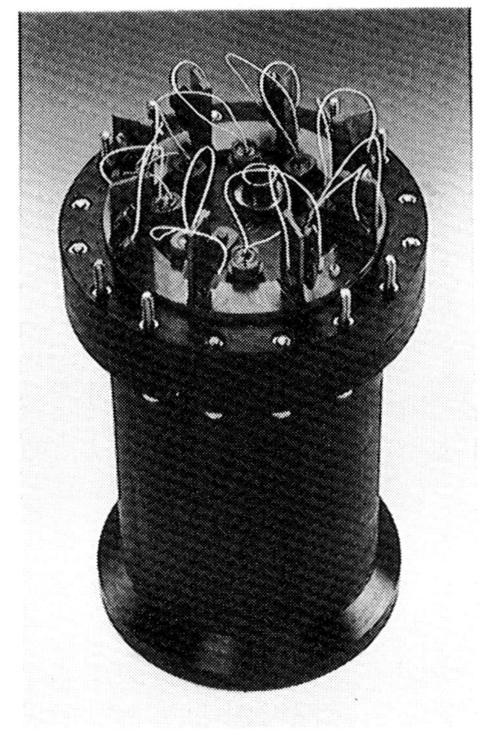

Figure 1. The Musselmonitor®.

Eight mussels are attached to the housing, and followed individually. This individual response detection allows discrimination between individual behaviour of the mussels, improving the reliability of the alarm detection. By providing the means to determine responses based on closure and activity behaviour, the Musselmonitor® may also detect, for example, dead organisms and low battery power. Various thresholds, like the number of mussels that should react, the time mussels should be closed, the increase in valve movement frequency, etc. can be changed. The monitor thus permits optimization of the system for specific requirements: an appropriate balance can be achieved between maximum sensitivity and minimum false alarm.

Since the system operates *in situ,* the natural environmental conditions minimize handling as food and oxygen are normally present in the water. Both in freshwater and the marine environment systems have been successfully used continuously for more than 3 months without replacement of the mussels.

The Dreissena Monitor® operates with two sets of 42 bivalves. Each mussel is followed using a magnet and reed relay glued to the shell surface. Only an arbitrarily set open/close response signal can be obtained.

Filter feeding mussels are considered best for use in a musselmonitor, as they are in maximum contact with the water column. Most applied bivalves so far, are for freshwater the zebra mussel *(Dreissena polymorpha)* or the painters' mussel (*Unio pictorum).* In the marine environment the blue or common mussel *(Mytilus spp.)* or oysters *(Ostrea spp.)* have been used in temperate climates, while the green mussel *(Perna viridis)* offers possibilities in tropical areas. In laboratory tests a number of other bivalves have been used (Manley & Davenport, 1979; Doherty *et al.,* 1987).

Under estuarine conditions the use of mussels in a BEWS is limited to estuarine species, as the salinity tolerance of *e.g. Mytilus* and *Dreissena* is about $> 10 \times 10^{-3}$ and $< 1 \times 10^{-3}$ respectively. True estuarine species, like *Mytilopsis leucophaeta (Congeria cochleata),* which are tolerant to the salinity regime offer here a better prospect.

3.3. Invertebrate monitors

Several monitors based on activity of small invertebrates or invertebrate larvae have been developed but of these, only the one based on *Daphnia spp.* have become commercially available and generally used for monitoring applications. The static Daphnia-test is a widespread standard for ecotoxicological work so it is not surprising that this should be developed for use in a continuous manner. Daphnia monitors are routinely used for monitoring on the River Rhine (Knie, 1982, 1988; Botterweg, 1989; Matthias & Puzicha, 1990). In systems like the Dynamic Daphnia Test® and the Daphnia Test® the swimming activity is monitored using infra-red light sources and detectors. Alarm limits are set directly on the impulse count, a pre-defined loss of activity or over-activity being used rather than one which is dynamically redefined, although such an approach could readily be adopted if it were demonstrated to confer any advantage. The method of monitoring small invertebrates is species independent and could be readily adapted for the prevailing conditions and species.

Voith (1979) described a system that was based on the activity measurement for various invertebrates (shellfish, crustacea) by the detection of the oxygen consumption. In principle this system can be applied to many small invertebrates.

Despite this, no marine/estuarine applications have been published yet.

3.4. Algal monitors

Monitors using unicellular algae offer a convenient means to provide a warning of the occurrence of *e.g.* herbicides (Whitton, 1991). Monitoring photosynthetic activity is achieved either by chlorophyll fluorescence or measuring oxygen production. Toxins interfering with the photosynthetic cycle may thus be rapidly detected. Other more general toxins may be detected as cell functionality declines, *e.g.* morphological changes or inhibition of growth rate, but such detection would be much slower than for substances which have a direct impact on photosynthesis. In algal monitors a culture of a suitable algal species (*e.g. Scenedesmus subspicatus, Microcystis aeruginosa, Chlamydomonas reinhardii)* is maintained in a culture chamber and measurements of oxygen production and/or fluoresence taken directly in the chamber or, more usually, in a sample drawn from the culture and mixed with the sample water.

There has been an interest in the use of fluorescence for monitoring photosynthetic activity in algae for many years with most of the literature being of German origin (see Lichtenhaler, 1988, for a review of applications). Chlorophyll molecules are usually detected by fluorescence spectroscopy, using an excitation wavelength of 430 nm, an emission wavelength of 665 nm. The emitted fluorescent light is a measure for the healthy living algal cells, compared to those that are damaged by algicidal compounds. This is utilised, for example in the commercially available Biosens® system which calculates a toxicity parameter from measurements of fluorescence of the sample water, the culture and the sample water and culture mixed (Noack, 1987, 1989; Benecke *et al.,* 1982; Schmidt, 1987).

If the excitation light is pulsed, a characteristic fluorescence curve can be obtained which can provide much more detailed information on the condition of the photosynthetic system. Fluorescence measured over time after a single light flash may be used to determine a variety of parameters (maximum, time to maximum, total and partial integrals, decay, delayed fluorescence, fluorescence quenching), changes to which are an indicator of interference to the photosynthetic system (Sayk & Schmidt, 1986; Ernst, 1986; Schmidt, 1987; Gerhard & Kretsch, 1989; Schreiber *et al.,* 1988). The DF-Algentest utilises the delayed fluorescence. Perhaps the most sophisticated system is the FluOx®-measurement system (Merschhemke, 1991) which simultaneously measures oxygen production and parameters derived from fluorescence curves determined from single and repetitive light flashes at different intensities. The parameters derived include fluorescence quenching which has been identified as a very sensitive indicator of disruption to photosynthesis.

Application of a so called "cage culture turbidostat" (Skipnes *et al.,* 1980) is proposed for continuous monitoring. The technique establishes a rate of growth for the test organism (*e.g.* the diatom *Phaeodactylum tricornutum)* in continuous culture with unpolluted medium, and then examines the changes in population growth rate with the introduction of the pollutant. Since the cultures are monitored continuously, immediate or short-term effects may readily be seen (Wangersky & Maass, 1990).

Algal monitors can operate with a wide variety of unicellular algal species and could therefore be readily adapted for use in coastal and estuarine conditions.

3.5. Bacterial monitors

As with algal monitors, bacterial monitors consist of a maintained culture of bacteria (either as free living mono- or mixed cultures, or attached to substrates) which is monitored by measurement of, usually, a metabolite (*e.g.* oxygen) rather than a product of metabolism. Many different micro-biotests have been developed (Blaise, 1991). The first such unit to be used in the field was developed by Holland & Green (1975) using *Nitrosomonas* bacteria and measuring their ability to metabolise dosed ammonia, but this system never achieved widespread use, owing to difficulties of establishing and maintaining the culture. Karube *et al.* (1977) described a system for continuous measurement of the Biological Oxygen Demand (BOD) by measuring the rate of oxygen consumption in a bacterial culture. Maintaining the culture again proved a problem, however. Control methods using microcomputers have now improved the ability to maintain bacterial cultures on a small scale and at least three on-line BOD monitors are now commercially available: Biomonitor®, BODyPoint®, BSB-Modul®.

Toxic substances will, of course, interfere with such a measurement and this has been used as the basis for development of BEWS for detection of toxic substances based on the same technique but measuring suppression of oxygen consumption (Martin, 1988; Pilz, 1990a,b). This has been adopted in the Stiptox® that monitors on-line the in- and outflow oxygen concentration of a bio-reactor containing activated sludge. Oxygen consumption may, of course, be used directly (Shieh & Yee, 1985; Clarke *et al.*, 1977), as may oxygen production (Solyom, 1977). Commercial systems using this technique are the ToxAlarm®, which measures either the consumption or production (with algae) of oxygen, the Toxiguard®, which is based on the respiration rate of a bacterial biofilm.

The on-line biomass monitoring of bacteria by capacitance measurement has recently been described (Fehrenbach *et al.*, 1992).

Careful choice of bacterial species should render adaptation of such monitors to saline conditions entirely feasible, but little information exists on their practical performance.

One method of toxicity measurement using bacteria which has been successfully exploited commercially for several years is the Microtox® test developed in 1979 (Bulich, 1979). This test uses the luminescent bacterium *Photobacterium phosphoreum*, and measures the suppression of luminescence when exposed to the sample water. The test was developed as a laboratory toxicity test but automated equipment for unattended field use has been developed: the Auto-Microtox® (Levi *et al.*, 1989; Henriet *et al.*, 1990; Rigaud *et al.*, 1992) and the BioLum®. There is a large body of literature reporting (toxicological) test results and comparisons with other static laboratory tests (*e.g.* Besson, 1983; Qureshi *et al.*, 1982; Vasseur *et al.*, 1984; Coleman & Qureshi, 1985, Nacci *et al.*, 1986). A data set on Microtox® EC_{50} data has been compiled by Kaiser & Palabrica (1991). Results are generally comparable to, or better than, those from static fish (96 hour LC_{50}) and daphnia (24 hour EC_{50}) tests but with greater speed and convenience and lower cost.

The Microtox® test with its use of a marine species of bacterium, can be used with saline samples without modification although some modification of automated equipment may be necessary to suit the conditions of estuarine and coastal monitoring. Recently a biosensor using these luminous bacteria has been described (Lee *et al.*, 1992).

A major advantage of the use of bacteria is the availability of different species and strains which can exhibit sensitivity to differing ranges of substances. For instance, a variation of the Microtox® test, the Mutatox®, uses a strain of *Photobacterium phosphoreum* which is sensitive to mutagens and DNA damaging agents, thereby providing the capability of detecting substances with chronic toxic effects (Kwan *et al.*, 1990). Bacteria are also easy to manipulate by genetic engineering and the lux gene of *Photobacterium* has been successfully transferred to *E. coli* thereby extending the range of bacteria to which the technique is applicable (Stewart *et al.*, 1989).

3.6. Biosensors

A biosensor is a chemical sensor in which a biological material is immobilised within, or on the surface of, an electrical or optical transducer. The biological elements may range from living organisms, tissues, cells, organelles, membranes, enzymes, etc., down to nucleic acids and organic molecules (Lowe, 1985; Turner *et al.*, 1987). The system is therefore an integral unit which may be dipped directly into the sample or contained within a system which presents it with the sample, after pre-conditioning as required. It is outside the scope of this chapter to discuss in detail the methods used for the development of biosensors. The reader is referred to standard texts such as Turner *et al.* (1987) for a detailed account of biosensors. However, biosensors are assuming an ever increasing importance and, although they have not yet achieved general use for environmental monitoring, the chapter would be incomplete without some discussion of them.

Research and development of biosensors has been largely inspired by the medical and food industries where the advantages of biosensors for rapid determination of defined biologically active organic substances were readily recognised. However, much of the development has focused on highly specific sensors utilising enzymes as the biological material, aiming at single components like glucose, alcohol, methane, ammonia etc (Clarke *et al.,* 1985; Karube, 1987). Although the need for specific toxic compound biosensors for environmental applications has been expressed (Rogers & Lin, 1992), their use seems limited to restricted applications where the nature of the toxicant is known or at least expected, *e.g.* in industrial effluents.

BEWS require a non-specific sensor that will respond to a range of substances, more or less limited according to requirements. Sensors using enzymes have been investigated for use in environmental monitoring, but the wide variation of environmental conditions coupled with the restricted range of operating conditions for an enzyme has largely led to serious practical problems. However, by using whole cells (bacteria or algae) as the biological material, the range of operating conditions is extended and the range of substances to which they will respond is much wider.

In theory, any of the methods cited for algal or bacterial monitors could be exploited in the development of a biosensor. However, such a simple statement belies the practical problems that beset such development. Notwithstanding, sensors monitoring the oxygen consumption of bacteria immobilised on the surface of a modified oxygen electrode have been developed for the determination of BOD (Riedel *et al.,* 1990).

The immobilisation of *Escherichia coli* at the surface of a CO_2 sensing electrode allowed the potentiometric detection of CO_2 production by the bacteria, and hence a reflection of the complex series of biochemical reactions which constitute the respiratory processes of the cells (Dorward & Barisas, 1984).

These examples employ the technique of measuring a metabolite giving an indirect measurement of metabolic processes within the cells. Such measurement may be prone to interference from changes in prevailing environmental conditions. A more sophisticated method of monitoring metabolic processes is to use a chemical mediator to transfer electrons directly from a metabolic process within the cell to the electrode, thereby giving a direct measurement of metabolic activity. The technique has been used to monitor photosynthesis in *Synechococcus*sp. and respiration in *E. coli* (Rawson *et al.,* 1987). Electrodes to determine the general microbial activity as measured by the electron transport system (ETS) were developed by several groups (Turner *et al.,* 1983; Ramsay *et al.,* 1986; Dobbs & Briers, 1988); the EuCyano bacteria electrode is an example.

Introduction of toxic substances which interfere with these metabolic processes cause a reduced output from the sensor within a few minutes. Whilst the technique shows promise (Evans *et al.,* 1986a), no practical monitor capable of sustained use in the field has yet been produced.

The microbial sensor using the luminous bacteria *Photobacterium phosphoreum,* already mentioned, uses the light emitted by the microorganisms which is detected with photomultiplier; a reduction in luminescence is observed under stress by toxic compounds (Lee *et al.,* 1992).

Biosensors offer a convenient and potentially inexpensive means of detecting toxic substances but they are as yet in an early stage of development and it will be some time before they find general use for monitoring applications in fresh or saline waters.

3.7. Multi-species monitors

Different species, even within similar phylla, will exhibit differing tolerances and susceptibilities to toxic substances. It will be clear that one BEWS based on a single species cannot, therefore, be expected to respond to any toxic substance at concentrations of interest (Diamond *et al.*, 1988). Intercomparisons between different BEWS have been reported, though not recently (Cairns & Gruber, 1980; Knie *et al.*, 1983). In Germany recently an intercomparison between some 20 different bio-monitoring systems has been caried out at the River Rhine. Unfortunately, no other than technical information has been published yet (Knie & Pluta, 1993).

The most effective way to extend the range of substances that can be detected is to use a number of different monitors, each employing a different species, or, ideally, to use several species within the same monitor. There are limitations to this latter approach since in many cases the monitoring technique is specific to, or at least tuned for, a particular species. However, avoidance and activity monitors do lend themselves to the possibility of incorporating several species in the same monitor. Such systems have been developed (Morgan *et al.*, 1987), but they are usually of a complex nature with a requirement for highly sophisticated means of data analysis and they are still in the developmental, or even research, stage and are not used in routine operation in the field.

Biosensors offer considerable scope for multi-species monitors, albeit that they are restricted to unicellular organisms. With the possibilities for using different bacterial strains and genetic engineering mentioned in the last section, this could lead to arrays of sensors each responding to a different, but overlapping, range of substances which would give some potential for actually identifying the type of toxicant present. Such developments are, however, in the future and await the development of biosensors that will operate reliably in the field.

For the present, the only means of operating multi-species systems is to use a number of different BEWS monitors together (Botterweg, 1989).

4. THE ROLE OF 'BEWS' IN COASTAL AND ESTUARINE MONITORING

For an overview of various BEWS systems that are on the market the reader is referred to Kramer & Botterweg (1991).

Most applications of biological early warning systems are in the various fields of water intake or discharge, and general water quality monitoring (see explanation in Figure 2):

- Influent (A)
  - drinking water plants, including desalination;
  - aquaculture;
  - public aquaria;
  - effluent treatment plants.
- Effluent (B)
  - industrial discharges, including off-shore structures;
  - domestic discharges;
  - water and sewage treatment plants;
  - cooling water;
  - stormwater runoff.
- Environmental monitoring (C)
  - rivers, estuaries (spill detection, storm surges);
  - coastal waters;
  - protection of oyster- and mussel cultures, fish farming.
- Anti-fouling control cooling water lines (D).

A BEWS will be most effective when a large gradient in the concentration of a toxic compound can be detected. When an alarming situation occurs, actions may be required like the closure of valves, a check upon proper functioning of the industrial installation, the automatic collection of a water sample.

BEWS have not so far found particular application in coastal and estuarine monitoring. However, their use may be valid, particularly in enclosed coastal areas where the residence time of waters within the system is high, an extreme example of such a situation being the Lagoon of Venice. Coastal and estuarine conditions pose particular problems to the application of BEWS. Firstly the high dilution factor that will commonly be present in coastal and estuarine systems. Whilst BEWS have been demonstrated to be capable of detecting toxic substances at concentrations likely to be acutely toxic and cause immediate damage to the environment, they are generally insufficiently sensitive to detect low concentrations which may be chronically toxic and cause long-term damage. Secondly, in estuarine conditions the sample is subject to rapid changes in salinity and other physico-chemical properties and, even if species could be found which adapt sufficiently quickly to these changes, they would undoubtedly exhibit a response of some sort which would trigger an alarm condition. In estuarine and coastal conditions, therefore, the use of BEWS seems in most instances limited to coastal waters. Locations close to effluent outfalls (pipelines, off-shore industry) or the effluent outfalls themselves.

Of the monitors reviewed in this chapter only two could be used directly in saline waters with no requirement for development work to adapt them. These are the Musselmonitor® and the Microtox® systems.

Those most likely to be readily adaptable are avoidance and activity monitors (BehavioQuant®), fish monitors based on rheotaxis and algal and bacterial monitors.

Considerations for choosing, installing and operating BEWS will be discussed under two distinct headings: practical and performance.

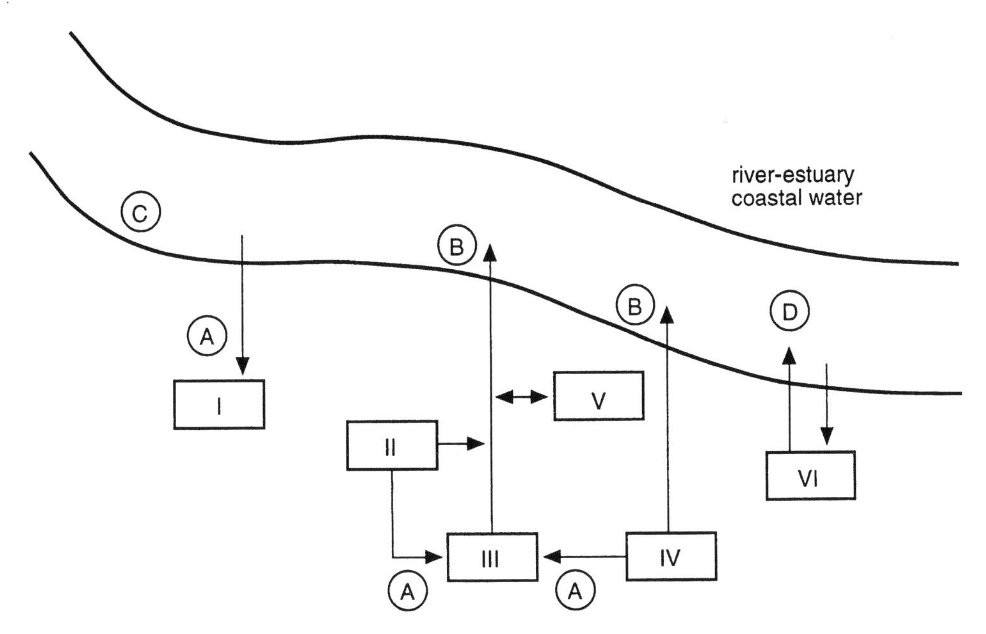

Figure 2. Overview of the various applications of biological early warning systems (BEWS).
I: drinking water inlet, aquaculture; II: industrial complex; III: treatment plant; IV: domestic sewage; V: calamity basin; VI: power plant (adapted from: Botterweg *et al.*, 1989)

4.1. Practical considerations

The first practical consideration is the species to be employed in the monitor. Whilst this may seem too obvious to mention, it is also so important that it cannot be overstressed. The physico-chemical conditions under which the monitor is to operate must be well understood, particularly the range of temperature, oxygen, and salinity; if an effluent is to be monitored then knowledge of other parameters such as pH and presence of low concentrations of toxic substances would also be needed. The conditions must fall comfortably within the survival conditions of the species to be used. It will be surprising how often technologists claim that their effluents are not harmful to biota, even at pH's 4 or 10, at temperatures exceeding 60 °C, etc. The chance is great that organisms will not survive in effluents proper. In these cases a pre-treatment of the effluent is necessary, *e.g.* cooling, pH adjustment, purging with air. Another option is the dilution of the effluent with the receiving water before entering the monitor, at a predetermined ratio. In the latter case the sensitivity of the system will undoubtedly be lowered.

The technical requirements for operation of the monitor are also of obvious importance. Most of the monitors reviewed, require a mains electricity supply and so, without some redevelopment, could not be used at remote locations away from such a service. Also the communication between the monitor and the control unit can be problematic when the distance becomes too large (see Appendix). As such they would only be of use as land-based monitors for monitoring effluents or on well equipped barges.

The size of the monitor and its housing requirements must also be taken into consideration. Again, most of the monitors reviewed are bulky and could not be installed on, for instance, a buoy so they may also be restricted to use on land or barges. The only exception is the Musselmonitor® that was specially designed to be used *in situ,* and future applications of field operable biosensors.

Maintenance requirements are also important: do reagents or the organisms themselves have to be changed at regular intervals? How frequently? How specialized should be the staff to do this, and what time does it take? All factors contributing to the cost of the operation of the system.

Siting of the monitor will be important if it is to draw a sample of the natural water rather than effluent. A good knowledge of currents and dispersion rates from the important effluent outfalls will be essential if the monitor is to be sited effectively.

It may be necessary to obtain subsidiary information, particularly with regard to physico-chemical conditions, in support of the information from the monitor itself. This will be particularly true if the monitor is known to be susceptible to interference from conditions that may, from time to time, be experienced. In such cases it may be important to measure the interferent continuously, or at least frequently, in order to rapidly identify responses caused by interference. Suitable equipment should, in such cases, be available which will operate under the required conditions.

Finally, but not least in importance, are the questions what is the purpose of the monitor and what should be done in the event of a response? If the purpose is research, then the monitor may be justified on the grounds of it being the only means of obtaining the information required. But if a monitor is being installed for a high level of routine surveillance for environmental protection, then the performance characteristics of the monitor must meet with the requirements for protection and it must be clear what actions can be taken in the event of a response. If an effluent is being monitored then the effluent may be stopped or redirected whilst the situation is investigated, which can be a costly process. In these situations the reliability of the BEWS (and of other sensors) should be high, false alarms are in principle not allowed.

If the natural waters are being monitored then rapid remedial action may not be possible in which case the monitor may serve no function other than to accumulate information for, for example, legislative action. It must be remembered, in such a case especially, that BEWS are, as implied in their title, warning systems; they do not provide hard and fast proof of the presence of a toxic pollutant and confirmation must be obtained through the more traditional methods of manual, or automatic, (BEWS directed) sampling and analysis in the laboratory. Of course, failure to confirm a response by laboratory analysis does not prove the response itself to be false since no laboratory analysis can be complete. In the River Rhine, a probable pollution event was detected by an alarm at two different locations employing two Musselmonitors. Samples were collected but no compound(s) could be identified by chemical analyses (de Zwart *et al.*, 1994).

4.2. Performance considerations

A thorough evaluation of a BEWS to determine its performance characteristics is a time consuming, lengthy and expensive process. The performance characteristics of a monitor may also be changed quite markedly by even apparently minor alterations to, for instance, the data analysis method. It is not surprising, therefore, that performance information for almost all the types of monitor reviewed here is very scarce and incomplete. The most rigorous evaluation for which results have been published is that undertaken on the WRc Fish Monitor® developed by Evans & Johnson (1985) (Baldwin *et al.*, 1993a,b).
Evaluation of a BEWS requires that the monitor be subjected to known concentrations of a wide range of toxic substances under laboratory conditions to determine the approximate concentrations to which the monitor will respond, what type of response may be expected and the typical response time. Ideally the effect of potential interferents, such as suspended particulate matter, food availability, vibrations or light climate, should also be assessed. Operation of the monitor under controlled laboratory conditions for an extended period, at least one year, is also necessary in order to establish the rate of false responses. If the response threshold is configurable, as it should be if the monitor is to be effectively tuned for prevailing conditions in the field, this information should be assessed in order to clearly establish the effect of altering the setting of the response threshold. This information should be supported by field trials undertaken, again for an extended period of at least one year, to confirm the performance thus established and to identify any further causes of interference. Reliable statements of concentrations of response and expected false response rate, whilst costly to obtain, are amply repaid when the cost of purchasing, installing and operating several monitors in the field is taken into consideration; it is of the utmost importance that such information is available in order to make an effective decision as to whether the monitor will meet the requirements of the application and to gain the confidence of the operators and permit them to maintain the system to its maximum effectiveness. Excessive false responses over and above that predicted from evaluatory trials would indicate that, in the absence of a fault, responses are valid and caused by the presence of either toxic substances or an unidentified interferent.

Another important aspect of performance is the ability, or lack of ability, of the monitor to detect low concentrations of toxic substance present for extended periods (chronic exposure). The requirement for this is very much dependent on the application and the performance of the monitor in this respect is often a function of the method used. In principle, BEWS will function best when relatively steep gradients are to be expected. Monitors that measure changes in physiological functions generally respond faster than those for which an absolute limit is defined on suppression or enhancement of some measured function. This is advanta
geous in most applications for which BEWS have been developed but the disadvantage is that they will not respond to greatly sub-acute concentrations of toxic substances, or very slow build-up of contaminant, until the normal functioning of the test organisms is severely impaired. If this is a priority consideration then it is probably more appropriate to select a BEWS which uses a method which is not based on determination of change since these may be more capable of responding to lower concentrations of toxicant but over much longer pe-

riods of exposure. Careful assessment of the performance characteristics outlined in the previous paragraph would, however, be necessary to confirm this.

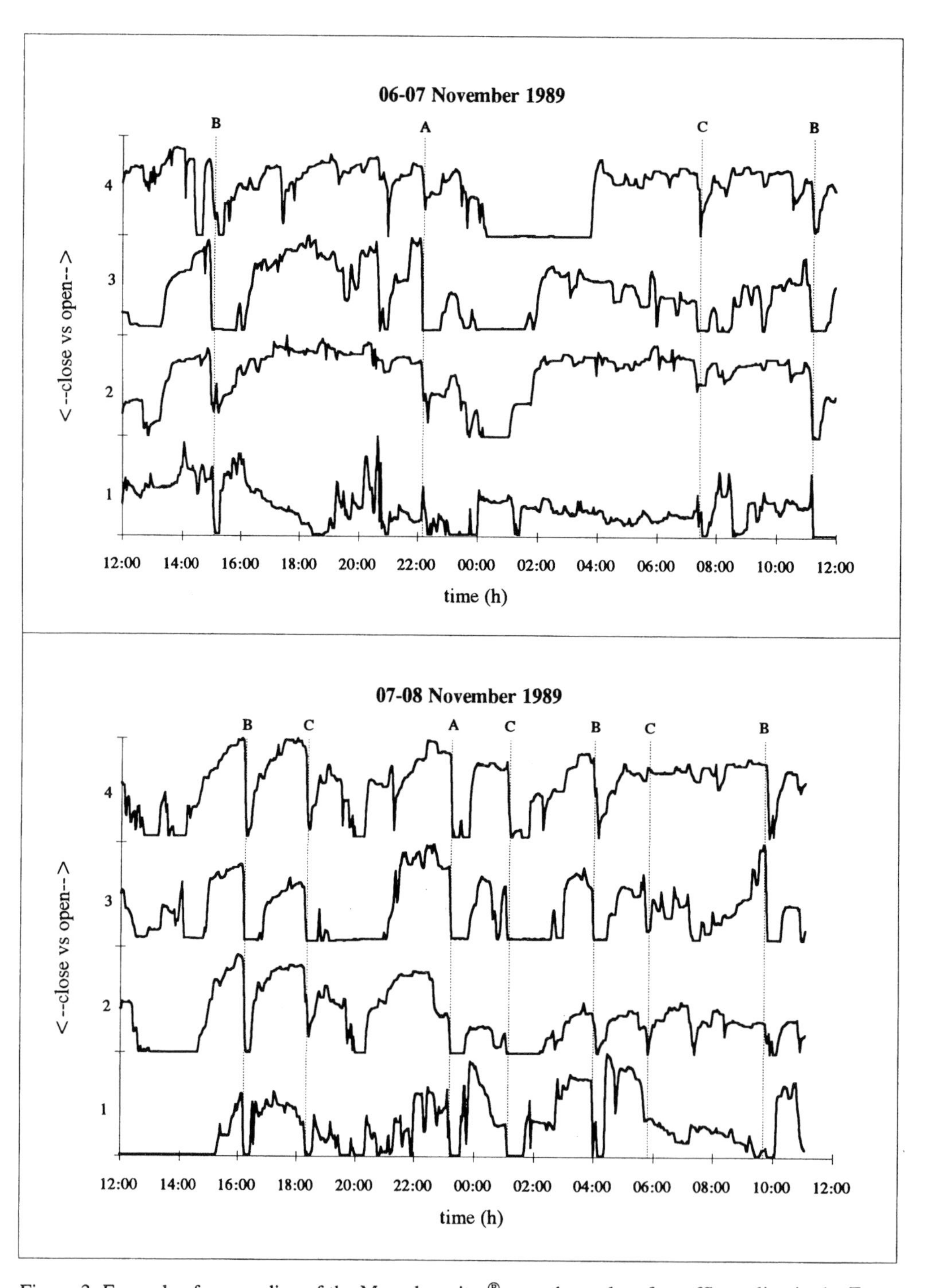

Figure 3. Example of a recording of the Musselmonitor® near the outlet of an effluent line in the Ems Estuary. Four different mussels show their closure behaviour (up = fully open, down = closed); A: alarm due to the start of a low water period; B: alarm; C: no alarm, but a visual response (After: Kramer & Foekema, 1990)

5. EXAMPLE OF AN ESTUARINE FIELD TRIAL

To our knowledge, so far only examples of one type of a BEWS, the Musselmonitor®, have been documented for the estuarine and marine environment.
The Musselmonitor® has been tested under natural conditions in the Ems/Dollard Estuary, The Netherlands (Kramer & Foekema, 1990). Here the effluent of an industrial complex was monitored in the receiving water for ten weeks to check upon the feasibility of the monitor under true field applications. For this study the same set of eight individuals of the common or blue mussel (*Mytilus edulis*) were used without any maintenance of the organisms. Although the tidal range caused some interference (mussels out of the water at spring tide low water), a number of the alarm situations corresponded with the increased free chlorine concentrations in the effluent (Figure 3).

CONCLUSIONS

BEWS are best when a relatively steep gradient in pollutant concentration is to be detected. In the marine environment concentrations are rather diluted except close to point resources. Under estuarine conditions the interference of the measurement with the salinity regime may result in a limited use of the systems here.
Nevertheless, BEWS will be useful for effluent monitoring in coastal marine waters (*e.g.* near industrial effluent lines, off shore discharges, and for the protection of aquaculture sites (fish, crustacea, shellfish) and of intake water of public aquaria/marinas. Their nature, in the potential biological detection of a wide spectrum of harmful substances that escape initially from chemical analysis, gives them a unique place in early warning networks, complementary to the physico-chemical detection systems.
Only two BEWS (Musselmonitor® and (Auto) Microtox®) are immediately applicable in the seawater environment, but several other systems could be developed by adapting systems operating under freshwater conditions. Algae and fish monitors (using rheotaxis) are considered to be most amenable to such adaptation. It is stressed that, whatever system used, a proper evaluation is carried out not only for sensitivity but also for the false alarm rate.

Note:

Aqua-Tox-Control® and Daphnia Test® are Registered Trademarks of Kerren Umwelt Technik GmbH, Viersen, Germany; Auto-Microtox® is a Registered Trademark of the Compagnie Générale des Eaux (CGE), Paris, France; BehavioQuant® is a Registered Trademark of GSF-Forschungszentrum, Neuherberg, Germany; BioMonitor® and ToxAlarm® are Registered Trademarks of LAR GmbH, Berlin, Germany; Biosens® is a Registered Trademark of Biosens GmbH, Hildesheim, Germany; Bio-Sensor® is a Registered Trademark of Biological Monitoring Inc., Blacksburg VA, USA; BODyPoint® is a Registered Trademark of Aucoteam mbH, Berlin, Germany; BSB-Modul® is a Registered Trademark of Medingen GmbH, Freital, Germany; Dynamischer Daphnientest® is a Registered Trademark of Elektron GmbH, Krefeld, Germany; Dreissena Monitor® is a Registered Trademark of Envicontrol, Cologne, Germany; Fish Monitor® is a Registered Trade Mark of WRc, Medmenham, UK; FluOx® and BioLum® are Registered Trademarks of Kolibri GmbH, Hattingen/Ruhr, Germany; Musselmonitor® is a Registered Trademark of Delta Consult, Capelle, The Netherlands; Microtox® and Mutatox® are Registered Trademarks of Microbics Corp. Carlsbad CA, USA; Stiptox® is a Registered Trademark of Stip Siepmann & Teutscher GmbH, Gross-Umstadt, Germany; Toxyguard® is a Registered Trademark of BTG Anlagen Technik GmbH, Bochum, Germany;

REFERENCES

Abel, P.D., 1976. Effect of some pollutants on the filtration rate of *Mytilus*. Mar. Pollut. Bull., 7: 228-231

Akberali H.B. & E.R. Treuman, 1985. Effects of environmental stress on marine bivalve molluscs. Adv. Mar. Biol. 22: 102-199

Baldwin, I.G., M.M.I. Harman, & D.A. Neville, 1993a. Performance characteristics of a fish monitor for the detection of toxic substances, part 1: laboratory trials. Wat. Res. (in press)

Baldwin, I.G., M.M.I. Harman & D.A. Neville, 1993b. Performance characteristics of a fish monitor for the detection of toxic substances, part 2: field trials. Wat. Res. (in press)

Belding, D.L., 1929. The respiratory movements of fish as an indicator of a toxic environment. Trans. Am. Fish. Soc. 59: 238-246

Benecke, G., W. Falke & C. Schmidt, 1982. Use of algal fluorescence for an automated biological monitoring system. Bull. Environm. Contam. Toxicol. 28: 385-395

Besch, W.K., A. Kemball, K. Meyer-Waarden & B. Scharf, 1977. A biological monitoring system employing rheotaxis of fish. In: Biological monitoring of water and effluent quality; J. Cairns (ed). ASTM STP 607, pp. 56-74

Besson, S., 1983. The control of toxicity in water by the Microtox. Eau, Ind. Nuis. 75: 67-68

Blaise, C., 1991. Microbiotests in aquatic ecotoxicology: characteristics, utility, and prospects. Environ. Toxicol. Wat.Qual. 6: 145-155

Borcherding, J., 1992. Another early warning system for the detection of toxic discharges in the aquatic environment based on valve movements of the freshwater mussel *Dreissena polymorpha*. In: The zebra mussel *Dreissena polymorpha;* D. Neumann & H.A. Jenner (eds). Gustav Fisher, Stuttgart, pp. 127-146

Botterweg, J., 1989. Evaluation of biological early warning systems (BEWS) in the River Rhine at Lobith (The Netherlands). Bioalarm project. Part 1. BKH consulting engineers in asignment of DBW/RIZA. Publications and reports of the project 'Ecological rehabilitation of the River Rhine, 8-89. DBW/RIZA nr 89.045, pp. 32

Botterweg, J., C. van der Guchte & L.W.C.A. van Breemen, 1989. Bio-alarm systems: a supplement to traditional monitoring of water quality. H_2O, 22: 778-794

Bulich, A.A., 1979. Use of luminescent bacteria for determining toxicity in aquatic environments. In: Aquatic toxicology; L.L. Marking & R.A. Kimerle (eds). ASTM STP 667, pp. 98-106

Cairns, J. & D. Gruber, 1980. A comparison of methods and instrumentation of biological early warning systems. Wat. Resourc. Bull. 16: 261-266

Cairns, J. & W.H. van der Schalie, 1980. Biological monitoring. Part I. Early warning systems. Wat. Res. 14: 1179-1196

Cherry, D.S. & J. Cairns, 1982. Biological monitoring. Part V. Preference and avoidance studies. Wat. Res. 16: 263-301

Clarke, A.N., W.W. Eckenfelder & J.A. Roth, 1977. The development of an influent monitor for biological treatment systems. Prog. Wat. Technol. J. 9: 103-107

Clarke, D.J., M.R. Calder, R.J.G. Carr, B.C. Blake Coleman, S.C. Moody & T.A. Collinge, 1985. The development and application of biosensing devices for bioreactor monitoring and control. Biosens. 1: 213-320

Coleman, R.N. & A.A. Qureshi, 1985. Microtox and *Spirilum volutans* tests for assessing toxicity of environmental samples. Bull. Environ. Contam. Toxicol. 35: 443-451

Davenport, J., 1977. A study of the effect of copper applied continuously and discontinuously to specimens of *Mytilus edulis* (L) exposed to steady and fluctuating salinity levels. J. mar. biol. Ass. U.K., 57: 63-74

de Zwart, D. & W. Slooff, 1987. Continuous effluent biomonitoring with an early warning system. In: Effluent and ambient toxicity testing in the Göta Älv and Viskan Rivers, Sweden; Bengston, Norberg-King & Mount (eds). Naturvardsverket report 3275

de Zwart, D., K.J.M. Kramer & H.A. Jenner, 1993. Practical experiences with the biological early warning system "Musselmonitor®. Environ. Toxicol. Wat. Qual. (in press)

de Graeve, G.M., 1982. Avoidance responses of rainbow trout to phenol. The Progressive Fish-culturalist, 44: 82-87

Diamond, J., M. Collins & D. Gruber, 1988. An overview of automated biomonitoring - past developments and future needs. In: Automated biomonitoring: living sensors as environmental monitors; D.S. Gruber & J.M. Diamond (eds). Ellis Horwood, Chichester, pp. 23-39

Dobbs, A.J. & M.G. Briers, 1988. Water quality monitoring using chemical and biological sensors. Anal. Proc. 25: 278-279

Doherty, F.G., D.S. Cherry & J. Cairns, 1987. Valve closure responses of the asiatic clam *(Corbicula fluminea)* exposed to cadmium and zinc. Hydrobiol. 153: 159-167

Dorward, E.J. & B.G. Barisas, 1984. Acute toxicity screening of water pollutants using a bacterial electrode. Environ. Sci. Technol. 18: 967-972

Drummond, R.A. & R.W. Carlson, 1977. Procedures for measuring the cough (gill purge) rates of fish. EPA-600/3-77-133. U.S. EPA, Washington D.C.

Ernst, D.E.W., 1986. Comments on fluorimetric chorophyll determinations in the field. Arch. Hydrobiol. 107: 521-527
Evans, G.P. & D. Johnson, 1984. Detection of pollution at drinking water intakes. In: Environmental protection: standards, compliance and cost; T. Lack (ed). Ellis Horwood, Chichester, pp. 239-254
Evans, G.P. & D. Johnson, 1985. Automatic detection of pollution at drinking water intakes using fish. Wat. Supply, 3: 179-186
Evans, G.P. & J.F. Wallwork, 1988. The WRc fish monitor and other biomonitoring methods. In: Automated Biomonitoring; D.S. Gruber & J.M. Diamond (eds). Ellis Horwood, Chichester, pp. 75-90
Evans, G.P., M.G. Briers & D.M. Rawson, 1986a. Can biosensors help to protect drinking water? Biosens. 2: 287-300
Evans, G.P., D. Johnson & C. Whithell, 1986b. Development of the WRc Mk 3 fish monitor: description of the system and its response to some commonly encountered pollutants. Water Research Center, Environment report TR 233, pp. 52
Famme, P., H.U. Riisgard & C.B. Jörgensen, 1986. On direct measurement of pumping rates in the mussel *Mytilus edulis*. Mar. Biol. 92: 323-327
Fehrenbach, R., M. Comberbach & J.O. Pêtre, 1992. On-line biomass monitoring by capacitance measurement. J. Biotechnol. 23: 303-314
Folmar, L.C., 1976. Overt avoidance reaction of rainbow trout fry to nine herbicides. Bull. Environ. Contam. Toxicol. 15: 509-514
Geller, W., 1983. A toxicity warning monitor using weakly electric fish, *Gnathonemus petersi*. Wat. Res. 18: 1285-1290
Gerhard, V. & G. Kretsch, 1989. Entwurf eines Biotestverfahrens zum Schadstoffnachweis mittels der verzögerten Fluoreszenz von Algen. In: Grundlagen und Anwendungsbereiche der Chorophyll-fluoreszenz; D. Ernst & C. Schmidt (eds). Dortmunder Beitrage zur Wasserforschung, pp. 87-95
Grace, A.L. & L.F. Gainey Jr., 1987. The effects of copper on the heart rate and filtration of *Mytilus edulis*. Mar. Pollut. Bull. 18: 87-91
Gruber, D.S. & J. Cairns, 1981. Industrial effluent monitoring incorporating a recent automated fish biomonitoring system. Wat. Air Soil Pollut. 15: 471-481
Gruber, D.S. & J.M. Diamond (eds), 1988. Automated biomonitoring: living sensors as environmental monitors. Ellis Horwood, Chichester, pp. 208
Gruber, D.S., J.M. Diamond & M.J. Parson, 1991. Automated biomonitoring. Environ. Audit. 2: ???
Hadjinicolaou, J. & L.D. Spraggs, 1988. Avoidance reactions of fish to an industrial effluent and its constituent toxic components. In: Automated Biomonitoring; D.S. Gruber & J.M. Diamond (eds). Ellis Horwood, Chichester, pp. 104-126
Hartwell, S.I., J.H. Jin, D.S. Cherry & J. Cairns, 1989. Toxicity verses avoidance response of golden shiner to 5 metals. J. Fish Biol. 35: 447-457
Henriet, C., Y. Levi & J.P. Coutant, 1990. Automatisation d'un test de toxicité aiguë utilisant des bactéries luminescentes. Eau, Ind. Nuis. 140: 80-82
Holland, G.J. & A. Green, 1975. Development of a gross pollution detector: laboratory studies. Wat. Treatm. Exam. 24: 81-99
Juhnke, I. & W.K. Besch, 1971. Eine neue Testmethode zur Früherkennung akut toxischer Inhaltsstoffe im Wasser. Gewässer Abwässer, 50/51: 107-114
Kaiser, K.L.E. & V.S. Palabrica, 1991. Photobacterium phosphoreum toxicity data index. Wat. Pollut. Res. J. Canada, 26: 361-431
Karube, I., 1987. Micro-organism based sensors. In: Biosensors: fundamentals and applications; A.P.F. Turner, I. Karube & G.S. Wilson (eds). Oxford Univ. Press, Oxford, pp. 13-29
Karube, I., S. Mitsuda, T. Matsunaga & S. Suzuki, 1977. A rapid method for estimation of BOD by using immobilised microbial cells. J. Ferment. Technol. B55B: 243-248
Kay, A.N. & J.W. Lewis, 1993. Fish monitors and the role of electric fish as potential indicators of water quality. J. Appl. Ichthyol. 9: 110-114.
Kleerekoper, H., A.M. Timms, G.F. Westlake, F.B. Davy, T. Malar & V.M. Anderson, 1970. An analysis of locomotor behaviour of goldfish *Carasius auratus*. Animal Behav. 18: 317-330
Knie, J., 1982. Der Daphnientest. Decheniana Beih. 26: 82-86
Knie, J., 1988. Der dynamische Daphnientest. Pracktische Erfahrungen bei der Gewässer-Wachung. Gewäss. Wasser Abwass. 68: 1-17
Knie, J. & H.-J. Pluta, 1993. Biomonitore zur kontinuierlichen Überwachung von Wasser und Abwasser. Ministry of Research and Technology (BMFT), Project BEO, Berlin, pp. 48
Knie, J., A. Halke, I. Juhnke & W. Schiller, 1983. Ergebnisse der Untersuchungen von chemischen Stoffen mit vier Biotests. Deutsche Gewässerkündl. Mitt. 27: 77-79

Kramer, K.J.M. & J. Botterweg, 1991. Aquatic biological early warning systems: an overview. In: Bioindicators and environmental management; D.W. Jeffrey & B. Madden (eds). Academic Press, London, pp. 95-126

Kramer, K.J.M. & E.M. Foekema, 1990. 'Mossel Monitors' getest: het Delfzijl experiment. MT-TNO report R 90/155, Delft, pp. 66

Kramer, K.J.M., H.A. Jenner & D. de Zwart, 1989. The valve movement response of mussels: a tool in biological monitoring. Hydrobiol. 188/189: 433-443

Kress, C. & W. Nachtigal, 1989. The monitoring of harmful substances in running water by means of behaviour-parameters in continuously swimming fish. Z. Wasser Abwasser Forsch. 22: 99-107

Kwan, K.K., B.J. Dutka, S.S. Rao, & D. Liu, 1990. Mutatox test: a new test for monitoring environmental genotoxic agents. Environ. Pollut. 65: 323-332

Lee, S. K. Sode, K. Nakanishi, J.L. Marty, E. Tamiya & I. Karube, 1992. A novel microbial sensor using luminous bacteria. Biosens. Bioelectr. 7: 273-277

Lemly, A.D. & R.J.F. Smith, 1986. A simple activity quotient for detecting pollution induced stress in fishes. Environ. Technol. Lett. 4: 173-178

Levi, Y., C. Henriet, J.P. Coutant, M. Lucas & G. Leger, 1989. Monitoring acute toxicity in rivers with the help of the Microtox test. Wat. Supply, 7: 25-31

Lewis, J.W., P.R. Campbell & I.P. Tems, 1990. Microcomputer based monitoring of the effect of phenol on electric organ activity in the mormyrid fish *Gnathonemus petersii.* Environ. Technol. 11: 571-584

Lichtenhaler, H. (ed), 1988. Application of chlorophyll fluorescence. Kluwer, Dordrecht

Lowe, C.R., 1985. An introduction to the concepts and technology of biosensors. Biosens. 1: 3-16

Manley, A.R., 1983. The effects of copper on the behaviour, respiration, filtration and ventilation activity of *Mytilus edulis.* J. mar. biol. Ass. U.K. 63: 205-222

Manley, A.R. & J. Davenport, 1979. Behavioral responses of some marine bivalves to heigtened seawater copper concentrations. Bull. Environ. Contam. Toxicol. 22: 739-744

Martin, J.V., 1988. Biomonitoring of polluted waters: three systems. In: Automated Biomonitoring; D.S. Gruber & J.M. Diamond (eds). Ellis Horwood, Chichester, pp. 173-181

Martin, J.M., F.M. Piltz & D.J. Reish, 1975. Studies on the *Mytilus edulis* community in Alamitos Bay, California. V. The effects of heavy metals on byssal thread production. Veliger 18: 183-188

Matthias, U. & H. Puzicha, 1990. Erfahrungen mit dem dynamischen Daphnientest - Einfluss von Pestiziden auf das Schwimmverhalten von *Daphnia magna* unter Labor und Praxisbedingungen. Z. Wasser Abwasser Forsch. 23: 193-198

Merschhemke, C., 1991. Development, testing and installation of automatic bioassays along the river Rhine. In: Use of algae for monitoring rivers; B.A. Whitton, E. Rott & G. Friedrich (eds.). Institut für Botanik, University of Insbruck, pp. 119-122

Morgan, W.S.G., 1979. Fish locomotor behaviour patterns as a monitoring tool. J. Wat. Pollut. Cont. Fed. 51: 580-589

Morgan, W.S.G., P.C. Kuhn, B. Allais & G. Wallis, 1982. An appraisal of the performance of a continuous automatic fish biomonitoring system at an industrial site. Wat. Sci. Technol. 14: 151-161

Morgan, E.L., R.C. Young, C.N. Crane & B.J. Armitage, 1987. Developing automated multispecies biosensing for contaminant detection. Wat. Sci. Technol. 19: 73-84

Nacci, D., E. Jackim, & R. Walsh, 1986. Comparative evaluation of three rapid marine toxicity tests: sea urchin early embryo growth test, sea urchin sperm cell toxicity test and Microtox. Environ. Toxicol. Chem. 5: 521-525

Noack, U., 1987. Quasi-kontinuierliche Chlorophyllfluoreszenz Messung in der Gewässerüberwachung. Arch. Hydrobiol. Beih. 29: 99-105

Noack, U., 1989. Algen-toximeter zur Gewässer und Abwasserüberwachung. In: Grundlagen und Anwendungsbereiche der Chorophyllfluoreszenz; D. Ernst & C. Schmidt (eds). Dortmunder Beitrage zur Wasserforschung, pp. 79-86

Petry, H., 1982. The 'motility test': an early warning system for the biological control of waters. Zentralb. Bakter. Mikrob. Hyg. 176: 391-412

Pilz, U., 1990a. Verbesserung der Betriebssicherheit und der Messempfindlichkeit des Bakterientoximeters. In: Grundlagen und Anwendungsbereiche der Chlorophyllfluoreszenz; D. Ernst & C. Schmidt (eds). Dortmunder Beitrage zur Wasserforschung, pp. 351-359

Pilz, U., 1990b. Verbesserung der Betriebssicherheit und der Messempfindlichkeit des Bakterientoximeters. Vom Wasser, 74: 351-359

Poels, C.L.M., 1977. An automatic system for rapid detection of acute high concentrations of toxic substances in surface waters using trout. In: Biological monitoring of water and effluent quality; J. Cairns (ed). ASTM STP 607: 85-95

Qureshi, A.A., K.W. Flood, S.R. Thompson, S.M. Janhurst, C.S. Inniss & D.A. Rokosh, 1982. Comparison of a luminscent bacterial test with other bioassays for determining toxicity of pure compounds and complex effluents. In: Aquatic toxicology and hazard assessment; J.G. Pearson, R.B. Foster & W.E. Bishop (eds). ASTM STP 766: 179-195

Ramsay, G., A.P.F. Turner, A. Franklin & I.J. Higgins, 1986. Rapid bioelectrochemical methods for the detection of living micro-organisms. In: Modelling and control of biotechnological processes; A. Johnson (ed). Pergamon, Oxford

Rawson, D.M., A.J. Willmer & M.F. Cardosi, 1987. The development of whole cell biosensors for on-line screening of herbicide pollution of surface waters. Tox. Assess. 2: 325-340

Redpath, K.J. & J. Davenport, 1988. The effect of copper, zinc and cadmium on the pumping rate of *Mytilus edulis* L. Aquat. Toxicol. 13: 217-226

Riedel, K., K.-P. Lange, H.-J. Stein, M. Kuhn, P. Ott & F. Scheller, 1990. A microbial sensor for BOD. Wat. Res. 24: 883-887

Rigaud, G., C. Henriet & J. Dowe, 1992. Une avancée décisive dans la détection automatique des pollutions: l'Auto-Microtox®. Point Sciences Techn. 3: 6-9

Roberts, D., 1975. The effect of pesticides on byssus formation in the common mussel, *Mytilus edulis*. Environ. Pollut. 8: 241-253

Rogers, K.R. & J.N. Lin, 1992. Biosensors for environmental monitoring. Biosens. Bioelectr. 7: 317-321

Sabourin, T.D. & R.E. Tullis, 1981. Effect of three aromatic hydrocarbons on respiration and heart rates of the mussel, *Mytilus californianus*. Bull. Environ. Contam. Toxicol. 26: 729-736

Salánki, J. & L. Varanka, 1978. Effect of some insecticides on the periodic activity of the fresh-water mussel *(Anodonta cygnea* L). Acta. Biol. Acad. Sci. Hung. 29: 173-180

Salánki, J., T.M. Turpaev & M. Nichaeva, 1991. Mussel as a test animal for assessing environmental pollution and the sub-lethals effect of pollutants. In: Bioindicators and environmental management; D.W. Jeffrey & B. Madden (eds). Academic Press, London, pp. 236-244.

Sayk, F. & C. Schmidt, 1986. Algen-Fluoreszenz-Autometer, eine computergesteuerte Biotest-Messapparatur. Z. Wasser Abwasser Forsch. 19: 182-184

Schmidt, C., 1987. Anwendungsbereiche und Ergebnisse des Algenfluoreszenztests. Arch. Hydrobiol. 29: 107-116

Schreiber, U., U. Schliwa & W. Bilger, 1988. Continuous recording of photochemical and non-photochemical chlorophyll fluorescence quenching with a new type of modulation fluorometer. Photosynth. Res. 10: 51-62

Shieh, W.K. & C.J. Yee, 1985. Microbial toxicity monitor for in situ continuous applications. Biotechnol. Bioeng. 27: 1500-1506

Skipnes, O., I. Eide & A. Jensen, 1980. Cage culture turbidostat: a device for rapid determination of algal growth rate. Appl. Environm. Microbiol. 40: 318-325

Slooff, W., D. de Zwart & J.M. Marquenie, 1983. Detection limits of a biological monitoring system for chemical water pollution based on mussel activity. Bull. Environ. Contam. Toxicol. 30: 400-405

Smith, E.H. & H.C. Bailey, 1988. Development of a system for continuous biomonitoring of a domestic water source for early warning of contaminants. In: Automated Biomonitoring; D.S. Gruber & J.M. Diamond (eds). Ellis Horwood, Chichester, pp. 182-205

Solyom, P., 1977. Industrial experiences with Toxiguard, a toxicity monitoring system. Progr. Wat. Technol. 9: 193-198

Spoor, W.A., T.W. Neiheisel & R.A. Drummond, 1971. An electrode chamber for recording respiratory and other movements of free-swimming anamals. Trans. Am. Fish. Soc. 100: 22-28

Sprague, J.B. & D.E. Drury, 1969. Avoidance reactions of salmonid fish to representative pollutants. Adv. Wat. Res. 1: 169-179

Stewart, G., T. Smith & S. Denyer, 1989. Genetic engineering for bioluminescent bacteria. Food Science Technol. Today, 3: 19-22

Turner, A.P.F., G. Ramsay & I.J. Higgins, 1983. Applications of electron transfer between biological systems and electrodes. Biochem. Soc. Trans. 11: 445-448

Turner, A.P.F., I. Karube & G.S. Wilson, 1987. Biosensors - Fundamentals and applications, Oxford. Univ. Press, Oxford, pp. 770

Vasseur, P., J.F. Ferard, J. Vial & G. Larbaigt, 1984. Comparison of the Microtox and daphnia tests for determining the toxicity of industrial wastewaters. Environ. Pollut. (A), 34: 225-235

Voith, J.M., 1979. Procédé et installation de mesure de la consommation d'oxygène par des animaux de rivière ou de mer, par example pour un test avertiseur à poissons. Patent application, publ. No. and date: FR24.27.604, 28.12.1979; DE28.24.435, 03.06.1978

Wangersky, P.J. & R.L. Maass, 1990. Bioavailability: the organism as sensor. Mar. Chem. 36: 199-213

Whitton, B.A., 1991. Use of algae for monitoring rivers. J. Appl. Phycol. 3: 287

Appendix 1

Methods of BEWS

ORGANISMS FOR USE IN A BEWS

If we limit ourselves to those groups of organisms that are of (potential) use in the marine or estuarine environment, only fish, bivalves and luminescent bacteria are considered.

Fish

It will depend on the type of fish BEWS (avoidance and activity, rheotaxis, ventilation) what species is to be selected. No typical marine species have been investigated so far.
Fish are rather sensitive organisms. Vibrations, light variation, etc. will disturb them. For most applications the fish should acclimatize for several days. Regular inspection is essential, while replacement has to occur after one to two weeks.

Bivalves

Preferred organisms are bivalves that are byssus forming filter feeders: they are in closest contact with the water column, and are sedentary. Species that are naturally attached to a hard substrate by byssus threads are preferred since the recording will then not be disturbed by movement through the sediment. Examples are the common or blue mussel *Mytilus edulis*, its representative from intermediate zones *Mytilus galloprovicialis* or the tropical green-lipped mussel *Perna viridis*. In estuaries *Mytilopsis leucophaeta* can be useful. Other bivalves have been used, like the oyster *Ostrea edulis*, or the cockle *Cerastoderma edule*. Of these species the sensitivity and practical application should be tested, however.
Bivalves for use in the Musselmonitor® should be of sufficient size to handle. A minimum of 3-4 cm is considered practical, due to the rounded shape of most mussels, larger sizes are no problem. The system is tuned to a given distance between the two sensors, but electronic corrections can be made to allow for other shell sizes.
Mytilus edulis has been used in Musselmonitor® tests for up to ten weeks without replacement or obvious malfunctioning. As the system operates *in situ*, feeding is not required. One may question the effects of adaptation over such long periods, and replace the mussels after *e.g.* two months.

Bioluminescent bacteria

For application in the Microtox® test, the bacteria *Photobacterium phosphoreum* are distributed by the supplier of the test system in freeze dried form. Proper instructions are given with the test.

STATISTICAL CONSIDERATIONS

In a time series, of which the fish ventilation frequencies or the valve movement response of mussels are examples, the successive observations are usually serially dependent and often non-stationary. Most statistical tests rely upon independence, and/or special symmetry in the distribution function, such as normality. In a strict sense they are therefore not applicable. In fact no standard statistical procedures for such data seems to exist. Evans *et al.* (1986b) investigated four different statistical approaches for treatment of the data from the WRc fish monitor: the Cusum method, Mann-Whitney U-test, Nonparametric regression and Percentile test. The latter method appeared to offer the best possibilities for use in BEWS where a number of organisms are monitored either parallel or as a group.
In Figure 4 the background of the method is illustrated for a data set considering the ventilation frequency of one fish. For each organism (in case of parallel detection) the present data are collected over the time period called 'inspection window' in a moving process. This window is here set to a practical 10 min interval. Separated by a 'gap' to separate present readings from true historical background values, the data are compared with the 'background inspection window'. From this interval the limits are adjusted to the 5% and 95% percentiles; the frequency may in- or decrease. In the case of a response to only one direction, such as for the Musselmonitor®, percentiles are determined to one side only. The data of the inspection window are compared with the range between upper and lower limits thus calculated. A distress value is generated from the ratio of the number of observed outsiders to the total possible number and a threshold set based on statistical probability, evaluatory data, or a combination of them. In the Musselmonitor® a warning level may also be set by considering the ratio which might be expected, based on practical experience, to exceed, for example, once a year by chance. The current version of the Fish Monitor® goes on from this initial analysis to provide a statistical result based on the binomial distribution. To overcome any serial correlation, a certain number of these consecutive incidents has to occur before an alarm may be generated. This number is empirically determined for the BEWS in question.

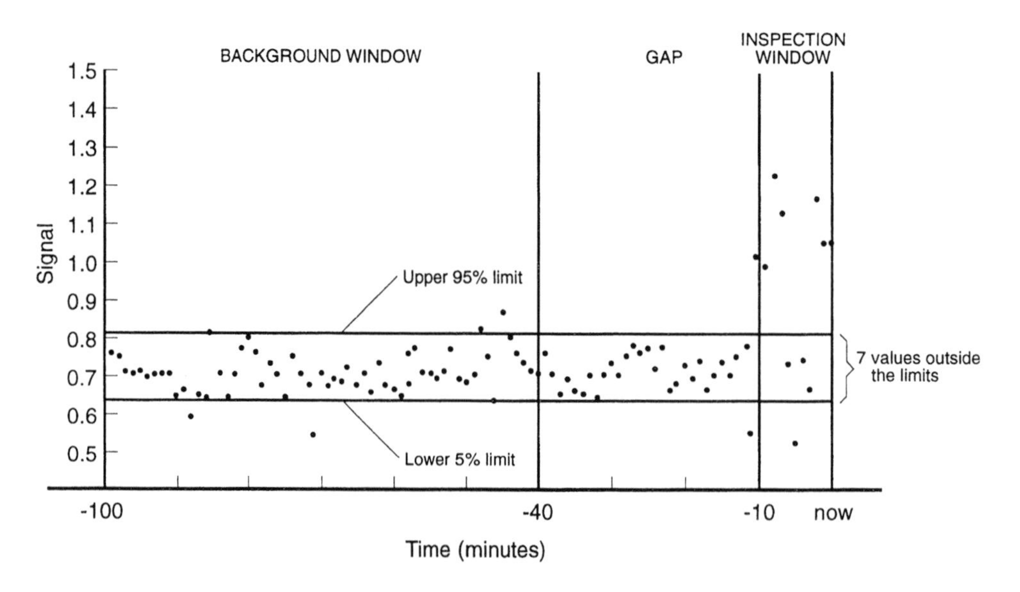

Figure 4. Example of the 'Percentile method' for the determination of a significant change in *e.g.* ventilation frequency of fish (After: Evans *et al.,* 1986)

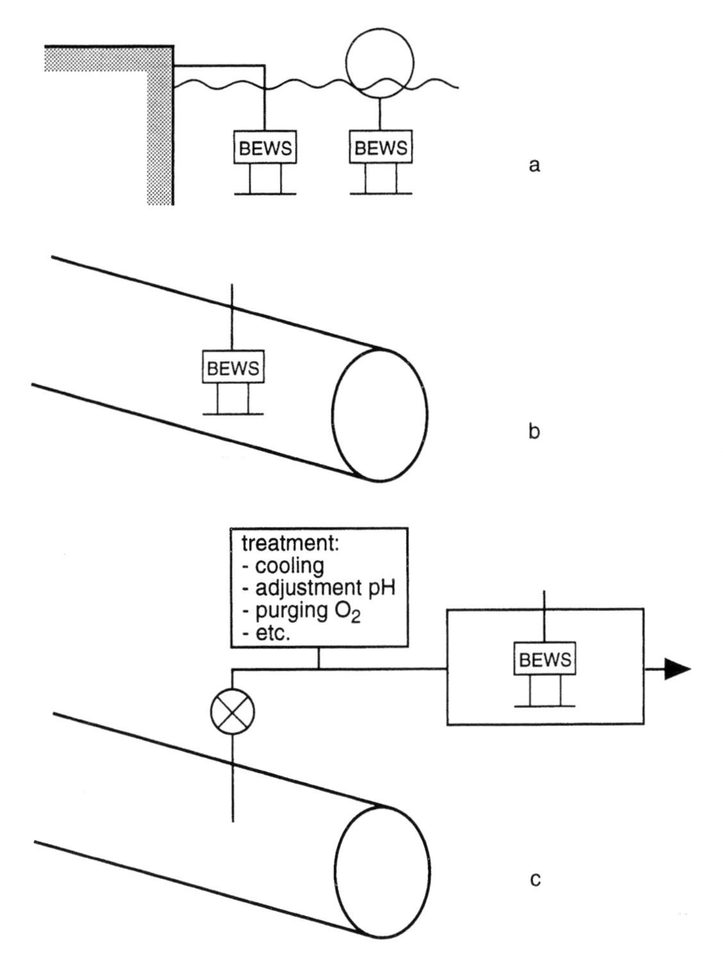

Figure 5. Lay-out of a BEWS installation involving a) *in situ* suspended from a jetty or buoy; b) in line operation; c) off-line operation in a flow-through tank; additional treatment of inflow water may be added to sustain life of the biota in the BEWS

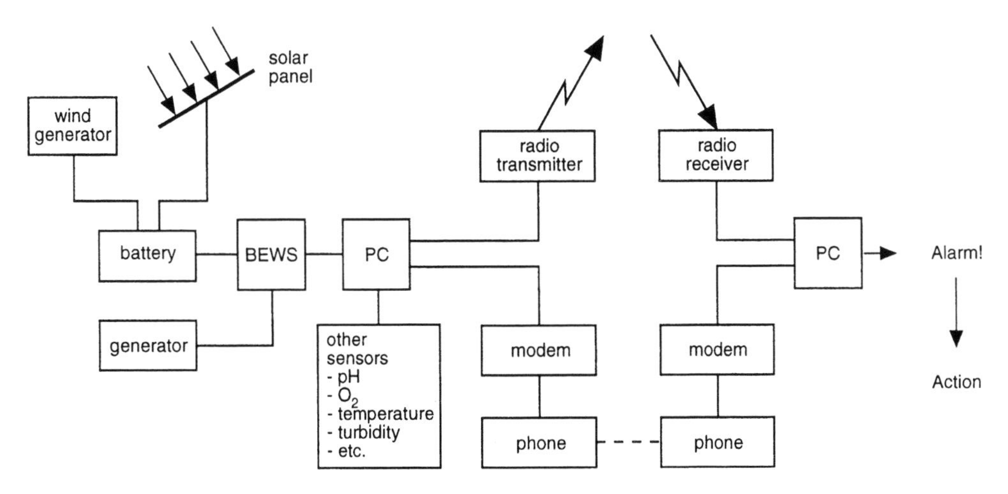

Figure 6. BEWS are often needed at rather remote locations (*e.g.* at the end of pipelines). Remote operation and control is feasible through wind generator or solar power supply and telephone/radio data transmission

LOGISTIC CONSIDERATIONS OF BEWS

In Figure 5 schematic diagrams are presented of the location of BEWS. BEWS may be placed *in situ* alongside a structure, suspended from a buoy (Figure 4a), or located directly in the effluent line (Figure 4b). Alternatively, and this will be the most occurring situation, the

BEWS is mounted in a flow-through system. In the latter case treatment of the supply becomes possible. Such a treatment may involve aeration, justification of pH and/or temperature, etc., in order to create an environment that allows the organisms to survive (Figure 4c) and to prevent interference.

In our opinion, BEWS are part of an integrated system that involves both physical, chemical and biological methods. The methods should be supplementary, and evaluation of several sensors may eliminate, or at least reduce false alarms. Triggering of an alarm could result not only in red lights or bells, but also in automated sampling, induced by the alarm, to provide physical/chemical proof of any adverse effect.
At remote locations (effluent pipes never end near a proper power supply) solar panels and radio transmission of data have been used (Figure 6). Power requirements should be rather low, however.

Chapter 2

Monitoring of Temporal Trends in Chemical Contamination by the NOAA National Status and Trends Mussel Watch Project

Thomas P. O'Connor, Adriana Y. Cantillo and **Gunnar G. Lauenstein**

Coastal Monitoring Branch
National Oceanic and Atmospheric Administration N/ORCA21
1305 East West Highway, Silver Spring, MD 20910, USA

ABSTRACT

Temporal trends in chemical contamination of the estuarine and coastal United States began being monitored in 1986 at more than 100 sites by the NOAA National Status and Trends Mussel Watch Project. The strategy of the project is to annually collect mussels or oysters from indigenous populations within a target size range at the same sites and times and to search for trends by application of non-parametric statistics to raw data. No attempt is made to adjust data for physiological or climatic conditions that could affect concentrations. Validity of identified trends is buttressed by finding common trends in groups of neighbouring sites. This strategy and its results are discussed in comparison with other approaches for detection of chemical trends through analysis of organisms.

1. INTRODUCTION

The Mussel Watch Project within the National Status and Trends (NS&T) Program of the United States National Oceanic and Atmospheric Administration (NOAA) began in 1986 with the goal of determining the status and trends of chemical contamination in the coastal and estuarine United States. While chemical concentrations were measured in surface sediment samples collected in the first two years of the programme, the main emphasis has been on the annual collection and analyses of mussels and oysters to follow annual trends in chemical concentrations. Organisms are collected for this purpose because on an annual scale they can reflect changes in chemical concentrations in their environment. Samples of surface sediment, on the other hand, respond to changes on time scales dictated by deposition and bioturbation rates, neither of which are usually known.

Biomonitoring of Coastal Waters and Estuaries. Edited by Kees J.M. Kramer.

Analyses could be made of water samples but the time scales of change could be as short as diurnal and, very importantly, extreme precautions would be necessary to avoid sample contamination.

Of particular applicability is the Mussel Watch Program in the United States sponsored by the U. S. Environmental Protection Agency that collected and analyzed mussels and oysters over the years 1976 to 1978 (Goldberg *et al.*, 1983). The state of California has been monitoring coastal waters through analyses of molluscs since 1977 (Martin, 1985).

Use of molluscs for monitoring chemical contamination is, in fact, a worldwide and relatively recent practice. Cantillo (1991) compiled a bibliography with more than 1000 entries on chemical measurements on mussels and oysters. Table 1, based on that bibliography, lists 54 countries from which there are reports. The earliest report was published in 1863 but between then and 1920 there were only 18 papers and there were fewer than 20 publications in each decade between 1920 and 1960. However in the decades of the 1960s and 1970s and since 1981 there were 75, 356, and 719 publications, respectively. For an overview of the field, the reader is referred to the bibliography. A very recent expansion of the mussel watch concept is the International Mussel Watch Program begun in 1992 with the collection of samples in South and Central America under the auspices of the Intergovernmental Oceanographic Commission of the United Nations (Tripp *et al.*, 1992).

Table 1. Number of papers, by country, reporting chemical analyses of mussels and/or oysters (Cantillo, 1991)

Country	Number of papers	Country	Number of papers
Algeria	2	Korea	4
Argentina	1	Kuwait	2
Australia	47	Malaysia	2
Bermuda	4	Melanesia	1
Brazil	6	Mexico	38
Canada	30	Monaco	2
Chile	3	Netherlands	18
China	1	New Zealand	26
Colombia	1	Nigeria	1
Croatia	8	Northern Ireland	3
Denmark	13	Norway	26
England	30	Pakistan	2
Fiji	1	Peru	1
Finland	7	Poland	1
France	39	Portugal	15
Germany	11	Scotland	15
Greece	8	South Africa	7
Greenland	2	Spain	25
Hong Kong	14	Sweden	9
Iceland	1	Taiwan	9
India	8	Tasmania	11
Indonesia	2	Thailand	8
Iraq	1	Turkey	1
Ireland	7	United States	237
Israel	1	USSR	1
Italy	34	Venezuela	1
Japan	21	Wales	12

2. METHODS

2.1. Species selection

Mussels and oysters are chosen as sentinel species because they are:

- sessile, therefore characteristic of their location;
- hardy enough to provide annual samples from each site;
- easily sampled.

Other organisms could have been chosen but none fit these three simple criteria so well. All fish, for example, are migratory to varying degrees and their chemical concentrations are not necessarily reflective of their location of capture. Furthermore, unlike molluscs, fish metabolize polycyclic aromatic hydrocarbons (PAHs), a group of compounds analyzed by the NS&T Program.

Other considerations being equal, there is added value to using organisms that have been used in the past and which are also currently being used in other parts of the world. Table 2, from Cantillo (1991), lists 81 species of mussels or oysters than have been used for chemical monitoring.

In particular, the species sampled by the NS&T Mussel Watch Project are, in clockwise progression around the continental United States, the mussel *Mytilus edulis* in sites from Maine to Delaware Bay, the oyster *Crassostrea virginica* from Delaware Bay south and through the Gulf of Mexico, and the mussels *M. edulis* and *M. californianus* on the West Coast. At sites in Hawaii the oyster *Ostrea sandvicensis.* has been collected. Figure 1 shows the distribution of sites around the continental United States.

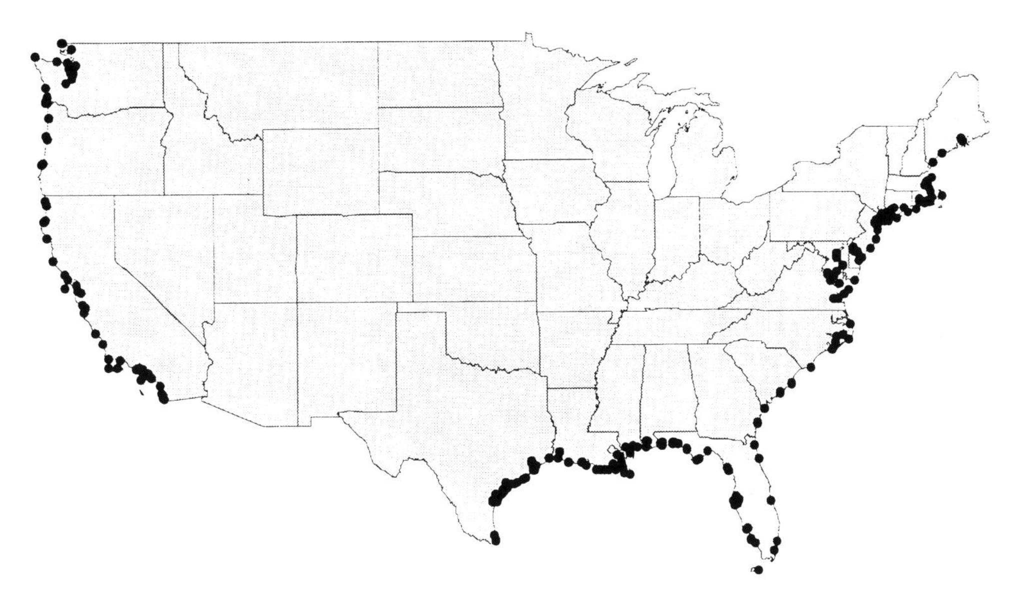

Figure 1. National distribution of NOAA NS&T Mussel Watch Sites

2.2. Chemical selection

The elements and compounds measured in the NS&T Program are listed on Table 3 and are quite representative of the chemicals reported in other analyses of mussels and oysters (Table 4). Not included on the NS&T list are silicon, which is measured in sediment but not molluscs, and antimony and thallium which were not detected in molluscs during the first two years of the programme and are no longer measured. Except for aluminium, iron and manganese, the elements in Table 3 are all possible contaminants in the sense that their concentrations in the environment have been altered by human activities (Nriagu, 1989). The mere existence of the chlorinated organic compounds and butyltins is due to human activities. Polycyclic aromatic hydrocarbons are similar to metals in the sense that they occur naturally.

Table 2. Number of papers reporting chemical analysis data of various mussel and/or oyster species (Cantillo, 1991)

Species	Common name	Number of papers
Arca zebra	Turkey wing	5
Aulacomya ater [1)]	Black ribbed mussel	1
A. maoriana	Maori mussel	2
Brachydontes variabilis	Mussel	1
B. demissus plicatulus [2)]	Ribbed mussel	1
Choromytilus meridionalis [1)]	Mussel	4
Crassostrea angulata	Portugese oyster	20
C. brasiliana [3)]	Oyster	1
C. commercialis[4)]	Rock oyster	11
C. corteziensis	Oyster	3
C. cucullata [4)]	Rock oyster	4
C. edulis [5)]	Edible oyster	1
C. gasar	Oyster	1
C. gigas [6)]	Pacific oyster	111
C. glomerata[7)]	Auckland rock oyster	4
C. laperousei [6)]	Oyster	1
C. lugubris	Oyster	1
C. madrasensis	Oyster	3
C. margaritacea	Cape rock oyster	3
C. rhizophorae [3)]	Caribbean edible oyster	3
C. rivularis	Oyster	1
C. virginica	Eastern oyster	153
Geukensia demissa [1)]	Ribbed mussel	1
Gryphaea angulata	Portugese oyster	3
Isognomon alatus	Flat tree-oyster	1
I. isognomon	Tree oyster	1
Malleus meridianus	Hammer oyster	1
Modiolus auriculatus	Mussel	1
M. barbatus	Bearded horsemussel	1
M. capax	Fat horsemussel	2
M. demissus [1)]	Ribbed mussel	6
M. demissus plicatulus [1)]	Ribbed mussel	1
M. modiolus	Northern horsemussel	19
M. neozelanicus	Mussel	1
Mytella falcata	Falcate swamp mussel	1
M. strigata	Mussel	1
Mytilus afer	Mussel	1
M. californianus	California mussel	69
M. coruscus	Hard shell mussel	4
M. edulis	Blue mussel	448
M. edulis aoteanus	Mussel	2
M. edulis chilensis	Mussel	2
M. edulis planulatus	Mussel	7
M. galloprovincialis	Mediteranean blue mussel	100
M. magellanicus	Mussel	1
M. minimus	Mussel	1
M. obscurus	Mussel	2
M. platensis	Mussel	1
M. smaradignus [8)]	Mussel	1

Table 2. (Cont'd)

Species	Common name	Number of papers
M. striagata	Mussel	1
M. trossulus	Mussel	1
M. viridis [8]	Green mussel	5
Ostrea angasi [9]	Oyster	6
O. angulata	Oyster	1
O. circumpicta	Oyster	1
O. edulis [5]	Edible oyster	63
O. equestris	Crested oyster	1
O. gigas [6]	Pacific oyster	7
O. heffordi	Oyster	1
O. lurida	Native Pacific oyster	6
O. lutaria [10]	Bluff oyster	3
O. pliculata	Oyster	2
O. sandvicensis	Hawaiian oyster	5
O. sinuata [9]	Pt. Lincoln oyster	4
O. spinosa	Oyster	1
Perna canaliculus	Channel mussel	3
P. indica	Mussel	1
P. perna	Brown mussel	5
P. viridis	Green-lipped mussel	26
Pinctada carchariarium	Shark Bay pearl oyster	1
P. fucata martensii	Japaneses pearl oyster	1
P. margaritifera	Black lipped pearl oyster	2
P. vulgans	Oyster	1
Saccostrea commercialis	Sydney rock oyster	5
S. cucullata [4]	Rock oyster	11
S. echinata	Spiny oyster	2
S. glomerata [7]	Oyster	4
S. iridescens	Iridescent oyster	1
Septifer bilocularis	Box mussel	1
Stavelia horrida	Hairy mussel	1
Tiostria lutaria [10]	Oyster	1
Oysters (unspecified species)		124
Mussels (unspecified species)		110

[1] *Choromytilus meridionalis* is a form of *Aulacomya ater;*
[2] *Modiolus demissus, M. demissus plicatulus* and *Brachydontes demissus plicatulus* are junior synonyms of *Geukensia demissa; Brachidontes* is the currently accepted spelling of *Brachydontes*;
[3] *Crassostrea brasiliana* and *C. rhizophorae* are probably the same species;
[4] *Crassostrea commercialis, C. cucullata*, and *Saccostrea cucullata* are the same species;
[5] *Crassostrea edulis* same as *Ostrea edulis*;
[6] *Ostrea gigas* and *Crassostrea laperousei* are junior synonyms for *Crassostrea gigas;*
[7] *Saccostrea glomerata* same as *Crassostrea glomerata*;
[8] *Mytilus smaradignus* probably same as *Mytilus viridis*;
[9] *Ostrea angasi* same as *Ostrea sinuata*;
[10] *Tiostria lutaria* same as *Ostrea lutaria.*

They are found in fossil fuels such as coal and oil and are produced during combustion of organic matter. Their environmental presence, however, is also attributable to humans because they are released during the use and transportation of petroleum products, while a multitude of human activities, from coal and wood burning to waste incineration, create PAH compounds in excess of those that exist naturally. Almost all the chemicals in Table 3 are also on the list of 127 Priority Pollutants created by the United States Environmental Protection Agency in the late 1970's (Keith & Teillard, 1979).

Table 3. Chemicals measured in the NOAA NS&T Program

DDT and its metabolites	Polycyclic aromatic hydrocarbons	Major elements	
2,4'-DDD	*2-ring*	Al	Aluminum
4,4'-DDD	Biphenyl	Fe	Iron
2,4'-DDE	Naphthalene	Mn	Manganese
4,4'-DDE	1-Methylnaphthalene		
2,4'-DDT	2-Methylnaphthalene		
4,4'-DDT	2,6-Dimethylnaphthalene		
	1,6,7-Trimethylnaphthalene		
		Trace elements	
Chlorinated pesticides other than DDT	*3-ring*		
	Fluorene	As	Arsenic
	Phenanthrene	Cd	Cadmium
Aldrin	1-Methylphenanthrene	Cr	Chromium
cis-Chlordane	Anthracene	Cu	Copper
trans-Nonachlor	Acenaphthene	Pb	Lead
Dieldrin	Acenaphthylene	Hg	Mercury
Heptachlor		Ni	Nickel
Heptachlor epoxide	*4-ring*	Se	Selenium
Hexachlorobenzene	Fluoranthene	Ag	Silver
Lindane (γ-HCH)	Pyrene	Sn	Tin
Mirex	Benz[*a*]anthracene	Zn	Zinc
	Chrysene		
Polychlorinated biphenyls		Tri- di- and mono-butyltin	
	5-ring		
PCB congeners 8, 18, 28, 44, 56, 66, 101, 105, 118, 128, 138, 153, 179, 180, 187, 195, 206, 209	Benzo[*a*]pyrene		
	Benzo[*e*]pyrene		
	Perylene		
	Dibenz[*a,h*]anthracene		
	Benzo[*b*]fluoranthene		
	Benzo[*k*]fluoranthene		
	6-ring		
	Benzo[*ghi*]perylene		
	Indeno[1,*2,3-cd*]pyrene		

2.3. Differences between species

The central premise to any chemical monitoring programme using biological samples is that chemical concentrations in the organism depend on concentrations in the water and food processed by that organism. The size of the Cantillo (1991) bibliography is testament to the empirical fact that chemical concentrations in mussels and oysters bear some relationship to their environment. Experimental evidence that this is the case for molluscs comes from laboratory experiments where conditions of exposure are controlled (*e.g.* Cunningham & Tripp, 1975; Fisher & Teyssie, 1986; Pruell *et al.*, 1987), and from field experiments where oysters or mussels are transplanted from areas of low contaminant concentrations to areas of high concentration and vice versa (Roesijadi *et al.*, 1984; Martin, 1985; Capuzzo *et al.*, 1989; Sericano *et al.*, 1992; see also de Kock & Kramer, this volume).

Molluscs are alive, growing, and reproducing. The simple fact that they are alive and are pumping ambient water through themselves makes them good monitors of change. It also means, however, that consideration must be given to the possibility that there are physiological influences on their concentrations of trace chemicals.

Table 4. Elements and groups of organic compounds found by Cantillo (1991) as having been listed in papers reporting analyses of mussels or oysters

Element	Number of papers	Organic group[a]	Number of papers
Ag	59	hydrocarbons	91
As	60	aromatic hydrocarbons	14
Cd	282	chlorinated hydrocarbons	19
Cr	90	chlorinated pesticides	80
Cu	325	DDT	110
Hg	156	dioxins	5
Ni	96	PCBs	164
Pb	210	pesticides	18
Sb	12	organomercury	8
Se	33	organotin	93
Sn	23		
Tl	3		
Zn	292		

[a] These groups are not mutually exclusive but represent the descriptions given in the original citations

The strongest physiological influence is simply that oysters and mussels are not equal in their ability to concentrate chemicals. At a few sites, two species have been collected by the NS&T Program. Figure 2, based on data from collections made in November 1989 at a site at the mouth of the Housatonic River in Long Island Sound, demonstrates that the trace elements silver, copper and zinc are highly enriched in the oyster, *C. virginica,* while chromium and lead are more than three times higher in the mussel *M. edulis*. For other elements and for organic compounds there is no strong species effect between mussels and oysters. This result

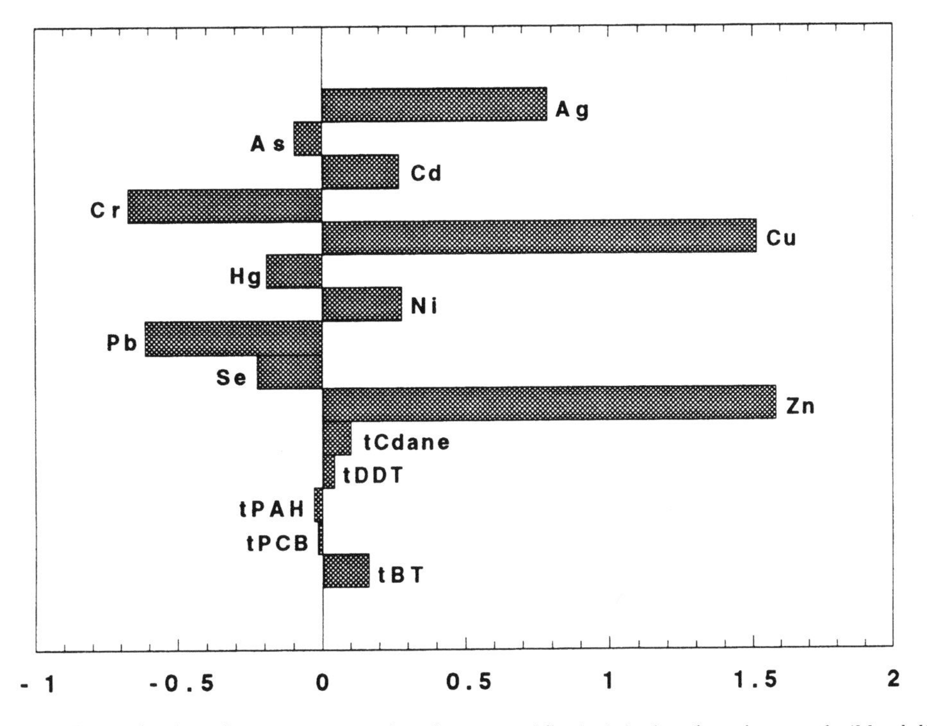

Figure 2. Logs of ratios of mean concentrations in oysters *(C. virginica)* to those in mussels *(M. edulis)* collected at the NOAA NS&T Housatonic River site in Long Island Sound in 1989

is not unique to that site or year. This same tendency for preferential concentration of Pb and Cr by mussels and Ag, Cu, and Zn by oysters was shown in 1987 and in 1988 at the Housatonic River site and in 1989 at two others sites in Long Island Sound. On the other hand, two species of mussels, *M. edulis* and *M. californianus*, collected at one site off the Columbia River showed no species preferences for any trace chemicals (O'Connor, 1993).

2.4. Physiological effects within a species

At least for some trace element concentrations, the species specificity precludes comparisons between sites with mussels and those with oysters. This is the reason why the NOAA NS&T Program bases its large scale national comparisons among sites on data obtained from analyses of sediments (O'Connor, 1990; O'Connor & Ehler, 1991). However, so long as the species remains constant, temporal trends in concentrations at any given site can be sought through annual collections and analyses of mollusc tissues.

Here the critical concern for the influence of physiology is on whether some attribute of the organism or the environment other than a change in chemical concentration can cause changes in the concentration in the organism. Phillips (1980) and Phillips & Segar (1986) discuss several possible reasons:

- organism size;
- location relative to tide;
- salinity;
- temperature;
- reproductive state.

The NOAA NS&T Program attempts to limit these influences by annually collecting organisms within:

- a set size range;
- at the same site;
- in the winter prior to the spawning season.

With these factors held constant by design, the NS&T approach to defining trends has been to look for statistically significant patterns in the directions of change of dry-weight based concentrations without any attempt to adjust those concentrations for physiological influences (O'Connor, 1992). It is, however, important to review other approaches to trend detection based on attempts to adjust measured concentrations in molluscs for factors other than external concentrations of chemicals.

2.5. Importance of season

Reproductive state has been considered an important determinant of chemical body-burden because in discharging eggs or sperm molluscs lose biomass and possibly trace chemicals. The remaining biomass should be more concentrated in chemicals not expelled with reproductive products and, possibly, less concentrated in other trace chemicals. The influence of reproductive state should be cancelled by annually sampling molluscs at the same time of year as done by the NS&T Program. Samples are collected in winter (November through March, but within 21 days of an annual sampling date at each site). Since spawning is keyed to temperature, this sampling strategy should insure that organisms are in always in a pre-spawning state. On the basis of examinations of gonadal state, that is generally true but, on occasion, molluscs have been found in the southern parts of the country to be in spawning condition in the winter.

Empirically, the importance of season on chemical concentrations can be estimated by comparing data from organisms collected at different times of the year. For trace elements the empirical data are inconsistent. Goldberg *et al.* (1983) collected mussels at a site in Bodega Bay, California, in 30 of 36 months from 1976 to 1978, and in 20 of those months from a site in Narragansett Bay, Rhode Island. They reported no systematic seasonality in changes in trace metal concentrations at the Bodega Bay site, but did find at the Narragansett Site that concentrations of lead, cadmium, zinc, and nickel (but not copper or silver) tended to be higher in winter than in summer. This type of result shows that seasonality may well affect measured concentrations, but not in a systematic manner that would allow raw data to be adjusted.

Since eggs and sperm are enriched with lipids, one would expect concentrations of lipophilic trace organic compounds (*e.g.* all of those measured by NS&T) to decrease after spawning. In principle, the way to account for this is to examine trace organic concentrations on a lipid basis rather than on the basis of whole body weight. In fact, adjusting data on the basis of lipid weight, adds rather than diminishes variability.

Capuzzo *et al.* (1989) transplanted mussels to the highly contaminated harbour in New Bedford, Massachusetts and, over a year, made biweekly measurements of PCB concentrations. They reported concentrations of three PCB congeners on both a dry-weight basis and on the basis of lipid content. After about day 50 the dry-weight based concentrations reached levels that remained steady within a factor of less than two for the remaining 300 days. The lipid-based concentrations, on the other hand, were much more variable and ranged by more than a factor of three. The central difficulty with lipid-based concentrations is that, within a given location, the amount of lipid itself varies more than does the amount of a trace organic compound. This is probably why, on a site-by-site basis, the variability in concentrations of organic compounds when expressed on a lipid basis has been found (NOAA, 1989) to be higher than for concentrations expressed on a dry-weight basis.

2.6. Adjustments for physiological effects

2.6.1. Relation with shell weight

Fischer (1983) proposed that for molluscs there was a useful relationship between total mass of cadmium in a mollusc and the weight of the mollusc shell such that:

$$\log[Cd] = a + b.\log[\text{shell weight}] \qquad (1)$$

where [Cd] is the mass in µg of cadmium in the whole soft parts of the organisms, and [shell weight] is the weight of shell in g. This was worked out in detail for *M. edulis* where large numbers of individuals (50 to 100) collected at various sites and in various seasons were analyzed. On log-log plots, the intercept (a) was found to range from -2.4 to -1.3, while the slope (b) ranged from 0.9 to 1.3 but was usually very close to 1.0. Since the slopes in these relationships were essentially the same, Fischer recommended that comparisons among cadmium concentrations at different sites be based on the intercepts of log-log plots of total cadmium in soft parts versus shell weight. The argument was essentially that site- and season-specific physiological factors causing shifts in the amount of soft weight in a mussel would alter concentrations based on body weight but that concentrations based on shell weight were independent of physiological influences. Trends at a given site (Fisher, 1989) would, in turn, be identified by changes in the intercept of annual log-log plots.

Rather than establish a relationship between chemical content and shell weight, the NS&T approach has been to collect composite samples of mussels and oysters within preselected size ranges of 50 to 80 mm and 70 to 100 mm, respectively. Developing a regression on shell weight would require expanding that range and, since the intercept (at shell weight = 1 g) is the critical parameter, it would be particularly important to analyze small specimens.

Rees & Nicholson (1989) recently applied the Fischer method to seek temporal trends in lead levels in the horse mussel *(Modiolus modiolus)* annually collected at two sites from 1984 to 1987. They found significant correlations between log of total lead and log of shell weight. Unlike Fischer (1983), they did not find nor did they consider the slopes (1.00 and 1.22) to be essentially the same at both sites. Fortunately, the slopes at a given site did not vary with year. Therefore, while the intercepts of log-log plots could not be compared between sites, they were the parameter that would reveal temporal trends at each individual site. No temporal trends were found.

2.6.2. Relation with shell length

Stronkhorst (1992) sought trends in concentrations of several trace elements and trace organic compounds in mussels *(M. edulis)* annually collected at one site in each of two Dutch estuaries (of the rivers Ems and Scheldt) from 1985 through 1990. He used the semi-log relationship:

$$\log[\text{Conc}] = a + b[\text{Length}] \tag{2}$$

where [Conc] is chemical concentration per unit weight of soft tissue. He tested for the temporal dependence of the slope at each location. If it was found to be independent of year, a term representing linear dependence of concentration on year was added to the equation. The justification for the semilog model of dependence of concentration on length was that concentrations were lognormally distributed and experience that the relationship was found to work. Nevertheless, despite the model, trends were also sought by plotting annual geometric means of unadjusted concentrations. With data for only 6 years, a strong linear pattern in concentration is needed to find a trend. There was no such pattern for Cd, Cu, Zn, Cr, As, Ni, Pb, or Hg concentrations on a wet-weight basis at either site. The organic chemicals: PCBs, DDE, lindane and dieldrin, expressed on a lipid weight basis, also showed no trends except for a decrease in dieldrin at one site. This same conclusion also resulted from sequential testing of the semilog model. While size did account for some of the overall variability, there were no trends except for dieldrin at the one site. The author stated that results for metals, while shown on a wet-weight basis, were essentially the same when examined on a dry-weight basis. For trace organic contaminants, the results, given on a lipid basis, were said to be the same if given on a wet-weight basis.

2.6.3. Adjustments to concentrations measured in fish

Adjustments for physiological factors prior to trend detection are also exemplified by attempts to detect trends in ambient concentrations of chemicals through annual analyses of fish. Stronkhorst (1992) in the same paper just discussed for mussels also sought trends in concentrations in flounder, *Platichthys flesus.* Data were plotted and, as with the mussel data, were also subjected to sequential testing of a semilog model. Plotted geometric means showed essentially the same result as did the model, *i.e.* no trends except for Cd increasing in the Ems Estuary. Interestingly, the PCB trend was only evident on a wet-weight basis. Since lipid content of livers also showed a decreasing trend, PCB concentrations on a lipid weight basis showed no trend.

Jensen & Cheng (1987) looked for trends in concentrations of Cd, Cu, Hg, Pb and Zn in livers of flounder from two areas along the Danish Coast sampled in 1976 through 1984. Data were tested, using the semilog model, for dependence of concentration on fish length, weight, sex, age and liver weight. Trends were sought after establishing that these dependencies differed between sites but, within a site, were independent of year. Jensen & Cheng (1987) also provided annual arithmetic means, to which we applied a simple non-parametric test (Spearman rank correlation, 0.05 level of significance). Results from the simple test were almost identical to those for the adjusted data: increasing trends for Cd and Zn concentrations in both areas and for Cu in one area. Differences were that the adjusted data indicated that males and females had to be treated separately and that at one site the increasing trend appeared only in males, while at the other only females showed the trend. At one site, where the non-parametric test found no trend in the Pb data, the semilog model did imply an increasing trend. However, Jensen & Cheng considered this an artifact produced by a very high mean in one year. Non-parametric tests, on the other hand, respond only to relative rankings of concentrations and not their absolute size.

Not all authors have adjusted data for physiological influences to identify trends. Suns *et al.* (1991), for example, analyzed organochlorine concentrations in whole specimens of young-of-the year (41 to 65 mm) spotted shiner, *Notropis hudsonius,* taken in September from six sites in Lake Ontario between 1975 and 1987. Not all sites were sampled in each year. Suns *et al.* (1991) found that examining data on a lipid-weight, rather than wet-weight, basis did not alter results. Parametric correlations between annual mean concentrations and year were significant and decreasing for both PCB and DDT at two sites. Application of Spearman correlation tests (0.05 level), by this author, to the means provided by Suns *et al.* (1991) also revealed those trends. In one case, a very high concentration in the first year led to a parametric trend that did not appear in the non-parametric analysis.

3. TREND DETECTION

3.1. Trends within single programmes

A common attribute of all the above-cited papers is that temporal trends were sought through parametric regressions between year and either raw concentrations or concentrations somehow adjusted for physiological factors. Such tests require comparing concentrations from normal populations with a common variance and that requirement is one justification for using logarithms of concentration. The non-parametric approach used by the NS&T Program (O'Connor, 1992) is independent of the distribution of the data. Also, rather than attempting to account for physiological influences at each site, in each year, and for each chemical, the NS&T approach identifies trends that can then be examined for possible physiological artifacts. In the absence of a trend in the raw concentrations, it is assumed that adjustments to the data will not reveal otherwise. Lastly, unlike the above-cited monitoring programmes, the NS&T Program annually samples more than 100 sites. This provides a basis for testing the validity of apparent trends by seeking similar behaviour among neighbouring sites.

The power of both parametric and non-parametric tests for trends is limited by the number of years for which there are data. Regardless of how many samples are taken at a site in a single year, when searching for annual trends, the number of cases (n) to be used for testing the significance of a correlation coefficient is the number of years. Taking multiple samples in a single year will increase confidence in how well a calculated concentration represents that year but it will not increase n for the overall correlation with year. If the objective is to identify differences between years, multiple analyses in each year will increase the ability to do that and to assign progressively smaller differences as "significant" as the numbers of analyses increases. However, that objective is separate from finding temporal trends. It is conceivable, for example, that strong correlations can be found between concentration and year (particularly if concentration is changing monotonically), but for there to be no significant difference between any two concentrations.

3.2. Long term trends from combined datasets

Lauenstein *et al.* (1990) used data from analyses of molluscs collected in different seasons and different decades at fifty sites common to the earlier and the current Mussel Watch Project in the United States. Means for the earlier programme were calculated from data on single composite samples collected in the summers and falls of 1976, 1977 and 1978. These were compared with means based on three composite samples collected in the winters of

1986, 1987 and 1988. Figure 3 displays the lead and cadmium data from that analysis. There are differences among sites and differences between species but the same tendencies for lead concentrations to be highest in mussels from urban areas and for cadmium to be higher on the Gulf of Mexico and Pacific, relative to the Atlantic Coast, appear in both sets of data, collected in different seasons a decade apart.

Lauenstein *et al.* (1990) concluded that lead had decreased between the decades because at 36 of the 50 sites the mean concentrations from the 1970's were greater than the more recent means. On the other hand, mean cadmium concentrations were higher in the 70s at 26 sites and higher in the 80s at 24 sites. Similarly, because differences were not predominantly in one direction or the other, no national decadal difference could be found for four other metals measured in both programmes: Ag, Cu, Ni and Zn. In essence, while trend data could be extracted from comparison of the two sets of data, the most common result, despite the different seasons and decades, was that concentrations had not changed.

4. CASE STUDIES

Results of trend detection on annual scales by Rees & Nicholson (1989) and by Stronkhorst (1992) have been described above. In both cases data were from two sites and the time scales were four to six years. Results described in this section are for more than 100 sites and, in the case of the French programme, for more than ten years at some sites.

4.1. Analysis of trends in the NS&T project

Temporal trends in molluscan concentrations of 14 chemicals annually measured by the NS&T Mussel Watch Project at 141 sites from 1986 through 1990 have been sought through two non-parametric statistical tests (O'Connor, 1992). The chemicals were ten trace elements.

Table 5. Results of a Sign test applied to mean chemical concentrations at the 141 sites sampled in at least four of the years between 1986 and 1990[a)]

Chemical	86-87	87-88	88-89	89-90	86-90
Ag	-	-	-	Inc	-
As	Dec	-	Dec	Inc	-
Cd	Dec	-	-	-	Dec
Cr	Dec	-	-	-	-
Cu	-	-	-	-	-
Hg	-	-	-	-	-
Ni	-	Dec	Dec	-	Dec
Pb	Inc	-	Dec	Inc	-
Se	-	Inc	Dec	-	-
Zn	Dec	-	-	Inc	-
ΣPCB	Dec	-	-	Dec	Dec
ΣDDT	-	Dec	-	-	-
ΣCdane	Inc	Dec	-	-	Dec
ΣPAH	b)	Inc	Dec	-	-[b)]
ΣBT				Dec	

a) Comparisons made on a year-to-year basis and between 1986 and 1990. A statistically significant (0.05 level) proportion of changes in the increasing (Inc) or decreasing (Dec) direction are indicated, as are cases with no significant direction of change (-). With 141 samples a direction of change is significant if it is common to at least 83 samples. Since it was not routinely measured in earlier years, ΣBT comparisons have been made for the 149 sites common to 1989 and 1990

b) ΣPAH data for 1986 were not used. The national geometric mean for that year was 430 $ng.g^{-1}$ which is 1.4 times the next highest annual geometric mean. Its high value may have been due to the analytical method used in that year for samples from the East and West Coasts. The last comparison for ΣPAH is an 87-90 comparison

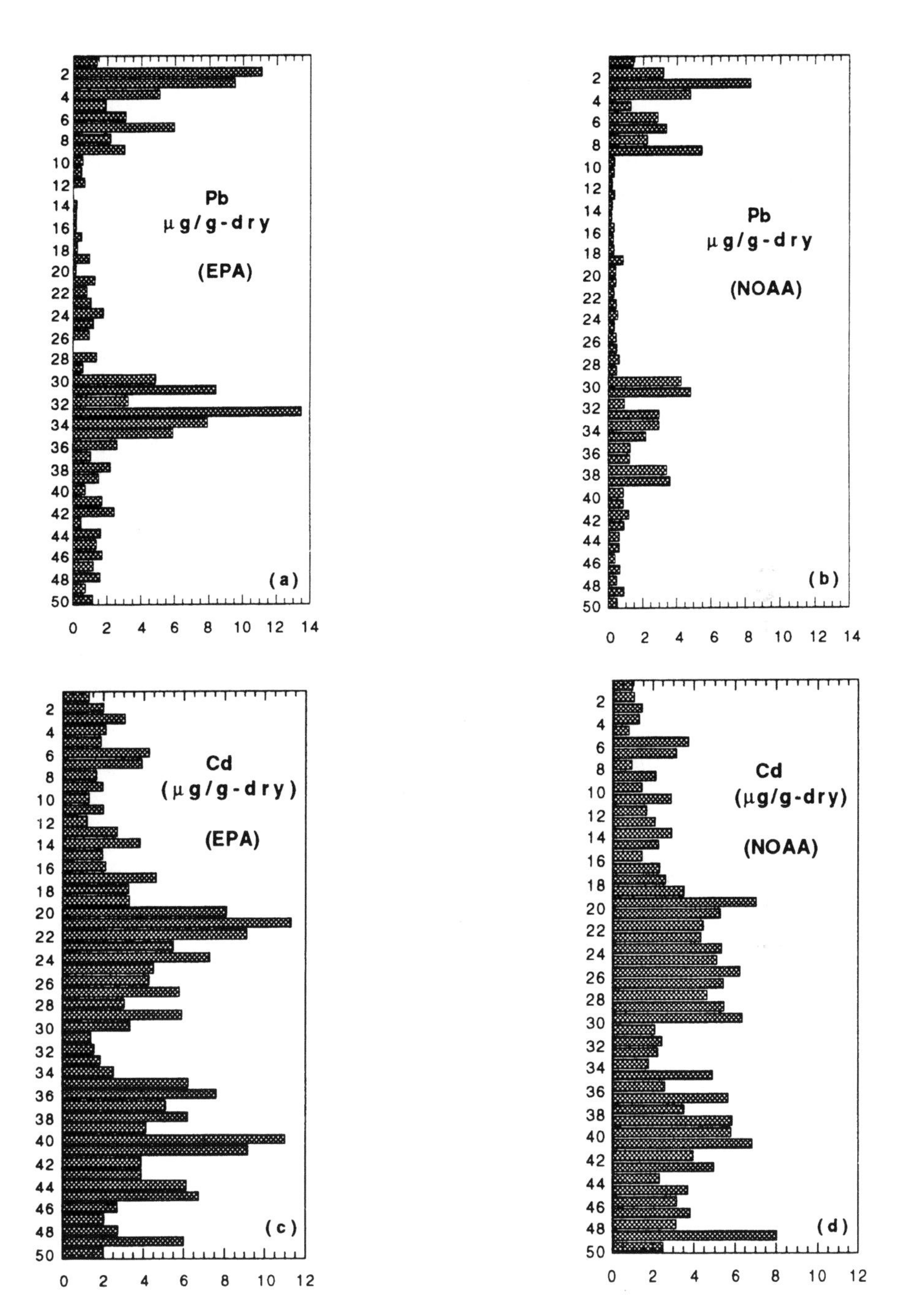

Figure 3. (a-d). Mean concentrations ($\mu g.g^{-1}$ dry weight) of lead (Figures a and b) and cadmium (Figures c and d) in whole soft parts of mussels collected at 50 sites by the EPA-sponsored program (Figures a and c) in 1976-1978 and by the NOAA NS&T Program in 1986-1988 (Figures b and d). Sites are listed clockwise around the United States. Sites 1 to 16 are from Maine to the East Coast of Florida, sites 17 to 29 extend from the West Coast of Florida to Western Texas, sites 30 to 50 are from Southern California to Puget Sound, WA. Oysters were collected at sites from 10 (Chesapeake Bay) through 29. Mean lead concentrations are higher in the 1970s at all but the following 14 sites; 1, 8, 9, 13, 16, 18, 20, 27, 37, 38, 39, 40, 43, 49. Cadmium is higher in the 1970s only 26 times out of 50 and cannot be considered different. Cadmium concentrations are statistically different (Kruskall-Wallis test, .05 level) in 12 cases; lead concentrations are statistically different in seven. All mean concentrations, site names, and site locations appear in Lauenstein *et al.* (1990)

(Ag, As, Cd, Cr, Cu, Hg, Ni, Pb, Se and Zn) and four aggregated groups of organic compounds (Σchlordane, ΣDDT, ΣPCB and ΣPAH). A fifth compound group, butyltins, began being measured only in 1989 (Table 3).

One test, the Sign test, is based on the fact that there are many sites with which to examine year-to-year changes. The other test, the Spearman rank correlation, examines the correlation between the ranks of concentration and year at each individual site.

Results of the Sign test in Table 5 show that, when viewed on a national scale, there usually was no particular direction of change between years in the concentrations of any chemical. Where between year changes occurred, they were most often a decrease and that over five years, from 1986 to 1990, the only changes were decreases. Decreases identified by this test are trends in the sense that a statistically significant majority of the sites showed the same behaviour. The Spearman test, on the other hand, based on correlations between concentration and time, searches for a statistically significant linear pattern at each site.

With only five years of data (n=5), a Spearman correlation coefficient must be at least 0.9 to be significant (at the 0.05 level), and only very strong trends are evident. Nevertheless, out of 1974 tests (14 chemicals x 141 sites) there were 239 chemical/site combinations with such strong trends: 152 decreasing and 87 increasing. However, as prescribed by the accepted level of significance, 5% or about 99 (0.05 x 1974) of the apparently strong trends can be nothing but random juxtapositions of concentrations and years. One way to assess which trends are most likely real is to find similar behaviour among groups of neighbouring sites.

Table 6 contains the annual mean concentrations of cadmium at the nine NS&T Mussel Watch sites in Long Island Sound (Figure 4) and their Spearman rank correlations with year. At seven of those sites there is a strong decreasing trend in cadmium. While the reason for this has not been determined, it is not a random result. It is not likely to be due to physiological factors such as lipid content or shell length because neither showed consistent and strong trends among the Long Island Sound sites. Nor is it likely that external factors such as salinity or temperature are responsible. These factors can affect concentrations (Phillips, 1980). They were measured at the time of mollusc collection and showed no trends. While climatic conditions at the exact time of collection may not be representative of general conditions at the site, if there were consistent trends in climatic conditions that affect molluscan concentrations of chemicals, they would have to be operating over spatial scales larger than Long Island Sound. Yet the trends observed in the Sound were not observed at sites to its north or south.

Table 6. Annual mean concentrations of Cd ($\mu g.g^{-1}$ dry weight) at nine NS&T Mussel Watch sites[a)] in Long Island Sound and Spearman rank correlation coefficients (r_s) between concentration and year

Year	LICR[a]	LIHH	LIHR	LIHU	LIMR	LINH	LIPJ	LISI	LITN
1986	3.1	3.2	5.7	3.0	3.4	2.8	3.1	2.2	4.6
1987	2.8	3.7	3.5	1.9	2.3	2.2	2.0	2.4	3.5
1988	2.4	2.6	2.0	2.0	2.0	1.7	2.1	1.7	2.3
1989	2.1	1.9	2.2	1.4	1.9	1.1	1.4	1.5	1.3
1990	2.1	1.7	2.3	1.7	1.3	1.6	1.5	1.5	1.2
r_s	-0.9	-0.9	-0.6	-0.8	-1.0	-0.9	-0.8	-0.9	-1.0

a) Sites are in Long Island (LI) Sound (Figure 4):
LICR = Connecticut River; LIHH = Hempstead Harbor; LIHR = Housatonic River; LIHU = Huntington Harbor; LIMR = Mamaroneck; LINH = New Haven; LIPJ = Port Jefferson; LISI = Sheffield Island; LITN = Throgs Neck

The French mussel watch programme, the National Network for the Observation of the Quality of the Marine Environment (Réseau National d'Observation, R.N.O.) published an analysis of secular trends over the period 1979 to 1990 (Claisse, 1989; RNO, 1992). Trends were calculated at 109 sites for concentrations Cd, Cu, Hg, Pb, Zn, PCBs, ΣDDT, α-HCH, γ-HCH (lindane), and PAHs. The molluscs collected have been mussels, *M. edulis* and *M. galloprovincialis* at 69 sites, and the oyster *Crassostrea gigas* at 35 sites. At five sites where collections ceased in 1986 or 1988, cockles *Cerastoderma (Cardium)* spp., were collected. Results in Table 7 show that, as the NOAA NS&T case, the most common result was no trend. Where trends were found decreases were much more common than increases. However, for one chemical group, ΣDDT, there were no increases and sites with decreases were three times more common than sites without trends. The original report lists trends not just for ΣDDT but for its components DDE, DDD and DDT as well, but the conclusion remains that there is an overwhelming decreasing trend.

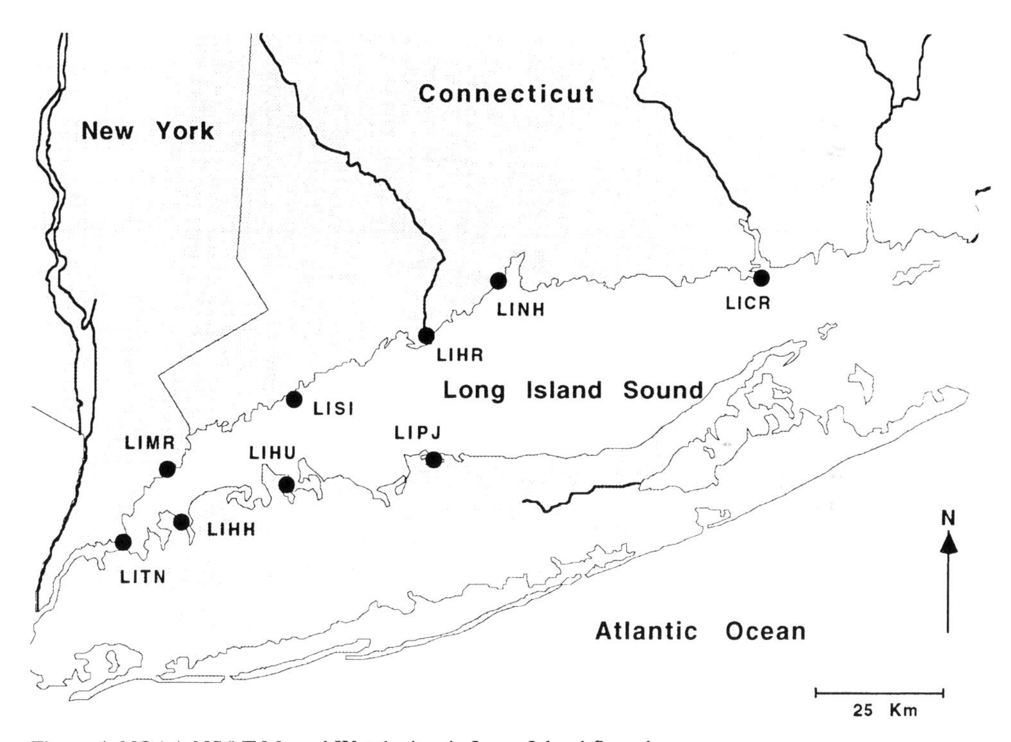

Figure 4. NOAA NS&T Mussel Watch sites in Long Island Sound

4.2. Mussel Watch in France

All concentrations were on a dry-weight basis. Trends were identified at sites where the slope on plots of concentration versus time was significantly different from zero at the 95% level of confidence. The standard procedure in the French programme is to collect composite samples at each site in each season, so that there are four samples per year. While details were not explicitly provided it appears from the figures in RNO (1992) that each sample was used in the correlation. In effect, if four samples were collected each year from 1979 through 1990, the correlation would be done with 48 concentrations over 12 years. The practice in the NS&T Program would have been to reduce the data to one value per year so that the "n" in the correlation corresponded to the number of years and to have examined the correlation non-parametrically. However, it is unlikely that doing so would have altered the main outcome.

Table 7. Trends found through analysis of concentrations measured in mussels and oysters collected at sites in France from 1979 through 1990 (RNO, 1992)

	numbers of sites with		
Chemical	**increasing trend**	**decreasing trend**	**no trend**
Hg	1	23	85
Cd	4	15	90
Pb	8	24	77
Zn	16	5	88
Cu	13	1	95
PCBs	1	19	89
ΣDDT	0	86	23
α-HCH	9	69	31
γ-HCH	2	31	76
PAHs	5	19	85

CONCLUSION

Annual sampling and chemical analysis of molluscs can reveal temporal trends in chemical contamination of the environment. Within the NOAA NS&T Mussel Watch Project, care is taken to be annually consistent in terms of sampling time and place and size range of organisms. However, analysis of trends is through application of non-parametric statistics to raw concentration, with no attempt to adjust those concentrations for physiological or external factors. Trends have been identified in only five years of data and their validity fortified by their occurrence among groups of neighbouring sites.

The primary conclusion with regard to the coastal United States is that, trends are usually not detected, but those that are found are predominantly decreases. Since trend detection will improve as time passes (and the "n" in the statistical tests increases), it is expected that trends that are now obscured by uncontrollable natural factors will emerge.

REFERENCES

Cantillo, A.Y., 1991. Mussel Watch worldwide literature survey - 1991. NOAA Technical Memorandum NOS ORCA 63. NOAA Office of Ocean Resources Conservation and Assessment, Rockville, MD, pp. 143

Capuzzo, J.M., J.W. Farrington, P. Rantamaki, C.H. Clifford, B.A. Lancaster, D.F. Leavitt & X. Jia, 1989. The relationship between lipid composition and seasonal differences in the distribution of PCBs in *Mytilus edulis* L. Mar. Environ. Res. 28: 259-264

Claisse, D., 1989. Chemical contamination of French coasts: the results of a ten-year Mussel Watch. Mar. Pollut. Bull. 20: 523-528

Cunningham, P.A. & M.R. Tripp, 1975. Factors affecting the accumulation and removal of mercury from tissues of the American oyster *Crassostrea virginica*. Mar. Biol. 31: 311-319

Fischer, H., 1983. Shell weight as an independent variable in relation to cadmium content of molluscs. Mar. Ecol. Prog. Ser. 12: 59-75

Fischer, 1989. Cadmium in seawater recorded by mussels: regional decline established. Mar. Ecol. Prog. Ser. 55: 159-169

Fisher, N.S. & J.-L. Teyssie, 1986. Influence of food composition on biokinetics and tissue distribution of zinc and americium in mussels. Mar. Ecol. Prog. Ser. 28: 197-207

Goldberg, E.D., M. Koide, V. Hodge, A.R. Flegal & J. Martin, 1983. U.S. Mussel Watch: 1977-1978 results on trace metals and radionuclides. Est. Coast. Shelf Sci. 16: 69-93

Jensen, A. & Z. Cheng. 1987. Statistical analysis of trend monitoring data of heavy metals in flounder *(Platichthys flesus)*. Mar. Pollut. Bull. 18: 230-238

Keith, L.H. & W.A. Teillard, 1979. Priority pollutants I: a perspective view. Environ. Sci. Technol. 13: 416-423

Lauenstein, G.G., A. Robertson & T.P. O'Connor, 1990. Comparisons of trace metal data in mussels and oysters from a Mussel Watch Programme of the 1970s with those from a 1980s Program. Mar. Pollut. Bull. 21: 440-447

Lauenstein, G.G., A.Y. Cantillo & S.S. Dolvin, 1993. Sampling and analytical methods of the NOAA National Status and Trends Program National Benthic Surveillance and Mussel Watch Projects. Vol. 1. Program overview and summary of methods. Technical Memorandum 71, NOAA Office of Ocean Resources Conservation and Assessment, Silver Spring, MD

Lauenstein, G.G. & A.Y. Cantillo (eds), 1993. Sampling and analytical methods of the NOAA National Status and Trends Program National Benthic Surveillance and Mussel Watch Projects 1984-1992: Vol. I-IV. Tech. memo. 71, NOAA/NOS/ORCA, Silver Spring, MD.

Martin, M., 1985. State Mussel Watch: Toxics surveillance in California. Mar. Pollut. Bull. 16: 140-146

NOAA, 1989. National Status and Trends Program progress report: a summary of data on tissue contamination from the first three years (1986-1988) of the Mussel Watch Project. NOAA Technical Memorandum NOS OMA 49. NOAA Office of Ocean Resources Conservation and Assessment, Rockville MD, pp. 22 & appendices

Nriagu, J.O., 1989. A global assessment of natural sources of atmospheric trace metals. Nature, 338: 47-49

O'Connor, T.P., 1990. Coastal environmental quality in the United States, 1990. Chemical contamination in sediments and tissues. A special NOAA 20th anniversary report, NOAA Office of Ocean Resources Conservation and Assessment, Rockville, MD, pp. 34

O'Connor, T.P., 1992. Recent trends in coastal environmental quality: Results from the first five years of the NOAA Mussel Watch Project. NOAA Office of Ocean Resources Conservation and Assessment, Rockville, MD, pp. 46

O'Connor, T.P., 1993. The NOAA National Status and Trends Mussel Watch Program: National monitoring of chemical contamination in the coastal United States. In: Environmental statistics, assessment and forecasting; C.R. Cothern & N.P. Ross (eds). Lewis Publ., Chelsea MI.

O'Connor, T.P. & C.N. Ehler, 1991. Results from NOAA National Status and Trends Program on distributions and effects of chemical contamination in the coastal and estuarine United States. Environ. Monit. Assess. 17: 33-49

Phillips, D.J.H., 1980. Quantitative aquatic biological indicators. Applied Science Publishers, London, pp. 488

Phillips, D.J.H. & D.A. Segar, 1986. Use of bio-indicators in monitoring conservative contaminants: programme design imperatives. Mar. Pollut. Bull. 17: 10-17

Pruell, R.J., J.G. Quinn, J.L. Lake & W.R. Davis, 1987. Availability of PCBs and PAHs to *Mytilus edulis* from artificially resuspended sediments. In: Oceanic processes in marine pollution; Vol. 1. Biological processes and wastes in the ocean; J.M. Capuzzo & D.R. Kester (eds). Krieger, Malabar FL, pp. 97-108

Rees, H.L. & M.D. Nicholson, 1989. Trends in lead levels in the horse mussel from the western North Sea. Mar. Pollut. Bull. 20: 86-89

RNO, 1992. Surveillance du milieu marin. Travaux du RNO, Edition 1991. Réseau National d'Observation de la Qualité du Milieu Marin. Report DEL/QM/B/67MJ/92. IFREMER, Brest

Roesijadi, G., J.S. Young, A.S. Drum & J.M. Gurtisen, 1984. Behavior of trace metals in *Mytilus edulis* during a reciprocal transplant field experiment. Mar. Ecol. Prog. Ser. 15: 155-170

Sericano, J.L., T.L. Wade & J.M. Brooks, 1992. The usefulness of transplanted oysters in biomonitoring studies. Proceedings of the Coastal Society Twelfth International Conference (in press)

Stronkhorst, J., 1992. Trends in pollutants in blue mussels *Mytilus edulis* and flounder *Platichthys flesus* from two Dutch estuaries, 1985-1990. Mar. Pollut. Bull. 24: 250-258

Suns, K., G. Hitchin & E. Adamek, 1991. Present status and temporal trends of organochlorine contaminants in young-of-the-year spottail shiner *(Notropis hudsonius)* from Lake Ontario. Can. J. Fish. Aquat. Sci. 48: 1568-1573

Tripp, B.W., J.W. Farrington, E.D. Goldberg & J. Sericano, 1992. International Mussel Watch: the initial implementation phase. Mar. Pollut. Bull. 24: 371-373

Appendix 2

Methods of NOAA NS&T Mussel Watch

This summary of the methods used by the NS&T Program for the collection and analysis of samples for the Mussel Watch Project was condensed from Lauenstein *et al.* (1993).

SITE SELECTION

The NS&T Mussel Watch Project is designed to describe chemical distributions over (US) national and regional scales, and its sampling sites need therefore to be representative of rather large areas. To avoid extremely local contamination, the project does not select sites near waste discharge points or poorly flushed industrialized waterways. A further requirement of the Mussel Watch Project is that indigenous mussels or oysters be collected for analysis. So Mussel Watch sites can only be located where large and apparently stable populations of mussels or oysters exist, and individuals must be of suitable sampling size (5-8 cm for mussels, 7-10 cm for oysters). Lastly, to allow comparisons of data with the earlier U.S. Mussel Watch Program (Goldberg *et al.,* 1983), its sites, when possible, were incorporated into the NS&T Program.

Sampling substrates are limited to natural substrates or structures containing them such as rock (including rip-rap and jetties), sand or mud. Collection of samples on buoys and preserved wooden structures could lead to artificially high results for the contaminants being quantified by the programme. The site must be suitable for follow-up sampling (*i.e.* it is not anticipated that the site will be physically disrupted by development or the molluscs population depleted by sampling).

Sites are not uniformly distributed along the coast of the United States. Within estuaries and embayments, they average about 20 km apart, while along open coastlines the average separation is 70 km. Almost half of the sites were selected in waters near urban areas (within 20 km of population centres in excess of 100,000 people). This choice was based on the assumptions that chemical contamination is higher, more likely to cause biological effects, and more spatially variable in these waters than in rural areas.

In 1986, 145 Mussel Watch sites were sampled. In 1988 a few sites were added on the East Coast to fill in large spatial gaps between sites, and twenty new sites were selected in the Gulf of Mexico for the specific purpose of gathering samples closer to urban centres. Results from the initial sampling showed that the highest chemical concentrations were near urban areas on the East and West Coasts, and that few sites in the Gulf of Mexico could be considered contaminated. Since urban centres along the Gulf are further inland than those on other coasts, an attempt was made to sample further inland. The major limitation in doing that, however, is that oysters are not found at salinities below about $10x10^{-3}$. By 1990, 234 sites had been sampled, with further additions made to test the representativeness of earlier sites and to expand the geographical coverage of the project.

SAMPLING PERIOD

Sampling for the Mussel Watch Project for all years occurs between mid-November and the end of March. While the entire sampling effort occurs during this time frame the criterion for the annual sampling of a Mussel Watch Project site once established requires that samples be collected within three weeks of the date the site was first sampled. The intention of sampling all sites in the time frame of late fall through early spring is to avoid the possible effect of spawning on chemical concentrations (Phillips, 1980).

SAMPLE COLLECTION

At each site, three composite samples of 20 oysters or 30 mussels have been collected (starting in 1992 this was reduced to one composite sample per site). Separate sets of composite samples were used for organic and trace element analyses.

A bivalve dredge was used to collect oysters or mussels in water deeper than 2 meters. The dredge bag was constructed of polypropylene mesh to minimize trace metal contamination from the steel chain used on other dredges of this type. In water less than 1 meter deep, bivalves were collected using a stainless steel pitch fork or quahog rake. In water 2 to 2.5 meters deep where the bottom is relatively soft, bivalves were sampled with stainless steel tongs. At some shoreline sites, the intertidal bivalve populations were collected from natural substrate by hand. Sometimes oysters were obtained directly from commercial oyster fishermen from privately leased sites.

The field team wore polythene, rubber or other non-contaminating gloves when removing the bivalves from the substrate or other handling. Bivalves were separated when found adhering to each other, and scrubbed with a nylon or natural fibre brush to remove adhering detritus.

FIELD MEASUREMENTS

For subtidal bivalve sites, *bottom depth* was determined by depth sounder or weighted line. At every site, *water temperature* was recorded using a surface-deployed temperature probe. Additionally, temperatures were measured from water samples collected with a Niskin bottle or equivalent using a portable digital thermometer or calibrated glass mercury thermometer. *Salinity* was measured with a refractometer or with a conductivity sensor attached to the temperature probe.

Tidal horizon data were incorporated into the Mussel Watch Project to ensure that the field team collects intertidal specimens from the same population so that contaminant concentrations found in molluscs are not a function of the amount of time that molluscs are able to filter feed.

SAMPLE HANDLING AND PACKAGING

Oysters and mussels were scrubbed free of mud and debris using pure bristle brushes and water from the collecting site.

For trace organic analyses, 30 clean mussels or 20 clean oysters were double wrapped in aluminum foil, placed in Ziploc bags. Because aluminum foil's high gloss side has been treated with organic compounds, foil used to package specimens was first solvent-rinsed in the laboratory. The field team placed the dull side (cleaned, untreated side) against the samples.

For trace elemental analyses, 30 clean mussels or 20 clean oysters were placed in Ziploc bag. This bag and a sample lable were placed inside another Ziploc bag.

Samples collected at East and West Coast sites were stored in dry ice and shipped to the laboratory, while samples from Gulf Coast locations were stored in ice chests until the day's sampling was complete, and then transferred to a mobile field laboratory for processing. All samples designated for trace element and trace organic analyses were frozen and shipped to the laboratory.

SAMPLE PREPARATION AND ANALYTICAL METHODS

While molluscs were sought within prescribed size-ranges, individual shell size was determined for each specimen collected. From 1986 through 1989 10 bivalves were collected at 3 stations per site for the determination of *gonadal index* on an organism by organism basis. These samples were placed in a fixative and returned to the laboratory for analysis. Starting in 1990, 15 bivalve specimens were required for the gonadal index sample from each site.

Tissue from 20 mussels or 30 oysters were shucked into a 500-ml pre-cleaned glass jar. Tissues were homogenized with a Polytron blender, or Tissumizer, for a minimum of 5 min. All tissue data were reported on a dry weight basis.

Sample *dry weight* was determined by drying a sub-sample of the bulk homogenized tissues. Approximately 1 g of wet tissue was weighed and then oven dried for 24 hours at 40 °C. The dried sample was cooled in a desiccator and reweighed. An alternate method was that samples were dried at 105°C, for two 12 hour periods and reweighed after each desiccation and cooling period. The resulting information was used to determine the percent moisture of the sample.

For *lipid determination* and the extraction of tissues for trace organic analysis, 50 g of sodium sulfate was added to approximately to 15 g of the bulk homogenized tissue. The tissue was then extracted three times with dichloromethane on a Tissumizer. Sample was removed from the extracted tissues with a centrifuge. Of the resulting 200 ml of sample approximately 10 ml was air dried in a pre-weighed aluminium pan, at room temperature for 24 hours. Lipid weight was determined by reweighing the aluminium pan and factoring in the amount this aliquot was of the original extract.

Methods for chemical analysis are not prescribed in the NS&T Mussel Watch Project because data quality is maintained through a performance-based Quality Assurance Project. Analytical methods can and have changed during this monitoring effort. The analytical methods vary among participating laboratories and are described in Lauenstein & Cantillo (1993) and in greater detail in the three accompanying volumes.

Preparation of tissues for trace element analyses included sample homogenization, freeze drying, and a second sample homogenization. Approximately 0.2 gram sample homogenate was completely digested in a 3 ml solution of $HN0_3$ and $HClO_4$ (3:1). Samples were initially allowed to digest, overnight, in loosely capped teflon bombs. These bombs were subsequently tightly capped and heated in either a microwave or standard oven to aid in complete sample dissolution. Subsequent analysis of the resulting solutions using flame or electrothermal atomic absorption spectroscopy (AAS) was commonly performed. In addition to AAS certain trace elements were quantified using x-ray florescence (XRF) and in 1991 inductively-coupled plasma mass spectrometery (ICP-MS) was incorporated as an analytical technique.

Oyster and mussel tissues and sediments to be quantified for organic contaminants were homogenized prior to extraction. Approximately 10 g of tissue was extracted with CH_2Cl_2. The extract was concentrated and purified using a silica gel/alumina column to remove matrix interferences. Further purification was performed using high performance liquid chromatography (HPLC). Characterization was carried out with gas chromatography (GC). Polycyclic aromatic hydrocarbons were primarily quantified with a mass spectrometer (MS) detector in the selected ion mode. Chlorinated contaminants were quantified by GC with an electron capture detector (ECD).

Chapter 3

Active Biomonitoring (ABM) by Translocation of Bivalve Molluscs

Willem C. de Kock and **Kees J.M. Kramer**

Laboratory for Applied Marine Research, IMW-TNO
P.O. Box 57, 1780 AB Den Helder, The Netherlands

ABSTRACT

'Active' biomonitoring (ABM) with organisms is based on the comparison of chemical and/or biological properties of samples which have been collected from one population and which have, after randomization and translocation, been exposed to different environmental conditions at monitoring sites. The technique offers the main advantage of reducing the inherent variability of analytical results usually encountered in field sampling programmes. A review is presented focusing on bivalve molluscs, especially mussels, discussing the advantages and disadvantages of the method, its applications and the treatment of possible (a)biotic processes that may interfere with the ABM technique. Case studies will demonstrate different applications in coastal waters and estuaries. Technical details on the performance of an ABM study will be presented in an Appendix.

1. INTRODUCTION

In the chemical monitoring of the aquatic environment for contaminants we should consider two reasons for the use of organisms:

- organisms are considered to live in equilibrium with the surrounding environment and may, thus, serve as an integrating 'sampling device';
- organisms may provide us with an insight into the relationships between internal pollutant concentrations and resultant biological effects.

This chapter is devoted mainly to the first reason, which relates to the more classical approach, *viz.* bio-monitoring applied in chemical 'Mussel Watch' type studies assuming that contaminants may be accumulated by aquatic biota to levels far above those found in the surrounding water. This long known phenomenon (Vinogradov, 1959; Eisler, 1981) has instigated the selection of certain tolerating indicator species for quantifying harmful

Biomonitoring of Coastal Waters and Estuaries. Edited by Kees J.M. Kramer.

substances resulting from anthropogenic inputs and has lead to the use of those species in field surveillance or monitoring programmes. The extensive literature on the use of organisms in estimating the degree of pollution in (a part of) the aquatic environment is, to a large extent, based on marine investigations (*e.g.* Butler, 1973; Goldberg *et al.*, 1978; Stephensen *et al.*, 1979; NAS, 1980; Phillips, 1980; Risebrough *et al.*, 1980; Farrington *et al.*, 1983; O'Connor, this Volume). Cantillo (1991) recently reviewed the literature on this subject as related to marine bivalves.

In a chemical sense, the biological availability of contaminants may be defined as the part of the total contaminant load that can be taken up by the organism, *i.e.* pass through cell membranes. This does not necessarily lead to bio-accumulation, however. A more dynamic view which comprises the physiological integrity of the organism, including its feed-back mechanisms, seems to be more appropriate here. The difference is based on the fact that the bio-accumulation of compounds is the net result of uptake and loss processes and not dependant on the uptake ability alone. We might even argue that toxic effects result from an organism's inability to cope with its internal contaminant load because of insufficient excretion/elimination or detoxification efficiencies. Significant accumulation does not at the same time necessarily mean that a compound is 'bioavailable', *i.e.* having easy access to physiological processes. From a biological viewpoint we should realize that, once the pollutant is inside the organism, it may not be available due to its storage in non-toxic forms (*e.g.* as granules or in fatty tissue reservoirs).

The establishment of a dose-effect relationship between a toxic contaminant and some biological response by a target organism requires not only that the type of response, but also that the dose of the contaminant be correctly defined. It is remarkable in the respect that most investigators in the literature on aquatic toxicology rely on the chemically defined ambient (total) exposure concentration of contaminants as the relevant dose parameter. The toxicological adequacy of using ambient concentrations should be doubted, however, because the response of an organism results from the presence of the toxic compound at sites of biochemical, cytological and physiological activity within the organism. De Kock & Bowmer (1993) stated earlier: "The actual toxic stress factor in the target organism is not some ambient external concentration but rather the bioavailable internal load of toxic elements or compounds accumulated by the organism from its immediate surroundings over a period of time".

Two different approaches have been developed for the use of biota in chemical monitoring programmes:

- Passive bio-monitoring (PBM), in which organisms are collected from their natural habitat at sites where a natural population exists; it is unavoidable that the samples will originate from different (sub)populations. 'Mussel Watch' type studies are examples (see O'Connor *et al.*, this Volume; Phillips, this Volume);
- Active bio-monitoring (ABM), in which samples are collected from one population at one location and translocated to the sites to be tested, ensuring comparable biological samples. It is concerned with a methodical way of working in the field in order to obtain accurate monitoring data on chemical stress.

The ABM technique offers the possibility of reducing the inherent variability of analytical results and, thus, improvement of data interpretation. In this chapter, the advantages and disadvantages will be discussed and the method of ABM will be illustrated by several case studies. Technical details on the performance of an ABM study, including the preparation and treatment of samples, and its logistic implications will be presented in an Appendix.

2. THE RATIONALE FOR ACTIVE BIO-MONITORING (ABM)

The internal concentration of a pollutant in an individual organism is not solely related to its external concentration, but also to internal biotic factors operating at the individual level. Such biotic factors are manifold as - according to Darwinian principles - variability is a basic prerequisite in the process of evolutionary development. Not surprisingly, a substantial variation in individual concentrations of *e.g.* trace metals has been measured in tissues of organisms sampled from a single population, even when the individuals were of the same length class (de Wolf, 1975; Goldberg *et al.*, 1978; de Kock, 1983), generally with a skewed distribution. For practical reasons it is imperative that natural variability due to biotic factors acting on the contaminant concentration in the animals' soft tissues is counteracted in the monitoring programme. Not only genetical differences govern such variability. Individual differences within one length group possibly arise from (among others) differences in excretory abilities related to age which in turn is related to individual growth rate. Especially in the tidal zone growth rates will greatly differ between individuals living at adjacent positions in the vertical plane leading to substantial age differences at similar lengths. As many 'Mussel Watch' monitoring programmes rely on samples taken from the tidal zone, a large variability in contaminant concentrations could be present in the pool of constituent animals. This would lead to a loss of spatial discrimination power.

It also follows that, notwithstanding the use of a suitable and well defined size class of bivalve samples in a 'passive monitoring' network (PBM), a large temporal variation may be observed at monitoring stations that is not easily explained by external factors (*e.g.* by seasonal influences).

For this reason a standardized, active biomonitoring method was developed by which the bio-accumulation (or bio-elimination) as well as estimated equilibrium levels of actual (bio-available) toxic stress factors could be established in the field. The method consisted of translocating batches of appropriately composed random samples of (*e.g.*) mussels of a selected length class from a sublittoral, less contaminated population to target positions on pollution gradients. Exposure of the samples, enclosed in polythene baskets or nets, was performed at defined locations (on gradients or transects) during defined periods.

As mentioned in earlier papers (de Kock, 1983, 1986; Smith *et al.*, 1986; Martincic *et al.*, 1992), the advantages of experimental exposure in so-called 'active bio-monitoring' (ABM) investigations through translocation experiments are fourfold:

- resolution power is optimized by employing statistically similar groups of organisms (with regard to population, size, age, pollution- and environmental history) for comparing chemical stress at different locations;
- the exposure period is known;
- monitoring stations may be selected independent of the natural (non) occurrence of the monitoring species;
- in case several species occur together (such as *M. edulis* and *M. californianus* in the San Francisco Bay, Smith *et al.*, 1986), experiments may be performed with selected species.

A draw-back of the method might be:

- its logistically more complicated nature, and
- it being subject to disturbances (or theft) of the equipment.

Translocation of organisms for ABM studies need not be limited to bivalves (such as mussels and clams of marine and fresh water origin: *e.g. Mytilus* spp., *Perna* spp., *Ostrea* spp, *Crassostrea* spp., and *Dreissena polymorpha*, *Anodonta* spp., *Unio* spp.). Successful attempts have also been described using *e.g.* fish (Fagerström & Åsell, 1976; Haasch *et al.*, 1993), macrophytes (*Fucus vesicolosus*, *Ascophyllum nodosum*; Myklestad *et al.*, 1978; Ho, 1984), barnacles (Wu & Levings, 1980) or gastropods (*Nucella lapillus*, see Gibbs & Bryan, this Volume).

Apart from their use in the determination of trace pollutants in biota, ABM experiments offer unique possibilities for the monitoring of biological effects. These may range from simple detection of the effects of eutrophication (Riisgård & Poulsen, 1981) and growth and fecundity testing (Wu & Levings, 1980), to more complex measurements of 'survival in air' during anaerobiosis (Eertman & de Zwaan, this Volume), acetylcholine esterase (Bocquené *et al.*, 1990) and histological defects (Gibbs & Bryan, this Volume; Bowmer *et al.*, 1991). Other examples include its use in studying physiological effects (Widdows *et al.*, 1980), scope for growth (Smaal *et al.*, 1991; Widdows *et al.*, 1990; Smaal & Widdows, this Volume), the induction of metallothionein (George & Olsen, this Volume) or cytochrome P450 (Haasch *et al.*, 1993). Advantages of ABM in effect studies are similar in terms of selection of the same population of organisms with a common life history, the defined exposure period and the possibility for exposing organisms at locations of the investigators' choice.

3. MUSSELS AS A MONITORING TOOL

Mussels, and some other selected bivalves, appear to comply to a high degree with what should be required of a useful biomonitoring organism. The basic requirements demanded for (chemical) monitoring studies with such species have been specified by different authors (Phillips, 1985; Bryan, 1980; de Kock, 1986) and include the following aspects: versatility, practicality, integrative ability and consistency.

Versatility: In coastal marine ecosystems, *Mytilus*-species are known as tolerators (de Kock & Kuiper, 1981) able to withstand the natural stress factors operating in the tidal zone. These stress factors include: exposure to the atmosphere during part of the tidal cycle which is the basis for variability in oxygen and food supply; surrounding temperature and degree of dessication; physical force exerted by wave action and biotic factors like predation pressure (Bayne, 1976). They are also well adapted osmotically to the variability of estuarine environments, tolerating salinities to ca. $10x10^{-3}$, or even down to $2x10^{-3}$ for short periods (Davenport, 1979). Moreover, substantial amounts of living and dead, inorganic and organic, suspended matter in the water column, as is usually found in such environments, can be dealt with because of an efficient mechanism for rejecting worthless particles as pseudofeces and by retaining valuable particles such as food (Jørgensen, 1990). They even tend to dominate some organically enriched polluted systems, being able to exploit food resources associated with untreated sewage impact (Bellamy, 1972).

Practicality: Species of the family Mytilidae, for example *Mytilus edulis* and *M. galloprovincialis*, *M. californianus*, *M. chilensis*, *Perna canaliculus* and *P. viridis* are abundant and widely distributed in many coastal regions of the world. For a more complete list of applied species one is referred to O'Connor *et al.* (this Volume). Many of the species mentioned are commercially exploited and cultivated and are easily collected and handled. They reach sizes and weights which are practical for chemical analytical purposes and may reach an age of several years (Phillips, 1980). The individuals are sedentary and, thus, representative of a certain location: they attach to a firm substrate (*e.g.* rocks, stones, buoys, jetties) by byssus threads. They are hardy enough to endure rough treatment, can be handled out of the water for a considerable time and, due to their protective shells, contamination of the soft parts during transport and storage is minimized.

Integrative ability: A large temporal variability usually exists in time series of dissolved contaminants at a single location. This is a normal characteristic of standard chemical water monitoring programmes, and may be the result of periodic processes on a scale varying from hours to seasons (tides, currents, monsoons), or originate from episodic events (storms). Whether sampling is carried out in rivers, estuaries or coastal waters, discrete samples collected by a water sampler will always provide only data that represent the water mass at the moment(s) of sampling. Submerged mussels and other aquatic organisms continuously interact with the water phase. In an environment in which mussels are exposed to a variable supply of contaminants, either dissolved or bound to (food) particles, two opposing processes are at work in the animal: *i.e.* uptake and loss of contaminants. The rate constants of both processes ensure a certain integrating capacity by which the mussel smoothes ambient variations of contaminant concentrations. In the light of an undesirable, temporal, inhomogenous contaminant distribution, mussels are practical integrant 'bio-samplers' of aquatic environments.

Consistency: An important requirement of an organism used in contaminant dose monitoring programmes is the existence of a simple and consistent relationship between external and internal concentrations of the compound that is investigated (Phillips, 1980; Bryan *et al.*, 1985). Supporting field data for this requirement are hardly available at all in the literature, however.

For practical reasons the (filterfeeding) blue mussel, *Mytilus edulis*, has been adopted for experimental application in the water column at our laboratory. Similar experiments in which we deployed common sediment dwelling bivalves, *e.g. Scrobicularia plana* (depositfeeder) and *Macoma balthica* (variable feeding behaviour), in the benthic environment of the Netherlands proved to be useful as well.

4. ACCUMULATION AND ELIMINATION IN THE FIELD

During the last decades, mussels have become known to be efficient bio-accumulators of trace elements and various organic micro-pollutants *e.g.* halogenated hydrocarbons (pesticides, polychlorinated biphenyls (PCBs)) and polycyclic aromatic hydrocarbons (PAHs). Like many other bivalve mollusc species, mussels seem to lack sufficient regulating capacity for preventing contaminating substances from building up in their tissues. The contaminants are usually not equally distributed over the various organs, some organs (*e.g.* kidney in the case of trace metals; gonads in the case of lipophilic compounds) often showing higher concentrations than others (*e.g.* foot and adductor muscle). See also section 5.1.

A mathematical description of the accumulation process for multi-compartment systems (organs), however, is somewhat outside the scope of this chapter as in the experimental field studies reported below the total organism is usually studied as a single compartment and treated accordingly.

4.1. The equilibrium concept

Neely *et al.* (1974) and Branson *et al.* (1975) developed a simple model for the bioaccumulation of organic contaminants in fish (*i.e.* trout muscle) at a constant concentration of the compound in the surrounding water. This model is kinetically described as concurrent first order uptake and loss processes (Bruggeman, 1983), leading to some steady state. The simultaneous occurrence of both processes, with different rate constants, measured in the laboratory also applies to a field situation, albeit that a true steady state may never be reached because of a continuous variable ambient contaminant concentration. Such a steady state may usually be approached, however, under conditions of chronic pollution operating at a scale of months to years. The internal concentration will then be the result of two continuously acting, opposed dynamic forces and will fluctuate within some band width of equilibrium values. The temporal variation of these values expressed as a variation coefficient is typically less than the variation found in ambient concentrations in the water column as the rate constant values of uptake and loss will guarantee a certain integrating capacity of the organism. The model of Neely *et al.* (1974) may easily be applied to mussels and other bivalve molluscs (as well as to other organisms with active excretion mechanisms), although the rate constants of the constituent processes in molluscs may differ from those in fish.

The general structure of such an accumulation-elimination model is based on a number of assumptions concerning the interaction of contaminating compounds with an organism:

- the external concentration (C_e) is constant with time;
- the internal concentration (C_i) reaches some equilibrium value $C_i(eq)$ at given C_e values;
- the uptake rate dC_i/dt into the organism is in linear proportion to C_e with rate constant k_u;
- the rate of loss dC_i/dt from the organism is in linear proportion to C_i with rate constant k_l.

It may then be written:

$$dC_i/dt = k_u.C_e - k_l.C_i \qquad (1)$$

On reaching the equilibrium state, $dC_i/dt = 0$, so that:

$$C_i(eq) = \{k_u/k_l\}.C_e \qquad (2)$$

In equation 2, the equilibrium value of the internal concentration is proportional to the external concentration. C_i and C_e are given in a weight on weight basis (*e.g.* $\mu g.kg^{-1}$), k_u and k_l are constants. The model parameter ratio k_u/k_l equals $C_i(eq)/C_e$, the latter known as the 'bioconcentration factor' (BCF).

The model is of a simplified nature. There are two contaminant sources in the field situation: dissolved and particulate contaminants, contained in water (w) and food (f) respectively. C_e should, therefore, be considered as consisting of two components: $C_e(w)$ and $C_e(f)$, so that equation (1) may be extended to:

$$dC_i/dt = k_u(w).C_e(w) + k_u(f).C_e(f) - k_l.C_i \qquad (3)$$

and the equilibrium value of C_i will then be:

$$C_i(eq) = \{k_u(w)/k_l\}.C_e(w) + \{k_u(f)/k_l\}.C_e(f) \qquad (4)$$

in which $k_u(w)$ and $k_u(f)$ are the rate constants for the uptake from water and food respectively. The parameter ratios $k_u(w)/k_l$ and $k_u(f)/k_l$ may be estimated by maximum likelihood analysis using series of coupled values of $C_i(eq)$, $C_e(w)$ and $C_f(f)$ which should have been measured over a range of concentrations. Such treatment of simultaneously collected field data is not known from various 'Mussel Watch' programmes around the world. The relative contribution of contaminants in food (*e.g.* phytoplankton) and water has been studied by a number of authors in the laboratory, however, and the general opinion is that most contaminants are accumulated predominantly from the water phase (Janssen & Scholtz, 1979; Borchardt, 1983; Riisgård *et al.*, 1987).

4.2. Accumulation *vs* elimination

Accumulation is defined as the net positive result of uptake and loss with time; elimination is the net negative result of uptake and loss with time. Accumulation and elimination are the usual result of increased and/or decreased bio-available contaminant concentrations in the organisms' immediate environment. Knowledge of the form of the accumulation or elimination curves implies that equilibrium levels of contaminants in living tissues of the organism can be estimated from the shape of the curve. It is the equilibrium level that characterizes the average supply of a bio-available contaminant fraction at a certain location after a certain exposure period has elapsed. The location may be situated on a pollution gradient which we wish to describe, the equilibrium period is, however, compound-specific and will depend on the combined rates of uptake and loss by the organism. These considerations are important for the correct application of mussels and other bivalve molluscs in active biomonitoring schemes. For monitoring purposes, routine establishment of pollution trends by means of repeated measurements of equilibrium levels in time at various locations should be tried. This is important from the viewpoint of water quality management, for estimating the functional dose of a toxic compound under field conditions.

In 1983, an attempt was made by our laboratory to measure the accumulation and elimination dynamics of cadmium and a number of higher PCB congeners (IUPAC numbers 101, 138 and 153) *in situ*. This could be managed logistically because the coexistence of different pollution regimes in the southwestern part of the Netherlands offers the opportunity to translocate batches of samples of *Mytilus edulis* originating from a relatively clean environment, in the Oosterschelde basin, to the polluted river Scheldt Estuary, and to translocate batches of mussel samples originating from the Scheldt Estuary *vice versa* (de Kock & van het Groenewoud, 1985). To construct accumulation/elimination curves, samples were retrieved from the experiment after 21 different exposure periods ranging up to 122 days.

In order to estimate the rates of accumulation and elimination under the assumptions mentioned above (equations 1 and 2), both uptake and loss obeying first order kinetics and the animal regarded as a 'one compartment' unit, the following simple model was applied. For accumulation:

$$C_{i\to w}(t) = \{C_{i\to w}(eq) + A\} - A.e^{-t/\tau(acc)} \quad (5)$$

and elimination:

$$C_{i\to w}(t) = C_{i\to w}(eq) + A.e^{-t/\tau(el)} \quad (6)$$

where $C_{i\to w}(t)$ = contaminant weight per mussel at time t, $C_{i\to w}(eq)$ = equilibrium contaminant weight per mussel under the least contaminated conditions (Oosterschelde), A = difference between contaminant weight per mussel at t = 0 and at t = ∞, $\tau(acc)$ = time constant for the accumulation process, a parameter for the rate at which the contaminant weight per mussel increases, representing the time in which an increase of 0.63 A occurs. $\tau(el)$ = similar time constant for the elimination process, representing the time in which a decrease of 0.63 A occurs. Figure 1 presents these parameters schematically. The model estimates the value of $C_{i\to w}(eq)$ (in ng. mussel^{-1}), τ (in days) and $C_{i\to w}(eq)$ + A (in ng.mussel^{-1}).

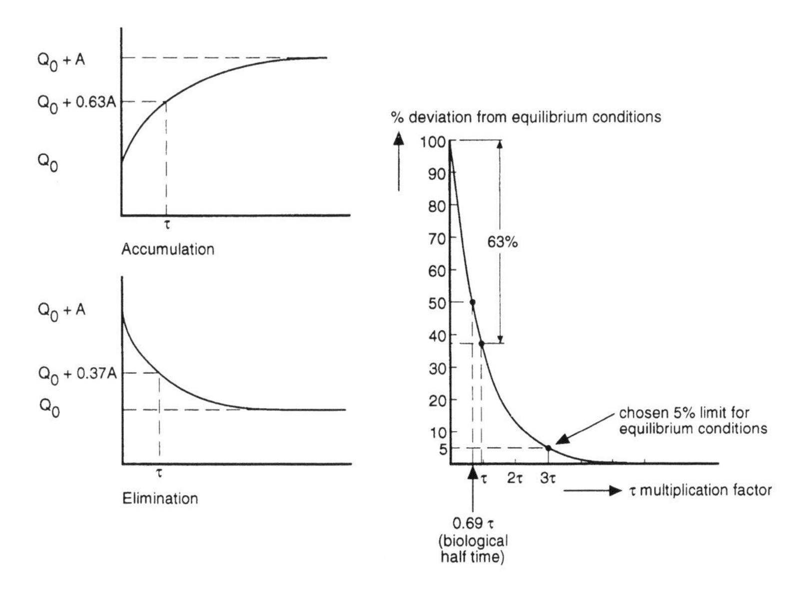

Figure 1. Presentation of theoretical accumulation and elimination curves, and the deviation from the equilibrium conditions as function of τ. The Q_0 is equal to $C_{i\to w}(eq)$ used in the test. (see text for details; After: de Kock & van het Groenewoud, 1985)

τ is defined as the time needed for attainment of a 63% pollutant weight difference from the initial pollutant weight in the mussel. This is illustrated in Figure 1. The relation between τ and the biological half-life of the contaminant (at a 50% difference body burden) is derived from $e^{-t/\tau} = 0.5$ or $t_{1/2} = 0.693\ \tau$. It is also seen that the time needed for equilibrium being approached to within 5% of the actual equilibrium value is about 3τ. This 5% concentration distance is already within the usual analytical variation range of various pollutants and greater accuracy in estimating accumulation or elimination times is not necessary.
The general picture of the experiment confirms the hypothesis that in polluted waters accumulation occurs, while in unpolluted waters loss may be expected. To compensate for weight irregularities resulting from variable mussel growth during the exposure period, contaminant concentrations (*i.e.* weight per unit tissue weight, C_i) were converted into contaminant weights per animal ($C_{i\rightarrow w}$). Table 1 summarizes the results and shows that both accumulation and elimination of Cd are slow processes with $\tau \approx 50$ days, implying that equilibrium is reached in ≈ 150 days. Also, $C_{i\rightarrow w}$(el) is higher than $C_{i\rightarrow w}$(acc). Complete elimination of Cd seems to be difficult once the metal has been taken up by the animal. There exists a fraction, however, that eliminates with a rate constant that is compatible with the rate constant for the accumulation process ($\tau \approx 50$ days). Compared with Cd, the dynamics for PCB congeners are more rapid with time not exceeding values of 12 days for both the accumulation and elimination process, leading to an equilibrium in ≈ 35 days (Table 1). Elimination of these compounds seems to be complete, down to initial equilibrium levels once the animals have been taken to the less contaminated environment (Oosterschelde). This concurs with Young *et al.* (1976) who observed similar findings for PCBs and DDT (see also Figure 8). No elimination of PCBs was observed, however, in a reciprocal field experiment between the Scheldt Estuary and the (clean) Oosterschelde (Hummel *et al.*, 1989). Literature data confirm the slow elimination rate of Cd. Borchardt (1983) estimated a biological half-life ($t_{1/2}$) of Cd to be 96-190 days; this long elimination time may be the reason why Luten *et al.* (1986) found no elimination for Cd. In contrast to cadmium, the kinetics of uptake and elimination for Cu, in *M. edulis*, are much faster (Elliot *et al.*, 1985). Hg reaches equilibrium in about 20 days, but elimination was found to be much slower (53-293 days) due to speciation effects (Riisgård *et al.*, 1985). Majori *et al.* (1978) investigated the accumulation and elimination of Hg and Pb, while Simpson (1979) studied Zn and Pb in a transplantation experiment (with *M. edulis*) between the Tyne and Coquet Estuaries, showing that elimination of Zn was not complete, perhaps due to seasonal factors. Other trace elements have hardly been studied,

Table 1. Accumulation and elimination data from an ABM experiment where mussels (*Mytilus edulis*) were transplanted from the clean Oosterschelde Basin to the polluted Scheldt Estuary, and *vice versa.* Cd and three PCB congeners were analyzed (concentrations in $\mu g.kg^{-1}$ AFDW) after exposure of up to 122 days. The $C_{i\rightarrow w}$(eq) gives the equilibrium contaminant weight per mussel under the least contaminated conditions (Oosterschelde), A is the difference between the contaminant weight at t=0 and t=∞, τ gives the time (days) needed to obtain 0.63A (see text for further explanation)

pollutant	**process**	**$C_{i\rightarrow w}$(eq)**	**$C_{i\rightarrow w}$(eq)+A**	τ	**$\{C_{i\rightarrow w}(eq)+A\}/C_{i\rightarrow w}(eq)$**
Cd	(acc.)	467	1671	55.0	3.58
	(el.)	1301	1866	40.0	1.43
PCB101	(acc.)	9.87	44.11	8.03	4.47
	(el.)	7.61	25.43	11.02	3.34
PCB138	(acc.)	15.20	43.68	9.06	2.87
	(el.)	11.02	27.91	8.01	2.53
PCB153	(acc.)	25.06	62.60	8.75	2.50
	(el.)	17.82	38.10	7.95	2.14

exept for a number of radiotracers such as ^{137}Cs (*e.g.* fast elimination; Clifton *et al.*, 1989) or the accumulation (Nolan & Dahlgaard, 1991) and elimination of ten different radiotracer elements (Pu, Am, Np, Eu, Ce, Ag, Tc, Zn, Co and Mn) in Baltic seawater, after accumulation under laboratory conditions (Dahlgaard, 1986).
In case of other organic compounds, rapid equilibrium was reached for PAHs (Narbonne *et al.*, 1992), while a fast elimination of naphtalene was found: $t_{1/2}$ = 25-50 h (Widdows *et al.*, 1983), thus interfering with any depuration.

4.3. Exposure

4.3.1. Sample size

In order to establish reliable mean levels of contaminant concentrations at a monitoring location, maximum effort is required to reduce the variance due to differing individual contaminant concentrations in the population. Or, according to Gordon *et al.* (1980) and Boyden & Phillips (1981) the sample should be of sufficient size for a reliable characterization of the pollutant levels. For this reason sufficient, randomly taken, individuals of a single length class (or age group) should be pooled in a sample. The larger the number of individuals, the better the coefficient of variation in sample series, the smaller the expected deviation from the population mean (Figure 2), as demonstrated by Baez & Bect (1989) for organic pollutants in mussels and Walker (1982) for trace metals in oysters. The (chemical) analysis of a large number of individuals will give excellent information on the natural variation, but usually the cost of these analyses will prohibit such approaches.

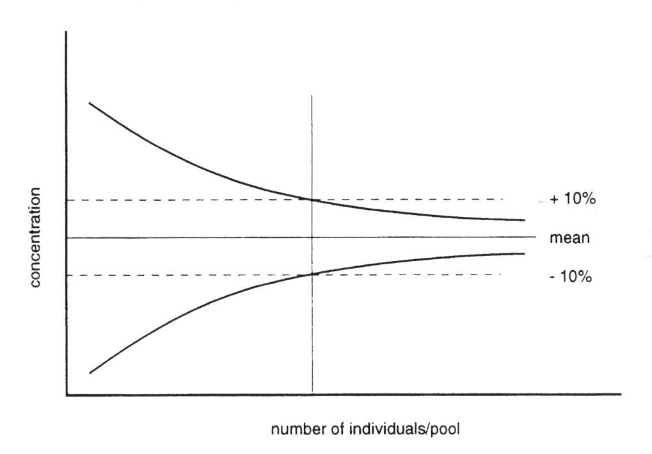

Figure 2. The higher the number of individuals per sample, the lower the (natural) variation in the analytical result

De Wolf (1975) observed that samples of at least 60 pooled *Mytilus edulis* individuals in the length class of 3-3.5 cm were required in order to obtain a measured Hg concentration within (+ or -) 10 % of the population mean, with a probability of 95 %. Comparable results were reported for Hg, Zn and Cu in *Macoma balthica* or the polychaete worm *Arenicola marina* (a minimum of 100 individuals; de Kock, unpublished results). For the freshwater zebra mussel, *Dreissena polymorpha*, it was found that at least 90 pooled animals of 1-1.5 cm length were needed in order to remain inside a range of analytical variation in standard samples, delimited by +/- 2 sd from the mean. Wright & Milhursky (1985) found that, depending on the location, 15-20 Cheasapeake Bay oysters were sufficient for measuring concentrations within 10% of the population mean.
Martin & Phelps (1979) plotted the relation between the number of samples and number of individuals of *M. edulis* per pool, giving a 90 % certainty of detecting differences between

means of Cu concentrations with p = 0.05 at various degrees of resolution (5, 10, 20, 50, 100 %) using a t-test. Extrapolation of these plots indicates that when a pool contains 100 individuals, only very few samples (possibly one) are required for detecting even a 5 % difference between means. Such plots are, however, not available for organic contaminants and most other trace metals. The variation found in monitoring data is the sum of natural variability and the analytical variation. This abiotic variation (expressed as a percentage) may be determined by analytical replicates for which a series of identical standard homogenates should be used. In Figure 3, an example of the contributions of analytical variation and the natural variation in PBM and ABM studies is depicted. The example relates to a case of mercury pollution in the Ems Estuary (see also section 6.1; Essink, 1988).

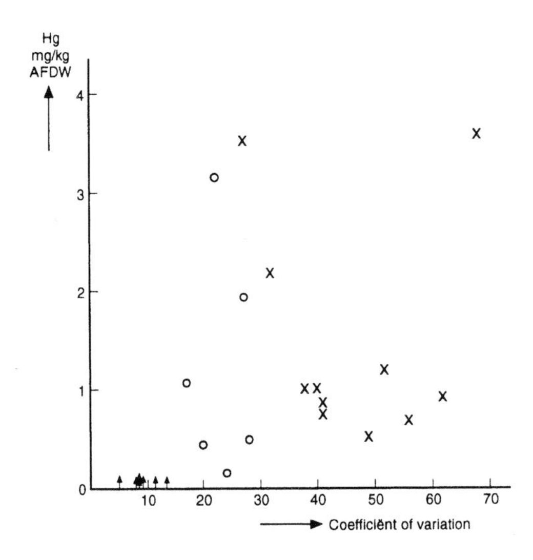

Figure 3. Natural variation in the observed results for the monitoring of Hg in the Ems Estuary, the Wadden Sea and the North Sea by PBM (x) or by ABM (o). The inherent variation of series of analytical measurements of standard samples is presented by (↑)(After: de Kock & Kuiper, 1981)

In many of our exposure experiments it was assumed that 100 pooled and randomly taken individuals per sample should constitute a reliable sample size. Discrimination power between spatially and/or temporally separated samples will in that case be almost limited by analytical reliability alone.

4.3.2. Exposure time

For general purposes, mussel exposure periods of about 6 weeks (40-50 days) have been used since 1989 in routine ABM progammes of estuarine and coastal contamination in the Netherlands. A practical balance has been sought between the need for sufficiently long exposure periods and the need for reducing maintenance efforts related to fouling of cages. Details of such ABM investigations will be given below (see section 6).

4.3.3. Depth of sampling and exposure

A single population has to be sampled for ABM programmes. As has already been mentioned, the tidal zone constitutes an important source of variation and this is not the appropriate area from which to collect an initial test population for conducting ABM experiments. Some authors (Nielsen, 1974; Lobel & Wright, 1982) observed height-dependant variation in tidal *M. edulis* populations in relation to physiological characteristics or trace element content (Zn). Tidal height also greatly influenced the Cd concentration in *M. californianus* from Monterey County, California (Martin & Severeid, 1984). Therefore it seems mandatory to use permanently submersed mussels as test animals in ABM exercises. This applies to the site of initial collection as well as to the sites of exposure. The exact depth of exposure depends on the monitoring purpose. Phillips (1976, 1985) suggests, in a general way, that identical water depths be selected and that sites above the halocline in estuaries and coastal areas that experience a large freshwater inflow be avoided. More specifically, however, it might be of interest to monitor near-bottom sites in relation to mobilization of contaminants from sediments or dumped, dredged materials (Koepp, *et al.*, 1987). Also, measurements at various depths taken at different sites may provide an insight into the spatial distribution of bio-available contaminants related to point sources. Figure 8 in section 6.5. presents a clear example of measuring DDT concentrations in *M. californianus* suspended at various depths in the water column (Young *et al.*, 1976).

4.3.4. Depuration

Sediment particles or suspended matter of inorganic and organic origin offer adsorption space and may, thus, become enriched in contaminants. The presence of small particles (either food or useless particles destined for rejection) in filterfeeding bivalve molluscs like mussels may interfere with the analysis for tissue contaminants (this applies even more to benthic depositfeeding organisms like polychaete worms). *M. edulis,* which has gills with well-developed laterofrontal cirri, is able to retain particles with a diameter of 3 µm with 90 % efficiency, whereas a particle diameter of 1 µm is still filtered with 50 % efficiency (Jørgensen, 1990). Depuration, during which the organism is allowed to rid itself of such particulates in clean seawater (*e.g.* for 24 hours) has, therefore, been recommended (Flegal & Martin, 1977).

The disadvantage, however, of depuration is the short biological half-life of various compounds, organic and inorganic, due to high elimination rates. An underestimation of the contaminant concentration will then be the result. It is unlikely that standardization of a depuration method will lead to useful results with the possible exception of applying an abundant supply of filtered water at the field exposure site before collecting the animals for analysis. As mussels contain only small amounts of particulate matter on their gills and in their intestine tracts relative to their own weight, only minor effects of non-depuration have been observed, and only so for elements that are closely related to clay fractions, such as Al, Fe and Mn (Latouche & Mix, 1982; Phillips, 1985). For this reason, depuration for mussels is advised against (NAS, 1980).

4.3.5. Expression of analytical results: units

As the ABM routine was developed to increase the resolution power of measurements of bio-available contaminants on estuarine gradients, proper attention should be paid to the unit used for presentation of concentration data in the technical literature in order to evade an unnecessary source of variability.

Two variables may impede the comparison of tissue contaminant concentrations at two (adjoining) monitoring sites: the water content per unit weight of soft tissue and the ash content per unit weight of dried soft tissue (*i.e.* the ash-free dry-weight/dry-weight ratio). These variables are determined by using subsamples of the homogenates.

Wet-weight (WW)

The analytical results should not be expressed on the basis of fresh- or wet-weight because the water percentage of homogenates may vary according to *e.g.* the animals' nutritional state (season dependant) and general condition (*e.g.* Popham & D'Auria, 1983) which may result in large variations in the concentrations.

Dry-weight (DW)

Usually dry-weight (of the total soft parts or of separate organs) is used (in mg) as the unit of expression in the literature.

Ash-free dry-weight (AFDW)
Using merely dry weight, however, does not exclude the ash percentage from being a source of variability. Mussels are highly isotonic with their external environment (*e.g. Mytilus edulis*, Potts, 1954) and their ash is largely composed of inorganic sea salt components. Isotony is maintained accross a broad range of salinities in estuarine areas. Analysis of a homogenate of *M. edulis* collected near our laboratory showed that, in this specific case, ca. 77% of the ash weight consisted of salts like those occurring in sea water. Significant linear correlations between ash content and water content were observed using different homogenates derived from various field stations.
Furthermore, such correlations have been found experimentally to shift within 3 hours after transfer of test animals to different salinities in the laboratory, pointing to a rapid response towards isotony.
In order to compensate for salinity differences when exposing ABM mussel samples in estuarine environments, concentrations of contaminants should be routinely expressed on the basis of ash-free dry weight of the homogenate. A source of variability may thus be counteracted.
Other ways of expression
Apart from discussions on the expression of results by WW, DW or AFDW, various other attempts have been made to normalize data in order to enhance the interpretation of results. The following procedures have been proposed: expression by body content, *i.e.* contaminant weight per individual (Simpson, 1979; Riisgård *et al.*, 1985), by shell-weight (Fisher, 1983), by size and DW (Cossa & Rondeau (1985), by lipid content (Capuzzo *et al.*, 1989; see also O'Connor this Volume, for a discussion on this unit of expression) and by condition index (condition index = DW/shell length) (Martincic *et al.*, 1987b). Nolan & Dahlgaard (1991) observed a linear relation between $\log_e$ [radiotracers] *vs* weight (shell+soft parts), hence a lognormal distribution.
No generally accepted normalization procedure has come forward yet, and the mixture of all these methods of expression leads to some confusion. To ensure the possibility of comparison between different investigations one is urged to express the results at least on the basis of DW or AFDW also.

5. COMPLICATING FACTORS

5.1. Total soft tissue or selected organs

In most monitoring studies, the chemical detection of contaminants is performed on whole soft tissue homogenates of the mussels. A number of authors, however, focussed on analysis of specific organs, the main argument being that some organs accumulate more pollutants than others, by which a better analytical performance would be accomplished (Amiard *et al.*, 1986a,b). It has also been observed that tissues with relatively low lipid levels eliminated xenobiotic lipophilic compounds at higher rates than lipid-rich tissues (Widdows *et al.*, 1983). Clark & Finley (1975) found higher elimination rates of petroleum hydrocarbons first accumulated in gill tissue from water as compared with elimination rates of the same compounds first accumulated from food by the hepatopancreas.
Up to eight different organs have been selected (viscera, mantle, gills, retractor muscle, adductor muscle, foot, labial palps and kidney), as described by Martincic *et al.* (1987a,b) for subsequent Cu and Zn analysis. A general sequence that is valid for all trace elements cannot be detected, however. Schulz-Baldes (1973) observed a sequence for Pb concentrations in *Mytilus edulis* as: kidney > intestine, adductor muscle > foot, gills, mantle and gonads, George & Coombs (1977) found likewise for Cd: kidney > viscera > gills > mantle > muscle,

which was confirmed in general by Everaarts (1990), while Lakshmanan & Nambisan (1989) found Hg, Cu, Zn and Pb concentrations in *Perna viridis* thus: gills > viscera > muscle > mantle. Martincic *et al.* (1987a,b) proposed that the foot be analysed for Cu and the mantle for Zn in *M. galloprovincialis*, while Odzak *et al.* (1993) prefer the foot and the gills for Cd and Pb respectively.

Other researchers have selected the shell as the target organ for chemical analysis (Rucker & Valentine, 1961; Koide *et al.*, 1982), thereby avoiding the problem of deciding on the application of a depuration period before analysis. Chemical analysis of the shells may serve in detecting a multi-year pollution history (Carrell *et al.*, 1987). Furthermore, the use of another secreted material, *i.e.* byssus threads, has been successfully tested in several *Mytilus* species (Coombs & Keller, 1981; Koide *et al.*, 1982). Still, their use for bio-monitoring purposes is very limited.

A disadvantage of the preparation of separate mussel tissues for chemical analysis is that it requires skilled staff to perform the dissection. Moreover, the operation reduces the amount of tissue for analysis and increases the number of samples with each organ selected, which involves more labour and costs in routine monitoring programmes. Since we observe no analytical problems with the analysis of the whole soft tissue, not all contaminants are proven to accumulate preferentially in the same organ and analytical funding is usually restricted, we prefer to use whole soft tissue of the mussels for our monitoring exercises.

5.2. Size and age

As is the case with most other organisms, the growth rate of mussels is dependant on the net energy available for the formation of somatic tissues and gametes. This in turn is strongly dependant on the available food supply and the energy needed by the animal to cope with adverse factors like stress in the tidal zone and high concentrations of inorganic suspended particles in the ambient water. As differences in growth rate lead to differences in size (length) at similar age, age related differences in contaminant concentrations could be expected between spatially separated samples of similarly sized animals. It is one of the advantages of ABM programmes that mussels of similar size and ecological history, derived from the same population, are used as test organisms.

In a fast growing, sublittoral population, small, young mussels show the highest concentrations of contaminants (weight per unit dry weight (DW) or, better, weight per unit ashfree dry weight, AFDW), as shown *e.g.* by Schultz-Baldes (1973) for Pb and by Cossa *et al.* (1980) for six elements in *Mytilus edulis*. The latter authors also observed in the St. Lawrence Estuary that the decrease of concentrations with size leveled off once maturity had been reached. No effect of length was observed in the Hg concentrations in *M. edulis planulatus* or the blacklip abalone, *Notohaliotis ruber* (Walker, 1982). Only minor changes seem to occur after maturity (Popham & d'Auria, 1983). It should be remarked that it is the concentration in the tissues (weight per unit weight) and not necessarily the body load (weight per individual) that decreases with size (Boalch *et al.*, 1981).

The reason for the size-related difference in a young population is presumably related to the surface:volume ratio of the animals' tissues, the surface structures (*e.g.* mantle, gill filaments) being the limiting factor for the uptake of contaminants. In a slow growing littoral or cold water population, where mussels may reach an age of 10 years or more, concentrations of persistant contaminants may increase with age as a result of diminishing excretion abilities.

5.3. Sex

Differences between contaminant concentrations in male and female mussels have been detected but are usually considered small *e.g.* in *Mytilus edulis* (Latouche & Mix, 1982) and

in *Choromytilus meridionalis* (Watling & Watling, 1976; Orren *et al.*, 1980), or not significant, *e.g.* in *Perna viridis* (Phillips, 1985).
Cossa *et al.* (1979) argue that sexual maturation is a major source of variation in cadmium concentrations (lipophilic xenobiotic compounds are not mentioned) and propose to use small (< 25 mm length), immature mussels (*M. edulis*) for marine monitoring purposes. Orren *et al.* (1980), because of slight differences observed between males and females, also plead for the use of immature mussels (black mussels, *C. meridionalis*).
Although the highest concentrations are found in juvenile organisms, most investigators use matured mussels of 1-3 years, taking care to avoid the reproductive period. In Dutch waters *M. edulis* individuals of 4-5 cm length are normally used, while practicality in *e.g.* Hong Kong dictates that the more easily collectable 6-9 cm size class of *P. viridis* has to be used there (K.K. pers. obs.).

5.4. Environmental variables

Data regarding interactions between contaminant concentrations in mussel tissue and common abiotic variables like season, temperature, salinity, dissolved oxygen levels, chemical speciation or biotic variables like quality and quantity of food particles is hardly available at all for most contaminants. Only a few trace metals among which Cd is most important, have been investigated in this respect (Nolan, 1991). The existing knowledge is summarized briefly below.

5.4.1. Seasonal variation

Seasonal variations in tissue contaminant concentrations have been observed by different authors and a number of explanations have been proposed. A plausible explanation may be derived from the negative regression of contaminant concentrations on individual weight within one year class. This would be consistent with the fact that highest concentrations are typically found in the late winter - early spring period in temperate areas (northern hemisphere: February - April; southern hemisphere: August - October) when temperatures and food availability have been low during extended periods and stored energy (*e.g.* glycogen) has been metabolized by the animal (*e.g.* for *Mytilus edulis*: Cossa *et al.*, 1980; Popham & D'Auria, 1983; Amiard *et al.*, 1986b; Borchardt *et al.*, 1988; for *Mytilus galloprovincialis*: Majori *et al.*, 1978). Such a mechanism would imply that algal food is a less important route of contaminant supply, that the mussels largely maintain their water processing capacity (although this is a temperature dependant process) and their steady state contaminant weights per animal (uptake equaling loss) at a seasonal time scale and that stored energy deposits are not a main reservoir for contaminants at the cellular level. It should be remarked, however, that some authors fail to find seasonal effects (*e.g.* for *Mytilus edulis*: Boalch *et al.*, 1981; and for *Mytilus galloprovincialis*: Martincic *et al.*, 1992).
Seasonal variations have also been attributed to the reproductive cycle (Boyden, 1974; Popham & D'Auria, 1983; Coimbra & Carraca, 1990). Conversely, Latouche & Mix (1982) found no seasonal effects supposedly because of the use of small, immature mussels that were not influenced by the reproductive cycle. In the case of trace metals, it has been observed that seasonal minima in metal concentrations of whole soft tissue homogenates coincided with pre-spawning periods of completed gonad development. The gametes contained relatively low concentrations of trace metals as compared to the remainder of the soft tissue mass (Phillips, 1976; Fowler & Oregioni, 1976). In the case of lipophilic organic contaminants, of which the concentrations are expressed on a lipid weight basis, less seasonal variation will be present as a result of the loss of spawning products. Seasonal variation, if present on the basis of lipid weight, might then relate to the application of pesticides in up-

stream agricultural areas and resultant seasonally increased environmental levels of such contaminants.
The effects of monsoon periods have also been observed, *e.g.* by elevated contaminant levels in the clams *Villorita cyprinoides*, *Meretrix casta* and *Perna viridis* in the Cochin Backwaters (India); the low summer concentrations were attributed to the high salinity regime at the time (Lakshmanan & Nambisan, 1983).
Another cause of seasonal variations might be ascribed to periodicity in hydrographic conditions such as regularly changing currents and upwelling by which ambient concentrations of contaminants co-vary (Ouelette, 1981).
Most authors agree that in longterm bio-monitoring studies it is best to perform the sampling and exposure of mussels during the late winter - early spring period. In order to minimize year to year differences, the same period should be selected within an interval of about two weeks (see also O'Connor *et al.*, this Volume). The spawning period should be avoided.

5.4.2. Food availability

The availability of suspended particulate matter (SPM, a mixture of micro-organisms, algal cells and silt particles among others) as a potential food source, may influence the physiological performance (water processing rate, assimilation efficiency, growth rate) of filter-feeding organisms. Jørgensen (1990), in a recent review of bivalve filterfeeding, elaborates on the unwarranted assumption that filtration rates are physiologically regulated in order to keep ingestion rates constant at varying levels of algal supply. Analysis of data collected by various researchers (Winter, 1978; Kiörboe *et al.*, 1981) indicated that, at natural concentrations of algal suspensions (0.1-0.3 mg dry weight.l^{-1}), addition of silt in concentrations amounting to 5-26 mg.l^{-1} enhanced clearance rates. Too high concentrations of SPM, however, are deleterious, although the levels at which unfavourable effects on growth occur (*e.g.* at ca. 50 mg.l^{-1} or higher, Prins & Smaal, 1989) will vary with the habitat.
In some natural environments, like oligotrophic offshore waters, low growth rates may prevail due to the limited availability of food particles. This factor may be of relevance for the expression of analytical results in ABM exposure experiments as steady state contaminant concentrations may vary inversely proportional to the ashfree dry weight of the animal. No problems are to be otherwise expected as more or less constant ingestion rates have been recorded at varying densities of algal food as long as the latter correspond to natural concentrations and suspended silt is largely absent.
Growth rates will decrease as a result of low availability of food. In PBM studies, selection of similarly sized animals from sites differing in food availability has resulted in results that are not comparable. This effect is minimized in ABM studies due to the limited exposure times.
An ongoing discussion exists about whether pollutants are taken up either from the water and/or from food. For the uptake of Cd in *M. edulis*, food does not seem to be important (Borchardt, 1983; Riisgård *et al.*, 1987), while the availability of food seems at least partly important for the uptake of Cu and Zn in *M. galloprovincialis* (Martincic *et al.*, 1987a,b). Schultz-Baldes (1974) estimated an equal importance of water and SPM for Pb.
Under starvation conditions *M. edulis* showed no elimination of Cd, Cu or Ag (Chou & Uthe, 1991), while the elimination of ten radionuclides was enhanced by the presence of food.

5.4.3. Salinity

Mussels may avoid the unfavourable osmotic pressure of a changing salinity regime by closing their shells. They can, thus, survive periodic low salinities down to 0.1-0.5 x 10^{-3} (Davenport, 1981). At less extreme situations they may function normally in isotony with their environment. Only a few elements have been tested for the effect of salinity upon the accumulation (rate). Cadmium has been investigated the most. A lower salinity leads to a

higher Cd concentration in the mussels (Jackim, *et al.*, 1977; George *et al.*, 1978; Fisher, 1986). Broman *et al.* (1991) found higher concentrations in the Gulf of Bothnia than along the Swedish coast, which was attributed to a salinity effect. Phillips (1976) found, in addition, that no effect could be observed for Zn while Pb was less accumulated at lower salinities. In an interesting study, Wright & Zamuda (1987), who kept the pCu constant during an experiment with variable salinities, found that the copper concentration in the oyster *Crassostrea virginica* and the clam *Mya arenaria* was inversely related to the salinity. Mussels that were allowed to accumulate ten radionuclides in the laboratory and transplanted to the natural environment of the Baltic showed an enhanced elimination at increased salinity (Dahlgaard, 1986; Nolan & Dahlgaard, 1991).
For the interpretation of biomonitoring results, one should be aware of the influence of salinity. In coastal waters and estuaries where a freshwater surface layer can be found, the ABM mussel samples should be exposed below this halocline to minimise salinity effects.

5.4.4. Temperature

Each organism will have an upper temperature limit above which it will not function properly (for *M. edulis* > 20 °C; Widdows, 1972). Diffusion is still a temperature dependant process below this threshold and sub-optimal temperature conditions may still influence metabolic rates. Changing temperatures may, therefore, have an influence upon the uptake and loss of pollutants by mussels. Everaarts (1990) found an increase in Cd uptake at higher temperatures (especially in the kidney and the gills), but others failed to see an effect after total tissue analysis for Cd (Jackim *et al.*, 1977; Fisher, 1986) or for Zn and Pb (Phillips, 1976). Dahlgaard (1986) found an increased elimination of Ag and Zn upon a temperature increase near a cooling water duct in winter time, but considers this to be only a minor temperature effect. Orren *et al.* (1980) attribute seasonal effects to differences in temperature for the black mussel (*Choromytilus meridionalis*).
In contrast to the mussel, other species that are a potential target for biomonitoring studies, *Mya arenaria* and *Mulinia lateralis*, did show a markedly increased Cd uptake when temperatures were elevated (Jackim *et al.*, 1977).

5.4.5. Dissolved oxygen

Fisher (1986) found no effect upon the *M. edulis* cadmium uptake at varying dissolved oxygen concentrations in the range of 2.5-6.5 $ml.l^{-1}$. For a discussion on the biological effects of even lower oxygen concentrations one is referred to Eertman & de Zwaan (this Volume).

5.4.6. Chemical speciation

Total (dissolved) trace element concentrations in the water column are often not indicative of the bio-availability or bio-reactivity and, hence, bio-accumulation. The form in which a trace metal is present, its distribution over various metal 'species', will largely determine the rates of accumulation (Creselius *et al.*, 1982; Sunda *et al.*, 1987; Batley, 1989).
The free ion concentration, *e.g.* [Cu^{2+}] or, analogous to pH: pCu, is especially of major importance. Zamuda & Sunda (1982) investigated the uptake of copper in the oyster *C. virginica* as a function of the pCu. In 14 days of experiments they found a strong correlation between uptake and pCu. Rapid accumulation occurred at a pCu above 10^{-10} M. In our laboratory, we found no difference in the equilibrium concentration of copper in the mussel *M. edulis* after exposure to a large range of total copper concentrations (1-500 $\mu g.l^{-1}$), where, in each case, the copper (Cu^{2+}) was buffered by EDTA to a constant pCu=11 (Kramer & Peereboom, unpublished data). Wright & Zamuda (1987) underline the importance of the free copper concentration in their salinity dependance experiments. The (rate of) elimination of Cd in *M. edulis* seems to be governed by complexation to EDTA, humics, etc., present in the ambient water (Guthiérrez-Galindo, 1980). Florence & Stauber (1986) studied the bio-

logical effects upon algae of copper complexed to a series of organic ligands. In the environment, natural ligands *e.g.* humic and fulvic acids may interfere in the same way. Martincic *et al.* (1987a,b) found a close correlation between the 'ionic' (= electrochemically labile) fraction in Adriatic seawater and the uptake in *M. galloprovincialis*. This relation, however, was not found for Zn. Del Castilho *et al.* (1984) found a close relation between the 'free' Cd fraction (passing a dialysis membrane) and the equilibrium concentration in the freshwater mussel *Dreissena polymorpha* in a river ABM experiment. Copper did not show this relationship, however.

6. CASE STUDIES

A large number of ABM studies has been performed with different objectives related to various pollutants. Active bio-monitoring has been performed for the detection of organic pollutants such as PCBs (Young *et al.*, 1976; Hervé *et al.*, 1988; Hummel *et al.*, 1990; Kannan *et al.*, 1989), pesticides (Young *et al.*, 1976; Green *et al.*, 1986) or PAHs (Boom, 1987; Kauss & Hamdy, 1991). Clifton *et al.* (1989) studied the accumulation of various radiotracers in several UK estuaries, and Nolan & Dahlgaard (1991) followed the elimination of radiotracers in Baltic waters following enrichment in the laboratory. Trace metals were studied by ABM *e.g.* in the Krka Estuary (Martincic *et al.*, 1992), and are routinely studied by ABM experiments in Dutch coastal waters. Also, the bioavailability of effluent mixtures has been quantified using the ABM technique, such as in the case of pulp mill effluents (Wu & Levings, 1980), treated effluents (Green *et al.*, 1986) or sewage outfalls (Anderlini, 1992). Furthermore, *Mytilus edulis* was used to monitor the levels of bio-available contaminants leaching from dredged material at a dump site (Koepp, *et al.*, 1987), while Narbonne *et al.* (1992) and Van het Groenewoud *et al.* (1988) used bivalves in the adjacent water column for the testing of sediment quality.

In the following sections, some examples are given that illustrate the various aspects of the ABM technique applied with mussels in coastal and estuarine environments by our laboratory and others.

6.1. Mercury in the Ems Estuary

De Wolf (1975), using intertidal *M. edulis* in an extensive 'passive' surveillance programme demonstrated the existence of a marine mercury pollution situation in the Dutch Ems Estuary which was related to (at the time only partly identified) industrial sources (a chlor-alkali plant and a pesticide factory) in the area near Delfzijl (Essink, 1988).

'Active Biomonitoring' with *M. edulis* samples began in 1974. Each sample consisted of 80 randomly selected individuals of 4-5 cm length derived from a population in the less contaminated North Sea environment. ABM was carried out at three exposure sites on a gradient from Delfzijl in the direction of the open sea (Figure 4). Total mercury concentrations in whole animal soft tissue homogenates measured at various intervals during a time series of about 250 days showed distinct and synchronous differences between the three sites (Figure 4A; de Kock & Kuiper, 1981).

The temporal variation of Hg concentrations, expressed as Pearson's coefficient of variation, was between 15% and 30% at all sites and this compared favourably with the temporal variation of between 25% and 70% found for 'passive' intertidal samples from nearby stations collected bi-monthly during the period March 1971 - March 1973 (Figure 3). The tissue samples were analysed by instrumental neutron activation analysis (INAA) and this analysis itself was responsible for a variation of not more than 15% as determined with various series of a standard homogenate kept in quartz capsules at -20 °C.

The spatial synchronicity indicates that some environmental factor was operational simultaneously at the three locations, one possibility being simultaneous loss or gain of weight of the animals (*e.g.* related to net available energy for growth in the area) by which Hg-concentrations could increase or decrease. Another possibility is that simultaneous changes occurred in external mercury supply, *e.g.* related to irregular industrial output or to changes in Hg mobility from sediments under the influence of turbulence in this shallow area.

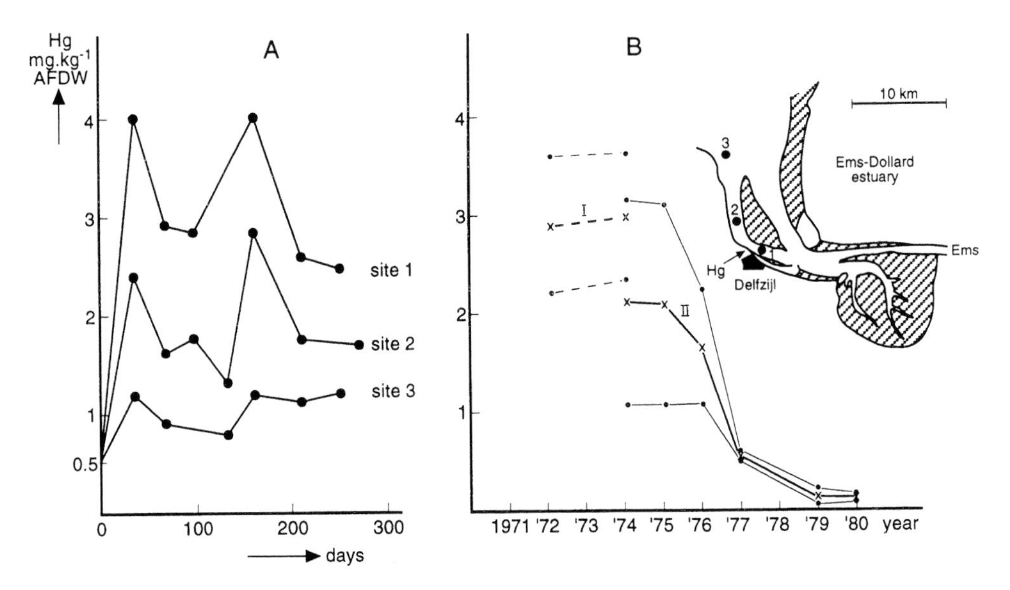

Figure 4. The monitoring of Hg in the Ems Estuary with the mussel *Mytilus edulis* at three stations along a transect away from the source (Delfzijl). In A an ABM study is presented as function of time of exposure (days) in 1974; in B the mean results over the years are plotted, where the dashed lines (I) indicate the initial PBM study, and the solid lines the ABM exercise (II). Clearly the tissue Hg concentrations follow a decrease in the amounts discharged (After: de Kock & Kuiper, 1981; 1986)

In order to monitor results of pollution abatement efforts by governmental water management authorities, repeated and similar exposure experiments at the same buoy locations followed the initial survey during the period 1974 -1980. Figure 4B shows results of the range of mean equilibrium Hg levels after at least 30 days of sublittoral exposure. After about 4 years the bio-available mercury in the water column dropped, following a S-shaped concentration curve towards more natural levels.

6.2. Cadmium in the river Scheldt Estuary

Two translocation experiments were performed covering a winter period (November 1979 - January 1980) and a summer period (June - September 1980). Temperatures in the sublittoral coastal waters of the Netherlands ranged from approximately 4-9 °C in winter and from 15-19 °C in summer at the time. For the exposure endeavour, *M. edulis* of 4-5 cm length was used, collected from a raft culture in pristine Atlantic waters off western Ireland (Killary Harbour, temperature 13-14 °C, salinity 30-35x10^{-3}) and divided into random samples of 100 individuals each. These samples were shipped to the Scheldt Estuary within 24 hours. Figure 5 depicts the locations used in the winter (closed symbols) and in the summer experiment respectively (open symbols). Samples were recollected after periods of (0), 10, 25 and 60 days. The figure clearly demonstrates a gradient of bio-available cadmium in the estuary. In view of the nearly linear accumulation pattern found during 60 days and the equilibration period of ≈150 days calculated above (see section 4.2), the eastern part of the estuary was considered to be substantially contaminated with bio-available Cd, as compared with North Sea values.

Following these experiments, ABM investigations of bio-available contaminants in coastal waters of the Netherlands by standardized exposure procedures with *M. edulis* have been conducted under the auspices of the Dutch government since 1989. These include three buoy exposure stations in the Scheldt Estuary (Figure 5b). The gradient in bioavailable Cd after 45 days of exposure is still present, but the general degree of Cd contamination seems to have declined between 1980 and 1990.

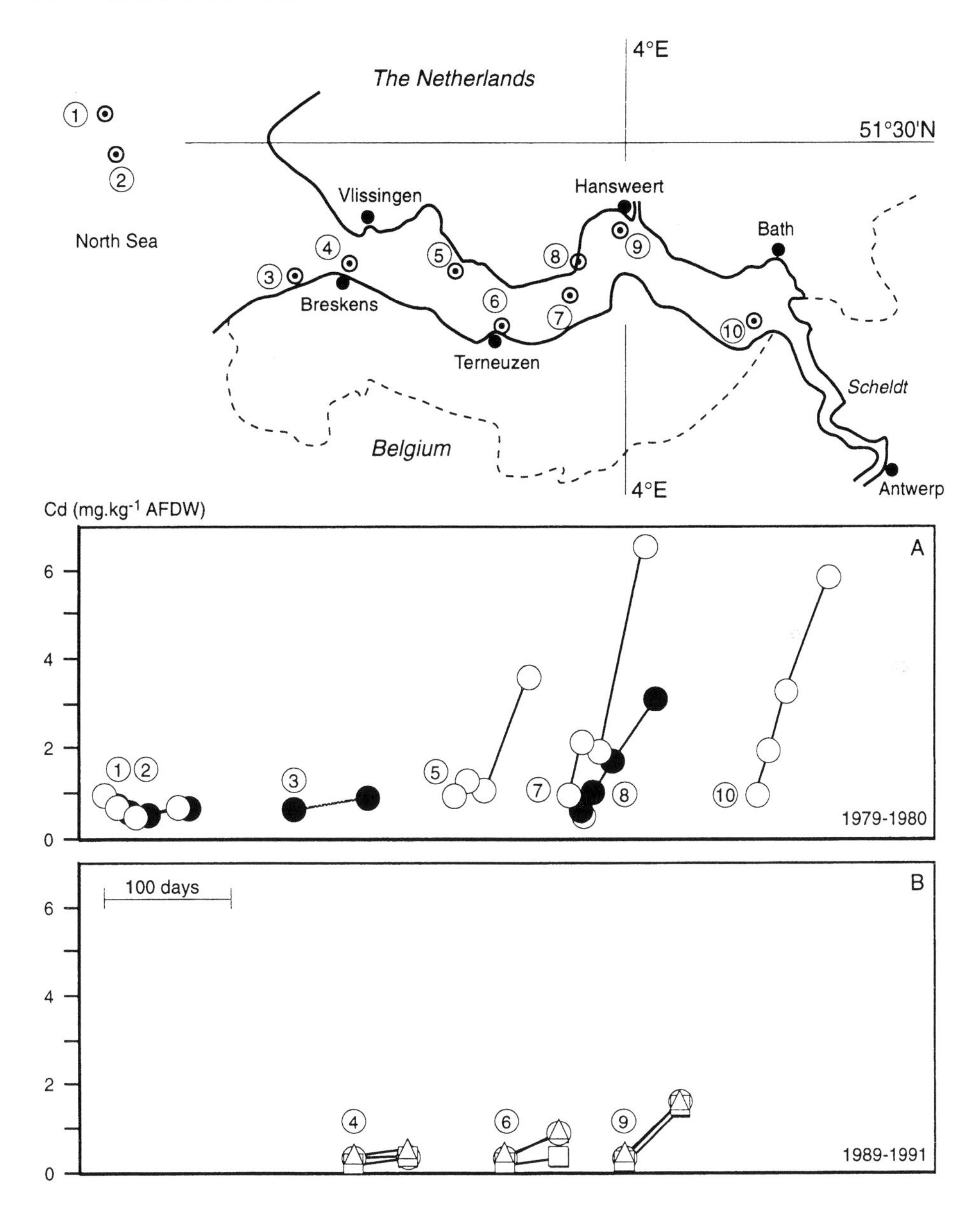

Figure 5. Active biomonitoring of Cd with *Mytilus edulis* in the river Scheldt Estuary in the period 1979-1980 (a) and 1989-1991 (b). In both cases the x-axis represents both time of exposure (days) and locations. The start of each curve marks the geographical position in the estuary, as corresponding to the map. Solid symbols refer to exposure in the winter, open symbols to exposure in the summer (After: de Kock, 1983; Dutch monitoring programme, non published)

6.3. ABM applied to detect offshore oil contamination

The use of 'oil based muds' (OBMs) in addition to 'water based muds' (WBMs) at platforms in the North Sea has steadily increased during the last decades. In recent years the practice of discharging drill cuttings to the sea bed has been prohibited in the Dutch sector of the North Sea but the practice unfortunately continues in other areas. As drill cuttings may contain substantial amounts of OBM adsorbed to the surface of the particles, the water column and sediments near offshore platforms may become contaminated by hydrocarbons (Bedborough *et al.*, 1987).

Two ABM experiments were conducted in the vicinity of platform site L-4-a in the North Sea in order to demonstrate the usefulness of *M. edulis* in monitoring oil contamination of the water column in relation to the leaching of oil from spilled drill cuttings.

The first exposure took place from September to November 1986, a period of 55 days during which wells were drilled using 'low-tox' OBMs. Mussel samples were exposed about 0.5 m above the sediment surface at distances of 250, 500, 1000 and 5000 m respectively from the platform in the direction of the residual current. Polythene baskets accommodating the samples were fastened on top of concrete anchor stones and provided with a marker buoy. An additional sample was exposed from the platform itself, about 3 m below the water surface.

The second exposure took place in April-May 1987 (42 days) at the same locations, several months after drilling activities were terminated. Only minor adjustments were made, *e.g.* the sample at the platform site (0 m point) was positioned at about 0.5 m above the sediment.

The chromatogrammes of mussel tissue extracts and original OBM showed a clear resemblance, a pronounced 'hill' of so-called unresolved carbon matter (UCM) found at 4-10 min. retention time, apart from identifiable n-alkanes (C8-C32) and aromatic compounds.

During the drilling operation, the UCM fraction in the mussels showed a clear gradient with concentrations decreasing with distance from the platform and still detectable at 5000 m during drilling periods (Figure 6). The second exposure showed significantly elevated UCM concentrations only at the platform site. As no direct correlations were found between UCM concentrations in the sediment at the different transect stations and the UCM concentrations in the mussel samples exposed at about 0.5 m above the sediment-water interface it was assumed that a desorbed soluble fraction of the OBM oil was mainly responsible for the observed bio-accumulation of UCM. The accumulation during the second exposure at the platform site probably resulted from leakage out of the bulk of discharged cuttings deposited directly near the platform (van het Groenewoud *et al.*, 1988).

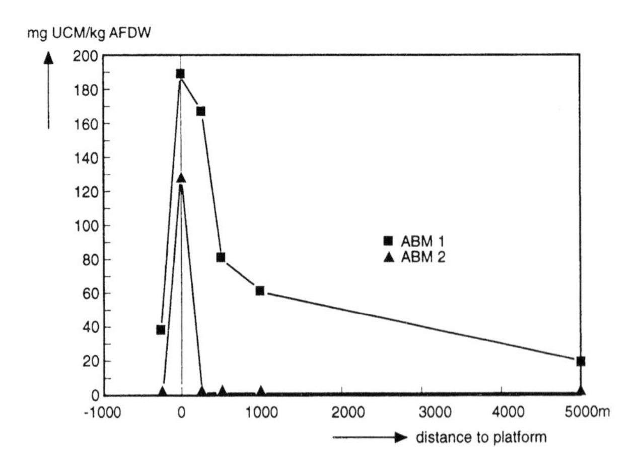

Figure 6. Oil concentration (expressed as UCM) in *Mytilus edulis* exposed in two ABM experiments at different distances from platform L-4-a in the North Sea; 1) during drilling; 2) several months after drilling was terminated. (after: van het Groenewoud *et al.*, 1988)

The case showed that ABM with mussels is a feasible option for quantifying bio-available contaminants at sites in the water column where bivalve filter feeders are normally lacking and that mussels may be used as a sensitive tool for monitoring the degree of contamination around offshore installations.

6.4. Comparison of trace metal concentrations in coastal waters

The water of the rivers Rhine and Meuse enters the North Sea close to Rotterdam and due to the local hydrodynamic situation, flows north as a residual current along the Dutch coast line. Hence, a salinity (and pollution) gradient may be observed from the coast to the open sea.
Mussels (*Mytilus edulis*) were collected from a commercial mussel farm, selected for similar size, randomly divided in test samples and stored in an intertidal area until use. At locations along a transect, perpendicular to the coast near Noordwijk, at 2, 4, 10, 20 and 60 km off shore, nine baskets with 50 mussels each were exposed at each station from bouys. Each following week after the start of the experiment one sample was retrieved and analysed for trace metal content.
The initial storage area was considered a 'clean' site, but proved to be contaminated. As a result, elimination of trace metals could be observed at all stations (Figure 7). All elimination curves tended to reach an equilibrium value. For copper and zinc this equilibrium was reached within several weeks, while cadmium (not shown) did not reach equilibrium within the 9 week exposure period.

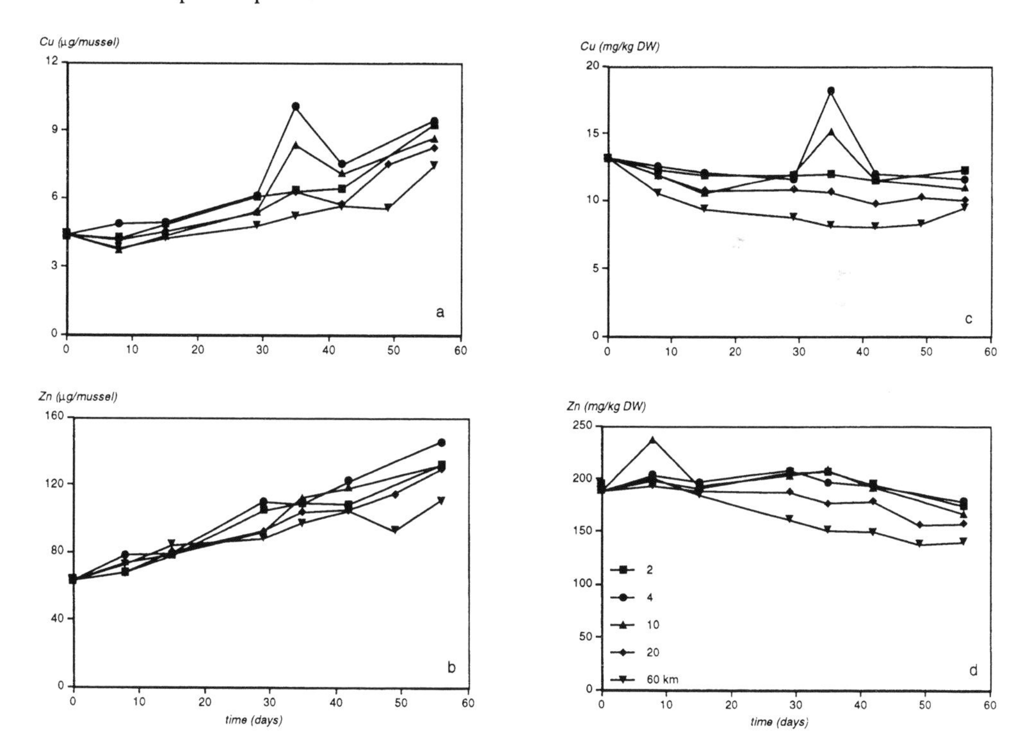

Figure 7. ABM exposure of mussels, *Mytilus edulis* along a transect perpendicular to the Dutch coast. Sub-samples were collected every week and analyzed for trace metals. Given are the total body burden and the concentrations based on DW for Cu and Zn. No explanation could be found for the two stations with a sudden increase in Cu at t=36 days (after: Kerdijk *et al.*, 1989)

Also clearly visible from the Figure are the differences between the coastal stations (2, 4 and 10 km), the intermediate 20 km station and the clean offshore station at 60 km, nicely reflecting the differences in trace metal concentrations dissolved in the water column (Kerdijk *et al.*, 1989).

6.5. Vertical profiles of DDT and PCBs in California waters

A similar ABM experiment was performed in Californian waters, where sub-samples were collected at different time intervals in a 13 week exposure experiment. Here, composite mussel samples (*M. californianus*) were exposed along a vertical line at different exposure depths: 0.5, 4, 15, 25 and near the bottom at 35 m (Young *et al.*, 1976). The samples were collected at bi-weekly intervals and analyzed for PCBs and DDT as a function of depth and exposure time (Figure 8).

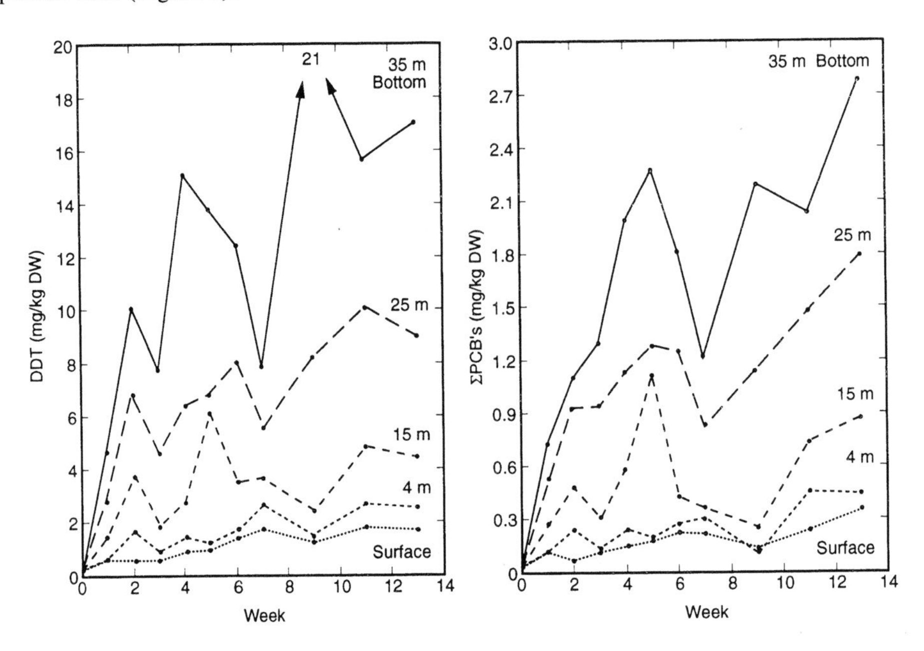

Figure 8. In an ABM experiment batches of mussels, *Mytilus californianus*, were exposed at different depths. Uptake curves are given for total DDT and total PCB in whole soft tissue (After: Young *et al.*, 1976)

Clearly, bio-accumulation of the compounds occurred and equilibrium levels were reached after about 4-6 weeks for both groups of organic pollutants. The differences in equilibrium concentration at the various water depths are interesting: the closer to the sediment, the higher the concentrations in the mussel tissue. The results clearly reflect differences in water quality between the surface and the bottom waters. The observed bio-accumulation pattern along the vertical was attributed to the influence of contaminated sediments and to the waste water plume that was trapped beneath the thermocline. The results are a good demonstration that exposure depth may affect the monitoring results, even in subtidal areas.

6.6. Comparison of trace metal concentrations in estuaries

In the period 1983-1987, ABM experiments were carried out in several north-western European estuaries and in adjacent coastal waters (Borchardt, 1988; Borchardt *et al.*, 1988). Selected mussels (*M. edulis*), either 80 large or 100 small individuals per sample, were transplanted to the estuaries of the rivers Ems, Jade and Elbe (all 1983-1987), to coastal areas in the German Bight (1985-1987) and the Danish Wadden Sea (1983-1986), as well as to offshore stations in the Southern Bight (1987) and the central North Sea (1985-1987). The samples were exposed in nets attached to specially designed frames (Figure 10a). For logistical reasons offshore stations were only sampled in the summer period. Upon retrieval, the mussels were left for 1-2 days in clean seawater to depurate. The tissue was analysed for Ag, Cd, Cu, Hg, Pb and Zn.

Table 2. Median concentrations of the six trace metals (in mg.kg^{-1} DW) in total soft tissue of mussels (*Mytilus edulis*) from ABM experiments performed over the period 1983-1987. The samples were exposed at different locations at the North Sea and in three estuaries (data after: Borchardt, 1988)

	Estuary:			**North Sea:**			
Element:	**Ems**	**Jade**	**Elbe**	**Danish Wadden Sea**	**German Bight**	**Southern Bight**	**Central**
Ag	0.10	0.11	0.15	0.23	0.27	0.28	0.23
Cd	0.9	1.3	2.7	0.7	1.4	1.0	1.5
Cu	6.9	7.6	8.0	6.3	6.8	11.9	7.1
Hg	0.27	0.28	0.72	0.20	0.26	0.10	0.19
Pb	1.3	1.8	1.5	0.7	1.3	1.9	2.0
Zn	95	129	130	103	131	116	106

A summary of the results, as median concentrations, is presented in Table 2. A general conclusion was, that in terms of trace metal pollution, the following sequence could be observed: Ems < Jade < Elbe (except for Pb), that the Ems was especially low in Ag and Zn and that all trace metals were highest in the Elbe (except for Pb). Surprisingly, Ag was found to be highest in coastal and open sea samples, Cu in the Southern Bight (summer values only) and Pb in the central North Sea.

CONCLUSION

With the presently available well defined procedures, ABM offers a good alternative for the more traditional collection of wild populations (PBM). The variability in the data is much reduced by the procedure, while comparability of different locations is improved. The application proves suitable not only for the chemical monitoring of biota, but also serves well where well defined exposure conditions are required for other bio-monitoring studies.

REFERENCES

Amiard, J.C., C. Amiard-Triquet, B. Berthet & C. Métayer, 1986a. Contribution to the ecotoxicological study of cadmium, lead, copper and zinc in the mussel *Mytilus edulis*. I. Field study. Mar. Biol. 90: 425-431.

Amiard-Triquet, C., B. Berthet, C. Metayer & J.C. Amiard, 1986b. Contribution to the ecotoxicological study of cadmium, copper and zinc in the mussel *Mytilus edulis*. II. Experimental study. Mar. Biol. 92: 7-13

Anderlini, V.C., 1992. The effect of sewage on trace metal concentrations and scope for growth in *Mytilus edulis aoteanus* and *Perna canaliculus* from Wellington Harbour, New Zealand. Sci. Total Environ. 125: 263-288

Baez, B.P.F. & M.S.G. Bect, 1989. DDT in *Mytilus edulis*: statistical considerations and inherent variability. Mar. Poll. Bull. 20: 496-499

Batley, G.E., 1989. Physicochemical separation methods for trace element speciation in aquatic samples. In: Trace element speciation: analytical methods and problems; G.E. Batley (ed). CRC Press, Boca Raton FL, pp. 43-76

Bayne, B.L., 1976. Marine mussels: Their ecology and physiology. Cambridge University Press, Cambridge, pp. 506

Bellamy, D.J., D.M. John, D.J. Jones, A. Starkie & A. Whittick, 1972. The place of ecological monitoring in the study of pollution of the marine environment. In: Marine pollution and sea life; M. Ruivo (ed). FAO, Publ. Fishing News (Books), pp. 421-425

Bedborough, D.R., R.A.A. Blackman & R.J. Law, 1987. A survey of inputs to the North Sea resulting from oil and gas developments. Phil. Trans. Roy. Soc. London B 316: 495-509

Boalch, R., S. Chan & D. Taylor, 1981. Seasonal variation in the trace metal content of *Mytilus edulis*. Mar. Pollut. Bull. 12: 276-280

Bocquené, G., F. Galgani & P. Truquet, 1990. Characterization and assay conditions for use of AChE activity from several marine species in pollution monitoring. Mar. Environ. Res. 30: 75-90

Boom, M.M., 1987. The determination of polycyclic aromatic hydrocarbons in indigenous and transplanted mussels (*Mytilys edulis* L.) along the Dutch coast. Intern. J. Environ. Anal. Chem. 13: 251-261

Borchardt, T., 1983. Influence of food quantity on the kinetics of cadmium uptake and loss via food and seawater in *Mytilus edulis*. Mar. Biol. 76: 67-76

Borchardt, T., 1988. Biologisches Monitoring in der zentralen und südlichen Nordsee: Schwermetall-belastung von Mischmuscheln (*Mytilus edulis* L.). Z. Angew. Zool. 75: 3-35

Borchardt, T., Burchert, S., Hablizel, H., Karbe, L. & Zeitner, R., 1988. Trace metal concentrations in mussels: comparison between estuarine, coastal and offshore regions in the southeastern North Sea from 1983 to 1986. Mar. Ecol. Prog. Ser. 42: 17-31

Bowmer, C.T., M. van der Meer & M.C.Th. Scholten, 1991. A histopathological analysis of wild and transplanted *Dreissena polymorpha* from the Dutch sector of the river Maas. Comp. Biochem. Physiol. 100C: 225-229

Boyden, C.R., 1974. Trace element content and body size in molluscs. Nature, 251: 311-314

Boyden, C.R. & D.J.H. Phillips, 1981. Seasonal variation and inherent variability of trace elements in oysters and their implications for indicator studies. Mar. Ecol. Prog. Ser. 5: 29-40

Branson, D.R., G.E. Blau, H.S. Alexander & W.B. Neely, 1975. Trans. Amer. Fish. Soc. 104: 785-792

Broman, D., L. Lindqvist & I. Lundbergh, 1991. Cadmium and zinc in *Mytilus edulis* L. from the Bothnian-Sea and the northern Baltic proper. Environ. Poll. 74: 227-244

Bruggeman, W.A., 1983. Bioaccumulation of PCBs and related hydrophobic chemicals in fish. PhD thesis, University of Amsterdam, pp. 160

Bryan, G.W., 1980. Recent trends in research on heavy metal contamination in the sea. Helgol. Wiss. Meeresunters. 33: 6-25

Bryan, G.W., W.J. Langston, L.G. Hummerstone & G.R. Burt, 1985. A guide to the assessment of heavy metal contamination in estuaries using biological indicators. Marine Biological Association of the United Kingdom, Special Publication No. 4

Butler, P.A., 1973. Organochlorine residues in estuarine mollusks, 1965-72. National Pesticide Monitoring Program. Pest. Monit. J. 6: 238-246

Cantillo, A.Y., 1991. Mussel Watch worldwide literature survey - 1991. NOAA Technical memorandum NOS ORCA 63, Rockville, MD, pp. 142

Capuzzo, J.M., J.W. Farrington, P. Rantamaki, C.H. Clifford, B.A. Lancaster, D.F. Leavitt & X. Jia, 1989. The relationship between lipid composition and seasonal differences in the distribution of PCBs in *Mytilus edulis*. Mar. Environ. Res. 28: 259-264

Carrell, B., S. Forberg, E. Grundelius, L. Henrikson, A. Johnels, U. Lindh, H. Mutvei, M. Olsson, K. Svaerdstroem & T. Westermark, 1987. Can mussel shells reveal environmental history? Ambio, 16: 2-10

Chou, C.L. & J.F. Uthe, 1991. Effect of starvation on trace metal levels in blue mussels (*Mytilus edulis*). Bull. Environ. Cont. Toxicol. 46: 473-478

Clark, R.C. & J.S. Finley, 1975. Uptake and loss of petroleum hydrocarbons by the mussel, *Mytilus edulis*. Fish. Bull. 73: 508-515

Clifton, R.J., H.E. Stevens & E.I. Hamilton, 1989. Uptake and depuration of 241Am, 239+240 Pu, 238Pu, 137Cs and 106Ru by *Mytilus edulis* under natural stress. Mar. Ecol. Prog. Ser. 54: 91-98

Coimbra, J. & S. Carraca, 1990. Accumulation of Fe, Cu and Cd during different stages of the reproductive cycle in *Mytilus edulis*. Comp. Biochem. Physiol. 95C: 265-270

Coombs, T.L. & P.J. Keller, 1981. *Mytilus* byssal threads as an environmental marker for metals. Aquat. Toxicol. 1: 291-300

Cossa, D., E. Bourget & J. Piuze, 1979. Sexual maturation as a source of variation in the relationship between cadmium concentration and body weight of *Mytilus edulis* L. Mar. Pollut. Bull. 10: 174-176.

Cossa, D., E. Bourget, D. Pouliot, J. Piuze & J.P. Chanut, 1980. Geographical and seasonal variations in the relationship between trace metal content and body weight in *Mytilus edulis*. Mar. Biol. 58: 7-14
Cossa, D. & J.G. Rondeau, 1985. Seasonal, geographical and size-induced variability in mercury content of *Mytilus edulis* in an estuarine environment: as reassessment of mercury pollution level in the Estuary in the Gulf of St. Lawrence. Mar. Biol. 88: 43-49
Creselius, E.A., J.T. Hardy, C.I. Gibson, R.L. Schmidt, C.W. Apts, J.M. Gurtisen & S.P. Joyce, 1982. Copper bioavailability to marine bivalves and shrimp: relationship to cupric ion activity. Mar. Environ. Res. 6: 13-26
Dahlgaard, H., 1986. Effects of season and temperature on long-term in situ loss rates of Pu, Am, Np, Eu, Ce, Ag, Tc, Zn, Co and Mn in a Baltic *Mytilus edulis* population. Mar. Ecol. 33: 157-165
Davenport, J., 1979. The isolation response of mussels (*Mytilus edulis* L.) exposed to falling sea-water concentrations. J. mar. biol. Ass. U.K. 59: 123-132
Davenport, J., 1981. The opening response of mussels (*Mytilus edulis*) exposed to rising sea-water concentrations. J. mar. biol. Ass. U.K. 61: 667-678
de Kock, W.Chr., 1983. Accumulation of cadmium and PCB's by *Mytilus edulis* transplanted from pristine water into polluted gradients. Can. J. Fish. Aquat. Sci. 40: 282-294
de Kock, W.Chr., 1986. Monitoring bio-available marine contaminants with mussels (*Mytilus edulis* L) in the Netherlands. Environm. Mon. Assessm. 7: 209-220
de Kock, W.C. & C.T. Bowmer, 1993. Bioaccumulation, biological effects and food chain transfer of contaminants in the zebra mussel. In: Zebra mussels: Biology, impacts and control; T.F. Nalepa & D.W. Schloesser (eds). Lewis Publ., Boca Raton Fl., pp. 503-533
de Kock, W.C. & J. Kuiper, 1981. Possibilities for marine pollution research at the ecosystem level. Chemosphere, 10: 575-603
de Kock, W.Chr. & H. van het Groenewoud, 1985. Modelling bioaccumulation and elimination dynamics of some xenobiotic pollutants (Cd, Hg, PCB, HCB) based on "*in situ*" observations with *Mytilus edulis*. TNO report R85/217, Netherlands Organisation for Applied Scientific Research, Delft, pp. 68
de Wolf, P., 1975. Mercury content of mussels from West European coasts. Mar. Poll. Bull. 6: 61-62
del Castilho, P., R.G. Gerritse, J.M. Marquenie & W. Salomons, 1984. Speciation of heavy metals and the in-situ accumulation by *Dreissena polymorpha:* a new method. In: Complexation of trace metals in natural waters; C.J.M. Kramer & J.C. Duinker (eds). Nijhoff/Junk Publ., The Hague, pp. 445-448
Eisler, R., 1981. Trace metal concentrations in marine organisms. Pergamon Press, New York, pp. 687
Elliott, N.G., R. Swain & D.A. Ritz, 1985. The influence of cyclic exposure on the accumulation of heavy metals by *Mytilus edulis planulatus* (Lamarck). Mar. Environ. Res. 15: 17-30
Essink, K., 1988. Decreasing mercury pollution in the Dutch Wadden Sea. Mar. Pollut. Bull. 19: 317-319
Everaarts, J.M., 1990. Uptake and release of cadmium in various organs of the common mussel, *Mytilus edulis* (L.). Bull. Environ. Contam. Toxicol. 45: 560-567
Fagerström, T. & B. Åsell, 1976. Caged fish for estimating concentrations of trace substances in natural waters. Health Phys. 31: 431-439
Farrington, J.W., E.D. Goldberg, R.W. Risebrough, J.H. Martin & V.T. Bowen, 1983. US "Mussel Watch" 1976-1978: an overview of the trace metal, DDE, PCB, hydrocarbon and artificial radionuclide data. Environm. Sci. & Technol. 17: 490-496
Fisher H., 1983. Shell weight as an independent variable in relation to cadmium content of molluscs. Mar. Ecol. Prog. Ser. 12: 59-75
Fisher, H., 1986. Influence of temperature, salinity and oxygen on the cadmium balance of mussels *Mytilus edulis*. Mar. Ecol. Prog. Ser. 32: 265-278
Florence, T.M. & J.L. Stauber, 1986. Toxicity of copper complexes to the marine diatom *Nitschia closterium*. Aquatic Toxicol. 8: 11-26
Fowler, S.W. & B. Oregioni, 1976. Trace metals in mussels from the NW Mediterranean. Mar. Pollut. Bull. 7: 26-29
George, S. & T. Coombs, 1977. The effects of chelating agents on the uptake and accumulation of cadmium by *Mytilus edulis*. Mar. Biol. 39: 261-267
George, S.G., E. Carpene & T.L. Coombs, 1978. The effect of salinity on the uptake of cadmium by the common mussel, *Mytilus edulis* (L.). In: D.S. McLusky & A.J. Berry (eds). Physiology and behaviour of marine organisms. Pergamon, Oxford, pp. 189-193
Goldberg, E.D., V.T. Bowen, J.W. Farrington, G. Harvey, J.H. Martin, P.L. Parker, R.W. Risebrough, W. Robertson, E. Schnieder & E. Gamble, 1978. The mussel watch. Environm. Conserv. 5: 101-125
Gordon, M., G.A. Knauer & J.H. Martin, 1980. *Mytilus californianus* as a bio-indicator of trace metal pollution: variability and statistical considerations. Mar. Poll. Bull. 11: 195-198

Green, D.R., J.K. Stull & T.C. Heesen, 1986. Determination of chlorinated hydrocarbons in coastal waters using a moored in situ sampler and transplanted live mussels. Mar. Pollut. Bull. 17: 324-329

Gutiérrez-Galindo, E.A., 1980. Etude de l'elimination du cadmium par *Mytilus edulis* en présence d'EDTA et de phosphate. Chemosphere, 9: 495-500

Haasch, M.L. Prince, R. Wejksnora, P.J. Cooper, K.R. Lech, J.J., 1993. Caged and wild fish - induction of hepatic cytochrome P-450 (CYP1a1) as an environmental biomonitor. Environ. Toxicol. Chem. 12: 885-895

Hervé, S., P. Heinonen, R. Paukku, M. Knuutila, J. Koistinen & J. Paasivirta, 1988. Mussel incubation method for monitoring organochlorine pollutants in water courses. Four-year application in Finland. Chemosphere, 22: 1945-1962

Ho, Y.B., 1984. Zn and Cu concentrations in *Ascophylum nodosum* and *Fucus vesicolosus* (Phaeophyta, Fucales) after transplantation to an estuary contaminated with mine wastes. Conserv. Recycling, 7: 329-337

Hummel, H., J.P. Uit Oude Groeneveld, J. Nieuwenhuize, J.M. van Liere, R.H. Bogaards & L. de Wolf, 1989. Relationship between PCB concentrations and reproduction in mussels *Mytilus edulis*. Mar. Environ. Res. 28: 489-493

Hummel, H., R.H. Bogaards, J. Nieuwenhuize, L. De Wolf & J.M. Van Liere, 1990. Spatial and seasonal differences in the PCB content of the mussel *Mytilus edulis*. Sci. Total Environ. 92: 155-163

Jackim, E., G. Morrison & R. Steele, 1977. Effects of environmental factors on radiocadmium uptake by four species of marine bivalves. Mar. Biol. 40: 303-308

Janssen, H.H. & N. Scholtz, 1979. Uptake and cellular distribution of cadmium in *Mytilus edulis*. Mar. Biol. 55: 133-141

Jørgensen, C.B., 1990. Bivalve filter feeding: hydrodynamics, bioenergetics, physiology and ecology. Olsen & Olsen, Fredensborg, pp. 140

Kannan, N., S. Tanabe, R. Tatsukawa & D.J.H. Phillips, 1989. Persistency of highly toxic coplanar PCBs in aquatic ecosystems: uptake and release kinetics of coplanar PCBs in green-lipped mussels (*Perna viridis* Linnaeus). Environ. Pollut. 56: 65-76

Kauss, P.B. & Y.S. Hamdy, 1991. Polycyclic aromatic hydrocarbons in surficial sediments and caged mussels of the St. Marys river, 1985. Hydrobiol. 219: 37-62

Kerdijk, H., K. Kramer, M. Eggens, B. van Eck & P. del Castilho, 1989. Biologische beschikbaarheid van contaminanten in het Nederlandse kustwater. RWS-DGW/MT-TNO/IB/WL report T17. Vol. I tekst/Vol. II tabellen & figuren. Ministry of Transport and Public Works, Tidal Waters Division, Middelburg, pp 88 + 130 (in Dutch)

Kiorboe, T., F. Mohlenberg & O. Nohr, 1981. Effect of suspended bottom material on growth and energetics in *Mytilus edulis*. Mar. Biol. 61: 283-288

Koepp, S.J., E.D. Santoro, R. Zimmer & J. Nadeau, 1987. Bioaccumulation of Hg, Cd and Pb in *Mytilus edulis* transplanted to a dredged-material dumpsite. In: Oceanic processes in marine pollution; Vol. I. Biological processes and wastes in the ocean. J.M. Capuzzo & D.R. Kester (eds). Krieger Publ., Malabar Fl, pp. 51-58

Koide, M., D.S. Lee & E.D. Goldberg, 1982. Metal and transuranic records in mussel shells, byssal threads and tissues. Est. Coast. Shelf Sci. 15: 679-695

Lakshmanan, P.T. & P.N.K. Nambisan, 1983. Seasonal variations in trace metal content in bivalve molluscs, *Villorita cyprinoides* var. cochinensis (Hanley), *Meretrix casta* (Chemnitz) & *Perna viridis* (Linnaeus). Indian J. Mar. Sci. 12: 100-103

Lakshmanan, P.T. & P.N.K. Nambisan, 1989. Bioaccumulation and depuration of some trace metals in the mussel *Perna viridis* (Linnaeus). Bull. Environ. Cont. Toxicol. 43: 131-138

Latouche, Y.D., & M.C. Mix, 1982. The effects of depuration, size and sex on trace metal levels in bay mussels. Mar. Pollut. Bull. 13: 27-29

Lobel., P.B. & D.A. Wright, 1982. Total body zinc concentration and allometric growth ratios in *Mytilus edulis* collected from different shore levels. Mar. Biol. 66: 231-236

Luten, J.B., W. Bouquet, M.M. Burggraaf & J. Rus, 1986. Accumulation, elimination and speciation of cadmium and zinc in mussels, *Mytilus edulis*, in the natural environment. Bull. Environ. Contam. Toxicol. 37: 579-586

Majori, L., G. Nedoclan, G.B. Modunutti & F. Daris, 1978. Study of the seasonal variations of some trace metals in the tissues of *Mytilus galloprovincialis*. Rev. Int. Oceanogr. Med. 49: 37-40

Martin, J.H. & D. Phelps, 1979. Bio-accumulation of heavy metals by littoral and pelagic marine organisms. In: EPA final report No. 600/3-79/038. U.S. Environmental Protection Agency, Washington D.C.

Martin, M. & R. Severeid, 1984. Mussel Watch monitoring for the assessment of toxic constituents in California marine waters. In: Concepts in marine pollution measurements; H.H. White (ed). Maryland Sea Grant College, Univ. of Maryland, pp. 291-323

Martincic, D., H.W. Nürnberg & M. Branica, 1987a. Bioaccumulation of metals by bivalves from Limski Kanal (North Adratic Sea) III. Copper distribution between *Mytilus galloprovincialis* (Lmk.) and ambient water. Sci. Total Environ. 60: 121-142

Martincic, D., M. Stoeppler & M. Branica, 1987b. Bioaccumulation of metals by bivalves from Limski Kanal (North Adratic Sea) IV. Zinc distribution between *Mytilus galloprovincialis*, *Ostrea edulis*, and ambient water. Sci. Total Environ. 60: 143-172

Martincic, D. Kwokal, Z. Peharec, Z. Margus, D. Branica, M., 1992. Distribution of Zn, Pb, Cd and Cu between seawater and transplanted mussels (*Mytilus galloprovincialis*). Sci. Total Environ. 119: 211-230

Myklestad, S., I. Eide & S. Melsom, 1978. Exchange of heavy metals in *Ascophylum nodosum* (L) Le Jol *in situ* by means of transplanting experiments. Environ. Pollut. 16: 277-284

Narbonne, J.F., D. Ribeira, P. Garrigues, M. Lafaurie & A. Romana, 1992. Different pathways for the uptake of Benzo[*a*]pyrene adsorbed to sediment by the mussel *Mytilus galloprovincialis*. Bull. Environ. Cont. Toxicol. 49: 150-156

NAS, 1980. The international Mussel Watch. National Academy of Sciences, Washington D.C., pp. 148

Neely, W.B., D.R. Branson & G.E. Blau, 1974. Partition coefficient to measure bioconcentration potential of organic chemicals in fish. Environ. Sci. Technol. 8: 1113-1115

Nielsen, S.A., 1974. Vertical concentration gradients of heavy metals in cultured mussels. N.Z. Jl. mar. Freshw. Res. 8: 631-636

Nolan, C., 1991. Trace metal accumulation in molluscs: the effect of variables and variability on sampling strategies. In: UNEP/FAO/IAEA reports, MAP technical reports series No. 59, UNEP, Athens, pp. 259-277

Nolan, C. & H. Dahlgaard, 1991. Accumulation of metal radiotracers by *Mytilus edulis*. Mar. Ecol. Prog. Ser. 70: 165-174

Odzak, N., D. Martincic, T. Zvonaric & M. Branica, 1993. Bioaccumulation rate of Cd and Pb in *Mytilus galloprovincialis* foot and gills. Mar. Chem. (in press)

Orren, M.J., G.A. Eagle, H.F-K.O. Hennig & A. Green, 1980. Variations in trace metal content of the mussel *Choromytilus meridionalis* (Kr.) with season and sex. Mar. Pollut. Bull. 11: 253-257

Ouellette, T.R., 1981. Seasonal variation of trace metals in the mussel *Mytilus californianus*. Environ. Conserv. 8: 53-58

Phillips, D.J.H., 1976. The common mussel *Mytilus edulis* as an indicator of pollution by zinc, cadmium, lead and copper. I. Effects of environmental variables on uptake of metals. Mar. Biol. 38: 59-69

Phillips, D.J.H., 1980. Quantitative aquatic biological indicators - their use to monitor trace metal and organochlorine pollution. Applied Science Publ., London, pp. 488

Phillips, D.J.H., 1985. Organochlorines and trace metals in green-lipped mussels (*Perna viridis*) from Hong Kong waters: a test of indicator ability. Mar. Ecol. Prog. Ser. 21: 251-258

Popham, J.D. & J.M. D'Auria, 1983. Combined effect of body size, season, and location on trace element levels in mussels (*Mytilus edulis*). Arch. Environ. Cont. Toxicol. 12: 1-14

Potts, W.T.W., 1954. The inorganic composition of the blood of *Mytilus edulis* and *Anodonta cygnea*. J. Exp. Biol. 31: 376-386

Prins, T.C. & A.C. Smaal, 1989. Carbon and nitrogen budgets of the mussel *Mytilus edulis* (L.) and the cockle *Cerastoderma edule* (L.) in relation to food quality. In: Topics in marine biology; J.D. Ros (ed). Scient. Mar. 53: 477-482

Riisgård, H.U. & E. Poulsen, 1981. Growth of *Mytilus edulis* in net bags transferred to different localities in a eutrophicated Danish fjord. Mar. Pollut. Bull. 12: 272-276

Riisgård, H., T. Kiorboe, F. Mohlenberg, I. Drabaek & P. Pheiffer-Madsen, 1985. Accumulation, elimination and chemical speciation of mercury in the bivalves *Mytilus edulis* and *Macoma balthica*. Mar. Biol. 86: 55-62

Riisgård, H.U., E. Bjornestad & F. Mohlenberg, 1987. Accumulation of cadmium in the mussel *Mytilus edulis*: kinetics and importance of uptake via food and sea water. Mar. Biol. 96: 349-353

Risebrough, R.W., B.R. de Lappe, W. Walker, B.R.T. Simoneit, J. Grimalt, J. Albaiges, J.A.G. Regueiro, A. Ballester, I. Nolla & M.M. Fernandez, 1983. Application of the mussel watch concept in studies of the distribution of hydrocarbons in the coastal zone of the Ebro delta. Mar. Pollut. Bull. 14: 181-187

Rucker, J.B. & J.W. Valentine, 1961. Salinity response of trace element concentration in *Crassostrea virginica*. Nature, 190: 1099-1100

Schulz-Baldes, M., 1973. Die Miesmuschel *Mytilus edulis* als Indicator für die Bleikonzentration in Weserestuar und in der Deutschen Bucht. Mar. Biol. 21: 98-102

Schulz-Baldes, M., 1974. Lead uptake from sea water and from food and lead loss in the common mussel *Mytilus edulis*. Mar. Biol. 25: 177-179

Simpson, R.D., 1979. Uptake and loss of zinc and lead by mussels (*Mytilus edulis*) and relationships with body weight and reproductive cycle. Mar. Pollut. Bull. 10: 74-78

Smaal, A.C., A. Wagenvoort, J. Hemelraad & I. Akkerman, 1991. Response to stress of mussels (*Mytilus edulis*) exposed in Dutch tidal waters. Comp. Biochem. Physiol. 100C: 197-200

Smith, D.R., M.D. Stephenson & A.R. Flegal, 1986. Trace metals in mussels transplanted to San Francisco Bay. Environ. Toxicol. Chem. 5: 129-138

Stephensen, M.D., R.M. Gordon & J.H. Martin, 1979. Biological monitoring of trace metals in the marine environment with transplanted oysters and mussels. In: Bioaccumulation of heavy metals by littoral and pelagic marine organisms, J.H. Martin (ed). US Department of Commerce, EPA report 600/3-79-038, National Technical Information Service PB-279-493, pp. 12-24

Sunda, W.G., P.A. Tester & S.A. Huntsman, 1987. Effects of cupric and zinc ion activities on the survival and reproduction of marine copepods. Mar. Biol. 94: 203-210

van het Groenewoud, H., L. van der Vlies, G. Hoornsman & T. Bowmer, 1988. A comparative study of monitoring techniques to establish distribution and biological effects of drilling muds around off-shore installations on the Dutch continental shelf (1986). MT-TNO report R88/405, Netherlands Organisation for Applied Scientific Research, Delft, pp. 68

Vinogradov, A.P., 1959. The geochemistry of rare and dispersed chemical elements in soils. Chapman and Hall, London

Walker, T.I., 1982. Effects of length and locality on the mercury content of blacklip abalone, *Notohaliotis ruber* (Leach), blue mussel *Mytilus edulis planulatus* (Lamarck), sand flathead *Platycephalus bassensis* Cuvier & Valenciennes, and long-noosed flathead *Platycephalus caeruleopunctatus* (McCulloch), from Port Phillip Bay, Victoria. Aust. J. Mar. Freshw. Res. 33: 553-560

Watling, H.R. & R.J. Watling, 1976. Trace metals in *Choromytilus meridionalis*. Mar. Pollut. Bull. 7: 91-94

Widdows, J. 1972. Thermal acclimation by *Mytilus edulis* L. Ph.D. thesis, University of Leicester, UK.

Widdows, J., D.K. Phelps & W. Galloway, 1980. Measurements of physiological condition of mussels transplanted along a pollution gradient in Narragansett-Bay. Mar. Environ. Res. 4: 181-194

Widdows, J., S.L. Moore, K.R. Clarke & P. Donkin, 1983. Uptake, tissue distribution and elimination of [1-^{14}C] naphthalene in the mussel *Mytilus edulis*. Mar. Biol. 76: 109-114

Widdows, J., K.A. Burns, N.R. Menon, D.S. Page & S. Soria, 1990. Measurement of physiological energetics (scope for growth) and chemical contaminants in mussels (*Arca zebra*) transplanted along a contamination gradient in Bermuda, North Atlantic Ocean. J. Exp. Mar. Biol. Ecol. 138: 99-118

Winter, J.E., 1978. A review on the knowledge of suspension-feeding in lamellibranchiate bivalves, with special reference to artificial aquaculture systems. Aquacult. 13: 1-33

Wright, D. & J. Mihursky, 1985. Trace metals in Chesapeake Bay oysters: Intra sample variability and its implications for biomonitoring. Mar. Environ. Res. 16: 181-197

Wright, D.A. & C.D. Zamuda, 1987. Copper accumulation by two bivalve molluscs: salinity effect is independent of cupric ion activity. Mar. Environ. Res. 23: 1-14

Wu, R.S.S. & C.D. Levings, 1980. Mortality, growth and fecundity of transplanted mussel and barnacle populations near a pulp mill outfall. Mar. Pollut. Bull. 11: 11-15

Young, D.R., T.C. Heesen & D.J. McDermott, 1976. An offshore biomonitoring system for chlorinated hydrocarbons. Mar. Pollut. Bull. 7: 156-159

Zamuda, C.D. & W.G. Sunda, 1982. Bioavailability of dissolved copper to the American oyster *Crassostrea virginica*. I. Importance of chemical speciation. Mar. Biol. 66: 77-82

Appendix 3

Active Bio-Monitoring (ABM) with mussels

SELECTION OF ORGANISMS; MONITORING SEASONS

Both the genera *Mytilus* (*M. edulis, M. galloprovincialis, M. californianus*) and *Perna* (*P. viridis*) have been succesfully used in respectively temperate and tropical chemical bio-monitoring programmes. These mussels are usually available in the area of investigation and are often cultivated in aquaculture industries.
For achieving optimal results, initial samples should be collected from a comparatively clean site as this will facilitate the best possible observation of accumulation processes in the field. In principle, animals of similar size (length) should be selected, preferably from a sublittoral population. The sampling location may consist of an aquaculture site or the site of a wild population below the tidal zone. Collection from polluted areas should be avoided as in an intentional depuration period before actual exposure, the elimination period may be too long for certain compounds (*e.g.* Cd). Test animals should not be bought from market places because of the many uncertainties involved as to their site of origin, treatment and contamination degree.
During spawning periods, mussels loose a considerable amount of biomass into the water column. As the spawning products are relatively rich in lipids, containing a large part of the lipophilic contaminant load such as PCBs and PAHs, spawning will strongly influence the monitoring exercise when chemical characterization of the bio-available contaminant load is the goal.
In relation to chemical monitoring with mussels, the periods of collection and exposure should preferably be chosen outside the spawning period (*e.g.* for the northern hemisphere: before April and after September; the local spawning periods should be used as a guide, however). In the case of one exposure per year, the late winter/early spring season should be used; in the case of two exposures per year (*e.g.* to investigate seasonal variation), the early autumn period should be included.

COLLECTION AND INITIAL STORAGE

For each of the exposure experiments, all mussels should be collected in one operation and at one site to avoid unnecessary inhomogeneity of the original test population. Mussels of similar size are selected from the population. The size of the mussels found is strongly dependant on the (sub)species concerned, the geographic area and site-specific growth characteristics. Hence large differences may be observed. The best size should be decided upon during sampling. As large numbers often need to be collected, a commonly available length class should be selected. Cultured mussels may offer the advantage of already belonging to a uniform size group. The length differences between the individuals should preferably be within 0.5 cm (*e.g.* range from 2.5-3.0 cm or from 5.0-5.5 cm, etc.). If necessary, 1 cm classes may be chosen. Each member of the sampling team should be provided with a simple length measuring device to provide a higher accuracy in selecting the size class concerned. An example is presented in Figure 10e. It is advisable to prepare a complete set of sizers for each person sampling.

The mussels should be collected by pulling them carefully off the hard substrate to which they are attached by byssus threads. Cutting the byssus threads with a pair of scissors is preferred as causing damage to the byssus gland by pulling must be avoided.

It has been shown above that between 50 and 100 individuals per sample are the minimum needed to reduce the effect of natural individual variability to acceptable levels; it is advised that samples of 100 individuals be used. This also allows for some loss during the experiment due to damage or mortality.

The mussels must be protected from overheating by the sun or from adverse high or low air temperatures. For this purpose they should, after their collection, be temporarily stored in moist, insulated boxes. Preferably, the fresh animals should be divided immediately into exposure samples at the collection site. If this is not feasible they should be transported to the laboratory, in which case temperature control of their environment is essential and insulated boxes with a cooling element (*e.g.* ice bags) should be used. The animals should not be immersed in water during transportation and contamination of the samples must, of course, be prevented.

PREPARATION OF EXPOSURE SAMPLES

First, the mussels must be cleaned of attached particles (sand, silt, pseudofaeces) by rinsing in water from their place of origin. Attached epibionts (*e.g.* Cirripedia, Hydrozoa, Bryozoa) must also be removed by careful scraping and brushing. It may be necessary to remove a surplus of byssus threads with a pair of scissors as well. Individuals that are not firmly closed or prove to be damaged (shell cracks) have to be discarded. Also, additional size checks are carried out and oversized and undersized animals have to be removed if present.

To ensure statistical similarity, the mussels have to be randomly distributed over the various exposure samples. To this end random tables should be used. A small computer programme (QBasic) is included (Figure 9) to assist in the preparation of these tables for each exposure experiment. An example of such a table is presented in Table 3.

Taking as an example 8 baskets each with 50 mussels, the 8 baskets are numbered 1-8. From the total batch of mussels, sub-samples are prepared of 5 individuals each. These sub-samples are to be picked randomly. In the present example the first sub-sample is placed in basket number 6, the second subsample in basket number 1, then number 8, etc., following the sequence of the first line of the random table. The random table has been designed in such a way that, in the end, all the baskets will contain (after following 10 lines) 10 subsamples, totalling 50 mussels. Two such baskets constitute an exposure sample of 100 individuals. In order to prevent confusion, one controller should be in charge, crossing out each number that has been used from the table.

The filled baskets should be kept in clean, natural seawater (*e.g.* at the site of origin) until use, giving the animals the opportunity to produce new byssus threads for attachment to the wall of their basket. This process might take one or two weeks; a longer wait should be avoided. To facilitate transportation to the exposure sites, the samples of mussels could also be transported in small net bags (taking up less space), after which the baskets can be filled and mounted at the exposure sites. However, exposure at turbulent buoy sites with wave

Table 3. Example result of the Random table generator given in Figure 9; number of samples (c) was selected 8, consisting of 10 sub-samples (d)

random table ABM *Mytilus edulis* - experiment 1							
6	1	8	7	4	3	5	2
7	6	5	2	8	1	4	3
6	2	8	1	7	3	5	4
6	5	3	4	8	2	7	1
6	3	5	8	1	4	7	2
2	4	8	1	7	6	3	5
5	4	2	3	1	8	7	6
2	7	5	4	1	3	8	6
3	7	4	6	2	5	1	8
5	2	8	4	1	6	3	7

```
OPTION BASE 1
10      CLS
20      OPEN "lpt1:" FOR OUTPUT AS #1
30      INPUT "What text do you want? "; ALG$
40      INPUT "How many samples and how many sub-divisions? ", c, d
50      DIM a(c), b(c)
60      PRINT #1, USING " random tabel   \                    \"; ALG$
70      FOR j = 1 TO d
80        FOR K = 1 TO c
90          a(K) = b(K) = 0
100     NEXT K
110     RANDOMIZE TIMER
120     b(1) = INT(c * RND + 1)
130     a(b(1)) = 1
140     FOR i = 2 TO c
150   150 :
160       b(i) = INT(c * RND + 1)
170       p = b(i)
180         IF a(p) = 1 THEN 150
190         a(b(i)) = 1
200     NEXT i
210     FOR K = 0 TO c - 1 STEP 10
220         IF (c - K) < 10 THEN 250
230         PRINT #1, USING "#### "; b(K + 1); b(K + 2); b(K + 3); b(K + 4);
240     NEXT K
250   250 :
260     Q = c - K
270     FOR n = 1 TO Q
280     PRINT #1, USING "#### "; b(K + n);
290     NEXT n
300     PRINT #1,
310     PRINT #1,
320   NEXT j
330   PRINT #1, CHR$(12)
340   CLOSE #1
350   END
```

Figure 9. Computer programme in QBasic for the preparation of a random table for subdividing (groups of) individuals (H. van het Groenewoud, pers. comm.)

action requires that the mussels are firmly attached (usually in clumps) to their baskets. The mussels should be kept moist and cool during transportation.

CONSTRUCTION OF EXPOSURE EQUIPMENT

Each batch of 50 mussels is contained in a polythene basket of sufficient size and strength. At our laboratory, tailor-made baskets from black UV resistant polythene are used. This material is free of contaminating constituents by which trace metal and organic micropollutant analysis of the mussel tissue samples might otherwise be disturbed. After checking for possible contamination (*e.g.* softeners in plastics), other holding devices (*e.g.* a pair of colanders or nylon fish netting) could be used. Realize, however, that the device will be subject to wave action for up to two months, which causes much strain on the construction. The pore size of the equipment should allow for sufficient exchange of water, even under severe fouling conditions. At our laboratory an average pore size of ca. 1.5 cm^2 is used, but maintenance (*i.e.* removing of fouling) is necessary during the summer season. Our baskets consist of two halves which should be bolted together and, if necessary, a separating disc may provide two compartments. The baskets are bolted into a stainless steel frame (Figure 10e). To prevent confusion, the baskets and their contents should be tagged for identification. The stainles steel exposure frames are welded together, made of about 8-10 mm diameter rods. A number of baskets could be accommodated in one frame (here 3, Figure 10e). On each vertical bar, flat supports with holes hold each basket in place. An identification of the frame (and its owner) may help future identification after accidental loss. The top of the frame should be provided with a revolving swivel to prevent curling of ropes.

The frames are usually exposed below buoys at 2-3 m below the water surface; other options are possible, however (Figure 10). The frames are suspended from the buoys by heavy duty nylon ropes (diameter 2.5 cm), protected against abrasion with polythene tubing (diameter 5.0 cm) and provided with eyelets in which connecting shackles fit. However, when no visible signs are permitted at the surface (*e.g.* in case of expected theft) the frames could be connected to the anchor chain of the buoy. It is important that all shackles used are bolted tight and that they are secured by using a split pin. Riding on the waves for weeks may otherwise loosen the bolts.

A number of exposure methods have been proposed (*e.g.* Young *et al.*, 1976; Riisgård & Poulsen, 1981; Smith *et al.*, 1986; Koepp *et al.*, 1987; Borchardt *et al.*, 1988), of which several are shown in Figure 10.

EXPOSURE TIME, RECOLLECTION, TRANSPORT AND STORAGE

In our laboratory, we routinely expose the frames with baskets for periods of 6-8 weeks, although much longer periods (up to 40 weeks) have been proven possible. Time series in regular monitoring schemes should relate to steady state concentrations of accumulated contaminants and repeated exposure runs are then necessary, *e.g.* twice every year.

We do not allow the mussels to depurate after collection as the influence of particulate matter is considered minimal and the analytical results may otherwise underestimate real concentration values for various compounds due to elimination during a depuration period at the laboratory. Immediately upon collection, the mussel samples are checked for dead individuals, which are counted and removed. Each sample is subsequently contained in either a cleaned glass jar or in aluminum foil for trace organic analysis, or in plastic bottles or heavy duty polythene zip-lock bags for trace metals. They are transported at a low temperature (< 4 °C) in an ice box and either dissected immediately in a 'clean bench' in the laboratory or stored deep frozen at at least - 20 °C (preferably less) until their further use in analysis.

DETERMINATION OF WW, DW, AFDW

The fresh homogenate should be prepared in a pre-cleaned glass (for trace organic pollutants) or plastic (for trace metals) beaker after removing all soft tissues (or specific organs of

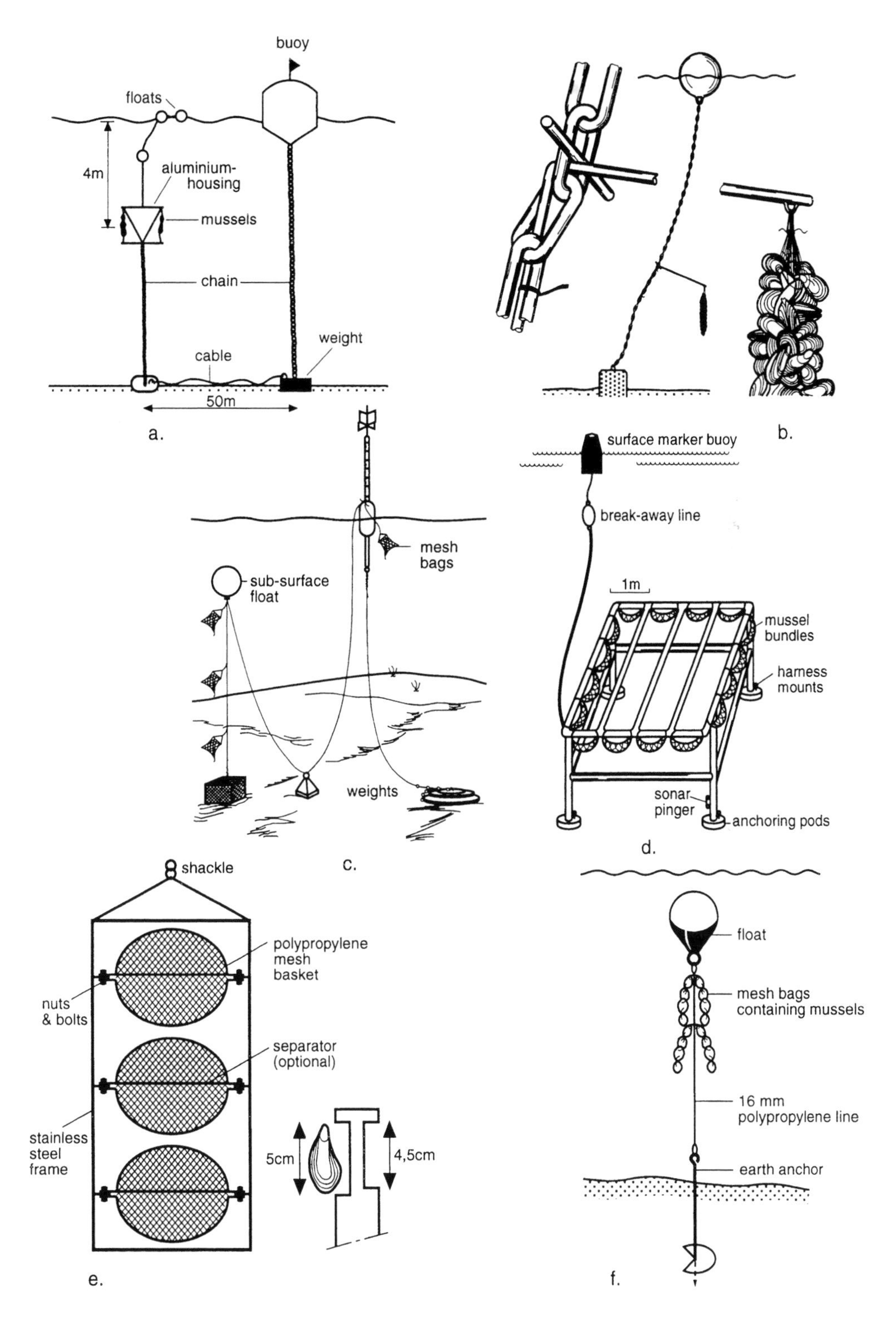

Figure 10. Examples of exposure equipment and methods for ABM experiments (After: a) Borchardt, 1988; b) Riisgård & Poulsen, 1981; c) Young *et al.*, 1976; d) Koepp *et al.*, 1987; e) design used by our laboratory, polypropylene baskets are mounted in a stainless steel frame; a simple measuring device for the proper estimation of range of the lengths of *e.g.* mussels is shown; f) Smith *et al.*, 1986)

interest) with a titanium scalpel. The authors always use a homogenizer with a titanium rotor shaft and cutting blades, in order to avoid contamination. Subsamples of the fresh homogenate are used for chemical analysis.

The water content of a fresh homogenate, or wet-weight, should not include 'attached' water contained within the two valves of the animal. The correct procedure is to sever the posterior adductor muscle and to drain interior water during a standard period (20 min) by placing the animal with open valves (ventral side down) on a slightly tilting, acid cleaned glass plate.

The *dry-weight* (DW) is determined by placing a pre-weighed subsample of the tissue homogenate (about 5 g) at 60 °C in an oven until constant weight (40-48 h), after which the sample is placed in a dessicator and allowed to cool. The sample is weighed to determine the DW.

The *ash-free dry-weight* (AFDW) is determined by initially following the procedure for DW. Following this, the sample is weighed and incinerated at about 550 °C in a muffle furnace for 2-4 h. After cooling to 80-90 °C, the residue is stored in a dessicator. After weighing, the AFDW is determined by subtraction.

Chapter 4

Macrophytes as Biomonitors of Trace Metals

David J.H. Phillips

Acer Environmental
Howard Court, Manor Park, Daresbury, Cheshire WA7 1SJ, U.K.

ABSTRACT

Aquatic macrophytes accumulate significant amounts of trace metals from ambient waters, and the concentrations of elements present in the tissues of these plants can provide information on the extent of contamination of estuarine and coastal waters by metals. Species of brown and green macroalgae have been employed most frequently in this fashion, but red algae and other macrophytes such as seagrasses have also been used as biomonitors of trace elements in particular circumstances.

Studies in both the laboratory and the field have provided valuable insights into the capacity of macrophytes to act as biomonitors of trace metals, and these are reviewed here. It is concluded that the analysis of macroalgae often provides useful information on the contamination of estuaries and coastal waters by trace elements. However, certain variables may significantly affect the accumulation of metals by such algae, interfering with their use as biomonitors. The most important of these variables are interactions between trace metals, and factors affecting plant growth. The latter include salinity and water temperature; light intensity and turbidity; and the abundance of nutrients and of certain trace contaminants.

Aspects of practical importance in sampling and preparing macroalgae for analysis are addressed, and the requirement for attention to plant taxonomy and to washing techniques is emphasised. In addition, the need is highlighted to standardise both the shore position of sampling of aquatic macrophytes used in biomonitoring surveys, and the part of the plant taken subsequently for the analysis of trace elements.

By comparison to macroalgae, much less is known about metal kinetics in seagrasses, although various seagrass species have been employed in the field as biomonitors, especially in tropical environments. Many of the parameters affecting trace element uptake and sequestration in macroalgae also appear to be active in seagrasses, but further studies are required to elucidate the impacts of variables such as growth rates and metal interactions in these species. Technical aspects of sampling and analysis are reviewed in the Appendix, these covering the various species of macrophytes employed to date as biomonitors of trace metals.

Biomonitoring of Coastal Waters and Estuaries. Edited by Kees J.M. Kramer.

1. INTRODUCTION

The net uptake of trace metals by aquatic organisms to concentrations which exceed those in their ambient waters is a widespread phenomenon, and most elements are now known to be accumulated to significant concentrations in the tissues of freshwater and marine biota (*e.g.* see Eisler, 1981). The metals of particular interest are those whose ions have been classified by Nieboer & Richardson (1980) on the basis of their Lewis acid properties as 'Class B' or 'Borderline' metal ions, as these are of significant toxicity in aquatic ecosystems (see also Hellawell, 1988).

A trend has emerged over the last two decades which involves the use of this capacity of organisms for the accumulation of trace elements in the monitoring of contamination of inland, coastal and open ocean waters. Thus, the concentrations of metals found in the tissues of aquatic organisms may be quantified relatively easily, and these may be employed as an estimate of the degree of contamination of such waters by the various elements of concern. Species which are employed in this fashion are known generically by several terms, including *bio-indicators*, *bioaccumulative indicators*, *biomonitors*, and *sentinel organisms* (see Goldberg *et al.*, 1978; Martin & Coughtrey, 1982; Hellawell, 1986; Phillips & Rainbow, 1993). None of these terms is considered perfect to generically describe such species, but *biomonitor* is employed here as it is considered the most accurate (and least mis-used) of the descriptors employed to date.

Pre-requisites for a species to act as an efficient biomonitor of trace contaminants in aquatic environments have been addressed by various authors (Butler *et al.*, 1971; Haug *et al.*, 1974; Bryan, 1976; Phillips, 1976; 1977; 1980; 1990a; Bryan *et al.*, 1980; 1985; Whitton *et al.*, 1981; Phillips & Segar, 1986; Phillips & Rainbow, 1989; 1993), and are not covered in detail here. Briefly, it may be noted that biomonitors should be strong net accumulators of trace contaminants and should reflect their ambient bioavailabilities. They should also exhibit widespread geographical distributions and year-round availability; be easy to sample and to identify taxonomically; and should preferably be sessile and tolerant of changes in salinity and turbidity. The latter is necessary to permit studies in estuarine areas, which are often sites of particular contamination.

Aquatic macrophytes conform to many of these pre-requisites, and have been widely employed as biomonitors of trace metals in estuarine and coastal environments. In most instances, species of brown or green macroalgae have been preferred, although red algae and seagrasses have also been employed in some studies. In freshwaters, vascular plants and aquatic mosses have been widely used as biomonitors of metals (*e.g.* see Harding *et al.*, 1981; Say *et al.*, 1981; Whitton *et al.*, 1981; 1982; Say & Whitton, 1983; Wehr & Whitton, 1983; Wehr *et al.*, 1983; Caines *et al.*, 1985; Mortimer, 1985), as has the red alga *Lemanea fluviatilis* (Harding & Whitton, 1981; Whitton *et al.*, 1981). These studies in freshwaters provide an interesting counterpart to investigations in estuarine and coastal environments, but are not reviewed in any detail here.

By contrast to their popularity as biomonitors of trace elements, macrophytes have been only rarely used to quantify the abundance of trace organic contaminants in aquatic ecosystems (*e.g.* see Parker & Wilson, 1975; Amico *et al.*, 1979; Coates *et al.*, 1986). This is probably a result of their low lipid contents, which reduce their capacity for the bioaccumulation of such hydrophobic compounds. As noted by Phillips & Rainbow (1993), there is evidence that the metabolism of trace organic contaminants in macrophytes is low, and this may provide scope for their further use in certain circumstances. However, these data are not reviewed here, as the present text concentrates on the information available for trace metals in aquatic plants.

The database supporting the use of macrophytes as biomonitors of trace elements in estuarine and coastal ecosystems consists of studies under both laboratory and field conditions, and these are discussed separately in the following text. The final section of the chapter discusses practical aspects of the sampling, preparation and analysis of aquatic macrophytes employed for biomonitoring studies, and technical information on such matters is provided in the Appendix to the chapter.

2. LABORATORY STUDIES AND THE EFFECTS OF ENVIRONMENTAL VARIABLES

The most comprehensive early laboratory studies on the accumulation of trace metals by macroalgae involved the brown alga *Laminaria digitata* (Bryan, 1969; 1971; 1976). This species was shown to respond principally to metals in solution, accumulating these in proportion to their ambient concentrations (Figure 1).

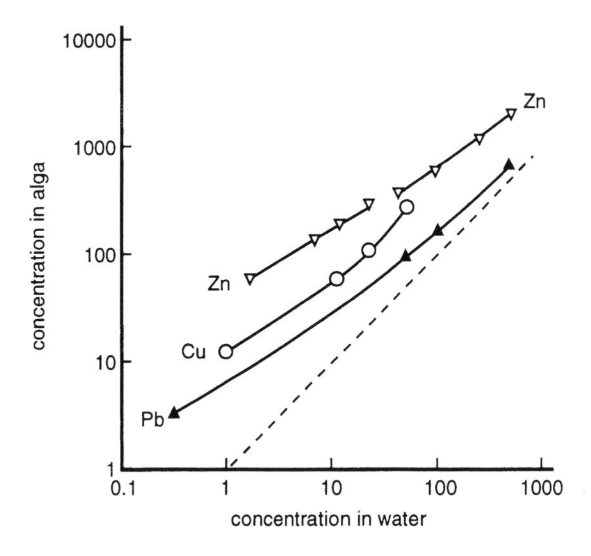

Figure 1. Relationships between the concentrations of metals in seawater ($\mu g.l^{-1}$) and those in the brown alga *Laminaria digitata* ($\mu g.g^{-1}$ dry weight) after exposure to the elements in solution under laboratory conditions for 30-31 days. Data for zinc represent two different experiments. Proportionality is shown by the broken line (After: Bryan, 1969; 1976)

These data suggest that macroalgae such as *L. digitata* satisfy the basic pre-requisite for use as biomonitors of trace elements in aquatic ecosystems, *i.e.* they reflect the ambient abundances (or more strictly, the ambient bioavailabilities) of metals. Similar results have been reported for other species also, including the brown algae *Fucus vesiculosus* and *Ascophyllum nodosum* and green algae of the genus *Enteromorpha* (Bryan *et al.*, 1985). The similarities in the responses of such species to increasing concentrations of metals in their ambient environment are evidenced by high correlations between metal levels in the various species at a range of locations (see Figure 2).

The mechanisms of uptake of trace elements by macroalgae, and of their subsequent sequestration in the tissues, have been investigated quite extensively. While the uptake of metals is believed to occur principally from solution (see Bryan, 1969; 1971; Phillips, 1979; 1980), some authors contend that elements may also be accumulated directly from sediments or suspended inorganic particulates (*e.g.* see Luoma *et al.*, 1982; Aulio, 1983; Barnett & Ashcroft, 1985; Say *et al.*, 1986a). This possibility is difficult to test, but there is certainly persuasive evidence for the contamination of the surface of macroalgae by suspended particulates and other material containing significant amounts of metals (see section 4 below).

Metals in solution in ambient waters bind to the cell walls of macroalgae through a process approximating ion exchange, and various authors have noted the high affinity of trace elements for polysaccharides such as alginates and carrageenin, which are present in the cell walls of brown and red algal species respectively (*e.g.* Haug, 1961; Haug & Smidsrød, 1965; Veroy *et al.*, 1980; Kuyucak & Volesky, 1989a,b; Maeda & Sakaguchi, 1990). The metal binding properties of alginates depend principally on the carboxylate groups present in guluronic and mannuronic acid residues, and have excited attention in particular for their affinity for strontium (Hesp & Ramsbottom, 1965; Tanaka *et al.*, 1971; Veroy *et al.*, 1980; Freitas *et al.*, 1988).

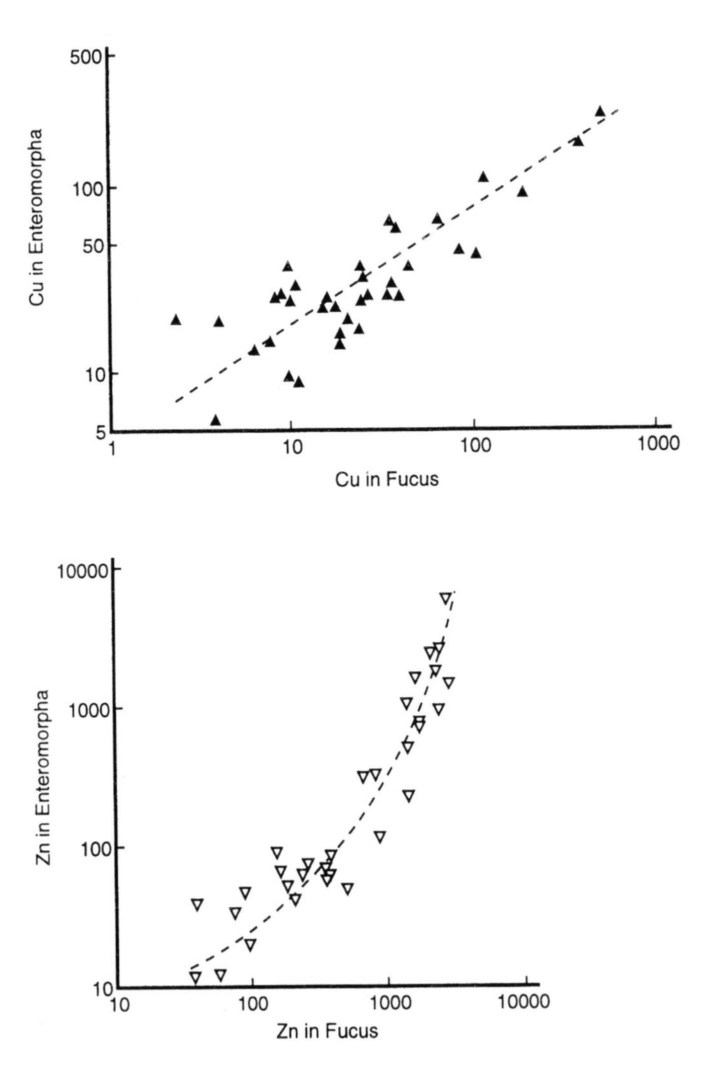

Figure 2. Comparisons between species of the genera *Fucus* and *Enteromorpha* in terms of their relative accumulation of copper and zinc at various study sites. All data are shown as $\mu g.g^{-1}$ dry weight (After: Bryan *et al.*, 1985)

While this process of adsorption to the cell walls of macroalgae undoubtedly accounts for a certain proportion of the total trace metal content of such species, the majority of the metals present are to be found within the cells, associated principally with polyphenols (Skipnes *et al.*, 1975; Ragan *et al.*, 1979; Bryan *et al.*, 1985). These polyphenols may be exuded into the surrounding seawater, not only providing a route for the loss of trace elements from macroalgae but also affecting the bioavailabilities of the metals once exuded (Ragan *et al.*, 1979). More recently, Grill *et al.* (1985; 1987) reported the existence of so-called phytochelatins in higher plants. Phytochelatins are small cysteine-rich peptides which are analogous in terms of their function to metallothioneins in invertebrate animals (see Roesijadi, 1980/81; 1992; Viarengo, 1985; 1989; George, 1990). They exhibit low molecular weights and a high affinity for trace metals, the latter depending essentially on thiolate coordination of the elements. The general structure of phytochelatins is $[\gamma\text{-Glu(-Cys)}]_n\text{-Gly}$, where n can range from 2 to 10. They have been documented to be present in a range of unicellular aquatic species, higher terrestrial plants, and certain yeasts (see Grill *et al.*, 1985; 1986; 1987; Kondo *et al.*, 1985; Gekeler *et al.*, 1988; Robinson, 1989), and a recent report from Bérail *et al.* (1991) has revealed their presence in an aquatic macrophyte, the brown seaweed *Cytoseira barabata f. repens*. Phytochelatins have been shown to be inducible by the presence of high concentrations of trace elements such as copper or cadmium (as documented also for metallothioneins; see also George & Olssen, this volume). They exhibit high affinities for a range of metals, these including cadmium, copper, lead, mercury and zinc. By contrast to metallothioneins, however, phytochelatins are not primary gene products, and are believed to be synthesised from glutathione (Grill *et al.*, 1986).

Thus, phytochelatins and metallothioneins constitute functionally analogous groups of compounds in plants and animals respectively, being widespread in occurrence and offering protection to organisms from the potential toxic effects of bioaccumulated trace metals. The protective mechanism is considered to involve the induction of the synthesis of phytochelatins and metallothioneins, and the preferential binding of metals to these peptides, hence reducing the "spillover" of elements into the general protein pool, in which non-specific binding may otherwise occur with toxic results (*e.g.* see George, 1990; Bérail *et al.*, 1991). Further work is required to fully document the biosynthetic routes for phytochelatins in plants; to establish their ubiquity amongst aquatic macrophytes and other plant species; and to clarify their relationship with polyphosphate bodies (*e.g.* see Jensen *et al.*, 1982), polyphenols and other cellular sites of metal sequestration in algae.

These various processes of metal accumulation and sequestration give rise to relatively high concentrations of trace elements in the tissues of aquatic macrophytes. Concentration factors for metals are frequently found to be in the range of 10^3 to 10^5, although these vary considerably between species and within a single species according to a variety of factors (*e.g.* see Foster, 1976; Saenko *et al.*, 1976; Seeliger & Edwards, 1977; Melhuus *et al.*, 1978; Bryan *et al.*, 1980; 1985; Förstner, 1980; Woolston *et al.*, 1982; Bryan & Gibbs, 1983; Bonotto *et al.*, 1986; Sawidis & Voulgaropoulos, 1986; Ho, 1987; Gouvea *et al.*, 1988; Catsiki *et al.*, 1991; Ganesan *et al.*, 1991; Kureishy, 1991).

The parameters which affect the bioaccumulation of trace elements in aquatic macrophytes include: salinity and water temperature; light intensity and growth rates of the species involved, the last of these producing seasonal changes in metal accumulation; and interactions between trace metals. These factors are addressed in greater detail below.

2.1. The effects of salinity and water temperature

Salinity is known to significantly affect the accumulation of metals by aquatic macrophytes in certain instances. Bryan *et al.* (1985) found that cadmium uptake by the brown alga *Fucus vesiculosus* was enhanced at lower salinities (see Figure 3), and this is believed to reflect the effects of salinity on the proportion of Cd^{2+} ions and hence on the levels of bioavailable cadmium in the ambient waters (see Zirino & Yamamoto, 1972; Cross & Sunda, 1985).

Bryan *et al.* (1985) also noted that the accumulation of copper and lead (but not that of zinc) by macroalgae was enhanced by lower salinities, although the presence of humic materials in natural freshwaters would be expected to significantly reduce the proportions of the bioavailable metal ions, tending to counteract such a salinity effect in field samples. Klumpp (1980) found no affect of salinity on the uptake of arsenic by *Fucus spiralis*, although it may be noted that arsenic kinetics in primary producers are now known to be a complex function of the accumulation of the element and its methylation, coupled to the synthesis of arseno-sugars (see Phillips, 1990b). Donard *et al.* (1987) reported that salinity influences the uptake and excretion of tin by species of the green algal genus *Enteromorpha*, which was demonstrated to be active in the methylation and demethylation of tin. Salinity is also known to affect the bioaccumulation of cobalt, manganese and zinc in *E. intestinalis* (Munda, 1984).

Water temperature may also exert significant effects on trace metal accumulation by aquatic macrophytes. Gutknecht (1961; 1963) showed that the accumulation of ^{65}Zn by the algae *Ulva lactuca* and *Porphyra umbilicalis* was heavily dependent on temperature, the rates of both uptake and excretion of the nuclide being more rapid at higher temperatures. Similar data were reported for zinc accumulation in *Fucus vesiculosus* and *Enteromorpha prolifera* from the Adriatic Sea by Munda (1979), and for technetium uptake in *F. serratus* by van der Ben *et al.* (1990).

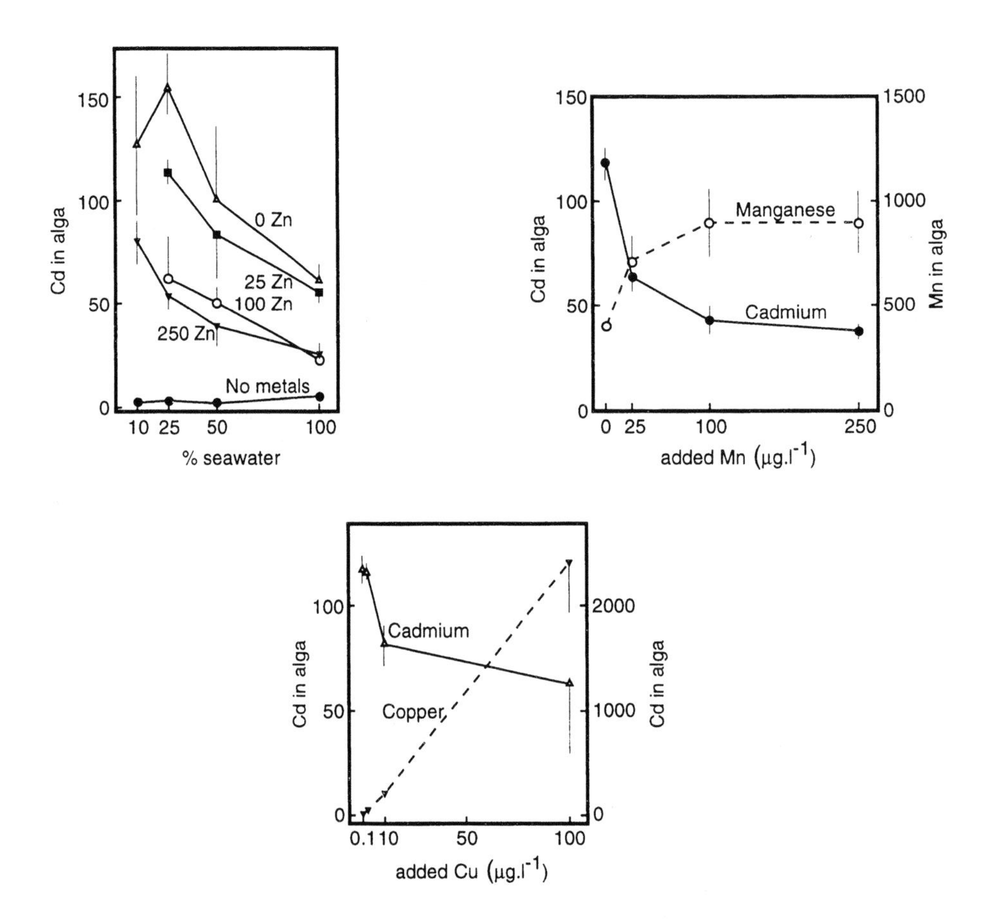

Figure 3. Effects of copper, manganese and zinc on the net uptake of cadmium by the brown alga *Fucus vesiculosus*. Exposure concentrations for cadmium were 5 $\mu g.l^{-1}$ for 16 days in each case; all concentrations of metals in the alga are shown as $\mu g.g^{-1}$ dry weight. The plot for zinc also shows the effects of salinity on metal uptake; added concentrations of zinc are shown in $\mu g.l^{-1}$. Vertical lines represent standard deviations (After: Bryan *et al.*, 1985)

2.2. Impacts of light intensity and growth rates

Many macrophyte species exhibit an enhanced net uptake of trace elements in the light compared to the dark (*e.g.* see Gutknecht, 1965; Bryan, 1969; Rice & Lapointe, 1981; Freitas *et al.*, 1988; van der Ben *et al.*, 1990; Wright & Weber, 1991), and this appears to be related to the need for metabolic energy in the process of absorption (as opposed to adsorption) of metals. Thus, for example, Wright & Weber (1991) have shown that only some of the uptake phases occurring in the accumulation of tin by *Fucus vesiculosus* and species of the genus *Enteromorpha* require energy, and that the processes of uptake differ between the two species, as shown in Figure 4.

The growth rates of aquatic macrophytes are also affected by light, and there is abundant evidence of the impacts of algal growth rates on the accumulation of trace metals (Bryan, 1969; 1971; Phillips, 1977; 1980; Rice & Lapointe, 1981). However, plant growth is affected not only by light conditions (including turbidity, which affects light penetration in water), but also by the ambient concentrations of nutrients and of trace metals, the former promoting growth and the latter sometimes inhibiting algal growth (*e.g.* see Bryan, 1969; Tewari *et al.*, 1990). These factors are important in relation to field studies of aquatic macrophytes, for two reasons. Firstly, the concentrations of elements found in aquatic macrophytes vary according to the part of the plant sampled, this being caused in most cases by the dilution of accumulated metals by plant growth. Secondly, the relationships between the metal concentrations found in an aquatic macrophyte and those in its ambient environment are a function of growth rates, and this influences the use of such species as biomonitors of metal contamination (see Phillips, 1977; 1980; Rice & Lapointe, 1981).

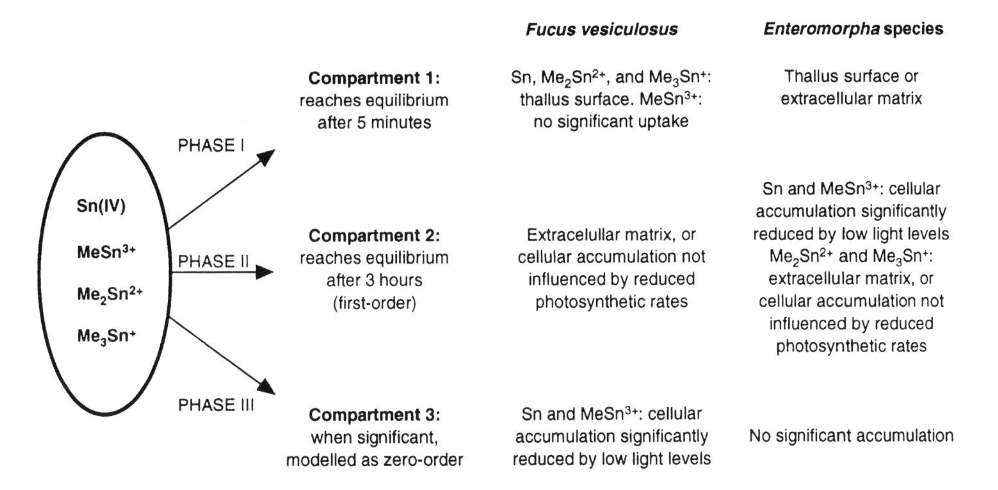

Figure 4. Models for the uptake of tin compounds from solution into different compartments of *Fucus vesiculosus* and *Enteromorpha* species (after Wright & Weber, 1991)

2.3. Seasonal influences and the sampling of aquatic plants

A further important factor defining the concentrations of trace elements in aquatic macrophytes is their season of collection. Many authors have reported significant variations in metal levels of such species with season (*e.g.* see Bryan & Hummerstone, 1973; Fuge & James, 1973; 1974; Haug *et al.*, 1974; Whyte & Englar, 1974; Burdon-Jones *et al.*, 1982; Sawidis & Voulgaropoulos, 1986; Say *et al.*, 1986a; Guner *et al.*, 1987; Ishikawa *et al.*, 1987; Ferreira & Oliveira, 1988; Geyer *et al.*, 1990; Rönnberg *et al.*, 1990; Miramand & Bentley, 1992). In most instances, such variations have been ascribed to the impacts of seasonal changes in growth of the plants, high growth rates in the warmer months of the year serving to "dilute" the accumulated metals and reduce their concentrations, with the reverse trend occurring in the winter. This need not be of particular concern in the use of aquatic macrophytes as biomonitors of trace metals, however, so long as growth rates of the samples taken do not vary significantly between sites, and plants at the locations of interest in biomonitoring surveys are sampled over a short period of time (Phillips, 1980).

Eide *et al.* (1980) noted in a field transplantation study that the uptake and release of cadmium and zinc was subject to seasonal changes, occurring at generally faster rates in summer than in winter. By contrast, the rate of accumulation of lead revealed few effects of season, while that of mercury was intermediate in this respect. They speculated that these differences between the metals might reflect the need for metabolic energy in the accumulation of cadmium and zinc, whereas lead was considered to be taken up through an ion exchange mechanism similar to that previously documented to exist for strontium.

With respect to seagrasses, it is known that trace elements are generally present at higher levels in the roots than the rhizomes of these species, and that significant age-related variations exist in metal concentrations in the stems and blades of the plants (as in macroalgae, with the stipe, thallus and tips varying in metal levels; see section 4 below). Elements appear to be taken up by the roots and rhizomes of seagrasses from interstitial waters or perhaps from the sediment surfaces themselves, and by the blades and (to some extent) the stems from ambient waters above the sediment surface. In most instances, there appears to be little translocation of metals between the different parts of the plant (*e.g.* see Brix & Lyngby, 1982).

2.4. Interactions between trace metals

It may be noted from Figure 1 that there is some evidence of the saturation of uptake at higher ambient metal concentrations, *i.e.* of diminished concentration factors for trace elements at higher exposure levels. This is thought to reflect the finite nature of binding sites in macroalgae for metals, and laboratory studies of trace element interactions support this concept. Bryan (1969) first showed that the presence of relatively high concentrations of cadmium, copper or manganese in solution reduced the net accumulation of ^{65}Zn by the brown alga *Laminaria digitata*, and additional examples of metal interactions have been reported since this early work. Thus, the accumulation of cadmium by the brown alga *Fucus vesiculosus* is known to be affected by copper, manganese and zinc (Figure 3), and Bryan *et al.* (1985) noted that this was evident not only under laboratory conditions but also in samples of the alga taken from highly contaminated locations in the field, such as Restronguet Creek in Cornwall, England (see section 3). Morris & Bale (1975) could not demonstrate any consistent correlation between the levels of manganese in *Fucus vesiculosus* from sites in the Bristol Channel and those in the ambient waters, although such correlations were present for cadmium, copper and zinc. Foster (1976) encountered similar problems in work on both *F. vesiculosus* and *A. nodosum* in Afon Goch in Wales, which was subjected to very considerable contamination by copper and zinc in particular. The high levels of these elements appeared to give rise to saturation of the available binding sites on the algae, and the accumulation of manganese was reduced as a result (see also Phillips, 1980).

Further examples of metal interactions in aquatic macrophytes have been reported more recently (*e.g.* see Zolotukhina *et al.*, 1990; Crist *et al.*, 1992), and this phenomenon has been considered by Phillips (1977; 1980; 1990a) to limit the usefulness of macroalgae as biomonitors of certain trace elements in field conditions.

It may be concluded that while the accumulation of trace elements by aquatic macrophytes is not subject to regulation as such (see Phillips & Rainbow, 1989; 1993), several parameters exert significant effects on this process. These effects may give rise to problems in field studies seeking to utilise aquatic macrophytes as biomonitors of the ambient abundances or bioavailabilities of trace metals, several examples of which are reviewed in the following section.

3. FIELD STUDIES

As noted previously, aquatic macrophytes have been widely employed in field studies of metal contamination, data for both macroalgae and seagrasses being reported from various estuarine and coastal ecosystems. The present section briefly reviews such data, emphasising studies which have provided information on the effects of variables which may influence the accumulation of trace elements in these species.

One important aspect of the use of biomonitors is their capacity to time-integrate the ambient levels of contaminants, *i.e.* to reflect long-term pollution trends, rather than simply the more recent abundance of contaminants (Phillips, 1980; Phillips & Segar, 1986). Laboratory-based investigations suggesting that aquatic macrophytes exhibit a considerable time-integration capacity for many elements (*e.g.* see Bryan, 1969; Young, 1975) have been followed more recently by field transplantation studies, providing an insight to the degree of time-integration offered by macroalgae in estuarine situations in particular.

Myklestad *et al.* (1978) transplanted the brown alga *Ascophyllum nodosum* from the metal-polluted Hardangerfjord to a relatively uncontaminated site in Tronheimsfjord further north in Norway. *A. nodosum* is of particular usefulness in such work, as it is easily aged due to its regular formation of vesicles during its growth. The concentrations of metals in the growing tips of the alga were found to approach those of native *A. nodosum* in the new location over a period of five months, whereas the levels of trace elements in the older parts of the transplanted plants altered relatively little over this period, suggesting the presence of metals in fractions exhibiting little exchange.

Later work by the same group involved the cross-transplantation of *A. nodosum* between a variety of polluted and relatively uncontaminated locations, extending the available database on metal kinetics in this species, in particular in relation to seasonal effects (see above and Eide *et al.*, 1980). Similar data were published by Ho (1984) involving copper and zinc in both *A. nodosum* and *Fucus vesiculosus*, these species being transplanted from the relatively unpolluted Torpoint in Cornwall, England, to Restronguet Creek, which is known to be highly contaminated by a range of trace elements derived principally from mining activities. Data from these studies are shown in Figure 5, and these reveal that metal levels in *F. vesiculosus* responded considerably more rapidly than those in *A. nodosum* to the new conditions. Similar results for other metals in the same two species were reported by Bryan & Gibbs (1983) and by Langston (1984). These data suggest the existence of a consistent difference in the time-integration capacities of the two species, perhaps due at least in part to their relative growth rates at the sites involved.

It should be noted here that the database available for trace elements in seagrass species is considerably less robust than that for macroalgae. In addition, the attention paid to laboratory studies of macroalgae to support their use as biomonitors in field situations has not been paralleled by similar investigations on seagrass species. Nevertheless, seagrasses of several genera have been employed as biomonitors of trace metals, with species of the genera *Posidonia* and *Zostera* being most commonly used (*e.g.* see Augier *et al.*, 1978; Drifmeyer *et al.*, 1980; Brix & Lyngby, 1982; 1983; Brix *et al.*, 1983; Wahbeh, 1984; Denton & Burdon-Jones, 1986; Nienhuis, 1986; Lyngby & Brix, 1987; Ward, 1987; Maserti *et al.*, 1988; Costantini *et al.*, 1991; Francesconi & Lenanton, 1992). Augier *et al.* (1978) reported a major source of mercury in the coastal zone near Marseille in France, this finding being based on the analysis of the roots, rhizomes and leaves of the seagrass *Posidonia oceanica.* This species is widespread in the Mediterranean, and would appear to offer scope for further studies.

The existence of limited data on metals in saltmarsh plants such as *Spartina alterniflora* and *Halimione portulacoides* may also be noted here in passing (*e.g.* see Gardner *et al.*, 1978; Drifmeyer & Redd, 1981; Newell *et al.*, 1982; Beeftink & Nieuwenhuize, 1986; Reboredo, 1992). Drifmeyer & Redd (1981) analyzed *S. alterniflora* from 16 sites along the Atlantic coast of the USA, revealing significant geographical variations in the levels of copper, iron and zinc, but not in those of manganese. Reboredo (1992) found cadmium to be accumulated rapidly through the roots of *H. portulacoides* and to be translocated efficiently to the leaves of this species. While the concentrations of cadmium present in the plant varied seasonally, it was nevertheless possible to demonstrate the existence of industrial pollution by cadmium in the River Sado estuary in southern Portugal through the analysis of cadmium levels in *H. portulacoides*.

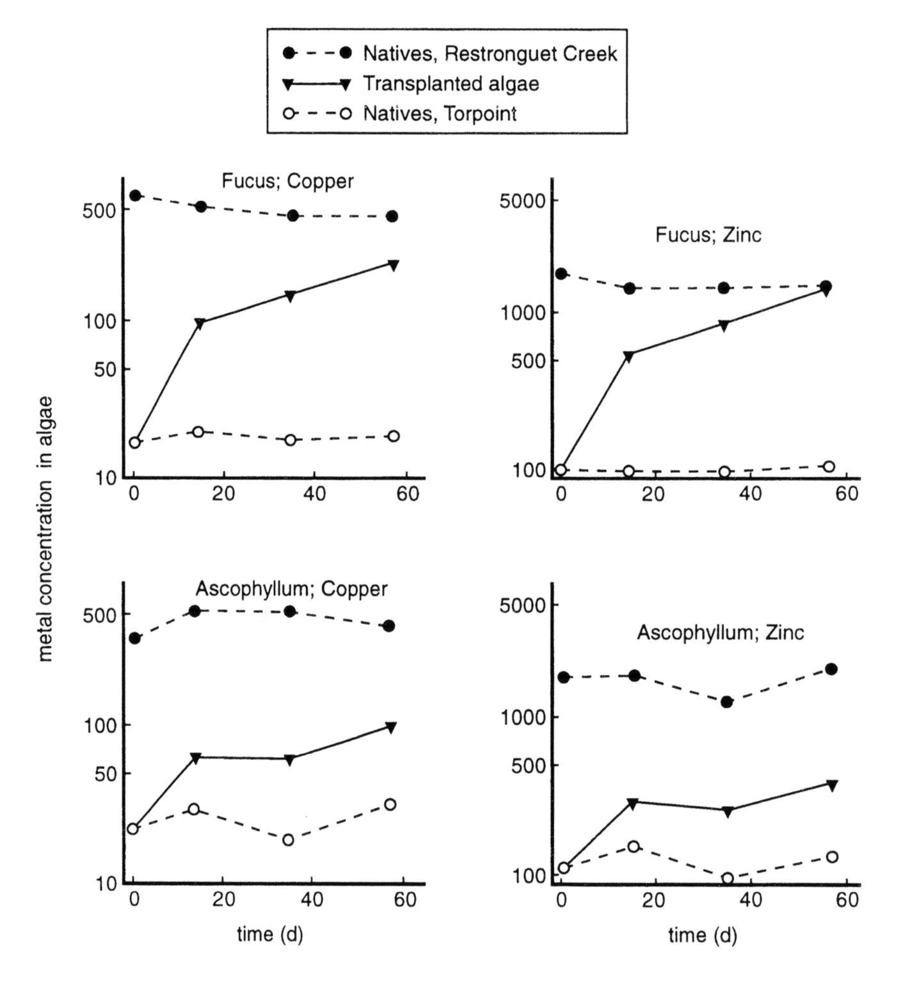

Figure 5. Temporal changes in the concentrations of copper and zinc in native *Fucus vesiculosus* and *Ascophyllum nodosum* from Torpoint and Restronguet Creek in Cornwall, England, and in the levels of metals in the same species transplanted from Torpoint to Restronguet Creek. All data for metal concentrations in the algae are shown as $\mu g.g^{-1}$ dry weight (After: Ho, 1984)

It may be concluded that field studies of macroalgae and seagrasses have generally supported the concept that these species act as efficient biomonitors of metals in estuarine and coastal waters (see also Say *et al.*, 1986a). However, in highly contaminated situations, metal interactions may interfere with attempts to employ aquatic macrophytes as biomonitors of trace elements, and caution must be exercised in the interpretation of field data for certain ele-

ments in particular. It appears likely that the metals most affected are those which exhibit relatively low affinity for uptake (either through adsorption or absorption processes) in algae, and these include cobalt, manganese and nickel in particular (Bryan, 1969; Bryan & Hummerstone, 1973; Ragan *et al.*, 1979; Bryan *et al.*, 1985). Phillips (1980) considered that the use of macroalgae to monitor manganese in estuaries could not be justified, due to the frequency of metal interactions affecting the net uptake of this element in such species.
Factors affecting plant growth are also of importance, and may influence the capacity of aquatic macrophytes to act as accurate biomonitors of metals in particular situations.

4. PRACTICAL ASPECTS OF SAMPLING, PREPARATION AND ANALYSIS

While the literature on the use of aquatic macrophytes as biomonitors of trace metals is extensive and the techniques involved can be considered in general to be robust, certain aspects merit particular attention. Principal among these are details of the sampling of the plants in the field, and their preparation for element analysis.
With respect to the sampling of aquatic macrophytes, attention should be paid to two aspects:

- plant taxonomy and identification; and
- the position of sampling on the shore (*i.e.* the site of the plants sampled, with reference to tidal fluctuations).

The preparation of macrophyte samples for analysis also involves two critical decisions:

- the selection of a method for washing of the plant material; and
- the choice of the part(s) of the plant to be taken for analysis.

4.1. Species identification

It is known from many studies that the concentrations of trace elements present in aquatic macrophytes vary significantly with species (*e.g.* see Black & Mitchell, 1952; Saenko *et al.*, 1976; Nienhuis, 1986; Ho, 1987; Gouvea *et al.*, 1988). All samples of macrophytes taken at various sites to support a biomonitoring survey of trace elements should therefore consist of the same species. In many instances, this is not problematical, as certain species (*e.g.* the commonly-used brown seaweeds of the genera *Laminaria, Fucus* and *Ascophyllum*) are easy to identify taxonomically and often dominate intertidal or near-subtidal communities. However, greater difficulties exist with other genera, perhaps the best example being the green algal genus *Enteromorpha*. Several authors have employed species of this genus in biomonitoring studies for trace elements (*e.g.* see Stenner & Nickless, 1974; Seeliger & Edwards, 1977; Seeliger & Cordazzo, 1982; Say *et al.*, 1986a,b). However, Say *et al.* (1986a,b) and others have noted that taxonomic identification is challenging in this unusually polymorphic genus, and that further work is required to enable the various species to be unequivocally identified (see also Koeman & van den Hoek, 1982a,b). Given the known differences in metal accumulation between even closely-related species of macrophytes, it is considered that genera such as *Enteromorpha* should not be employed as biomonitors of trace elements unless a reliable system of taxonomic identification is available.

4.2. Shore level of collection

Several authors have reported that the concentrations of trace metals present in aquatic macrophytes vary according to their position on the shoreline (*e.g.* see Nickless *et al.*, 1972; Bryan & Hummerstone, 1973; Fuge & James, 1973; 1974; Bryan *et al.*, 1985). This may be related to differential growth rates among macrophytes according to their shore position and degree of submergence over the tidal cycle (Phillips, 1980), or to other factors. Whatever the cause of such a phenomenon, researchers should take account of such variation by sampling at similar shore levels with respect to tide marks at all monitoring sites. The mid-shore level is generally considered to be the most appropriate (Bryan *et al.*, 1985; Say *et al.*, 1986a,b). It may also be noted here that the sampling techniques employed in studies of algal biomass (*e.g.* see Sfriso *et al.*, 1991) could be adapted for use in the collection of macrophytes for trace element analysis in biomonitoring investigations.

4.3. Surface contamination of macrophytes

The selection of washing techniques employed for collected samples is particularly important, as aquatic macrophytes inevitably exhibit surface contamination by a variety of organic and inorganic materials.

4.3.1. Organic coatings

Patrick & Loutit (1977) have shown that epiphytic bacteria and other organisms such as diatoms and protozoa present on the surfaces of the freshwater plant *Alisma plantago-aquatica* contribute very significantly to the total metal concentrations present at analysis, accounting for up to 50% of certain elements and being of particular importance in older plants. Given the capacities of bacterial communities for metal accumulation (*e.g.* see Remacle, 1981), this can hardly be considered surprising. Wehr *et al.* (1983) found that the metal concentrations present in the aquatic moss *Rhynchostegium riparioides* varied with the type of washing of the moss samples prior to analysis, and this trend may also reflect contamination of the sample surfaces by organic material containing significant amounts of trace elements. Recent data from Holmes *et al.* (1991) have extended these observations to the marine red algal species *Gracilaria sordida*, the concentrations of zinc in this species being affected very significantly by epiphytic bacteria on its surface.

4.3.2. Inorganic particulates

Similar concerns exist with respect to the presence of small inorganic particulates adhering to the surfaces of macrophytes. For example, Barnett & Ashcroft (1985) reported that the great majority of the site-dependent variations in the concentrations of chromium, iron and lead in *Fucus vesiculosus* from the Humber Estuary in England were a function solely of the differential contamination of samples by inorganic particulates derived from sediments. Such problems would be anticipated to be most extreme in instances where the concentrations of metals in sediments significantly exceed those in the macrophyte employed as a biomonitor, and this situation may be encountered more frequently for elements which favour the particulate phase (Bryan & Hummerstone, 1973; Barnett & Ashcroft, 1985). It remains possible that the correlations reported by certain authors between metal concentrations in macrophytes and those in nearby sediments (*e.g.* see Luoma *et al.*, 1982; Aulio, 1983; Say *et al.*, 1986a) are due in part at least to the contamination of the plants by fine particulates exhibiting high levels of trace elements. Beeftink & Niewenhuize (1986) reported data for metals in unwashed salt marsh plants which clearly reflect the impacts of deposited suspended particulate matter on the total metal concentrations of the plant.

It is also notable here that surface contamination by particulates tends to vary between different species of macroalgae. Thus, for example, Bryan *et al.* (1985) reported this factor to be less of a problem for *Ascophyllum nodosum* than for most other species of brown algae or for members of the genus *Enteromorpha*. Such differences between species may relate to their tendencies to produce surface slimes or mucus (see Patrick & Loutit, 1977); to the leaf area available for deposition of the particulates, as noted for salt-marsh species (Beeftink & Niewenhuize, 1986); or to more sophisticated aspects of the chemistries or microstructure of their surfaces.

Reliable measures to deal with this problem of the surface contamination of macrophytes have not been developed to date. While certain authors have recognised this factor as a possible source of interference in the use of macrophytes as biomonitors, no standard technique exists which appears sufficiently robust to account for such problems. Bryan *et al.* (1985) proposed the washing of macroalgae such as *Fucus vesiculosus* in diluted seawater, coupled to brushing of the fronds to remove adhering particles. Other authors have preferred shaking samples with deionised water or seawater, or (in a few instances) the use of sonication techniques to remove particulates (Phillips, 1979; Woolston *et al.*, 1982; Freitas *et al.*, 1988). Say *et al.* (1986b) reviewed methods employed to that time, emphasising the lack of any standardised technique to remove either epiphytes or adhering inorganic particulates. Updated information on this aspect is shown in the Appendix to this chapter.
Bryan *et al.* (1985) also noted that the maximum contribution of metals bound to inorganic particulates could be calculated using aluminium or iron as a basis for corrections, as these elements are known to associate intimately with inorganic particulates. Similar methods have been employed to account for contamination of invertebrate animals by suspended particulates in the gut (see Flegal & Martin, 1977; NAS, 1980; Brumbaugh & Kane, 1985; Lobel *et al.*, 1991). The use of such correction factors should certainly be considered in biomonitoring surveys employing macrophytes as biomonitors, which should in any event incorporate a standardised washing technique for all samples. Further work is clearly required with respect to the development of the most appropriate technique for the washing of macrophyte samples destined for trace element analysis.

4.4. The part of the plant sampled

Decisions are also required in biomonitoring studies employing macrophytes as to which part of the plant should be taken for analysis. The older portions of both macroalgae and seagrasses generally contain higher concentrations of trace elements (*e.g.* see Bryan, 1969; Bryan & Hummerstone, 1973; Haug *et al.*, 1974; Brix & Lyngby, 1983; Bryan *et al.*, 1985). Bryan *et al.* (1985) concluded that the younger tissue exhibits a shorter time-integration period for trace metals, whereas the older parts of aquatic plants reflect contamination conditions over long periods of time and are generally less responsive to short-term changes in metal abundance. Researchers may use such differences (perhaps coupled to transplantation techniques, in appropriate cases) to select the period of most interest in biomonitoring surveys. However, care must be taken to standardise the techniques for preparation of macroalgae, such that tissues of similar age are taken from all samples. This tends to be simplest in species such as *Ascophyllum nodosum*, which can be easily aged by the presence of vesicles formed at regular periods throughout its growth.

4.5. Digestion and analysis techniques

By contrast to the problems discussed above, the digestion and analysis of aquatic macrophytes present relatively few concerns. Wet digestion techniques are generally employed for most elements, using nitric acid or a mixture of nitric and perchloric acids (see the review by Say *et al.*, 1986b). This may be followed by analysis using conventional atomic absorption spectrophotometric (AAS) methods. For the determination of arsenic and tin concentrations in macrophytes, dry ashing is recommended, prior to hydride generation analysis (HGAAS; see Phillips, 1990b). Mercury should be analyzed by the cold vapour technique (CVAAS) following digestion with a mixture of nitric and sulphuric acids and the subsequent addition of hydrogen peroxide (Bryan *et al.*, 1985). The multi-element technique employing inductively coupled plasma and mass spectrometry (ICP-MS) may become advantageous.

CONCLUSION

It may be concluded that the analysis of macrophytes as biomonitors has provided useful data on the abundances and distributions of trace metals in estuaries and coastal waters. The state of the art is further advanced at present for macroalgae (particularly for brown and green algae) than for seagrasses or salt marsh species. Similarly, much more is known of the accumulation of trace elements by macrophytes inhabiting temperate zones than for those species found in tropical locations. Given the geographical distributions of the various species, it appears probable that future work will tend to concentrate on macroalgae in temperate regions (extending the database to species other than those employed to date), but on seagrasses in warmer waters. There is a need for considerable additional research on the effects of environmental variables on the net uptake of metals by macrophytes. However, practical aspects which affect the use of such species as biomonitors should not be ignored, and the present difficulties with taxonomic identification and washing techniques are of particular note.

REFERENCES

Amico, V., G. Orient, M. Piattelli & C. Tringali, 1979. Concentrations of PCBs, BHCs and DDTs residues in seaweeds of the east coast of Sicily. Mar. Pollut. Bull. 10: 177-179

Augier, H., G. Gilles & G. Ramonda, 1978. Recherche sur la pollution mercurielle du milieu maritime dans la région de Marseille (Méditerranée, France): Partie 1. Degré de contamination par le mercure de la phanerogame marine *Posidonia oceanica* Delile á proximité du complexe portuaire et dans la zone de réjet du grand collecteur d'égouts de la ville Marseille. Environ. Pollut. 17: 269-285

Aulio, K., 1983. Heavy metals in the green alga *Cladophora glomerata* as related to shore types in the Archipelago Sea, SW Finland. Mar. Pollut. Bull. 14: 347-348

Barnett, B.E. & C.R. Ashcroft, 1985. Heavy metals in *Fucus vesiculosus* in the Humber Estuary. Environ. Pollut. (B), 9: 193-213

Beeftink, W.G. & J. Niewenhuize, 1986. Monitoring trace metal contamination in salt marshes of the Westerschelde Estuary. Environ. Monit. Assess. 7: 233-248

Bérail, G., P. Prudent, C. Massiani & M. Pellegrini, 1991. Isolation of heavy metal-binding proteins from a brown seaweed *Cytoseira barbata f. repens* cultivated in copper or cadmium enriched seawater. In: Metal compounds in environment and life, Vol. 4; E. Merian & W. Haerdi (eds). Science and Technology Letters, Middlesex and Science Reviews Inc., Wilmington, USA. Special Supplement to Chemical Speciation and Bioavailability, pp. 55-62

Black, W.A.P. & R.L. Mitchell, 1952. Trace elements in the common brown algae and sea water. J. mar. biol. Ass. U.K. 30: 575-584

Bonotto, S., R. Kirchmann, J. van Baelen, C. Hurtgen, M. Cogneau, D. van der Ben, C. Verthe & J.M. Bouquegneau, 1986. Behaviour of technetium in marine algae. In: Speciation of fission and activation products in the environment; R.A. Bulman & J.R. Cooper (eds). Elsevier, Amsterdam, pp. 382-390

Brix, H. & J.E. Lyngby, 1982. The distribution of cadmium, copper, lead, and zinc in eelgrass (*Zostera marina* L.). Sci. Total Environ. 24: 51-63

Brix, H. & J.E. Lyngby, 1983. The distribution of some metallic elements in eelgrass (*Zostera marina* L.) and sediment in the Limfjord, Denmark. Estuar. Coast. Shelf Sci. 16: 455-467

Brix, H., J.E. Lyngby & H.-H. Schierup, 1983. Eelgrass (*Zostera marina* L.) as an indicator organism of trace metals in the Limfjord, Denmark. Mar. Environ. Res. 8: 165-181

Brumbaugh, W.G. & D.A. Kane, 1985. Variability of aluminum concentrations in organs and whole bodies of smallmouth bass (*Micropterus salmoides*). Environ. Sci. Technol. 19: 828-831

Bryan, G.W., 1969. The absorption of zinc and other metals by the brown seaweed *Laminaria digitata*. J. mar. biol. Ass. U.K. 49: 225-243

Bryan, G.W., 1971. The effects of heavy metals (other than mercury) on marine and estuarine organisms. Proc. Roy. Soc. Lond., Ser. B. 177: 389-410

Bryan, G.W., 1976. Heavy metal contamination in the sea. In: Marine pollution; R. Johnston (ed). Academic Press, London, pp. 185-302

Bryan, G.W. & P.E. Gibbs, 1983. Heavy metals in the Fal Estuary, Cornwall: a study of long-term contamination by mining waste and its effects on estuarine organisms. Occ. Publ. mar. biol. Ass. U.K. 2: 1-112

Bryan, G.W. & L.G. Hummerstone, 1973. Brown seaweed as an indicator of heavy metals in estuaries in south-west England. J. mar. biol. Ass. U.K. 53: 705-720

Bryan, G.W., W.J. Langston & L.G. Hummerstone, 1980. The use of biological indicators of heavy-metal contamination in estuaries; with special reference to an assessment of the biological availability of metals in estuarine sediments from south-west Britain. Occ. Publ. mar. biol. Ass. U.K. 1: 1-73

Bryan, G.W., W.J. Langston, L.G. Hummerstone & G.R. Burt, 1985. A guide to the assessment of heavy metal contamination in estuaries using biological indicators. Occ. Publ. mar. biol. Ass. U.K. 4: 1-92

Burdon-Jones, C., G.R.W. Denton, G.B. Jones & K.A. McPhie, 1982. Regional and seasonal variations of trace metals in tropical Phaeophyceae from North Queensland. Mar. Environ. Res. 7: 13-30

Butler, P.A., L. Andrén, G.J. Bonde, A. Jernelöv & D.J. Reisch, 1971. Monitoring organisms. In: Food and Agricultural Organisation technical conference on marine pollution and its effects on living resources and fishing, Rome, 1970. Supplement 1: Methods of detection, measurement and monitoring of pollutants in the marine environment; M. Ruivo (ed). Fishing News (Books), London, pp. 101-112

Caines, L.A., A.W. Watt & D.E. Wells, 1985. The uptake and release of some trace metals by aquatic bryophytes in acidified waters in Scotland. Environ. Pollut. (B), 10: 1-18

Catsiki, V.A., E. Papathanassiou & F. Bei, 1991. Heavy metal levels in characteristic benthic flora and fauna in the central Aegean Sea. Mar. Pollut. Bull. 22: 566-569

Coates, M., D.W. Connell, J. Bodero, G.J. Miller & R. Back, 1986. Aliphatic hydrocarbons in Great Barrier Reef organisms and environment. Estuar. Coast. Shelf Sci. 23: 99-113

Costantini, S., R. Giordano, L. Ciaralli & E. Beccaloni, 1991. Mercury, cadmium and lead evaluation in *Posidonia oceanica* and *Codium tomentosum*. Mar. Pollut. Bull. 22: 362-363

Crist, R.H., K. Oberholser, J. McGarrity, D.R. Crist, J.K. Johnson & J.M. Brittsan, 1992. Interaction of metals and protons with algae. 3. Marine algae, with emphasis on lead and aluminum. Environ. Sci. Technol. 26: 496-502

Cross, F.A. & W.G. Sunda, 1985. The relationship between chemical speciation and bioavilability of trace metals to marine organisms - a review. Proc. Int. Symp. Utilization of coastal ecosystems: planning, pollution and productivity, 21-27 Nov 1982, Rio Grande, Brazil. N.L. Chao & W. Kirby-Smith (eds). Vol. 1 Editora da Furg, Rio Grande Brazil, pp. 169-182

Cullinane, J.P. & P.M. Whelan, 1982. Copper, cadmium and zinc in seaweeds from the south coast of Ireland. Mar. Pollut. Bull. 13: 205-208

Denton, G.R.W. & C. Burdon-Jones, 1986. Trace metals in algae from the Great Barrier Reef. Mar. Pollut. Bull. 17: 98-107

Donard, O.F.X., F.T. Short & J.H. Weber, 1987. Regulation of tin and methyltin compounds by the green alga *Enteromorpha* under simulated estuarine conditions. Can. J. Fish. Aquat. Sci. 44: 140-145

Drifmeyer, J.E. & B. Redd, 1981. Geographic variability in trace element levels in *Spartina alterniflora*. Estuar. Coast. Shelf Sci. 13: 709-716

Drifmeyer, J.E., G.W. Thayer, F.A. Cross & J.C. Zieman, 1980. Cycling of Mn, Fe, Cu and Zn by eelgrass, *Zostera marina* L. Amer. J. Bot. 67: 1089-1096

Eide, I., S. Myklestad & S. Melsom, 1980. Long-term uptake and release of heavy metals by *Ascophyllum nodosum* (L.) Le Jol. (Phaeophyceae) *in situ*. Environ. Pollut. (A), 23: 19-28

Eisler, R., 1981. Trace metal concentrations in marine organisms. Pergamon, New York, pp. 685

Ferreira, J.G. & J.F.S. Oliveira, 1988. Spatial and temporal variation of mercury levels in macrophyte algae from the Tagus Estuary. In: Heavy metals in the hydrological cycle; M. Astruc & J.N. Lester (eds). Selper, London, pp. 241-248

Flegal, A.R. & J.H. Martin, 1977. Contamination of biological samples by ingested sediment. Mar. Pollut. Bull. 8: 90-92

Förstner, U., 1980. Inorganic pollutants, particularly heavy metals in estuaries. In: Chemistry and biogeochemistry of estuaries; E. Olausson & I. Cato (eds). Wiley, New York, pp. 307-348

Foster, P., 1976. Concentrations and concentration factors of heavy metals in brown algae. Environ. Pollut. 10: 45-53

Francesconi, K.A. & R.C.J. Lenanton, 1992. Mercury contamination in a semi-enclosed marine embayment: organic and inorganic mercury content of biota, and factors influencing mercury levels in fish. Mar. Environ. Res. 33: 189-212

Freitas, A.C.S., J.R.D. Guimarães, V.A. Gouvea & E. Penna Franca, 1988. Strontium-85 bioaccumulation by *Sargassum* spp. (brown seaweed) and *Galaxaura marginata* (calcareous seaweed). Sci. Total Environ. 75: 225-233

Fuge, R. & K.H. James, 1973. Trace metal concentrations in brown seaweeds, Cardigan Bay, Wales. Mar. Chem. 1: 281-293

Fuge, R. & K.H. James, 1974. Trace metal concentrations in *Fucus* from the Bristol Channel. Mar. Pollut. Bull. 5: 9-12

Ganesan, M., R. Kannan, K. Rajendran, C. Govindasamy, P. Sampathkumar & L. Kannan, 1991. Trace metals distribution in seaweeds of the Gulf of Mannar, Bay of Bengal. Mar. Pollut. Bull. 22: 205-207

Gardner, W.S., D.R. Kendall, R.R. Odom, H.L. Windom & J.A. Stephens, 1978. The distribution of methyl mercury in a contaminated salt marsh ecosystem. Environ. Pollut. 15: 243-251

Gekeler, W., E. Grill, E.-L. Winnacker & M.H. Zenk, 1988. Algae sequester heavy metals via synthesis of phytochelatin complexes. Arch. Microbiol. 150: 197-202

George, S.G., 1990. Biochemical and cytological assessments of metal toxicity in marine animals. In: Heavy metals in the marine environment; R.W. Furness & P.S. Rainbow (eds). CRC Press, Boca Raton, pp. 123-142

Geyer, A.L.M., J.C. Moreira, J.F. Faigle, R.E. Bruns & A.J. Curtius, 1990. Local and temporal variations in essential elements and agar of the marine algae *Pterocladia capillacea*. Hydrobiol. 194: 143-148

Goldberg, E.D., V.T. Bowen, J.W. Farrington, G. Harvey, J.H. Martin, P.L. Parker, R.W. Risebrough, W. Robertson, E. Schneider & E. Gamble, 1978. The mussel watch. Environ. Conserv. 5: 101-125

Gouvea, R.C., M.E. Castelo Branco, P.L. Santos & V.A. Gouvea, 1988. Accumulation of ^{210}Po by benthic marine algae. Environ. Technol. Lett. 9: 891-897

Grill, E., E.-L. Winnacker & M.H. Zenk, 1985. Phytochelatins: the principal heavy metal-complexing peptides of higher plants. Science, 230: 674-676

Grill, E., E.-L. Winnacker & M.H. Zenk, 1986. Synthesis of seven different homologous phytochelatins in metal-exposed *Schizosaccharomyces pombe* cells. FEBS Letts, 197: 115-120

Grill, E., E.-L. Winnacker & M.H. Zenk, 1987. Phytochelatins, a class of heavy-metal-binding peptides from plants, are functionally analogous to metallothioneins. Proc. Natl. Acad. Sci. USA, 84: 439-443

Guner, H., V. Aysel, S. Ozelsel & A. Sukatar, 1987. Periodical variation of trace element accumulations in some algae found in the Bay of Izmir. Rev. Int. Océanogr. Méd. 85/86: 52-55

Gutknecht, J., 1961. Mechanism of radioactive zinc uptake by *Ulva lactuca*. Limnol. Oceanogr. 6: 426-431

Gutknecht, J., 1963. ^{65}Zn uptake by benthic marine algae. Limnol Oceanogr. 8: 31-38

Gutknecht, J., 1965. Uptake and retention of cesium 137 and zinc 65 by seaweeds. Limnol. Oceanogr. 10: 58-66

Harding, J.P.C. & B.A. Whitton, 1981. Accumulation of zinc, cadmium and lead by field populations of *Lemanea*. Wat. Res. 15: 301-319

Harding, J.P.C., I.G. Burrows & B.A. Whitton, 1981. Heavy metals in the Derwent Reservoir catchment, northern England. In: Heavy metals in northern England: Environmental and biological aspects; P.J. Say & B.A. Whitton (eds). Department of Botany, University of Durham, U.K., pp. 73-86

Haug, A., 1961. The affinity of some divalent metals to different types of alginates. Acta Chem. Scand. 15: 1794-1795

Haug, A. & O. Smidsrød, 1965. Fractionation of alginates by precipitation with calcium and magnesium ions. Acta Chem. Scand. 19: 1221-1226

Haug, A., S. Melsom & S. Omang, 1974. Estimation of heavy metal pollution in two Norwegian fjord areas by analysis of the brown alga *Ascophyllum nodosum*. Environ. Pollut. 7: 179-192

Hellawell, J.M., 1986. Biological indicators of freshwater pollution and environmental management. Elsevier, Amsterdam, pp. 546

Hellawell, J.M., 1988. Toxic substances in rivers and streams. Environ. Pollut. 50: 61-85

Hesp, R. & B. Ramsbottom, 1965. Effect of sodium alginate in inhibiting uptake of radiostrontium by the human body. Nature, Lond. 208: 1341

Ho, Y.B., 1984. Zn and Cu concentrations in *Ascophyllum nodosum* and *Fucus vesiculosus* (Phaeophyta, Fucales) after transplantation to an estuary contaminated with mine wastes. Conserv. Recycling, 7: 329-337

Ho, Y.B., 1987. Metals in 19 intertidal macroalgae in Hong Kong waters. Mar. Pollut. Bull. 18: 564-566

Holmes, M.A., M.T. Brown, M.W. Loutit & K. Ryan, 1991. The involvement of epiphytic bacteria in zinc concentration by the red alga *Gracilaria sordida*. Mar. Environ. Res. 31: 55-67

Ishikawa, M., G. Izawa, T. Omori & K. Yoshihara, 1987. Annual variations of elemental quantities in brown sea algae Hijiki *Hizikia fusiforme*. Nippon Suisan Gakkaishi, 53: 853-859

Jensen, T.E., M. Baxter, J.W. Rachlin & V. Jani, 1982. Uptake of heavy metals by *Plectonema boryanum* (Cyanophyceae) into cellular components, especially polyphosphate bodies: an X-ray energy dispersive study. Environ. Pollut. (A), 27: 119-127

Klumpp, D.W., 1980. Characteristics of arsenic accumulation by the seaweeds *Fucus spiralis* and *Ascophyllum nodosum*. Mar. Biol. 58: 257-264

Koeman, R.P.T. & C. van den Hoek, 1982a. The taxonomy of *Enteromorpha* Link, 1820, (Chlorophyceae) in the Netherlands. I. The section *Enteromorpha*. Archiv. f. Hydrobiol. Suppl. 63: 279-330

Koeman, R.P.T. & C. van den Hoek, 1982b. The taxonomy of *Enteromorpha* Link, 1820, (Chlorophyceae) in the Netherlands. II. The section *Proliferae*. Cryptogamie Algol. 3: 37-70

Kondo, N., M. Isobe, K. Imai & T. Goto, 1985. Synthesis of metallothionein-like peptides cadystin A and B occurring in a fission yeast, and their isomers. Agric. Biol. Chem. 49: 71-83

Kureishy, T.W., 1991. Heavy metals in algae around the coast of Qatar. Mar. Pollut. Bull. 22: 414-416

Kuyucak, N. & B. Volesky, 1989a. Accumulation of cobalt by marine alga. Biotechnol. Bioeng. 33: 809-814

Kuyucak, N. & B. Volesky, 1989b. The mechanism of cobalt biosorption. Biotechnol. Bioeng. 33: 823-831

Langston, W.J., 1984. Availability of arsenic to estuarine and marine organisms: a field and laboratory evaluation. Mar. Biol. 80: 143-154

Lobel, P.B., S.P. Belkhode, S.E. Jackson & H.P. Longerich, 1991. Sediment in the intestinal tract: a potentially serious source of error in aquatic biological monitoring programs. Mar. Environ. Res. 31: 163-174

Luoma, S.N., G.W. Bryan & W.J. Langston, 1982. Scavenging of heavy metals from particulates by brown seaweed. Mar. Pollut. Bull. 13: 394-396

Lyngby, J.E. & H. Brix, 1987. Monitoring of heavy metal contamination in the Limfjord, Denmark, using biological indicators and sediments. Sci. Total Environ. 64: 239-252

Maeda, S. & T. Sakaguchi, 1990. Accumulation and detoxification of toxic metal elements by algae. In: Introduction to applied phycology; I. Akatsuka (ed). Academic Publ., The Hague, pp. 109-136

Martin, M.H. & P.J. Coughtrey, 1982. Biological monitoring of heavy metal pollution: land and air. Applied Science Publ., London, pp. 475

Maserti, B.E., R. Ferrara & P. Paterno, 1988. *Posidonia* as an indicator of mercury contamination. Mar. Pollut. Bull. 19: 381-382

Melhuus, A., K.L. Seip, H.M. Seip & S. Myklestad, 1978. A preliminary study of the use of benthic algae as biological indicators of heavy metal pollution in Sørfjorden, Norway. Environ. Pollut. 15: 101-107

Miramand, P. & D. Bentley, 1992. Heavy metal concentrations in two biological indicators (*Patella vulgata* and *Fucus serratus*) collected near the French nuclear reprocessing plant of La Hague. Sci. Total Environ. 111: 135-149

Morris, A.W. & A.J. Bale, 1975. The accumulation of cadmium, copper, manganese and zinc by *Fucus vesiculosus* in the Bristol Channel. Estuar. Coast. Mar. Sci. 3: 153-163

Mortimer, D.C., 1985. Freshwater aquatic macrophytes as heavy metal monitors - the Ottowa River experience. Environ. Monit. Assess. 5: 311-323

Munda, I.M., 1979. Temperature dependence of zinc uptake in *Fucus vesiculosus* (Don.) J. Ag. and *Enteromorpha prolifera* (O.F. Müll.) J. Ag. from the Adriatic Sea. Bot. Mar. 22: 149-152

Munda, I.M., 1984. Salinity dependent accumulation of Zn, Co and Mn in *Scytosiphon lomentaria* (Lyngb.) Link and *Enteromorpha intestinalis* (L.) from the Adriatic Sea. Bot. Mar. 27: 371-376

Myklestad, S., I. Eide & S. Melsom, 1978. Exchange of heavy metals in *Ascophyllum nodosum* (L.) Le Jol. *in situ* by means of transplanting experiments. Environ. Pollut. 16: 277-284

NAS, 1980. The international Mussel Watch. National Academy of Sciences, Washington D.C., pp. 248

Newell, S.Y., R.E. Hicks & M. Nicora, 1982. Content of mercury in leaves of *Spartina alterniflora* Loisel. in Georgia, U.S.A.: an update. Estuar. Coast. Shelf Sci. 14: 465-469

Nickless, G., R. Stenner & N. Terrille, 1972. Distribution of cadmium, lead and zinc in the Bristol Channel. Mar. Pollut. Bull. 3: 188-190

Nieboer, E. & D.H.S. Richardson, 1980. The replacement of the nondescript term 'heavy metals' by a biologically and chemically significant classification of metal ions. Environ. Pollut. (B), 1: 3-26

Nienhuis, P.H., 1986. Background levels of heavy metals in nine tropical seagrass species in Indonesia. Mar. Pollut. Bull. 17: 508-511

Parker, J.G. & F. Wilson, 1975. Incidence of polychlorinated biphenyls in Clyde seaweed. Mar. Pollut. Bull. 6: 46-47

Patrick, F.M. & M.W. Loutit, 1977. The uptake of heavy metals by epiphytic bacteria on *Alsima plantago-aquatica*. Wat. Res. 11: 699-703

Phillips, D.J.H., 1976. The common mussel *Mytilus edulis* as an indicator of pollution by zinc, cadmium, lead and copper. I. Effects of environmental variables on uptake of metals. Mar. Biol. 38: 59-69

Phillips, D.J.H., 1977. The use of biological indicator organisms to monitor trace metal pollution in marine and estuarine environments - a review. Environ. Pollut. 13: 281-317

Phillips, D.J.H., 1979. Trace metals in the common mussel, *Mytilus edulis* (L.), and in the alga *Fucus vesiculosus* (L.) from the region of the Sound (Öresund). Environ. Pollut. 18: 31-43

Phillips, D.J.H., 1980. Quantitative aquatic biological indicators: Their use to monitor trace metal and organochlorine pollution. Applied Science Publ., Barking, pp. 488

Phillips, D.J.H., 1990a. Use of macroalgae and invertebrates as monitors of metal levels in estuaries and coastal waters. In: Heavy metals in the marine environment; R.W. Furness & P.S. Rainbow (eds). CRC Press, Boca Raton, pp. 81-99

Phillips, D.J.H., 1990b. Arsenic in aquatic organisms: a review, emphasizing chemical speciation. Aquatic Toxicol. 16: 151-186

Phillips, D.J.H. & P.S. Rainbow, 1989. Strategies of trace metal sequestration in aquatic organisms. Mar. Environ. Res. 28: 207-210

Phillips, D.J.H. & P.S. Rainbow, 1993. Biomonitoring of trace aquatic contaminants. Elsevier, London, pp. 371

Phillips, D.J.H. & D.A. Segar, 1986. Use of bio-indicators in monitoring conservative contaminants: programme design imperatives. Mar. Pollut. Bull. 17: 10-17

Ragan, M.A., O. Smidsrød & B. Larsen, 1979. Chelation of divalent metal ions by brown algal polyphenols. Mar. Chem. 7: 265-271

Reboredo, F., 1992. Cadmium accumulation by *Halimione portulacoides* (L.) Aellen. A seasonal study. Mar. Environ. Res. 33: 17-29

Remacle, J., 1981. Cadmium uptake by freshwater bacterial communities. Wat. Res. 15: 67-71

Rice, D.L. & B.E. Lapointe, 1981. Experimental outdoor studies with *Ulva fasciata* Delile. II. Trace metal chemistry. J. exp. mar. Biol. Ecol. 54: 1-11

Robinson, N.J., 1989. Algal metallothioneins: secondary metabolites and proteins. J. Appl. Phycol. 1: 5-18

Roesijadi, G., 1980/81. The significance of low molecular weight, metallothionein-like proteins in marine invertebrates: current status. Mar. Environ. Res. 4: 167-179

Roesijadi, G., 1992. Metallothioneins in metal regulation and toxicity in aquatic animals. Aquatic Toxicol. 22: 81-114

Rönnberg, O., K. Adjers, C. Ruokolahti & M. Bondestam, 1990. *Fucus vesiculosus* as an indicator of heavy metal availability in a fish farm recipient in the northern Baltic Sea. Mar. Pollut. Bull. 21: 388-392

Saenko, G.N., M.D. Koryakova, V.F. Makienko & I.G. Dobrosmyslova, 1976. Concentration of polyvalent metals by seaweeds in Vostok Bay, Sea of Japan. Mar. Biol. 34: 169-176

Sawidis, T. & A.N. Voulgaropoulos, 1986. Seasonal bioaccumulation of iron, cobalt and copper in marine algae from Thermaikos Gulf of the northern Aegean Sea, Greece. Mar. Environ. Res. 19: 39-47

Say, P.J. & B.A. Whitton, 1983. Accumulation of heavy metals by aquatic mosses. 1: *Fontinalis antipyretica* Hedw. Hydrobiol. 100: 245-260

Say, P.J., J.P.C. Harding & B.A. Whitton, 1981. Aquatic mosses as monitors of heavy metal contamination in the River Etherow, Great Britain. Environ. Pollut. (B), 2: 295-307

Say, P.J., I.G. Burrows & B.A. Whitton, 1986a. Use of estuarine and marine algae to monitor heavy metals. Final Report from Northern Environmental Consultants Ltd. to the Department of the Environment, U.K., Two Volumes. Northern Environmental Consultants Ltd., Consett, Co. Durham, pp. 26 (Vol. 1) and 125 (Vol. 2)

Say, P.J., I.G. Burrows & B.A. Whitton, 1986b. *Enteromorpha* as a monitor of heavy metals in estuarine and coastal intertidal waters. A method for the sampling, treatment and analysis of the seaweed *Enteromorpha* to monitor heavy metals in estuaries and coastal waters. Occasional Publication No. 1, Northern Environmental Consultants Ltd., Consett, Co. Durham, pp. 25

Seeliger, U. & C. Cordazzo, 1982. Field and experimental evaluation of *Enteromorpha* sp. as a quali-quantitative monitoring organism for copper and mercury in estuaries. Environ. Pollut. (A), 29: 197-206

Seeliger, U. & P. Edwards, 1977. Correlation coefficients and concentration factors of copper and lead in seawater and benthic algae. Mar. Pollut. Bull. 8: 16-19

Sfriso, A., S. Raccanelli, B. Pavoni & A. Marcomini, 1991. Sampling strategies for measuring macroalgal biomass in the shallow waters of the Venice Lagoon. Environ. Technol. 12: 263-269

Skipnes, O., T. Roald & A. Haug, 1975. Uptake of zinc and strontium by brown algae. Physiol. Plant. 34: 314-320

Stenner, R.D. & G. Nickless, 1974. Distribution of some heavy metals in organisms in Hardangerfjord and Skjerstadfjord, Norway. Water Air & Soil Pollut. 3: 279-291

Tanaka, Y., A.J. Hurlburt, L. Angeloff & S.C. Skoryna, 1971. Application of algal polysaccharides as *in vivo* binders of metal pollutants. Proc. Intl. Seaweed Symp. 7: 602-604

Tewari, A., S. Thampan & H.V. Joshi, 1990. Effect of chlor-alkali industry effluent on the growth and biochemical composition of two marine macroalgae. Mar. Pollut. Bull. 21: 33-38

van der Ben, D., M. Cogneau, V. Robbrecht, G. Nuyts, A. Bossus, C. Hurtgen & S. Bonotto, 1990. Factors influencing the uptake of technetium by the brown alga *Fucus serratus*. Mar. Pollut. Bull. 21: 84-86

Veroy, R.L., N. Montaño, M.L.B. de Guzman, E.C. Laserna & G.J.B. Cajipe, 1980. Studies on the binding of heavy metals to algal polysaccharides from Philippine seaweeds. I. Carrageenan and the binding of lead and cadmium. Botan. Mar. 23: 59-62

Viarengo, A., 1985. Biochemical effects of trace metals. Mar. Pollut. Bull. 16: 153-158

Viarengo, A., 1989. Heavy metals in marine invertebrates: mechanisms of regulation and toxicity at the cellular level. CRC Crit. Revs. Aquatic Sci. 1: 295-317

Wahbeh, M.I., 1984. Levels of zinc, manganese, magnesium, iron and cadmium in three species of seagrass from Aquaba (Jordan). Aquatic Bot. 20: 179-183

Ward, T.J., 1987. Temporal variation of metals in seagrass *Posidonia australis* and its potential as a sentinel accumulator near a lead smelter. Mar. Biol. 95: 315-321

Wehr, J.D. & B.A. Whitton, 1983. Accumulation of heavy metals by aquatic mosses. 2: *Rhynchostegium riparioides*. Hydrobiol. 100: 261-284

Wehr, J.D., A. Empain, C. Mouvet, P.J. Say & B.A. Whitton, 1983. Methods for processing aquatic mosses used as monitors of heavy metals. Wat. Res. 17: 985-992

Whitton, B.A., P.J. Say & J.D. Wehr, 1981. Use of plants to monitor heavy metals in rivers. In: Heavy metals in northern England: Environmental and biological aspects; P.J. Say & B.A. Whitton (eds). Department of Botany, University of Durham, U.K., pp. 135-145

Whitton, B.A., P.J. Say & B.P. Jupp, 1982. Accumulation of zinc, cadmium and lead by the aquatic liverwort *Scapania*. Environ. Pollut. (B), 3: 299-316

Whyte, J.N.C. & R.J. Englar, 1974. Elemental composition of the marine alga *Nereocystis luetkeana* over the growing season. Fisheries and Marine Service, Technical report No. 509, Environment Canada, Vancouver, pp. 29

Woolston, M.E., W.G. Breck & G.W. VanLoon, 1982. A sampling study of the brown seaweed, *Ascophyllum nodosum* as a marine monitor for trace metals. Wat. Res. 16: 687-691

Wright, P.J. & J.H. Weber, 1991. Biosorption of inorganic tin and methyltin compounds by estuarine macroalgae. Environ. Sci. Technol. 25: 287-294

Young, M.L., 1975. The transfer of ^{65}Zn and ^{59}Fe along a *Fucus serratus* (L.) $\rightarrow$ *Littorina obtusata* (L.) food chain. J. mar. biol. Assoc. U.K. 55: 583-610

Zirino, A. & Yamamoto, S., 1972. A pH-dependent model for the chemical speciation of copper, zinc, cadmium, and lead in seawater. Limnol. Oceanogr. 17: 661-671

Zolotukhina, E., E.E. Gavrilenko & K.S. Burdin, 1990. Interactions among metal ions during uptake by marine macroalgae. Gidrobiologecheskii Zhurnal, Kiev, 26: 46-52 (In Russian with an English summary)

Appendix **4**

Detailed Aspects of the Sampling and Washing of Macrophytes Prior to Analysis, as Reported by Various Key Authors

(Adapted and updated from Say *et al,* 1986b).

Species	Sampling details	Washing technique	Author(s)
Fucus vesiculosus	Whole vertical range sampled. Stipe, thallus and tips separated.	Shaken with water of appropriate salinity; some samples brushed.	Bryan & Hummerstone, 1973
Fucus vesiculosus	Mostly fronds taken from several plants over entire vertical range.	Epifauna/epiphytes removed manually; samples washed in distilled water.	Fuge & James, 1974
Fucus vesiculosus; Ascophyllum nodosum	20 samples taken from mid-tidal range.	Epifauna/epiphytes removed; samples washed in distilled water.	Foster, 1976
Zostera marina	Quadruplicate samples separated into live and dead blades, roots, and rhizomes.	Sediment and macroscopic fauna removed; epiphytes not removed.	Drifmeyer *et al.*, 1980
Spartina alterniflora	Whole plants used, both live and standing dead, clipped at level of marsh surface.	Rinsed with filtered seawater to remove attached sediment particles.	Drifmeyer & Redd, 1981
Various macroalgae	Whole plants used, minus holdfast.	Three washes in distilled-deionised water.	Cullinane & Whelan, 1982
Spartina alterniflora	Leaves collected of senescent and dead plants; these cut at leaf-sheath junction.	Rinsed twice with filtered marsh water and once with distilled water.	Newell *et al.*, 1982
Ascophyllum nodosum	Material from three plants combined; 10-15cm tips taken with no fruits or injured parts.	Ultrasonicated to remove particulates and epiphytes; dipped in 0.1M nitric acid; rinsed twice in distilled water.	Woolston *et al.*, 1982
Cladophora glomerata	Collected at 0.5-1.0m depth.	Washed with seawater, then with tap water. Epiphytes not removed.	Aulio, 1983
Zostera marina	Samples separated into roots, rhizomes, stems, and leaves of different ages.	Epiphytes and other adherent material removed by scraping and washing.	Brix & Lyngby, 1983; Brix *et al.*, 1983
Fucus vesiculosus; Ascophyllum nodosum	Ten shoots taken for each species. Mature frond taken for *F. vesiculosus*; nodes of varying age sampled for *A. nodosum*.	Washed with clean seawater; rinsed with tap water.	Ho, 1984
Fucus vesiculosus	Fronds taken from several plants at high water neap level.	Scrubbed with nylon brush to remove particulates; rinsed; fruiting bodies, bladders, tips removed; thallus analysed.	Barnett & Ashcroft, 1985
Various macroalgae	Collected mostly subtidally over a wide range. Holdfast and older portions discarded.	Rinsed with clean seawater.	Denton & Burdon-Jones, 1986
Enteromorpha species	Collected at mid-shoreline, at least five locations at each site of study.	Washed under tap water on nylon sieve; rinsed in deionised water.	Say *et al.*, 1986a
Enteromorpha species	Samples taken from healthy, non-reproductive parts of plants.	Rinsed sequentially in seawater from the collection site and artificial seawater.	Donard *et al.*, 1987
Various macroalgae	Stratified random sampling technique used.	Washed in field at collection site, and again in laboratory to remove particles.	Ferreira & Oliveira, 1988
Sargassum spp.; *Galaxaura marginata*	Whole plants used, collected manually.	Hand-cleaned and ultrasonicated to remove epiphytes and particulates.	Freitas *et al.*, 1988
Fucus serratus	Fronds of 10 plants used, taken from low water neap level.	Washed with seawater at collection site; epifauna/epiphytes removed manually.	Miramand & Bentley, 1992
Halimione portulacoides	Whole plants collected by hand.	Rinsed with deionised water to remove particulates and epiphytes.	Reboredo, 1992

Chapter 5

Physiological Biomarkers and the Trondheim Biomonitoring System

Tore Aunaas[1] and **Karl Erik Zachariassen**[2]

[1] SINTEF Applied Chemistry
Environmental Technology Group
N-7034 Trondheim, Norway
[2] ALLFORSK
Department of Ecotoxicology
N-7055 Dragvoll, Norway

ABSTRACT

A system based on biochemical or physiological biomarkers may compensate for many of the shortcomings of the classical ecological methods involving monitoring of populations and species diversity. Physiological parameters may be ecologically relevant in the sense that they are closely related to the normal function and health of organisms. Whenever such parameters are displaced beyond their normal range, it indicates that the organisms are under considerable stress. Ecological effects including injury and death of organisms are likely to occur if the stress persists.

Physiological and biochemical processes linked to the normal functioning of the cell, *e.g.* metabolism and the regulation of transmembrane ionic gradients are suggested for use in environmental bio-monitoring. Changes and fluctuations in such parameters might function as an early warning (alarm) of pollutants and provide information on the nature of the pollutant and the toxicity stress caused by it.

A monitoring concept, the Trondheim Biomonitoring System (TBS), based upon pollutant induced changes in physiological and biochemical parameters, is outlined and discussed.

1. INTRODUCTION

The marine environment is the recipient of a wide variety of chemical waste products from industry and other sources. The possible environmental impacts of these chemicals are causing considerable concern. The main reason for this concern is the possibility that the chemicals may cause ecological changes in the environment.

Biomonitoring of Coastal Waters and Estuaries. Edited by Kees J.M. Kramer.

The impact of a chemical in an environment depends mainly on the concentration of the chemical, the period of time during which the chemical is present and the sensitivity of the organisms that live in the environment to the chemical. The presence of a chemical in an organism or an environment does not in itself imply that the organisms or the environment is affected or threatened by the chemical. Thus, in order to evaluate or predict the impact of a chemical on an environment it is necessary to operate in terms of effects at the level of the organism or at the ecological level.

Since the main concern behind environmental monitoring is the fear of effects that may occur at ecosystem level, the classical ecological approach involving monitoring of populations and species diversity is of great importance in the monitoring of the environment. However, the ecological approach has serious limitations, particularly with regard to the detection of effects at an early stage and with regard to the substantiated prediction of effects. Furthermore, the ecological approach is rather slow and retrospective, in that it provides no information until substantial ecological changes have taken place. In addition, such changes are difficult to substantiate, partly because large natural variations exist between different locations and within one location over a period of time. Finally, it can provide only very limited information as to the cause of observed disturbances. Even when manifest ecological damage has occurred in a system and a pollutant is shown to be present, the damage is not necessarily caused by that pollutant. There may be other chemicals present in the same environment and, in order to draw more substantiated conclusions about causal relationships, it is necessary to know the bioactive concentration ranges of the pollutants with respect to the particular organisms and, preferably, also to its biological effect characteristics.

A system based on biochemical or physiological biomarkers can compensate for many of the shortcomings of the classical ecological methods. Biomarkers may respond quickly to an environmental pollutant before ecological effects can be seen. It may be difficult to ascertain the ecological relevance and hazard associated with changes in physiological parameters. However, this does not imply that all physiological parameters are of low ecological relevance. Physiological parameters may be ecologically relevant in the sense that they are closely related to the normal function and health of the organisms. Whenever such parameters are displaced beyond their normal range, it indicates that the organisms are under considerable stress. Ecological effects including injury and death of organisms are likely to occur if the stress persists.

The combined response of a number of biomarkers may also reflect the mechanisms of action of the respective pollutants. This gives rise to the possibility that such parameters can be used to form a pollutant specific effect fingerprint that can be used to identify unknown pollutants.

Thus, the use of biochemical and physiological biomarkers may provide different types of information that are of great interest when it comes to environmental monitoring. However, it is important to select parameters that are appropriate for each particular purpose. The term 'biomarker' is used in different ways, and often no clear distinction is made between changes that reflect tolerable stress and those that reflect a direct hazard. To differentiate between these concepts a more complex terminology is required. This article presents sets of concepts and terms which distinguish between the various monitoring functions and categories of biomarkers. An integrated monitoring system called the 'Trondheim Biomonitoring System' (TBS) is also presented. Used as an early warning system, it is based on physiological parameters and great care is taken to assign the different physiological parameters to an appropriate biomarker function.

Some physiological parameters are well regulated and do not respond easily to stress imposed on the organism. Obviously, such parameters do not possess the high sensitivity required by an 'early warning' parameter, but the fact that they are effectively regulated implies that they are of vital importance to general physiological function. If the regulation

of such parameters should fail, it may have serious consequences for the function and health of the organism. The moderate sublethal changes of such parameters in exposed organisms can be used to detect serious physiological strain at an early stage.
The processes involved in maintaining the regulated parameters at their optimal level will usually display quick and pronounced compensatory responses when regulated parameters are under stress. The rapid response of these processes makes them suitable for detection of stress before the regulated parameters yield to that stress. Regulatory parameters which are to be used for the early warning of pollutants should be the most sensitive parameters which can be registered continuously in a field situation.
If the rapid response of a regulatory process secures that the regulated parameter is maintained at an optimal level, the change will not imply that the stress is harmful to the organism. However, if the stress exceeds the compensatory capacity of the regulatory process, the regulated parameter will yield to the stress, and damage will occur. Thus, in order to ascertain the degree to which an organism is under a health threatening stress when an alarm parameter has responded, it is necessary to measure the status of the corresponding regulated parameter. The displacement of a regulated physiological parameter when an organism is under stress by a pollutant is termed the 'toxicity strain'.
Fingerprint parameters should be a set of physiological parameters which combined are under the influence of a variety of regulatory processes. This will ensure that they respond to pollutant interactions with a great number of different processes and enzyme sites. The number of parameters included in the alarm parameter set should be as low as possible, but should still be sufficiently high so as to provide the required specificity.

2. BIOMARKERS AND STRESS INDICATORS

Many tests for biomarkers and other stress indicators have been developed over the last decade. Most have been tested and applied only under controlled laboratory conditions. We may expect, however, that at least some will find their application in biomonitoring of the natural environment. In this section an overview will be given of those methods which are considered to be the most important at present. Without pretending to be complete, Table 1 gives an overview of the most important physiological/biochemical techniques that may be useful in (aquatic) biomonitoring studies, subdivided according to their biological level of organisation. Indices are added which give an indication of the ease at which the measurements are performed (resource index), and the (potential) applicability in the field.
Only a few techniques enable measurement directly in the field. Organisms for sampling, come from either their natural populations at specific locations or from exposure in cages (active biomonitoring, ABM). Analyses are carried out in the laboratory.
In the following sections several methods will be treated briefly and their fundamentals explained. Those techniques used in the Trondheim Biomonitoring System (TBS, see later) are explained in more detail, while the methods and techniques covered extensively by other chapters in this volume, such as biological early warning systems (Baldwin & Kramer), metallothioneines (George & Olsson), P450 and related enzymes (Förlin & Goksøyr), survival in air of mussels (Eertman & de Zwaan), histological characteristics (Yevich & Yevich) and genotoxic effects (Kotelevtsev *et al.*) will not be further explained.

Table 1. Parameters within different organisation levels for practical application in bio-monitoring and toxicity evaluation in field studies and in the laboratory. The resource index indicates the resources required in order to achieve the necessary data of the parameter and is indicated on a scale from 1 to 3 (1: easily measured; 2: routine analyses; 3: specialised analyses). The applicability of the parameters for field use is evaluated (1: suitable for automatic registrations in field; 2: suitable for field investigations, but not automatically registered; 3: not suited for field investigations).

Organisation level	Parameter	Parameter in biomonitoring	Resource index	Field index
Individual organism	Scope for growth*	Toxicity stress	2-3	2
	Survival in air*	Toxicity strain	1	1
	Valve movement response*	Alarm	1	1
	Oxygen consumption rates	Alarm	1	1
	Body condition	Toxicity stress	2	2
	Growth	Toxicity strain	2	2
Cell tissue	Gonad index	Toxicity strain	2	2
	Glycogen content	Toxicity strain	2	2
	Extracellular space	Toxicity strain	2-3	2
	Transmembrane distribution of free amino acids	Toxicity stress/ Fingerprint	2-3	2
	Transmembrane distribution of inorganic ions	Toxicity strain	2-3	2
	Histological characteristics*	Toxicity strain	2-3	2-3
Sub-cellular	Metallothioneine*	Toxicity stress	2-3	2
	Superoxide dismutase	Toxicity stress	2-3	2-3
	P450 and related enzymes*	Toxicity stress	2-3	2
	Glutathione S-transferase activity	Toxicity stress	2-3	2
	Metabolic end products	Toxicity stress	2-3	2
	Adenylic energy charge	Toxicity stress	3	2
	Energy rich phosphates	Toxicity stress	3	2
	ATPases	Toxicity stress	2-3	2
	Metabolic intermediates	Toxicity stress	2-3	2
	Genotoxic effects*	Toxicity stress	2-3	2-3
	Neurotoxic effects	Alarm?/ Toxicity stress	1-3	1-2

* These parameters are treated extensively in other chapters of this Volume.

2.1. Rate of oxygen consumption

Mussels often close their shells as a response to pollutants in their environment. This feature has been used to establish an automatic system for monitoring (Kramer *et al.*, 1989; de Zwart *et al.*, 1994). Pollutants which induce a closure of the siphon and without affecting the shell opening will probably fail to be detected by the "Musselmonitor®" (Baldwin & Kramer, this Volume). It is possible, however, that such compounds can be detected by a system using either the pumping rate or the rate of oxygen consumption as variables (Nordtug *et al.*, 1991b). Shell closure or other effects that alter the rate of oxygen consumption are induced by a number of environmental pollutants. The latter method has been used in the Trondheim monitoring unit (see section 3). In mussels the rate of oxygen consumption will be affected by any pollutant that enters the body fluids and interferes with the metabolic system. Figure 1

shows how the oxygen consumption rates vary in the mussel *Mytilus edulis* exposed to elevated concentrations of various organic and inorganic chemicals. The respiration rate of *M. edulis* is sensitive to organic pollutants like formaldehyde and benzene down to the low $mg.kg^{-1}$ range and it is also sensitive to heavy metals such as cadmium, copper and mercury in the $\mu g.kg^{-1}$ range (Nordtug & Olsen, 1993).
The rate of oxygen consumption can be continuously registered, at low cost, with an automatic system, enabling its use as alarm parameter.

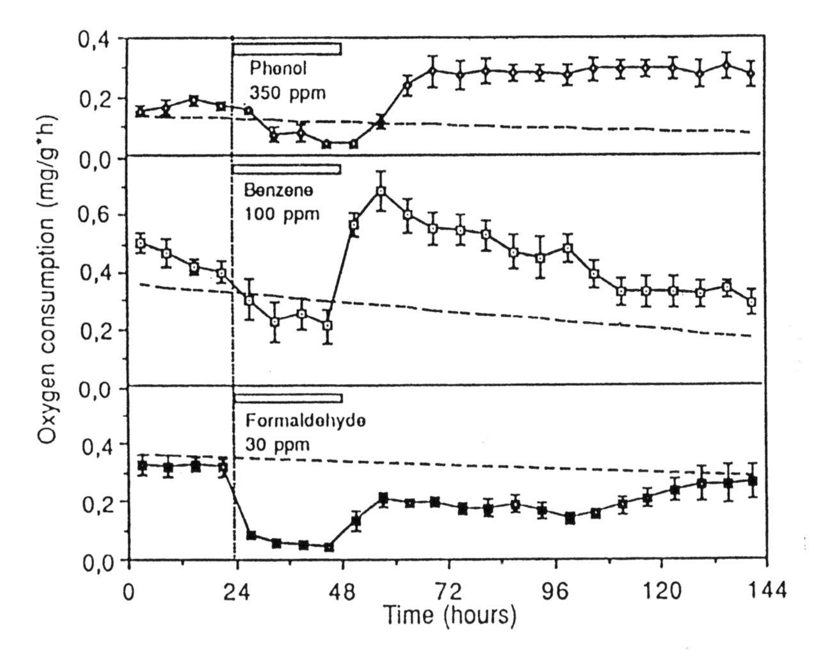

Figure 1. Oxygen consumption rates in *Mytilus edulis* during acclimation, exposure to phenol, formaldehyde and benzene and subsequent recovery in $34.5x10^{-3}$ S seawater at 10 °C (From: Nordtug *et al.*, 1991b, with permission)

2.2. Energy-rich phosphates

Energy rich phosphates such as ATP, phospho-arginine and creatine phosphate act as regulatory physiological parameters and are produced and consumed according to the requirements of the organism. They also provide energy for a variety of cellular processes. While ATP is the direct source of energy for a number of physiological processes, phospho-arginine is the major phosphagen in the tissues of marine invertebrates (Beis & Newsholme, 1975). Since these compounds take part in a wide variety of processes in the organisms (Figure 2) they are subject to alterations by various stress factors influencing the organism. The energy rich phosphates are, thus, well suited to serve as an early warning parameter in environmental monitoring.

An inhibition of ATP production will cause a reduction in the levels of energy-rich phosphates and, in turn, in the activity of energy-requiring processes in the cells. The concept of adenylate energy charge (AEC) has been used to express the energy available for cellular processes. The AEC is defined as:

$$AEC = \frac{ATP + 1/2\ ADP}{(ATP + ADP + AMP)} \qquad (1)$$

This parameter has great potential as a biomarker for environmental pollution (Giesy *et al.*, 1981, 1983). It responds to any pollutant that inhibits the oxidative metabolism and to a stressor that increases the energy turnover in the cell substantially.
To be able to measure the levels of energy-rich phosphates by means of NMR Aunaas *et al.* (1991b) developed a phosphate index to express the energy state of an organism. The phosphate index (PI) is defined as:

$$PI = \frac{ATP \times P\text{-arginine}}{P_i^2} \tag{2}$$

The phosphate index is a useful way to express the energy state since ADP is not readily measured by phospho-NMR.

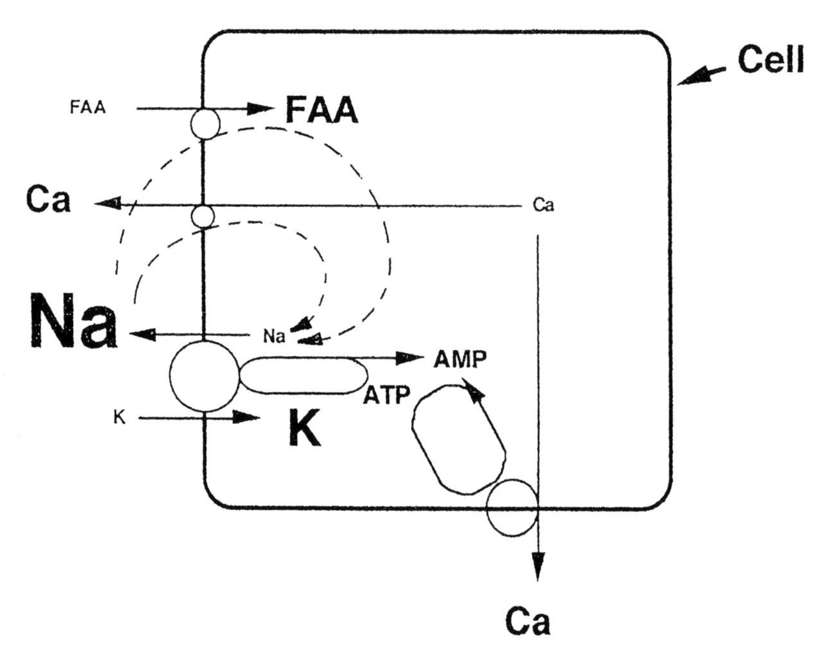

Figure 2. Mechanisms related to the creation and utilization of the electrochemical potential difference of sodium across cell membranes

The amounts of energy rich phosphates have been determined *in vivo* by ^{31}P NMR (Higashi *et al.*, 1989; van den Thillart *et al.*, 1990; Aunaas *et al.*, 1991b). While NMR techniques allow energy-rich phosphates to be determined easily for live mussels in the laboratory, other techniques are required in order to carry out measurements on animals collected in the field. Since no NMR technique is available for measuring phosphates in the field, the phosphate determination for field organisms has to be made on dissected tissue and treated in such a manner that the phosphate levels remain constant during transportation to the laboratory.
For routine use, methods based on high performance liquid chromatography (HPLC) should be applied when phosphates are used as biomarkers in practical environmental monitoring. A method described by Sellevold *et al.* (1986) allows for the determination of ATP, ADP, AMP and phospho-arginine in one single HPLC run. Enzymatic methods for the determination of energy-rich phosphates in animal tissues have also been widely used.
Changes in the amounts of intracellular energy-rich phosphates are shown to occur during exposure to both organic pollutants and heavy metals (Aunaas *et al.*, 1991b). Figure 3 shows changes in ATP and phospho-arginine amounts in *M. edulis* following 96 hours of exposure in the laboratory to different chemicals. As demonstrated for formaldehyde (Figure 4), there

seems to be a correlation between the chemical dose and the effect on the phosphates. The threshold for the observed effect of formaldehyde at this exposure temperature is between 1 and 10 $mg.kg^{-1}$. Benzene does not affect the amounts of ATP and phospho-arginine, suggesting that the toxic effects of this chemical acts through other mechanisms that do not influence the phosphate amounts.

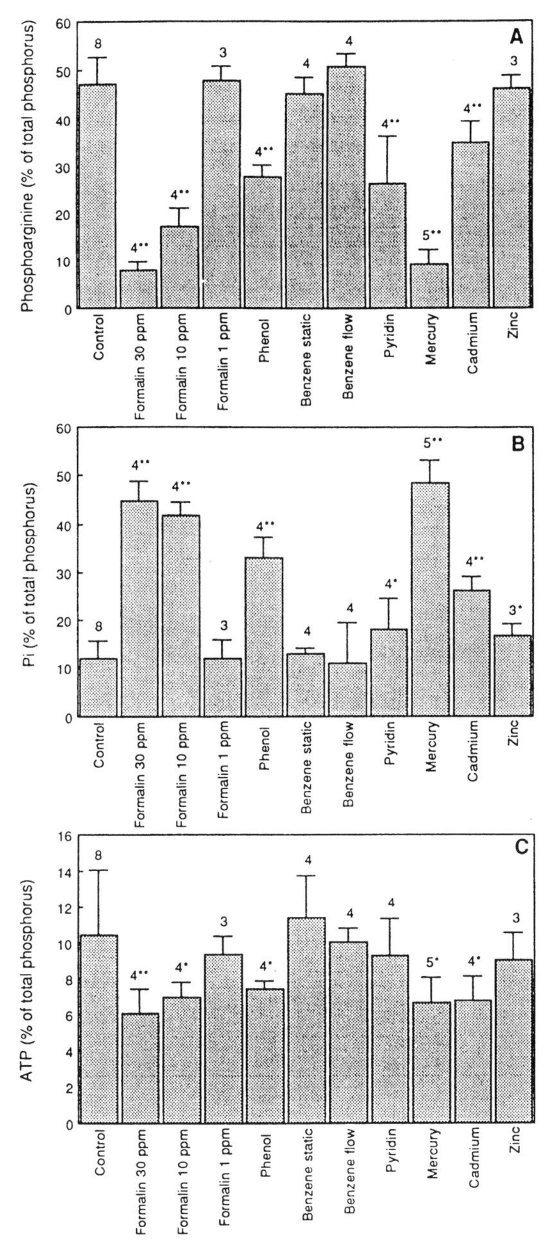

Figure 3. Changes in levels of (A) phospho-arginine, (B) inorganic phosphate, and (C) ATP in *Mytilus edulis* exposed to formaldehyde (1, 10 and 30 $mg.l^{-1}$), phenol (100 $mg.l^{-1}$), benzene (100 $mg.^{-1}$), pyridine (100 $mg.l^{-1}$), mercury (0.52 $mg.l^{-1}$), cadmium (1.84 $mg.l^{-1}$) and zinc (2.4 $mg.l^{-1}$) for 96 hours at 10 °C in 34.5×10^{-3} S sea water. The metabolite levels are expressed as the percentage of the total phosphorus signal (phospho-monoesters + inorganic phosphate + phospho-diesters + phospho-arginine + ATP) and obtained by measuring the respective peak heights in ^{31}P-NMR *in vivo* spectra. Values are presented as mean ± SD. The number of individuals in each group is given above the bars; $*=P<0.01$ and $**=P<0.001$ with respect to the control (From: Aunaas *et al.*, 1991b, with permission)

The energy-rich phosphates could respond to chemical exposure within minutes and the phosphate levels continues to drop as long as the chemical stress is present (Figure 4). If the metabolic apparatus is not damaged, phosphate levels recover when the chemical stress is removed.

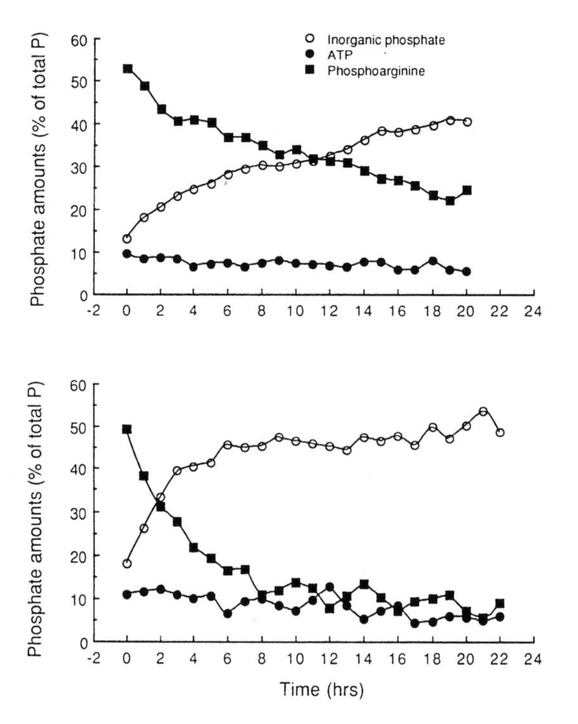

Figure 4. Sequence of changes in levels of phospho-arginine, inorganic phosphate and ATP in *M. edulis* exposed to 10 (top) and 30 mg.l^{-1} (bottom) formaldehyde at 10 °C in 34.5x10^{-3} sea water. The metabolite levels are expressed as in Figure 3.

2.3. Strombine

Strombine is a N-carboxymethyl compound (2,2-iminopropioacetic acid) which belongs to a family of naturally occurring amino acid related substances referred to as opines (de Zwaan & Zurburg, 1981, Sato *et al.*, 1988). It was first isolated from *Strombus gigas* (Sangster *et al.*, 1975). Opines are common end-products of anaerobic glycolysis in marine invertebrates including mussels (Zurburg *et al.*, 1981; de Zwaan *et al.*, 1983; Eertman & de Zwaan, this Volume). Strombine is produced from glycine by reductive condensation of the amino acid, in a reaction considered as a redox control mechanism (Dando *et al.*, 1981; Kreutzer *et al.*, 1989).

Strombine formation has been demonstrated only in chemical exposures which caused visible or clinical injury to the mussels (Olsen & Skjærvø, 1991). Thus, like the effects on the sodium gradient and cellular concentrations of calcium, strombine formation seems to be associated with pollutant stress which eventually may threaten the normal physiological function of the organism.

A HPLC method for analyzing strombine and free amino acids in haemolymph and tissue samples from *M. edulis* has been described (Olsen *et al.*, 1991a).

Olsen & Skjærvø (1991) demonstrated in laboratory experiments that strombine formation is also induced when mussels (*M. edulis)* are exposed to formaldehyde and other organic pollutants (Figure 5) and they suggested that alterations in haemolymph or cellular concentrations of strombine induced by chemicals or other stressors could indicate an interference by the compounds with the energy turnover and ATP synthesis of the organism. The observed formation of strombine indicates a mobilisation of anaerobic production of ATP from glycine and pyruvate.

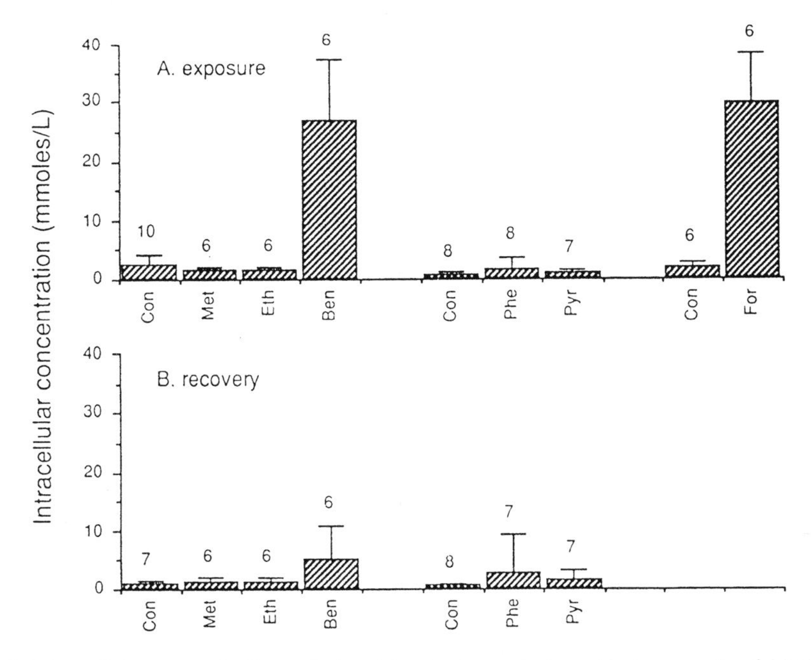

Figure 5. Intracellular concentrations of strombine in the posterior adductor muscle of *Mytilus edulis* after exposure for 96 hours to A: control, clean seawater (Con), formaldehyde (For, 30 $mg.l^{-1}$), benzene (Ben, 100 $mg.l^{-1}$), phenol (Phe, 275 $mg.l^{-1}$), methanol (Met, 5000 $mg.l^{-1}$), ethylene glycol (Eth, 5000 $mg.l^{-1}$) and pyridine (Pyr, 100 $mg.l^{-1}$) and B: after a subsequent 96 hour recovery period in clean seawater. Values are presented as mean ± SD of 6-10 individuals (After: Olsen & Skjærvø, 1992)

2.4. Free amino acids

A great number of free amino acids and substances similar to amino acids are present in the body fluids of marine animals, intracellular concentrations being 100 to 1000 times higher than extracellular concentrations. The free amino acids are involved in a wide variety of biochemical and physiological processes, and are, therefore, likely to be sensitive to a number of different pollutant-induced disturbances. Attempts have been made to use changes in levels of individual amino acids in *Mytilus edulis* (Briggs, 1979) and concentration quotients between different amino acids in the hard clam, *Mercenaria mercenaria* (Jeffries, 1972) as biomarkers for environmental pollution. In both cases the natural variations were found to be so substantial that the parameters were considered unfit for use as biomarkers for pollutants.
Olsen *et al.* (1991b) suggested the use of transmembrane concentration quotients as pollutant specific biomarkers. The high intracellular concentration is the result of a sodium coupled symport uptake of the amino acids from the haemolymph, implying that the amino acid accumulation takes place by the use of energy from the sodium energy gradient. Accordingly, a reduction in the transmembrane sodium gradient will affect the transmembrane distribution of free amino acids. The free amino acids are also involved in a great number of other cellular processes, such as anaerobic metabolism, pH regulation and isosmotic cell volume regulation. Consequently, the amino acids are likely to respond to a wide variety of pollutant-induced impacts.
A fingerprint can also include the transmembrane distribution of a number of substances similar to amino acids. The total number of amino acids and comparable compounds in the common mussel is close to 40.

The changes in the concentration quotients of free amino acids were found to be specific for a number of organic pollutants. Further studies are required in order to establish whether or not they can be used as a pollutant specific biomarker in environmental monitoring.

2.5. Na^+/K^+-ATPase

Several enzyme systems are involved in the regulation of ions across cell membranes. Marine fish are hyposmotic regulators and maintain their body fluids hyposmotic by continuously pumping monovalent cations out of the body fluids across the gill membranes. The anisosmotic regulation and pumping of ions across the gills of freshwater fish is primarily conducted by the Na^+/K^+-ATPase. Pollutants which have a direct impact on the performance of this enzyme cause malfunction in the osmoregulatory system of the fish and will cause death if the stress persists. Thus, investigations into this enzyme system are ecologically relevant and could be fitted into a biomonitoring system in the evaluation of the toxicity stress imposed on the organism.

Marine teleost fish are hyposmotic to the ambient seawater and the concentration difference of the major osmolytes sodium and chloride is maintained by an active extrusion of ions across the gill epithelium. Na^+/K^+-ATPase plays an essential role in this extrusion mechanism. The regulatory system in the gill epithelium is, thus, in many respects similar to that of animal cell membranes.

Results from acid rain research (Leivestad & Muniz, 1976) and other studies have revealed that the system is sensitive to pollutants in the same way as is the sodium transport system of cell membranes in the adductor muscle cells of *Mytilus edulis*.

Staurnes *et al.* (1984) have shown that freshwater fish gill Na^+/K^+-ATPase activity is inhibited by the presence of aluminum, which was correlated with the clinical effects in the fish.

2.6. Electrochemical potential difference of sodium

The electrochemical potential difference of sodium ($\Delta\mu_{Na}$ or sodium energy gradient) is the energy source for a number of physiological processes, *e.g.* cellular accumulation of free amino acids and cellular extrusion of Ca^{2+} (Ganong, 1987). The energy gradient is made up by a concentration difference of sodium across the cell membrane and a high negative membrane potential. The electrochemical potential difference for sodium is given by the formula:

$$\Delta\mu_{Na} = -Z_{Na} \times F \times E_m + R \times T \times \ln([Na]_e/[Na]_i) \quad (3)$$

where Z_{Na} is the electrical charge of sodium ions (+1), F the Faraday constant (96500 coulomb.mol^{-1}), R the universal gas constant (8.3 J.K^{-1}.mol^{-1}), T the absolute temperature (K), E_m the membrane potential (mV), $[Na]_e$ the haemolymph concentration of sodium and $[Na]_i$ the intracellular concentration of sodium.

The sodium gradient is maintained by the membrane-bound ATP-consuming sodium/potassium pump, and Florey (1966) estimated that in resting frog muscle cells a substantial fraction of the energy turnover is required to maintain $\Delta\mu_{Na}$. The maintenance of the ATP-dependent sodium gradient depends on a number of factors, such as an intact metabolic system for the production of ATP, an adequate supply of oxygen, a functional Na^+/K^+-ATPase and a low membrane permeability to sodium. Pollutants that reduce the ATP production of the organism, influence membrane permeabilities or block the sodium pump are likely to affect the sodium gradient.

The close correlation between sodium energy gradient and sodium content gives rise to the possibility of using the tissue sodium content as a substitute for the sodium gradient in environmental monitoring. Investigations have revealed that the sodium gradient is altered by exposure to both organic pollutants and heavy metals (Denstad, 1989; Børseth *et al.*, 1992;

Nordtug & Olsen, 1993). The results from exposures of *M. edulis* to formaldehyde (Figure 7) indicate that there is a certain limit (6.5 kJ.mol^{-1}) to which the sodium gradient could be reduced by the pollution stress without causing major damage to the organism. If the gradient is further depressed the animals are unable to recover and they die from the stress. The lack of ability to maintain the sodium gradient at normal levels is probably associated with serious disturbances of vital biological processes. Thus, the sodium gradient has the properties required by a biologically relevant toxicity parameter and is well suited for evaluating the toxicity strain on the organism.

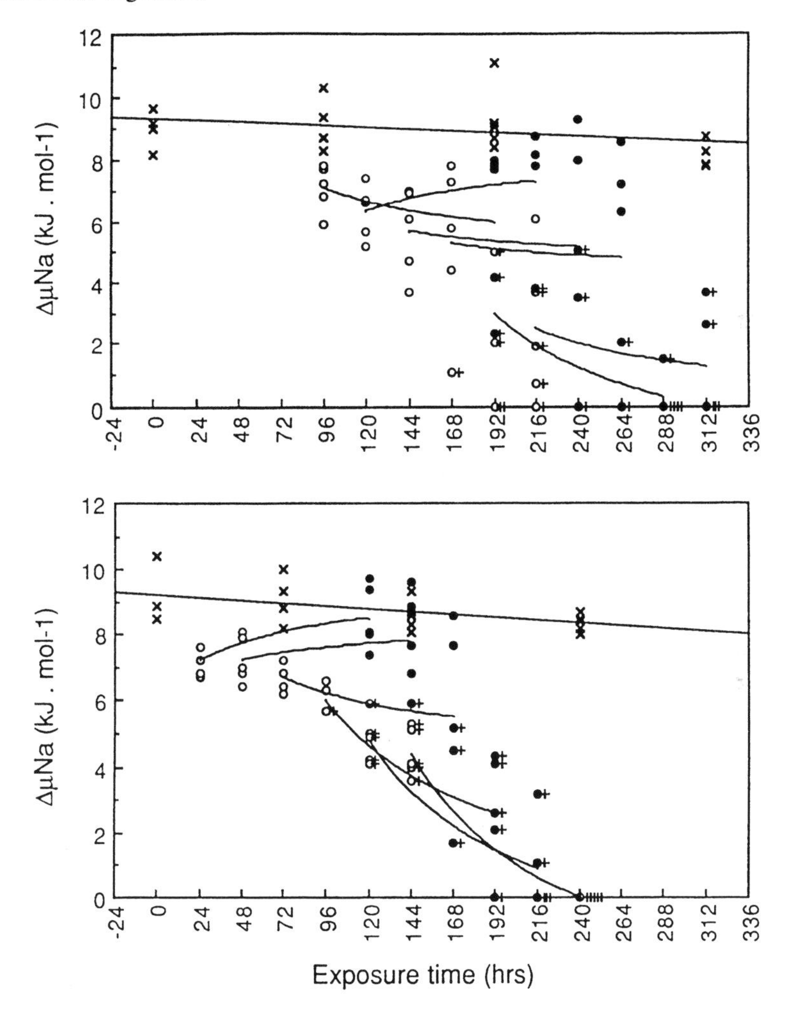

Figure 6. Transmembrane electrochemical potential difference of sodium ($\Delta\mu_{Na}$) in the posterior adductor muscle of *Mytilus edulis* during exposure to 30 mg.l^{-1} (top) and 60 mg.l^{-1} (bottom) formaldehyde (open symbols) and subsequent recovery (closed symbols) in 34.5x10^{-3} S seawater at 10 °C. Values representing individual mussels and lines between the mean of corresponding exposed and recovery groups are shown. The straight line is the regression line of control groups (x). Dead or moribund individuals are indicated to the right of respective values (+) (From: Zachariassen *et al.*, 1991; with permission)

The determination of the sodium energy gradient requires the membrane potential and the intracellular and extracellular concentrations of sodium to be known. The intracellular concentration of sodium and the membrane potential are not readily determined and, ideally the use of intracellular ion-selective micro electrodes is required. This technique is available only in a few laboratories, and is therefore not feasible for practical environmental monitoring. However, Børseth *et al.* (1992) have described a method by which the extra-cellular space and membrane potential across the cell membrane in the posterior adductor muscle could be calculated on the basis of analyses of haemolymph and total tissue concentrations of chloride, potassium and sodium. Intracellular concentrations of sodium could then be calculated from haemolymph and total tissue sodium concentrations, and $\Delta\mu_{Na}$ can be calculated from formula 3.

The reduction in the transmembrane energy gradient of sodium is associated with a comprehensive net influx of sodium to the intracellular compartments, leading to an increase in intracellular sodium concentration (Nordtug *et al.,* 1991a). This influx is also expressed as an increase in the total tissue concentration of sodium and there is a close correlation between the sodium energy gradient and the tissue concentration of sodium (Figure 7). While the determination of the sodium gradient requires the measurement of seven variables (tissue water content and tissue contents of sodium, potassium and chloride and extracellular concentrations), the determination of tissue sodium concentration only requires the measurement of two variables (tissue water content and tissue sodium content). Thus, the use of tissue sodium concentration as a substitute for the sodium gradient offers a considerable methodological simplification, which would be a great advantage when used in routine environmental monitoring.

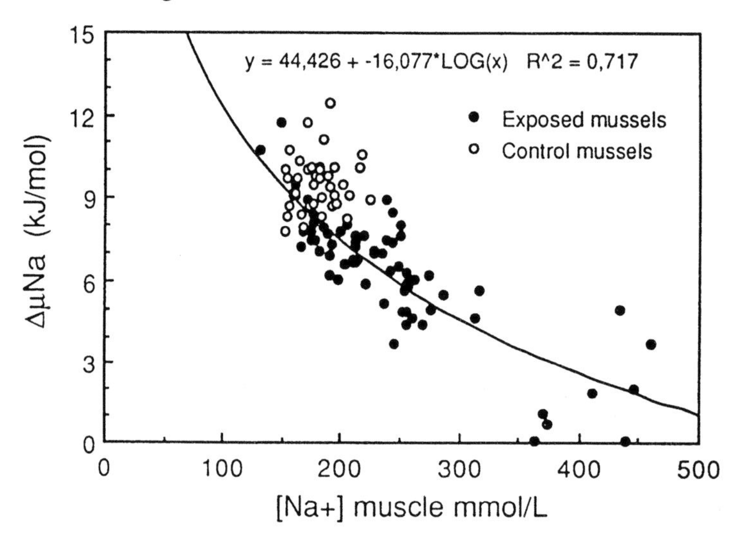

Figure 7. Relationship between the transmembrane electrochemical potential difference ($\Delta\mu_{Na}$) of sodium and tissue concentration of sodium in the posterior adductor muscle of *Mytilus edulis*

2.7. Cellular calcium

Calcium ions are involved in the regulation of a multitude of physiological processes (Rasmussen & Barrett, 1984). The major amount of cell associated calcium is bound to proteins or compartmentalized in the cell, therefore, free calcium concentrations in cytosol are in the low micro-mole range. Intracellular calcium content is well regulated and the homeostasis is maintained by a number of extrusion and compartmentalization systems (Carafoli, 1987). Calcium is extruded from cells by means of active pumping by the membrane-bound enzyme Ca^{2+}-ATPase and by an antiport mechanism coupled to sodium influx to the cell. The energy for the latter mechanism is obtained from the energy gradient for sodium. In *M. edulis* the sodium dependent extrusion is probably more important than the Ca^{2+}-ATPase.

Pollutants may inhibit the calcium extrusion and compartmentalization processes and disturb the intracellular calcium homeostasis (Nicotera *et al.*, 1989). Malfunction in these processes will cause an increase in free cellular calcium, which binds to enzymes in cytosol and thereby causes malfunctioning of physiological mechanisms. The role of Ca^{2+} in cytotoxicity has been reviewed by Viarengo & Nicotera (1990).

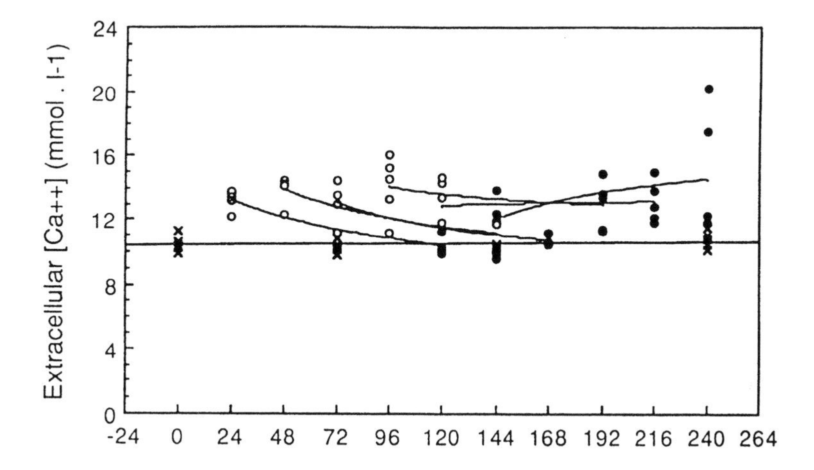

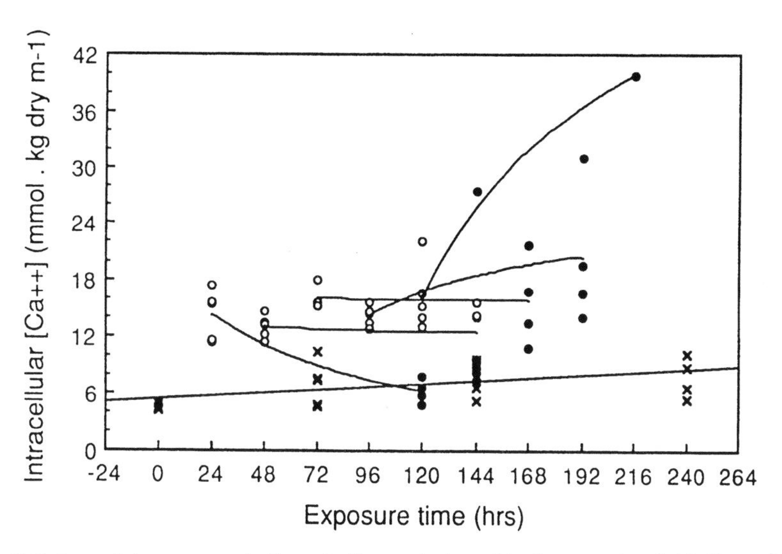

Figure 8. Cellular calcium concentrations in the posterior adductor muscle of *Mytilus edulis* during exposure to 30 $mg.l^{-1}$ (top) and 60 $mg.l^{-1}$ (bottom) formaldehyde (open symbols) and subsequent recovery (closed symbols) in $34.5x10^{-3}$ S sea water at 10 °C. See Figure 6 for details. (After: Aunaas *et al.*, 1991a)

Because of the very low cytosol concentrations and the compartmentalization of the ion, the intracellular concentrations of free Ca^{2+} ions are not readily available by standard analysis techniques. Analyzing the total cellular amounts of calcium will yield values three orders of magnitude higher than the values of the free cytosolic calcium. Despite this, total cellular amounts of calcium could serve as an indication of the cellular calcium stress.

Aunaas *et al.* (1991a) reported the effects on cellular calcium in *M. edulis* exposed to sublethal concentrations of formaldehyde and other organic pollutants. In mussels exposed to toxic concentrations of chemicals the total cellular concentration of calcium increased, while no effects were observed during exposure to concentrations tolerated by the mussels (Figure 8). The increased cellular calcium concentration in mussels exposed to formaldehyde was also correlated to a reduction in the transmembrane energy gradient of sodium (Figure 6). Thus, the cellular calcium concentration is indicative of both clinical toxic effects and of a depression of the sodium gradient. Consequently the cellular calcium concentration appears to be a feasible toxicity strain parameter.

2.8. Glutathione S-transferase

Glutathione S-transferases (GSTs) are inducible isoenzymes involved in the detoxification of contaminants. They have been isolated in a large number of organisms, including marine species (Ketterer *et al.*, 1988; Clark, 1989) and freshwater and marine molluscs (Looise, 1982). GSTs bind the endogenous compound glutathione (γ-glutamylcysteinylglycine) to the electrophilic (thus reactive) molecules enlarging the water solubility and enhancing their excretion. The electrophilic substrates can be of endogenic or exogenic origin. The isoenzymes are specific to both substrate and inducibility. GST could have an additional function in performing a covalent binding to electrophilic molecules (Mannervik & Danielson, 1988).

The analysis of GST activity is relatively simple and has been described by Habig *et al.* (1974).

The activity of the enzyme system (and the occurrence of the different isoenzymes) is dependent upon *e.g.* species, tissue, season and environmental conditions, including the effects of pollutants. A number of contaminants are proven inducers of GSTs in fish, *e.g.* pesticides, PAHs and PCBs. Lee (1989) found compounds associated with GST in the crab *Callinectes sapidus* which had been exposed to labelled benzo[*a*]pyrene and tributyltin (TBT). Examples of field applications are given in the case studies, section 4.

2.9. Enzyme systems linked to the glycolysis and Krebs' cycle

Different chemicals with different reactive groups are likely to affect different enzymes. The glycolysis and Krebs' cycle involve a great number of enzymatic reactions. As outlined by Förlin & Goksøyr (this Volume), the induction of the cytochrome P-450 monooxygenase liver enzyme system in marine organisms shows that the organisms have been exposed to organic pollutants, *i.e.* PAHs and/or PCBs. Similarly, different pollutants may affect different enzymes and pinpointing of the affected enzymes can be used to obtain an understanding of the mechanisms by which the pollutants act. When an enzyme is blocked by a pollutant, the amount of substrates for the particular enzyme will probably increase and the amount of product decrease. In this way the effect sites of pollutants on the different reactions and enzymes in the glycolysis and Krebs' cycle could be investigated. By observing the levels of substrate and product from various enzyme reactions, the affected enzyme could be identified.

Thus, identification of the enzyme affected by a pollutant could also provide data that could be included in a physiological fingerprint system.

3. THE TRONDHEIM BIOMONITORING SYSTEM (TBS)

This section deals with the various categories of biomarkers and how they can be used to establish a consistent and practically operative integrated system for biomonitoring of marine environments. At least three different aims can be attached to biomonitoring of the environment. These are:

- To *detect* the presence of environmental pollutants at the earliest possible stage, so that countermeasures can be taken before ecological damage develops;

- To *evaluate* whether or not the organisms or ecosystem is seriously stressed or strained by an environmental pollutant;
- To *identify* the pollutant that is threatening the organisms or ecosystem or that has caused manifest ecological damage.

In order to make sure that the monitoring system meets the biomonitoring objectives in an adequate manner, it is recommended that three different categories of monitoring parameters to be established, each of which corresponds to one of the aims listed above.
The biomonitoring related parameters are:

- **Alarm parameter**
  An alarm parameter is a parameter which reveals that something abnormal, perhaps pollutant induced, is taking place in the environment. An ideal alarm parameter responds quickly to low concentrations of a broad spectrum of environmental pollutants. The main feature of an alarm parameter is high sensitivity, but a change in such a parameter does not necessarily imply that the organism or ecosystem is threatened.
- **Toxicity parameter**
  It is recommended that two categories of toxicity parameters be proceeded with.
  A *toxicity stress* parameter is a parameter which reflects the stress imposed on an organism by a pollutant. The stress parameter should display high sensitivity to pollutants and preferably respond quickly to the chemical stress. Displacement of the toxicity stress parameter value indicates that this compensatory mechanism is mobilised to handle the stress.
  A *toxicity strain* parameter is a well regulated and, thus, a normally stable parameter that is closely related to the normal function of an organism. When a toxicity strain parameter is brought to bear on the stress imposed by an environmental pollutant, it indicates that the organisms are about to be seriously affected by the pollutant and that the pollutant is about to have ecological effects.
- **Fingerprint parameter**
  A fingerprint parameter is a single physiological or biochemical parameter or a combination of such parameters which, alone or combined, reflect the physiological impacts of environmental pollutants on the organism in a specific manner. If the physiological fingerprints of various pollutants are known from previous laboratory experiments, the fingerprint parameters can be used to identify a pollutant in cases where the pollutant is unknown. Chemical analysis should then serve as verification of the presence of the fingerprinted pollutant in the environment.

The Trondheim Biomonitoring System (TBS) is a monitoring approach designed to meet the aims of the biomonitoring described above. It combines various biomonitoring approaches into one integrated concept. The procedure is divided into three operational phases, as presented in Figure 9.
The first phase is the alarm phase, which must be based on the continuous monitoring of a sensitive biological parameter, *e.g.* shell closure of mussels as applied in the 'Mussel-monitor® system (see Baldwin & Kramer, this Volume), oxygen consumption rates, or other regulatory physiological or behavioural parameters.
The second phase is a toxicity evaluation phase, in which the physiological stress imposed on an organism by a pollutant is evaluated by the use of regulatory physiological parameters, and the physiological strain or health hazard is evaluated by using regulated physiological parameters.
The third and final phase is the effect fingerprint determination phase, which aims at identifying the active pollutant in an affected environment by using a set of physiological parameters. The effect fingerprint obtained in this phase must be compared with typical effect fingerprints of a variety of pollutants that should be available in a previously established fingerprint library.

A number of methods, as summarized in section 2, may be fit for use during the different phases of the TBS. A system meeting the requirements of an automated biomonitoring system was developed in our Department. The system is designed to monitor acute toxicity but, with some modifications, it can also be used to monitor chronic pollution.

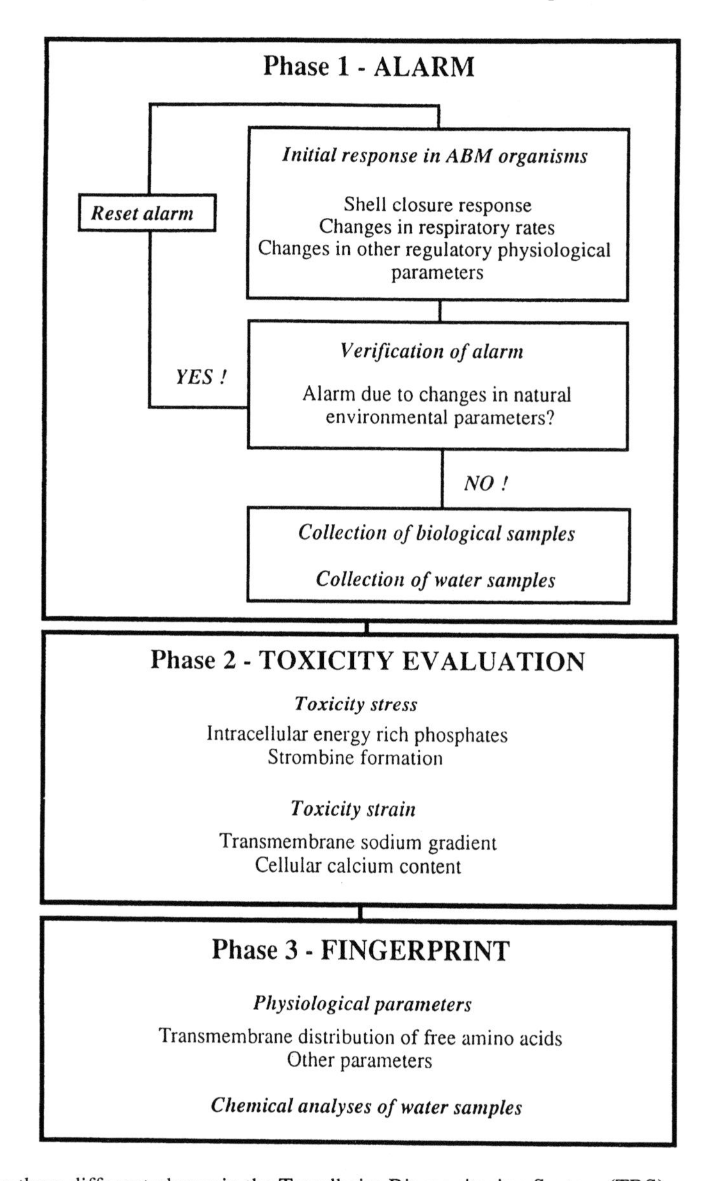

Figure 9. The three different phases in the Trondheim Biomonitoring System (TBS)

The TBS is focused on physiological parameters related to the maintenance and utilisation of the energy contained in the electrochemical potential difference (energy gradient) of sodium across the cell membranes of the posterior adductor muscle of the mussel, *Mytilus edulis*. The system involves monitoring of *oxygen consumption rates* and the levels of *energy-rich phosphates* (ATP and phospho-arginine), which are all elements in processes that serve to maintain the transmembrane sodium energy gradient, and the *transmembrane transport of calcium* and *free amino acids*, which utilise energy from the energy gradient. The sodium gradient and the intracellular calcium concentration appear to be regulated physiological parameters, whereas the oxygen consumption rate and energy rich phosphates are parameters which are likely to vary substantially even under natural conditions. The integrated regulatory system is illustrated in Figure 2.

While the Trondheim TBS system has so far been based on the use of *M. edulis* as a monitoring organism, other species may also be suitable. Teleost fish of a size that fits into the exposure chambers would offer the advantage of using the plasma sodium concentration as a monitoring parameter. Toxic pollutants may cause the plasma sodium concentration of marine teleosts to increase in a manner similar to the intracellular sodium concentration of *Mytilus edulis*. The use of fish as test animals would allow the rate of oxygen consumption to be used as a monitoring parameter in the same way as for blue mussels, but the alarm signal would probably be different. The toxicity stress and strain could be evaluated by the use of plasma sodium concentration or plasma osmolality. Since an acute increase in the plasma osmolality would lead to a shrinking of blood cells, even haematocrit could be used as a parameter for stress evaluation.

The technical set-up consists of ten exposure chambers, each containing one mussel, and mounted in an apparatus as illustrated in Figure 10. Seawater is pumped through the exposure chambers, so that the mussels are continuously exposed to water from the ambient medium. The apparatus also serves as a respirometer unit, in that the rate of oxygen consumption of the individual mussels is continuously recorded and displayed on a computer (see Appendix). When desired, the mussels can be removed from the chambers, and samples of haemolymph and mussel tissue taken for determination of other parameters. The instrumentation can be land-based or mounted in a buoy with radio transmission of the alarm parameter signal to a land station. The system has been tested extensively under laboratory conditions and in one case it has been operated in the field.

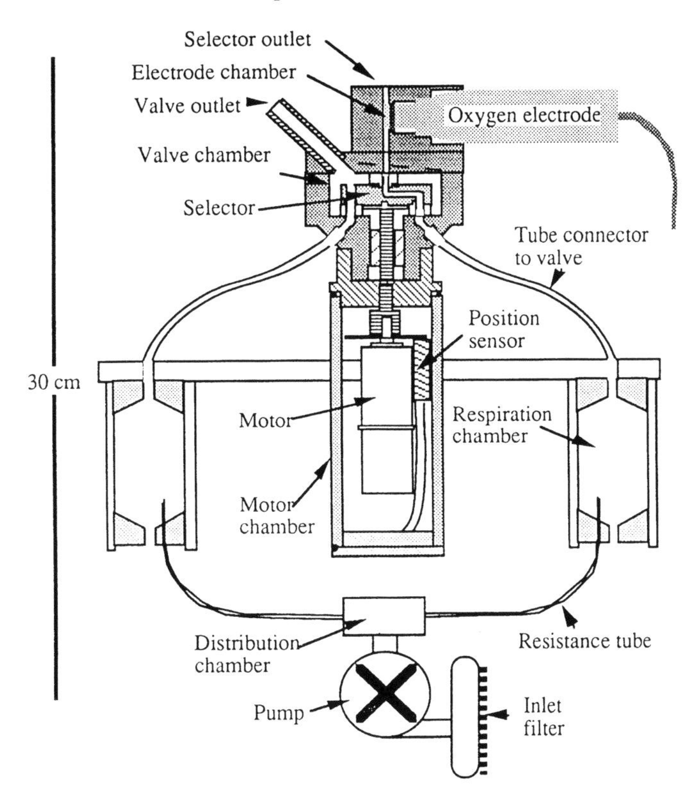

Figure 10. Transected view of a submersible system for automatic registration of the oxygen consumption in mussels and other aquatic animals (From: Nordtug & Olsen, 1993)

3.1. Phase 1 - The alarm phase

3.1.1. The alarm signal

The first phase in the TBS system is an early warning based on the rate of oxygen consumption of the mussels as an alarm parameter. The respiratory rates are continuously registered by a self-calibrating oxygen electrode and abrupt changes in the oxygen consumption rate are used as the alarm signal. A computer linked to the TBS instrumentation continuously registers, stores and transfers data from the monitoring site.

The rate of oxygen consumption of *M. edulis* responds to a wide variety of chemical and physical changes, partly as a result of shell closure and partly as the result of interactions with the oxygen transport system or with enzymes of the energy metabolic system. Shell closure leads to a sudden and substantial drop in oxygen consumption.

3.1.2. Verification of the alarm signal

Shell closure with a subsequent drop in oxygen consumption may also be induced by sudden changes in the salinity (Schumway, 1977) or temperature of the medium. Furthermore, food particle concentrations in the water will influence ventilation rates and the oxygen extraction from the water. Davenport (1979a, b) believed that the flow rate of water through the mantle cavity and subsequently the amount of oxygen extracted from the water was mainly based on the closure of the exhalant siphon of the mussels. Thus, there seems to be a close correlation between the water flow through the mantle cavity and respiratory rates of *M. edulis*.

Since variations in natural environmental parameters might influence the ventilation rates of the mussels, it is possible that the alarm signal can also be triggered by natural variations. The fluctuations in natural environmental parameters have to be taken into consideration when the respiratory rate is to be used as an alarm parameter. In the Trondheim TBS system sensors for these parameters should be linked to the computer, and the entire system could be set to ignore alarms caused by abrupt changes in natural variables.

3.1.3. Sampling

When natural variations cannot explain the alarm signal displayed by the system, water samples should be collected automatically from the environment. The water collection could be initiated by the TBS immediately after the alarm signal is verified. This ensures that water is obtained while the pollutants are still present in the environment. The computer linked to the alarm set-up could initiate water collection automatically and immediately and it could be programmed to collect samples at defined time intervals following the alarm. By chemical analysis of the water it will be possible to reconstruct the sequence of the event and thereby evaluate the total pollution load to the environment.

Samples of the TBS organisms for analysis of the other parameters must be collected manually in the field (see Appendix). Immediately after dissection and wet weight determination, the samples must be frozen and stored on liquid nitrogen.

Haemolymph samples are primarily preserved for analysis on inorganic ions and free amino acids, but should be prepared in such a manner that makes analysis of other parameters possible. Different tissue samples might be collected from *M. edulis*. Since the Trondheim TBS system has focused on the transmembrane processes in the posterior adductor muscle, samples of the muscle are primarily preserved for analysis of the content of intracellular energy-rich phosphates, inorganic ions and free amino acids.

3.2. Phase 2 - Evaluating the toxic stress and strain

The purpose of the second phase of the Trondheim TBS system is to establish whether or not the pollutant which caused the alarm has imposed a health threatening stress or strain on the organisms.

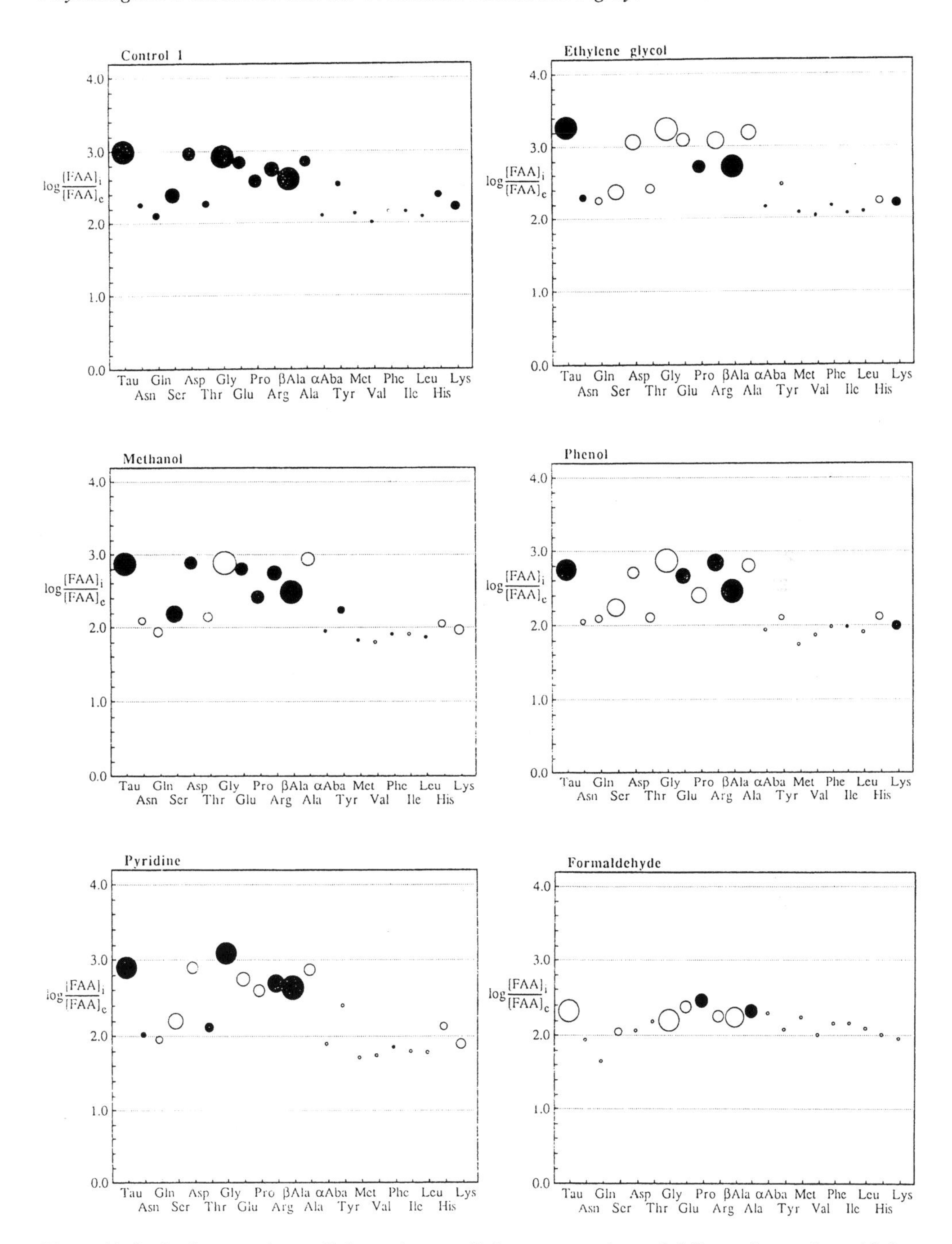

Figure 11. Ratios between intracellular and extracellular concentrations of different free amino acids in the posterior adductor muscle of blue mussels (*Mytilus edulis*). The size of the spots represents intracellular concentration. Open spots represent ratios markedly different from control values (From: Olsen *et al.*, 1991b)

The physiological *stress* imposed on an organism by a pollutant can be evaluated in terms of changes in regulatory parameters in the TBS system. When a pollutant has caused an inhibition of the aerobic pathways of the energy metabolism, the organism seeks to compensate by mobilizing energy-rich phosphate from phospho-arginine and by activating anaerobic pathways for production of ATP. The anaerobic metabolism is expressed as an accumulation of succinic acid, propionic acid and strombine.

The degree to which a pollutant has caused a health threatening physiological *strain* is evaluated in terms of changes in the transmembrane energy gradient of sodium and the calcium content of the adductor muscle cells. Both the sodium gradient and the calcium content are regulated physiological parameters, which will be the first to undergo substantial changes when the stress imposed by the pollutant exceeds the compensatory capacity of the organism. Since both the sodium gradient and the cellular calcium content are of great importance to a variety of cellular processes, failure in the regulation of these parameters will have serious consequences for the normal functioning of the organisms. Therefore, these parameters are well suited for the evaluation of a pollutant-induced strain threatening the organism.

3.3. Phase 3 - Fingerprinting of pollutants

The purpose of the third phase is to identify the pollutant by means of a physiological fingerprint. The rationale behind this phase is the fact that pollutants affect organisms by interfering chemically with the structure and function of proteins, membranes and other components in the organism. The chemical impact of the pollutants must be related to their chemical structure and properties. Since pollutants have chemical properties which are characteristic of each particular pollutant specimen, the biochemical and physiological impact of pollutants is also likely to be pollutant specific.

Accordingly, a physiological fingerprint system must be based on the use of a multitude of physiological parameters. The combined set of parameters must respond to a variety of direct chemical impacts, so that different pollutants, according to their particular sites of impact, can produce different combined response patterns.

A physiological fingerprint can be presented in different ways: solute concentrations and rates of processes; concentrations and rates in relation to a control value or in more sophisticated manners *e.g.* ratios between intracellular and extracellular concentrations.

The latter is the correct manner for expressing the FAA based fingerprint in the TBS system. An example of FAA fingerprints of various chemicals based on this form of presentation is shown in Figure 11. The graphs reveal a clear specificity for the various chemicals. The specificity can be increased by including more parameters in the fingerprint. As fingerprints based on a high number of parameters are established for a high number of chemicals, it might be necessary to apply multivariate analysis to operate the fingerprint system.

4. CASE STUDIES

A number of the physiological and biochemical parameters that have already been applied in coastal waters and estuaries are treated extensively in the various other chapters. Most other variables discussed in this chapter have been tested only in controlled laboratory experiments in order to evaluate the effects of possibly toxic compounds, and have, so far, not been incorporated into many field investigations in marine environments.

In situ registrations of respiratory rates, for example, which is a more integrated response by the organism, have so far not been applied in field investigations for biomonitoring purposes due to logistical restrictions, but a large number of laboratory experiments indicate that this parameter is well suited as an *in situ* early warning of both organic and inorganic pollutants in seawater (Nordtug *et al.*, 1991a,b).

In the following sections a number of the biochemical parameters, not treated in other chapters, are discussed using a few practical field applications that have been reported in the literature.

4.1. Glutathione S-transferase (GST)

Studies of glutathione S-transferase (GST) in marine mussels, crustaceans and fish carried out along a pollution gradient in the Langesund Fjord, Norway, indicated that the GST activity in crab (*Carcinus maenas*) hepatopancreas was significantly higher at two polluted sites than at more pristine sites. Such differences were not found for mussel (*M. edulis*) and winkle (*Littorina littorea*) digestive glands, despite large differences in tissue concentrations of PAHs and PCBs (Lee, 1988; Suteau *et al.,* 1988). GST levels in flounder (*Platichthys flesus*) intestines from the same fjord were generally higher at the more polluted sites. However, the only significant differences in activities ran counter to the anticipated pollution gradient, suggesting that intestinal GST activity may not be a sensitive bioindicator under field conditions (van Veld & Lee, 1988).

In another study, GST activity identified in the mussel, *M. edulis*, was significantly elevated in mussels from the polluted Cork Harbour as compared to mussels from the less polluted Bantry Bay (both in Ireland) or in depurated mussels from Cork Harbour, suggesting the potential use of GST in mussels (Sheenan *et al.*, 1991).

4.2. Biochemical parameters in dab (*Limanda limanda*) from the German Bight

In the Bremerhaven workshop intercomparison of biochemical techniques (edited by Stebbing *et al.*, 1992) several methods were tested along a pollution gradient. Samples were collected from a transect that stretched from the (polluted) inner German Bight, near the mouths of the Rivers Elbe and Weser, to the off-shore Dogger Bank. Apart from the parameters discussed in other chapters of this volume (*e.g.* metallothionein, MFO-induction, histological and pathological techniques), several other biochemical techniques were applied. They involved the detection of antioxidant enzymes, cholinesterase activity and changes in branchial Na^+/K^+-ATPase in dab (*Limanda limanda*).

Antioxidant enzymes

At the 7 stations along the decreasing pollution gradient from Heligoland to the Dogger Bank dab liver was sampled. The activities of catalase and glutathione peroxidase were higher at the more polluted sites. However, catalase and superoxide dismutase (SOD) activity were also high at the location near the Dogger Bank, the most pristine location. The reason for this (which was also found to be the case with several other methods) is not yet established (Livingstone *et al.*, 1992).

Cholinesterase activity

The (butyryl-)cholinesterase activities were determined in dab muscle along the transect. The activity of both enzymes was depressed in near-shore stations. No variation was observed in relation to sex or size (Galgani *et al.*, 1992).

Na^+/K^+-ATPase

In *L. limanda* gill tissue the Na^+/K^+-ATPase activity was determined along the pollution gradient investigated at the Bremerhaven Workshop. In general the ATPase activity was not significantly different at the five stations studied, although there were small differences which followed the contamination gradient, *i.e.* higher in the inner than the outer German Bight, but (again) elevated at the Dogger Bank station. It was considered likely that changes in the levels of ouabain binding reflect modulation of branchial Na^+, K^+-ATPase along the contaminant gradient (Stagg *et al.*, 1992).

CONCLUSION

Outside the more commonly applied methods (such as cytochrome P450, metallothioneines) only relatively few biochemical techniques have so far been tested in real field applications.

They include parameters like glutathione S-transferase, anti-oxidant enzymes, cholinesterase activity and Na^+/K^+-ATPase. It is anticipated, however, that several other methodologies that have so far been investigated only in laboratory experiments, will become available for field application.

Within this context, the Trondheim biomonitoring system (TBS) offers a concept that incorporates the three phases which form the basis of any biomonitoring strategy: early biological effect detection, evaluation of stress or strain that is imposed upon the organism or ecosystem and the identification of the pollutant(s) through fingerprinting, by means of a set of physiological and biochemical parameters.

REFERENCES

Aunaas, T., J.F. Børseth, T. Nordtug, A. Olsen, G. Skjærvø & K.E. Zachariassen, 1991a. Physiological parameters reflecting acute toxicity. In: Effects of organic chemicals on the physiology of the blue mussel (*Mytilus edulis*): Perspectives for toxicity testing and environmental monitoring; K.E. Zachariassen (ed). Tapir Trykk, Trondheim, Norway, pp. 168

Aunaas, T., S. Einarson, T.E. Southon & K.E. Zachariassen, 1991b. The effects of organic and inorganic pollutants on intracellular phosphate compounds in blue mussels (*Mytilus edulis*). Comp. Biochem. Physiol. 100C: 89-93

Beis, I. & E.A. Newsholme, 1975. The contents of adenine nucleotides, phosphagens and some glycolytic intermediates in resting muscles from vertebrates and invertebrates. Biochem. J. 152: 23-32

Briggs, L.B.R., 1979. Effects of cadmium on the intracellular pool of free amino acids in *Mytilus edulis*. Bull. Environ. Contam. Toxicol. 22: 838-845

Børseth, J.F. T. Aunaas, S. Einarson, T. Nordtug, A.J. Olsen & K.E. Zachariassen, 1992. Pollutant-induced depression of the transmembrane sodium gradient in muscles of mussels. J. exp. Biol. 169: 1-18

Carafoli, E., 1987. Intracellular calcium homeostasis. Ann. Rev. Biochem. 56: 395-433

Clark, A.G., 1989. The comparative enzymology of the glutathione S-transferases from non-vertebrate organisms. Comp. Biochem. Physiol. 92B: 419-446

Dando, P.R., K.B. Storey, P.W. Hochachka & J. Storey, 1981. Multiple dehydrogenases in marine molluscs: Electrophoretic analysis of alanopine dehydrogenase, strombine dehydrogenase, octopine dehydrogenase and lactate dehydrogenase. Mar. Biol. Lett. 2: 249-257

Davenport, J., 1979a. The isolation response of mussels (*Mytilus edulis* L.) exposed to falling seawater concentrations. J. mar. biol. Ass. U.K. 59: 123-132

Davenport, J., 1979b. Is *Mytilus edulis* a short term osmoregulator? Comp. Biochem. Physiol. 136: 53-59

Denstad, J.-P., 1989. Effects of water soluble fractions and water emulsions of oils and chemicals on inorganic ions, electrochemical potential difference of sodium and content of free amino acids in blue mussels, *Mytilus edulis*. In: Biological effects of chemical treatment of oilspills at sea; K.E. Zachariassen (ed.). Tapir Trykk, Trondheim, Norway, pp. 330

de Zwaan, A. & W. Zurburg, 1981. The formation of strombine in the adductor muscle of the sea mussel *Mytilus edulis* L. Mar. Biol. Lett. 2: 179-192

de Zwaan, A., A.M.T. de Bont, W. Zurburg, B.L. Bayne & D.R. Livingstone, 1983. On the role of strombine formation in the energy metabolism of adductor muscle of a sessile bivalve. J. comp. Physiol. 149: 557-563

de Zwart, D., K.J.M. Kramer & H.A. Jenner, 1994. Practical experiences with the biological early warning system "Musselmonitor®". Environ. Toxicol. Wat. Qual. (in press)

Einarson, S., B. Josefsson & S. Lagerkvist, 1983. Determination of amino acids with 9-fluorenylmethyl chonoformate and reversed-phase high-performance liquid chromatography. J. Chromatogr. 282: 609-618

Florey, E., 1966. An introduction to general and comparative animal physiology. W.B. Saunders, Philadelphia, pp. 713

Galgani, F., G. Bocquené & Y. Cadiou, 1992. Evidence of variation in cholineesterase activity in fish along a pollution gradient in the North Sea. Mar. Ecol. Prog. Ser. 91: 77-82

Ganong, M.F., 1987. Review of medical physiology, 13th Edn., Appelton & Lange, Prentice-Hall Int., Englewood Cliffs, NJ, pp. 357

Giesy, J.P., S.R. Denser, C.S. Duke & G.W. Dickson, 1981. Phosphoadenylate concentrations and energy charge in two freshwater Crustaceans: Responses to physical and chemical stressors. Verh. Int. Verein. Limnol. 21: 205-220

Giesy, J.P., C.S. Duke, R.D. Bingham & G.W. Dickson, 1983. Changes in phosphoadenylate concentrations and adenylate energy charge as an integrated biochemical measure of stress in invertebrates. The effects of cadmium on the freshwater clam *Corbicula fluminea*. Toxicol. Environ. Chem. 6: 259-295

Habig, W.H., M.J. Pabst & W.B. Jakoby, 1974. Glutathione S-transferases. The first enzymatic step in mercapturic acid formation. J. Biol. Chem. 249: 7130-7139

Higashi, R.M., T.W. Fan & J.M. MacDonald, 1989. Monitoring of metabolic responses of intact *Haliotis* (Abalones) under salinity stress by ^{31}P surface-probe localized NMR. J. exp. Zool. 249: 350-356

Jeffries, H.P., 1972. A stress syndrome in the hard clam, *Mercenaria mercenaria*. J. Invert. Pathol. 20: 242-251

Ketterer, B., D.J. Meyer & A.G. Clark, 1988. Soluble glutathiome transferase isoenzymes. In: Glutathione conjugation, mechanisms and biological significance; H. Sies & B. Ketterer (eds). Academic Press, London, pp. 74-137

Kramer, K.J.M., H.A. Jenner & D. de Zwart, 1989. The valve movement response of mussels: a tool in biological monitoring. Hydrobiologia, 188/189: 433-443

Kreutzer, U., B.R. Siegmund & M.K. Grieshaber, 1989. Parameters controlling opine formation during muscular activity and environmental hypoxia. J. comp. Physiol. 159B: 617-628

Lee, R.F., 1988. Glutathion S-transferase in marine invertebrates from Langesundfjord. Mar. Ecol. Prog. Ser. 46: 33-36

Lee, R.F., 1989. Metabolism and accumulation of xenobiotics within hepato-prancreas cells of the blue crab, *Callinectes sapidus*. Mar. Environ. Res. 28: 93-97

Leivestad H. & I.P. Muniz, 1976. Fish kill at low pH in a Norwegian river. Nature (Lond). 259: 391-392

Livingstone, D.R., S. Archibald, J.K. Chipman & J.W. Marsh, 1992. Antioxidant enzymes in liver of dab *Limanda limanda* from the North Sea. Mar. Ecol. Prog. Ser. 91: 97-104

Looise, B.A.S., 1992. Glutathion S-transferasen in bivalven als biomarker. IMW-TNO report R92-433, TNO, Delft, pp. 45 (in Dutch)

Mannervik, B. & U.H. Danielson, 1988. Glutathione transferases -Structure and catalytic activity. CRC Crit. Rev. Biochem. 23: 283-337

Neff, J.M. & P.D. Boehm, 1985. Petroleum contamination and biochemical alterations in oysters (*Crassostrea gigas*) and plaice (*Pleuronectes platessa*) from bays impacted by the Amoco Cadiz crude oil spill. Mar. Environ. Res. 17: 281-283

Nicotera, P., D.J. McConkey, J.M. Dypbukt, D.P. Jones & S. Orrenius, 1989. Ca^{2+}-activated mechanisms in cell killing. Drug. Metab. Rev. 20: 193-201

Nordtug, T., T. Aunaas, J.F. Børseth, A. Olsen, G. Skjærvø & K.E. Zachariassen, 1991a. Dynamics of solute redistribution in muscular tissue during exposure to formaldehyde. In: Effects of organic chemicals on the physiology of the blue mussel (*Mytilus edulis*): Perspectives for toxicity testing and environmental monitoring; K.E. Zachariassen (ed). Tapir Trykk, Trondheim, Norway, pp. 168

Nordtug, T., J.F. Børseth, A. Olsen & K.E. Zachariassen, 1991b. Measurements of oxygen consumption in *Mytilus edulis* during exposure to, and recovery from, high sublethal concentrations of formaldehyde, benzene and phenol. Comp. Biochem. Physiol. 100C: 85-87

Nordtug, T. & A. Olsen, 1993. A submersible system for continous *in situ* monitoring of the rate of oxygen consumption in aquatic animals. In: Effects of organic chemicals on the physiology of the blue mussel (*Mytilus edulis*): Developing a method for environmental monitoring; K.E. Zachariassen (ed). Tapir Trykk, Trondheim, Norway, pp. 148

Olsen, A. & G. Skjærvø 1991. Induction of strombine formation. In: Effects of organic chemicals on the physiology of the blue mussel (*Mytilus edulis*): Perspectives for toxicity testing and environmental monitoring; K.E. Zachariassen (ed). Tapir Trykk, Trondheim, Norway, pp. 168

Olsen, A., T. Aunaas, J.F. Børseth, T. Nordtug, G. Skjærvø & K.E. Zachariassen, 1991a. Exposure, sampling and analyses. In: Effects of organic chemicals on the physiology of the blue mussel (*Mytilus edulis*): Perspectives for toxicity testing and environmental monitoring; K.E. Zachariassen (ed). Tapir Trykk, Trondheim, Norway, pp. 168

Olsen, A., T. Aunaas, J.F. Børseth, T. Nordtug, G. Skjærvø & K.E. Zachariassen, 1991b. Physiological effects as fingerprint parameters in environmental monitoring. In: Effects of organic chemicals on the physiology of the blue mussel (*Mytilus edulis*): Perspectives for toxicity testing and environmental monitoring; K.E. Zachariassen (ed). Tapir Trykk, Trondheim, Norway, pp. 168

Prosser, C.L., 1986. Adaptational biology: Molecules to organisms. Wiley, New York, pp. 784

Rasmussen, H. & P.Q. Barrett, 1984. Calcium messenger system: an integrated view. Physiol. Rev. 64: 938-984

Sangster, A.W., S.E. Thomas & N.L. Tingling, 1975. Fish attractants from marine invertebrates. Arcamine from *Arca zebra* and strombine from *Strombus gigas*. Tetrahedron, 31: 1135-1137

Sato, M., Y. Suzuki, H. Yasuda, H. Kawauchi, N. Kanno & Y. Sato, 1988. Quantitative HPLC analysis of acidic opines by phenylthiocarbamyl derivatization. Anal. Biochem. 174: 623-627

Schumway, S.E., 1977. The effects of fluctuating salinity on the osmotic pressure and Na^+, Ca^{++} and Mg^{++} concentrations in the haemolymph of bivalves. Mar. Biol. 41: 153-177

Sellevold, O.F.M., P. Jynge & K. Aarstad, 1986. High performance liquid chromatography: a rapid isocratic method for determination of creatine and adenine nucleotides in myocardial tissue. J. Mol. Cell Cardiol. 18: 517-527

Sheenan, D., K.M. Crimmins & G.M. Burnell, 1991. Evidence for glutathione-s-transferase activity in *Mytilus edulis* as an index of chemical pollution in marine estuaries. In: Bioindicators and environmental management; D.W. Jeffrey & B. Madden (eds). Academic Press, London

Stagg, R., A. Goksøyr & G. Rodger, 1992. Changes in branchial Na^+,K^+-ATPase, metallothionein and P450 1A1 in dab *Limanda limanda* in the German Bight: Indicators of sediment contamination? Mar. Ecol. Prog. Ser. 91: 105-115

Staurnes, M., T. Sigholt & O.B. Reite, 1984. Reduced carbonic anhydrase Na-K-ATPase activity in gills of salmonids exposed to aluminum-containing acid water. Experientia, 40: 226-227

Stebbing, A.R.D., V. Dethlefsen & M. Carr, 1992. Biological effects of contaminants in the North Sea. Mar. Ecol. Progr. Ser. (special volume), 91: 1-361

Suteau, P., M. Daubeze, M.L. Migaud & J.F. Narbonne, 1988. PAH-metabolizing enzymes in whole mussels as biochemical tests for chemical pollution monitoring. Mar. Ecol. Prog. Ser. 46: 45-49

van den Thillart, G., A. van Waarde, H.J. Muller, C. Erkelens & J. Lugtenberg, 1990. Determination of high-energy phosphate compounds in fish muscle: ^{31}P-NMR spectroscopy and enzymatic methods. Comp. Biochem. Physiol. 95B: 789-795

van Veld, P.A. & R.F. Lee, 1988. Intestinal glutathion S-transferase activity in flounder *Platichtys flesus* colleceted from contaminated and reference sites. Mar. Ecol. Prog. Ser. 46: 61-63

Viarengo, A. & P. Nicotera, 1991. Possible role of Ca^{2+} in heavy metal cytotoxicity. Comp. Biochem. Physiol. 100C: 81-84

Zacheriassen, K.E., T. Aunaas, J.F. Børseth, S. Einarson, T. Nordtug, A. Olsen & G. Skjærvø, 1991. Physiological parameters in ecotoxicology. Comp. Biochem. Physiol. 100C: 77-79

Zurburg, W., A.M.T. de Bont & A. de Zwaan, 1981. Recovery from exposure to air and the occurrence of strombine in different organs of the sea mussel *Mytilus edulis* L. Mol. Physiol. 2: 135-147

Appendix 5

Methods for Physiological Biomarkers

THE TBS RESPIRATORY RATE REGISTRATION UNIT

The respiratory rate registration unit consists of twelve registration chambers, in which the ABM organisms stay during the registration period (Figure 10). Two of these chambers are left vacant and serve as control chambers during the registration. Water is pumped through the chambers at a constant flow rate during the registration period by a low power pump. The flow through the chambers is controlled by resistance tubes from the pump into each chamber and could be regulated by either changing the tube diameter or the tube length. A multi-valve system leads water from each chamber to an oxygen electrode. This system is electronically and automatically operated and the electrode is self-calibrating during operation by the flushing of water from the control chambers over the electrode between each registration of the oxygen tension in water entering from the occupied chambers. Each registration period lasts for two minutes to allow the electrode signal to be stabilized before the value is recorded.

The respiratory rate of the organism is calculated from the deviation between the oxygen tension in the water entering the respiration chamber, which is the oxygen tension in water entering the electrode from the control chamber and the oxygen tension in the water leaving the occupied chambers. The flow rate through the chambers is regulated to give about 5% extraction of oxygen from the water while it remains in the chambers.

BIOLOGICAL SAMPLING FOR BIOCHEMICAL METHODS

For the various biochemical analyses, mantle fluid/haemolymph samples and tissue samples have to be prepared. The method described, is based on our methods for the preparation of samples from mussel species.

Mantle fluid and haemolymph samples

Before sampling the haemolymph, mantle fluid should be drained from the mussels in order to prevent mixing of mantle fluid into the haemolymph. The draining can be done by inserting a small screwdriver between the shells, forcing the shell halves apart and draining the water through the siphons. If the mantle fluid is to be collected and stored for analysis, the draining of the fluid is performed by inserting a canyla between the shells from the ventral side, inserted 3-4 mm into the mantle cavity and sucking the water into a syringe. The tubes containing the mantle fluid should be frozen in an upright position and stored for later analysis.

The haemolymph samples are obtained by inserting a thin canyla between the shells and into the central part of the posterior adductor muscle. The haemolymph is sucked out of the mussel by means of a 1 ml disposable syringe. Normally there is no problem in collecting 0.5-1.0 ml of haemolymph from mussels with a length of 40-60 mm. Haemolymph samples are easily distinguishable from mantle fluid by the occurrence of haemocytes in the former. The haemolymph samples are kept in 2 ml centrifuge tubes and should be stored on ice (0 °C) during the collection period. Thereafter they are centrifuged for 5 minutes at 8000 rpm in order to remove the cells from the fluid.
For later analysis of free amino acids, 100 µl of cell free haemolymph is transferred to convenient tubes and added 100 µl 10% TCA for precipitation of proteins. The cell free haemolymph is transferred to marked tubes and, along with the samples for analyses of free amino acids, the tubes are frozen in an upright position and stored at -20 °C.

Tissue samples

Samples from the posterior adductor muscle are obtained after collecting mantle fluid and haemolymph samples. The muscle is cut loose from one half of the shell by sliding a scalpel knife along the inner side of the shell. Thereafter the shell is bent open and other tissue is cut loose from the muscle before the whole muscle is cut loose from the other shell half. Surface water is removed from the muscle by means of lint free paper. The muscle is transferred to 4 ml tarred and marked cryo tubes and the muscle wet weight is determined. The tubes are immediately put in liquid nitrogen and the samples can be stored in this manner. However, it is recommended that after 24 hours the samples should be taken from the nitrogen tank and freeze dried. After freeze drying the tubes are capped and the dry weight of the samples determined. Freeze dried samples are mortared by a rounded glass rod while inside the tube and 4 ml (preferably not less than 10 times the volume of the sample) of cooled 10% TCA is added. The tissue samples should then be stirred for 24 hours at 0-5 °C in order to extract inorganic ions and free amino acids while proteins are precipitated. Tissue extracts are stored at -20 °C in a deep freezer. Before volumes are used for analysis on the different parameters the samples are centrifuged for 5 minutes at 8.000 rpm in order to deposit particulate matter. Other tissue samples from the organism should be treated in the same way as the muscle samples.

ANALYSES

Energy rich phosphates

A method for the determination of energy-rich phosphates in muscle tissue by the use of HPLC is described by Sellevold *et al.* (1986).

Inorganic ions

The haemolymph and tissue concentrations of sodium, potassium and calcium are analyzed by means of flame photometry or atomic absorption spectrophotometry (AAS) according to standard procedures. The determination of chloride concentrations is conducted on a chloride titrator.

Free amino acids and strombine

The supernatants from haemolymph and tissue extracts are diluted with supra pure water in order to obtain concentrations appropriate for HPLC analysis. As a rule, one part of the haemolymph/TCA mixture is diluted with 4 parts of water. Tissue extracts are diluted with

TCA in a 1:1 ratio and are further diluted with 7 parts water. In order to achieve ca pH 8.0 in the samples, 100 µl of a borate buffer (pH 12.3) is added to 400 µl of the diluted samples.
In order to make fluorescent derivatives, 500 µl FMOC-Cl dissolved in acetone is mixed with the buffered solutions. The reaction is allowed to continue at room temperature for 3 minutes before 10 µl of the samples are injected on the HPLC column. The samples are eluted according to a method described by (Einarson *et al.*, 1983) and detected by a fluorescence detector.

Chapter 6

Cytochrome P450 Monooxygenase as Indicator of PCB/Dioxin like Compounds in Fish

Lars Förlin[1], **Anders Goksøyr**[2] and **Astrid-Mette Husøy**[2]

[1] Department of Zoophysiology, University of Göteborg
Medicinaregatan 18, S-413 90 Göteborg, Sweden

[2] Laboratory of Marine Molecular Biology, University of Bergen
HIB, N-5020 Bergen, Norway

ABSTRACT

In this chapter different methods for measuring the induction of fish cytochrome P450 monooxygenase are described and discussed. The induction is usually established by measuring the ethoxyresorufin O-deethylase (EROD) activity. Immunochemical probes employing specific antibodies to cytochrome P450 1A are also used.

This induction response is used in environmental monitoring in order to indicate exposure of fish to certain organic contaminants, including planar halogenated PCBs and dioxins. Examples of these techniques are presented and illustrated in several case studies of fish from the aquatic environment.

Details of the methods are treated in detail in a technical Appendix.

1. INTRODUCTION

Biomonitoring of aquatic chemical contaminants includes the monitoring of biological effects in fish. Environmental control of the health of fish is relevant given its importance both commercially and as a source of food. This concern has resulted in more knowledge being gathered about contaminant effects in fish than in other organisms of the aquatic ecosystem.

Its large size is another advantage when it comes to monitoring fish as this facilitates the analysis of effects in different tissues. In addition, since they comprise a relatively high trophic level, toxic effects of contaminants on fish may integrate effects occurring at lower levels in the ecosystem.

Effects of chemical contaminants can be studied at many different levels of biological organization, from the subcellular to the population and ecosystem levels. It is obvious that the

Biomonitoring of Coastal Waters and Estuaries. Edited by Kees J.M. Kramer.

direct effect of a contaminant occurs at the biochemical or molecular level. The more severe this effect is, the more likely it is that the effect is 'transmitted' to higher levels of biological organization. Biochemical methods can, therefore, serve as early warning indicators (biomarkers) in the monitoring of environmental contaminants (see Haux & Förlin, 1988; Stegeman *et al.*, 1992).

Biotransformation of lipophilic chemicals to water soluble compounds facilitates detoxification and excretion from the fish through urine or bile or over the gills. Certain steps in the biotransformation are, however, involved in the activation of xenobiotics to reactive intermediates that ultimately result in toxicity, carcinogenicity and other effects (see Varanasi, 1989).

The first step in the biotransformation process is often an oxidative step (phase I reaction), catalyzed by the cytochrome P450 monooxygenase system. In the second step (phase II reaction) larger endogenous molecules are conjugated to the oxygenated xenobiotics. This reaction is catalyzed by different transferase enzymes, *e.g.* UDP glucuronyl transferase and sulphotransferase, thereby transforming the xenobiotic to polar products that can be excreted from the organism (see Varanasi, 1989).

The aim of this chapter is to review briefly the cytochrome P450 (P450) monooxygenases system and its responses in fish to environmental contaminants to describe the catalytic measurement of EROD (ethoxyresorufin-O-deethylase) activity and the immunochemical measurement of P450 1A in fish tissues and, finally, to describe and discuss the measurement of EROD activity and P450 1A protein content in the monitoring of aquatic pollution. For more detailed information of biochemistry, toxicology and other functional aspects and regulation of the biotransformation enzymes in fish the reader is referred to other recent reviews (Varanasi, 1989, Stegeman, 1989; Goksøyr & Förlin, 1992; Andersson & Förlin, 1992). Some practical details based on the authors' experiences in their studies of the P450 system are presented in the Appendix.

2. THE CYTOCHROME P450 SYSTEM

2.1. Definitions

The cytochrome P450s belong to a superfamily of structurally and functionally related hemoproteins. In this chapter our nomenclature of P450 genes, gene transcripts and proteins will be according to that recommended by Nebert *et al.* (1991). Cytochrome P450 is denoted CYP and the following Arabic number stands for the P450 family, a letter indicating the subfamily and finally an Arabic numeral for the individual gene or protein, for example (CYP1A1). However, at the protein level CYP or P450 can be used.

The P450 protein is partly embedded in the endoplasmatic reticulum. The P450 functions with several other enzymes in the monooxygenase system in a coupled electron transport system in the endoplasmatic reticulum. Electrons are transferred from NADPH (and sometimes NADH) through a flavoprotein (NADPH cytochrome P450 reductase, and sometimes NADH cytochrome b_5 reductase) to P450 (see Guengerich, 1991; Porter & Coon, 1991). By these means P450 inserts one atom of oxygen into the substrate and reduces the second oxygen atom to water.

Many reactions are catalyzed by the P450 monooxygenase system (Table 1). The basal characteristics of these reactions are the same but there are large differences in the chemistry of the substrates and products. The measurement seems to reflect the catalytic activity of individual P450 proteins for very few of these reactions. Benzo[*a*]pyrene hydroxylase (aryl hydrocarbons hydroxylase, AHH) and EROD (ethoxyresorufin-O-deethylase) seem to be rather specific for P450s in the CYP1A subfamily. Therefore these catalytic probes are therefore important tools in the studies of responses of environmental contaminants (see below).

Table 1. Cytochrome P450 monooxygenase reactions catalyzed with xenobiotics (from Goksøyr & Förlin, 1992)

reactions	**example xenobiotics**
aromatic epoxidation and hydroxylation	Benzo[*a*]pyrene and other PAHs
aliphatic hydroxylation	n-propylbenzene
N-dealkylation	aminopyrine, ethylmorphine
O-dealkylation	ethoxyresorufin, 7-ethoxycoumarin
S-dealkylation	some thioethers like methylmercaptan
N-oxidation	aniline, amphetamine
S-oxidation	thioethers in general, insecticides
P-oxidation	trisubstituted phosphines
desulphuration and breaking of ester bonds	parathion, insecticides
oxidative deamination	
oxidative dehalogenation	
reductive dehalogenation	

2.2. Induction of the P450 1A system

Many different chemicals induce *'de novo'* synthesis of P450. The inductive response is a classical process by which a chemical stimulates the rate of gene transcription, resulting in increased levels of messenger RNA and synthesis of P450 protein. The induction response of the CYP1A subfamily occurs via an intracellular receptor (Ah receptor), translocation of the receptor complex to the nucleus and transcriptional activation of the genes. Each step in the inductive response that is mRNA, protein and catalytic activity can be analyzed with a probe suitable for the detection of induction (Table 2).

Table 2. Induction of cytochrome P450 1A can be measured at several levels

level	**type**	**assay**
enzyme	catalytic activity	EROD, AHH
protein	immunochemical	Western blot, ELISA, immunocytochemistry
mRNA	molecular biology	Northern blot

The induction process is not fully characterized in fish but it is well known that P450 1A can be induced by polyaromatic hydrocarbons (PAH)-type inducers (Stegeman, 1981). The presence of an Ah-receptor in fish was not demonstrated until recently (Hahn *et al.*, 1992). Measurement of AHH and EROD activities appears to be the most sensitive means of determining the inductive response in fish. These enzyme activities occur at low and sometimes undetectable levels in control or untreated fish.

The fact that many of the inducers of fish EROD and AHH activities are well known aquatic pollutants has greatly stimulated research of the CYP system in fish. Environmental contaminants of major concern are PAHs, polychlorinated dioxins and furans, polyhalogenated biphenyls (PCBs) and other halogenated organic compounds such as some pesticides (Table 3). Also, complex mixtures such as Aroclor and Clophen (commercial PCBs), petroleum hydrocarbons and industrial effluents, *e.g.* from bleached kraft mills, have been shown to have inducing properties because these mixtures contain specific inducers.

Table 3. Examples of xenobiotics and mixtures with proven inducing properties on the cytochrome P450 system in fish

Polyaromatic hydrocarbons (PAHs):	**Polychlorinated biphenyls (PCBs):**
benzo[*a*]pyrene	3,3',4,4'-tetrachlorobiphenyl
3-methylcholanthrene	3,3',4,4',5-pentachlorobiphenyl
ß-naphthoflavone	
Isosafrol	
Chlorinated dioxins and furanes:	**Industrial effluents:**
2,3,7,8-tetrachlorodibenzo-p-dioxin	bleached pulp mill effluents
2,3,7,8-tetrachlorodibenzofurane	effluents from oil refineries

2.3. Factors modulating the induction

The induction response in fish can be influenced by many biotic and abiotic factors such as species, sex, reproductive stage and water temperature. For example, it is known that certain gonadal steroids affect induction by xenobiotics, but glucocorticoids may also play a regulatory role (for a review see Andersson & Förlin, 1992). In biomonitoring of the environmental induction responses of pollutants, the biotic and abiotic factors must be taken into account. We are, however, far from understanding the mechanisms by which these factors affect the CYP system and we still need to characterize the magnitude and timing of the changes. Therefore, these interferences must be standardized among sampling sites in field studies.
In addition, the susceptibility of the P450 monooxygenase system to degradation, *e.g.* during sampling and storage, requires the use of well characterized and properly functioning techniques when monitoring P450 induction in field studies.

3. MEASUREMENT OF THE P450 1A SYSTEM

Induction of cytochrome P450 can be analyzed with suitable probes at enzyme, protein and/or mRNA level. The catalytic activity (AHH or EROD) is the most commonly used measurement of P450 1A induction in environmental monitoring of aquatic pollution. The measurement of P450 1A protein and mRNA can, however, supplement in the analysis of the induction process. For example, induction of an organ can be studied with catalytic assays and it is possible to study P450 1A induction of specific cell types in an organ with immunocytochemistry.
The choice of method is a matter of cost with costs increasing from catalytic activity to mRNA level analysis. Induction analysis also requires well equipped laboratories and experienced personnel and access to specific antibodies that are not yet commercially available. In the following sections the collection and storage of samples, the catalytic and immunochemical assays and a brief treatment of mRNA analysis for CYP1A induction studies will be discussed.

3.1. Collection and storage of samples

The activity levels of the P450 monooxygenase in fish are influenced by the previous handling of the liver. Erroneous results are obtained if great care is not given to the sampling and handling of the sample. The membrane-bound P450 is susceptible to inactivation during sampling (time after death of the fish), freezing before preparation of the enzyme fractions, enzyme fractionation (homogenisation and centrifugation) and storage of the enzyme fraction before enzyme analysis. It is recommended to investigate the influence of these factors on each species examined. The following general recommendations should be followed. Suggestions for proper sample handling are presented in the Appendix.

3.2. Catalytic probes

The activity of cytochrome P450 monooxygenases in fish can be measured by an array of different substrates. Since different forms of P450s often show overlapping substrate specificity it is important to choose the substrate with care and with knowledge of the specificity of the chosen substrate.

The best substrates for the catalytic measurement of P450 1A activity in fish are ethoxyresorufin (EROD assay) and benzopyrene (AHH assay). Different methods exist for each substrate. The O-deethylation of ethoxyresorufin (7-ER) is catalyzed by specific inducible cytochrome P450 (P450 1A isoenzymes) and results in the production of resorufin which can be readily estimated fluorometrically. Burke & Mayer (1974) described the first fluorometric technique for EROD-activity. The appearance of resorufin is measured at an excitation wavelength of 535 nm and at an emission wavelength of 585 nm. The resorufin formation is dependent upon the presence of enzyme, substrate, oxygen (O_2), and cofactor (NADPH). An alternative technique was described by Pohl & Fouts (1980). In this procedure the EROD reaction is stopped by adding methanol and the sample fluorescence is measured after centrifugation and filtration. A spectrophotometric assay was introduced by Klotz *et al.* (1984). In this method the product resorufin is monitored continuously at 572 nm. Galgani *et al.* (1991) describe an EROD procedure where hepatic EROD activity is measured using a fluorescence plate-reader. This method is time saving compared to fluorescence spectrophotometry (Burke and Mayer, 1974) and UV/VIS spectrophotometry (Klotz *et al.*, 1984).

Several laboratories now measure fish hepatic EROD activity in order to indicate environmental contamination. The ICES-IOC arranged an EROD inter-calibration workshop in Aberdeen, Scotland (4-6 September, 1991) to evaluate the different methods. The procedure presented in the Appendix is based on the method of Burke & Mayer (1974) and the results from the EROD intercalibration workshop, and carried out by the methods used by the authors. However, many variants have been published. We recommend the reader to consult the original paper for further information on the assays (Burke and Mayer, 1974). We also recommend the report by Hodson *et al.* (1991a) where detailed protocols for measuring P450 1A activity and protein of fish liver are given.

There are many aspects to be taken into account when making the choice of assay. One has to consider substrate purity (source), the cost of the assay, equipment needed and the ease and speed of the assay. Also, sensitivity and reliability of the assay have to be considered. After the choice has been made the assay conditions must be controlled for each species studied. This includes studies of the optimal concentrations of the substrate, protein and NADPH, and incubation temperature as well as pH for the assay. It is recommended that known samples are always run in parallel as a reference, in order to assure the consistency of results obtained with unknown samples.

3.3. Immunochemical probes

In the immunochemical techniques the amount of protein cross-reacting with a specific antibody is measured. Polyclonal and monoclonal antibodies were prepared to cytochrome P450 1A1. These antibodies bind to epitopes on cytochrome P450 1A1; the immunocomplex formed is an excellent target for labelled secondary antibodies. The principle of detection is to use a secondary antibody conjugated with an enzyme, radioactive or fluorescent group. Immunodetection systems are powerful tools with high sensitivity, detectability and specificity.

This analysis of P450 1A protein is a good complement to the catalytic measurement of the P450 1A system. These techniques are not dependent upon biologically active samples, they can have high sensitivity and detectability and can be developed into time saving techniques in which a large number of samples can be assayed.
Three immunochemical techniques were applied:

- Immuno-blotting (Western blot);
- ELISA (enzyme linked immuno sorbent assay);
- Immunocytochemistry.

The detection and quantization of antigens by antigen-specific antibodies using Western blotting, an indirect enzyme-linked immunosorbent assay (ELISA) and immunohistochemical localization, are described in the Appendix.
Western blotting combines gel electrophoresis with the specificity of immunological detection. Denaturated proteins are separated by SDS-gel electrophoresis and transferred to a membrane support. After blocking for nonspecific binding sites, specific antibodies will bind to the antigen and the protein can be detected. In this technique the antibodies only recognize epitopes on the antigen that were not destroyed by denaturation.
Western blotting is suitable when few samples are analyzed and should always be used to study cross reactivity of the P450 1A antibody in new species.
The indirect P450 1A1 ELISA was recently described by Goksøyr (1991) and Celander & Förlin (1991). In the ELISA procedure, antigen is immobilized on a microtiter plate and antigen specific antibodies are allowed to bind to the antigen. Labelled antibodies (enzyme-antibody conjugate) bind specifically to the first antibodies and the immuncomplex can be detected. The conjugated enzyme cleaves a substrate to generate a coloured reaction product that can be detected spectrophotometrically. The absorbance of the coloured solution in individual microtiter wells is proportional to the amount of antigen, relative to total sample protein. Some technical information on the ELISA method is presented in the Appendix. ELISA works best when a large amount of samples are assayed.
Immunohistochemical techniques demonstrate both the presence and cellular localization of the antigen. Double-labelling techniques permit simultaneous detection of two antigens. These forms of study of P450 1A1 induction have demonstrated that only specific cell groups (endothelium) are capable of being induced in the different organs, thereby resulting in a dilution effect when homogenized tissue is used as in catalytic measurements, Western blotting and ELISA (Smolowitz *et al.*, 1991; Stegeman & Lech, 1991; Husøy *et al.*, 1992). In field studies, immunohistochemistry has the advantage of not requiring sophisticated equipment or reagents (like liquid N_2), only containers and the fixative.
The immunochemical techniques require specific antibodies. Several laboratories have developed their own antibodies, many of which have been shown to be suitable in the immunochemical detection of P450 1A in different species of fish (Williams & Buhler, 1984; Goksøyr, 1985; Park *et al.*, 1986; Goksøyr *et al.*, 1991; Celander & Förlin, 1992; Förlin & Celander, 1993). Unfortunately there are as yet no fish specific antibodies commercially available.

3.4. Gene transcripts

Besides catalytic and immunochemical assays, mRNA analysis can be applied for measuring CYP1A expression. After the first fish CYP1A1 gene was cloned a cDNA probe was used in order to follow the temporal relationship between mRNA levels, immunodetectable protein and catalytic activities in fish. These studies show that mRNA levels return rather rapidly to control levels whereas P450 1A protein levels stay elevated for extended periods. The mRNA analysis is more complicated than the catalytic and immunochemical techniques and includes purification of RNA (susceptible to rapid degradation) and the use of labelled cDNA probe. Therefore this analysis is not recommended (at present) as a routine method. However, analysis of mRNA could be valuable, for example in situations where gene activation is suspected but the enzyme is inhibited or degraded. In addition, more studies are needed in order to assess the response at this level in chronic exposure situations.

4. FIELD APPLICATIONS

The interpretation of the results from field studies requires a reference site located in an unpolluted area with a given fish species of similar age and sex. In addition, the interpretation relies upon the fact that the response is relatively specific with respect to the inducing chemical. One has, thus, to assume that the environmental contaminant is structurally similar to chemicals known to induce the enzyme system.

4.1. Early studies

In the work by Payne & Penrose (1975) an increased hepatic P450 1A activity (AHH) was observed for the first time in brown trout (*Salmo trutta*) taken from a small urban lake with a history of hydrocarbon contamination in Newfoundland. Subsequently, through a number of studies carried out during the last 15-20 years it has become evident that P450 1A dependent monooxygenase activities (EROD and AHH) are often elevated in fish from waters polluted with oil hydrocarbons, industrial and municipal effluents containing aromatic and/or chlorinated hydrocarbons (for reviews see Payne, 1984; Lindström-Seppä *et al.*, 1985; Payne *et al.*, 1987; Haux & Förlin, 1988; Vindimian & Garric, 1989; Goksøyr & Förlin, 1992).

4.2. Dioxin/PCB studies

Polyhalogenated hydrocarbons are among the most persistent and ubiquitous contaminants and have been shown to accumulate in virtually all biota worldwide, including the oceanic and polar environments. The inductive effects of dioxins and PCBs on the fish cytochrome P450 system has been demonstrated in a number of laboratory experiments using specific congeners or technical mixtures (*e.g.* van der Weiden *et al.*, 1992; Boon *et al.*, 1992; Skaare *et al.*, 1991; Gooch *et al.*, 1989). However, pronounced species differences appear to exist in this response. Atlantic cod *(Gadus morhua)* is one species that seems quite unresponsive to PCB congeners and technical mixtures (Beyer *et al.*, 1992). Whether this is caused by genetic factors, or by the high lipid content of cod liver (50-80%), which could be "hiding" the lipophilic PCB compounds from contact with the Ah receptor, is not known. Also, differences in the inducing response of perch, *(Perca fluviatilis),* have been observed during recent studies (Förlin, unpublished). In these studies, the perch caught at a polluted site were non-responsive to a PCB congener (IUPAC 77) whereas the fish taken from a reference site showed a "normal" inducing response. It is not known if this difference is due to genetic factors or if chronic exposure of fish to certain pollutants may mask or alter the inducing response of the CYP system. However, such species or population differences have to be taken into account when choosing indicator species for field studies. Great caution should be taken if a species used in the field has not been studied in controlled exposure situations.
Consistent with laboratory experiences, field studies show that elevated levels of P450 1A often occur in fish from environments contaminated with organochlorines such as dioxins and PCBs (Addison & Edwards, 1988; Stegeman *et al.*, 1988; Elskus *et al.*, 1989; Goksøyr *et al.*, 1991; Förlin *et al.*, 1992; Eggens *et al.*, 1992; Goksoyr *et al.*, 1992; Renton & Addison, 1992). Good correlations have been reported between *e.g.* PCB, rates of EROD or AHH activities and P450 1A levels in flounder *(Platichthys flesus)* from the Freijerfjord, Norway (Stegeman *et al.*, 1988; Addison & Edwards, 1989), and TCDD equivalents and liver EROD

activities or P450 1A levels in pike from Lake Vänern (Förlin *et al.*, 1992). It was reported recently that EROD activity in dab (*Limanda limanda*) liver (Figure 1) was highly correlated with the concentrations of individual PCB congeners in the liver (Eggens *et al.*, 1992). In this study EROD activity was measured in dab caught along a gradient in the German Bight during the ICES/IOC Bremerhaven workshop on biological effects of contaminants (Stebbing & Dethlefsen, 1992). Also, higher monooxygenase activity at higher PCB levels were reported by Stegeman *et al.* (1986) in rattail *(Coryphanoides armatus)*, a fish species typical of the deep sea below 1000 m. Caution should be taken, however, because polycyclic aromatic hydrocarbons (PAHs), and a number of other chemicals which are not normally measured, often tend to be present in the same environments and, in these cases, may contribute significantly to the observed effect.

In some cases, effects of naturally contaminated or spiked sediments on fish under laboratory conditions have also been investigated. Again, strong induction is seen in some species (van der Weiden *et al.*, 1992), whereas others, like cod, is more refractory to induction (Beyer *et al.*, 1992).

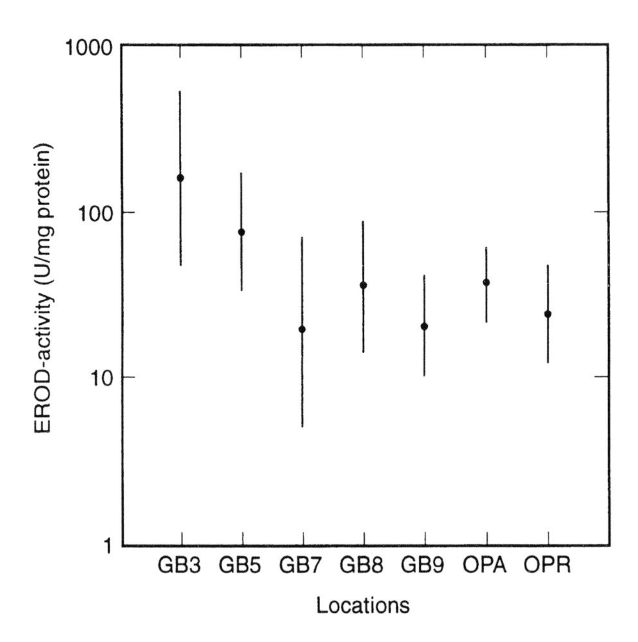

Figure 1. Dab (*Limanda limanda*) EROD activities (with 95% confidence limits) for stations along a gradient in the North Sea in the German Bight, extending from the mouths of the Rivers Elbe and Weser out to the Dogger Bank. Station GB3 is located closest to the coast and GB9 close to the Dogger Bank. Stations OPA and OPR are from oil drilling sites in the North Sea. The EROD activities were measured in the liver postmitochondrial fraction (From: Eggens *et al.*, 1992; with permission)

4.3. Studies on bleached kraft mill effluents (BKME)

In Sweden recently, an extensive trial was carried out in order to examine the usefulness of biochemical and physiological responses as health indicators in fish exposed to effluents from pulp mills, and to use the responses to elucidate the nature and extent of sublethal toxic effects on fish populations exposed to pulp mill effluents. The work included both longterm laboratory studies on fish exposed to pulp mill effluents and field investigations on fish in the receiving waters of different effluents (Förlin *et al.*, 1985; Andersson *et al.*, 1988; Södergren, 1989; Förlin *et al.*, 1991).

These investigations showed that perch *(Perca fluviatilis)* was strongly affected by the effluents (Andersson *et al.*, 1988; Södergren, 1989), with induced EROD activities as the strongest signal (Figure 2; Andersson *et al.*, 1988). Also, biochemical and physiological effects including a profound induction of EROD activity near the source of BKME were observed in other Scandinavian studies and in Canadian field studies of fish (Lindström-Seppä & Oikari, 1990; Munkittrick *et al.*, 1991; Hodson *et al.*, 1991b).

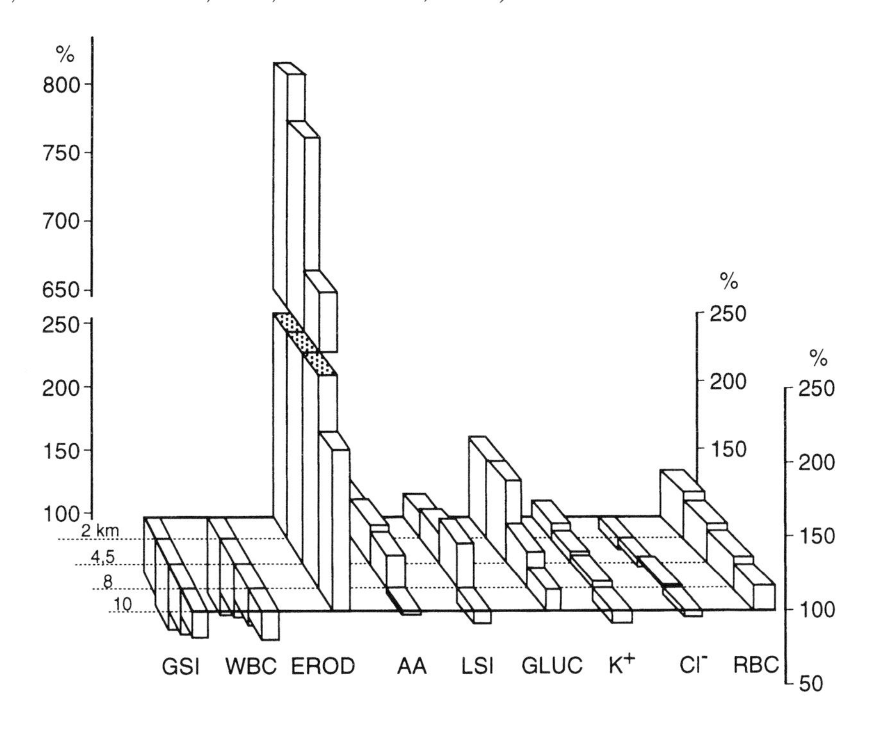

Figure 2. Physiological and biochemical variables in perch (*Perca fluviatilis*) captured at different distances from the effluent outlet of a pulp bleaching plant on the Swedish east coast. The data are expressed as percentages of the control site (120 km away, 100%).
GSI: gonad somatic index; WBC: total white blood cell count; EROD: 7-ethoxyresorufin O-deethylase; AA: ascorbic acid; LSI: liver somatic index; GLUC: blood glucose; K^+: plasma potassium; Cl^-: plasma chloride; RBC: red blood cell count (From: Andersson *et al.*, 1988; with permission)

In the studies near the source of BKME, the induction of EROD activity or P450 1A is amongst the easiest and most sensitive responses to detect. Therefore, from the biochemical and physiological techniques applied today, the measurement of EROD induction in fish appears to be the most sensitive biomonitoring method for representing an area exposed to BKME (see also Hodson *et al.*, 1991a). Near the source of a BKME, EROD induction may also be indicative of an area where the health of fish is affected. This assumption is supported by the simultaneous occurrence of other adverse biological effects, including effects on fish population (see Södergren, 1991). Therefore, if induction is detected in fish captured near sources of a BKME then more detailed studies are needed of, for example, reproduction, growth and survival, and chemical analyses of suspected inducers that might have bioaccumulated in the fish populations.
The above mentioned studies of fish, captured near sources of BKME, clearly indicated that bleached kraft mill effluents caused serious disturbances in vital biochemical and physiological functions, which affected the health status and also, presumably, the survival of the fish (Andersson *et al.*, 1988; Förlin *et al.*, 1991). The studies also showed that measurement of EROD activity can serve as an early warning signal (biomarker) in the monitoring of BKME effects in fish, although no causal links are established between EROD induction and, for example, reproduction, growth or survival.

CONCLUSION

Suitable probes at catalytic, protein and mRNA levels are available for the measurement of the induction of fish cytochrome P450 1A monooxygenase by pollutants. General application of this response in biomonitoring programmes requires access to standardized techniques. For ethoxyresorufin O-deethylase (EROD), this process has started in inter-laboratory workshops and suitable protocols for this assay have been suggested. This work will continue. For a wide application of immunochemical analysis, standardised techniques must wait until antibodies become commercially available. For gene transcript analysis the relatively high costs may make them unsuitable for inclusion in routine biomonitoring programmes.

REFERENCES

Addison, R.F. & A.J. Edwards 1988 Hepatic microsomal monooxygenase activity in flounder, *Platichthys flesus* from polluted sites in Langesundfjord and from mesocosms experimentally dosed with diesel oil and copper. Mar. Ecol. Prog. Ser. 46: 51-54

Andersson, T., L. Förlin, J. Härdig & Å. Larsson, 1988. Physiological disturbances in fish living in coastal water polluted with bleached kraft mill effluents. Can. J. Fish Aq. Sci. 45: 1525-1536

Andersson, T. & L. Förlin, 1992. Regulation of the cytochrome P450 enzyme system in fish. Aquat. Toxicol. 24: 1-20

Beyer, J., Serigstad, B., Wilhelmsen, S., Klungsøyr, J. & A. Goksøyr. 1993 Toxicokinetic and effects of PCB on the cytochrome P450 system of juvenile Atlantic cod (*Gadus morhua*). Mar. Environ. Res. 35: 206

Bradford, M. 1976. A rapid and sensitive method for quantization of microgram quantities or protein utilizing the principle of protein-dye binding. Anal. Biochem. 72: 248-251

Boon, J.P., J.M. Everaarts, M.T.J. Hillebrand, M.L. Eggens, J. Pijnenburg & A. Goksøyr, 1992. Changes in the levels of hepatic biotransformation enzymes and haemoglobin levels in female plaice (*Pleuronectes platessa*) after oral administration of a technical polychlorinated biphenyl mixture (Clophen A40). Sci. Total Environ. 114: 113-133

Burke, M.D. & R.T. Mayer, 1974. Ethoxyresorufin: Direct fluorometric assay of microsomal dealkylation which is preferentially inducible by 3-methylcholanthrene. Drug Metab. Disp. 2: 583-588

Celander, M. & L. Förlin, 1991. Catalytic activity and immunochemical quantification of hepatic cytochrome P-450 in ß-naphthoflavone and isosafrole treated rainbow trout (*Oncorhynchus mykiss*). Fish Physiol. Biochem. 9: 189-197

Eggens, M., F. Galgani, J. Klungsøyr & J. Everts, 1992. Hepatic activity in dab *Limanda limanda* in the German Bight using an improved plate-reader method. Mar. Ecol. Prog. Ser. 91: 71-75

Elskus, A.A., J.J. Stegeman, L.C. Susani, D. Black, R.J. Pruell & S.J. Fluck, 1989. Polychlorinated biphenyl concentration and cytochrome P-450E expression in winter flounder from contaminated environments. Mar. Environ. Res. 28: 25-30

Förlin, L., 1980. Effects of Clophen A50, 3-methylcholanthrene, pregnenolone-16αq-carbonitrile, and phenobarbital on the hepatic microsomal cytochrome P-450-dependent monooxygenase system in rainbow trout, *Salmo gairdneri*, of different age and sex. Toxicol. Appl. Pharmacol. 54: 420-430

Förlin, L., T. Andersson, B.-E. Bengtsson, J. Härdig & Å. Larsson, 1985. Effects of pulp bleach effluents on hepatic xenobiotic biotransformation enzymes in fish: laboratory and field studies. Mar. Environ. Res. 17: 109-112

Förlin, L., L. Balk, M. Celander, S. Bergek, M. Hjelt, C. Rappe, C. de Witt & B. Jansson, 1992. Biotransformation enzyme activities, and PCDD and PCDF levels in pike caught in a swedish lake. Mar. Environ. Res. 34: 169-173

Förlin, L., T. Andersson, L. Balk & Å. Larsson, 1991. Biochemical and physiological effects of pulp mill effluents in fish. In: Environmental fate and effects of bleached pulp mill effluents; A. Södergren (ed). SEPA report 4031, Stockholm, Sweden, pp. 235-243

Förlin, L. & M. Celander, 1993. Induction of cytochrome P450 1A in teleosts: environmental monitoring in Swedish fresh, brackish and marine waters. Aquat. Toxicol. 26: 41-56

Galgani, F., Bocquene, G., Lucon, M., Grzebyk, D., Letroui, F. & Claisse, D. 1991. EROD measurements in fish from the northwest part of France. Mar. Pollut. Bull. 22: 494-500.

Goksøyr, A., 1985. Purification of hepatic microsomal cytochromes P-450 from ß-naphthoflavone-treated Atlantic cod (*Gadus morhua*), a marine teleost fish. Biochim. Biophys. Acta, 840: 409-417

Goksøyr, A., 1991. A semi-quantitative cytochrome P-450IA1 ELISA: A simple method for studying the monooxygenase induction response in environmental monitoring and ecotoxicological testing of fish. Sci. Total Environ. 101: 255-262

Goksøyr, A., T. Andersson, D.R. Buhler, J.J. Stegeman, D.E. Williams & L. Förlin, 1991a. An immunological comparison of ß-naphthoflavone-inducible microsomal cytochrome P-450 in different fish species and rat. Fish Physiol. Biochem. 9: 1-13

Goksøyr, A., A.-M. Husøy, H.E. Larsen, J. Klungsoyr, S. Vilhelmsen, A. Maage, E.M. Brevik, T. Andersson, M. Celander, M. Pesonen & L. Förlin, 1991b. Environmental contaminants and biochemical responses in flatfish from the Hvaler Archipelago in Norway. Arch. Environ. Contam. Toxicol. 21: 486-496

Goksøyr, A. & L. Förlin, 1992. The cytochrome P450 system in fish, aquatic toxicology and environmental monitoring. Aquat. Toxicol. 22: 287-312

Goksøyr, A., H.E. Larsen, S. Blom & L. Förlin, 1992. Detection of cytochrome P450 1A in North Sea dab liver and kidney. Mar. Ecol. Prog. Ser. 91: 83-88

Gooch, J.W., A.A. Elskus, P.J. Kloepper-Sams, M.E. Hahn & J.J. Stegeman, 1989. Effects of ortho- and non-ortho polychlorinated biphenyl congeners on the hepatic monooxygenase system in scup *(Stenotomus chrysops)*. Toxicol. Appl. Pharmacol. 98: 422-433

Guengerich, F.P., 1991. Reactions and significance of cytochrome P-450 enzymes. J. Biol. Chem. 266: 10019-10022

Hahn, M.E., A. Poland, E. Glover & J.J. Stegeman, 1992. The Ah receptor in marine animals: phylogenetic distribution and relationship to cytochrome P4501A inducibility. Mar. Environ. Res. 34: 87-91

Haux, C. & L. Förlin, 1988. Biochemical methods for detecting effects of contaminants on fish. Ambio, 6: 376-380

Hodson, P.V., P.J. Kloepper-Sams, K.R. Munkittrick, W.L. Lockhart, D.A. Metner, L. Luxon, I.R. Smith, M.M. Gagnon, M. Servos & J.F. Payne, 1991a. Protocols for measuring mixed function oxygenases of fish liver. Can. Tech. Rep. Fish. Aquat. Sci. 1829, pp. 49

Hodson, P.V., D. Bussieres, M.M. Gagnon, J.J. Dodson, C.M. Couillard & J.C. Carey, 1991b. Review of biochemical, physiological, pathological and population responses of white sucker *(Catostomus commersoni)* to BKME in the St. Maurice river, Quebec. In: Environmental fate and effects of bleached pulp mill effluents; A. Södergren (ed). SEPA report 4031, Stockholm, Sweden, pp. 261-269

Husøy, A.M., Willis, M.J., Myer, M.S., Collier, T.K. & A. Goksøyr. 1993. Immunohistochemical localization of cytochrome P450 1A1 in different organs of Atlantic cod (*Gadus morhua*). Sci. Tot. Environ., (in press)

Klotz, A.V., J.J. Stegeman & C. Walsh, 1984. An alternative 7-ethoxyresorufin-O-deethylase activity assay: a continuous visible spectrophotometric method for measurement of cytochrome P-450 monooxygenase activity. Anal. Biochem. 140: 138-145

Lindström-Seppä, P., U. Koivusaari, O. Hänninen & H. Pyysalo, 1985. Cytochrome P-450 and monooxygenase activities in the biomonitoring of aquatic environment. Pharmazie, 40: 232-234

Lindström-Seppä, P. & A. Oikari, 1990. Biotransformation activities of feral fish in waters receiving bleached pulp mill effluents. Environ. Toxicol. Chem. 9: 1415-1424

Lowry, O.H., N.J. Rosebrough, A.L. Farr & R.J. Randall, 1951. Protein measurement with the Folin phenol reagent. J. Biol. Chem. 193: 265-275

Munkittrick, K.R., C.B. Portt, G.J. van der Kraak, I.R. Smith & D.A Rokosh, 1991. Impact of bleached kraft mill effluent on population characteristics, liver MFO activity and serum steroid levels of a Lake Superior white sucker *(Catostomus commersoni)* population. Can. J. Fish. Aquat. Sci. 48: 1371-1380

Nebert, D.W., D.R. Nelson, M.J. Coon, R.W. Estabrook, R. Feyereisen, Y. Fujii-Kuriyama, F.J. Gonzalez, F.P. Guengerich, I.C. Gunsalus, E.F. Johnson, J.C. Loper, R. Sato, M.R. Waterman & D.J. Waxman, 1991. The P450 superfamily: Update on new sequences gene mapping, and recommended nomenclature. DNA Cell Biol. 10: 1-14

Payne, J.F. & W.R. Penrose, 1975. Induction of aryl hydrocarbon (benzo(a)pyrene)hydroxylase in fish by petroleum. Bull. Environ. Cont. Toxicol. 14: 112-116

Payne, J.F., 1984. Mixed-function oxygenases in biological monitoring programs: review of potential usage in different phyla of aquatic animals. In: Ecotoxicological testing for the marine environment; G. Persone, E. Jasper & C. Claus (eds). Vol. I. State University of Ghent and Institute for Marine Scientific Research, Bredene, Belgium, pp. 625-655

Payne, J.F., L.L. Fancey, A.D. Rahimtula & E.L. Porter, 1987. Review and perspective on the use of mixed-function oxygenase enzymes in biological monitoring. Comp. Biochem. Physiol. 86(C): 223-245

Pohl, R.J. & J.R. Fouts, 1980. A rapid method for assaying the metabolism of 7-ethoxyresorufin by subcellular fractions. Anal. Biochem. 107: 150-155

Porter, T.D. & M.J. Coon, 1991. Cytochrome P-450: multiplicity of isoforms, substrates, and catalytic and regulatory mechanisms. J. Biol. Chem. 266: 13469-13472

Renton, K.W. & R.F. Addison, 1992. Hepatic microsomal mono-oxygenase activity and P450IA mRNA in North Sea dab *Limanda limanda* from contaminated sites. Mar. Ecol. Prog. Ser. 91: 65-69

Skaare, J.U., E. Gram Jensen, A. Goksøyr & E. Egaas, 1991. Response of xenobiotic metabolizing enzymes of rainbow trout *(Oncorhynchus mykiss)* to the mono-ortho substituted polychlorinated PCB congener 2,3',4,4',5-pentachlorobiphenyl, PCB-118, detected by enzyme activities and immunochemical methods. Arch. Environ. Contam. Toxicol. 20: 349-352

Smolowitz, R.M., M.E. Hahn & J.J. Stegeman, 1991. Immunohistochemical localization of cytochrome P-4501A1 induced by 3,3',4,4'-tetrachlorobiphenyl and by 2,3,7,8-tetrachlorodibenzofuran in liver and extra hepatic tissues of the teleost *Stenotomus chrysops* (Scup). Drug Metab. Disp. 19: 113-119

Södergren, A., 1989. Biological effects of bleached pulp mill effluents. National Swedish Environmental Protection Board, Report 3558, Stockholm, pp. 139

Södergren, A. (ed), 1991. Proceedings: Environmental fate and effects of bleached pulp mill effluents. Swedish Environment Protection Agency, report 4031, Stockholm, Sweden, pp. 394

Stebbing, A.R.D. & V. Dethlefsen, 1992. Introduction of the Bremerhaven Workshop on biological effects of contaminants. Mar. Ecol. Prog. Ser. 91: 1-8

Stegeman, J.J., 1981. Polynuclear aromatic hydrocarbons and their metabolism in the marine environment. In: Polycyclic hydrocarbons and cancer; P.O.P. Ts'o (ed). Academic Press, New York, pp. 1-60

Stegeman, J.J., P.J. Kloepper-Sams & J.W. Farrington, 1986. Monooxygenase induction and chlorobiphenyls in the deep-sea fish *Coryphanoides armatus*. Science, 231: 1287-1289

Stegeman, J.J., 1989. Cytochrome P-450 forms in fish: catalytic, immunological and sequence similarities. Xenobiotics, 19: 1093-1110

Stegeman, J.J., B.R. Woodin & A. Goksøyr, 1988. Apparent cytochrome P-450 induction as an indication of exposure to environmental chemicals in the flatfish *Platichthys flesus*. Mar. Ecol. Prog. Ser. 46: 55-60

Stegeman, J.J. & J.J. Lech, 1991. Cytochrome P-450 monooxygenase systems in aquatic species: Carcinogen metabolism and biomarkers for carcinogen and pollutant exposure. Environ. Health Perspect. 90: 101-109

Stegeman, J.J., M. Brouwer, R.T. Di Guilio, L. Förlin, B.A. Fowler, B.M. Sanders & P.A. van Veld, 1992. Molecular responses to environmental contamination: Enzyme and protein systems as indicators of chemical exposure and effects. In: Biomarkers: biochemical, physiological, and histological markers of anthropogenic stress; R.J. Hugget, R.A. Kimerle, P.M. Mehrle & H.L. Bergman (eds). SETAC Special Publications Series, Lewis Publishers, Boca Raton, Fl, pp. 235-335

Towbin, H., T. Staehelin & J. Gordon, 1979. Electrophoretic transfer of proteins from polyacrylamide gels to nitrocellulose sheets: procedure and some applications. Proc. Nat. Acad. Sci. USA, 76: 4350-4354

van der Weiden, M.E.J., M. Celander, W. Seinen, M. van den Berg, A. Goksøyr & L. Förlin, 1992. The use of immunochemical determination of P450IA1 and EROD activity assay in rainbow trout and carp treated with 2,3,7,8-TCDD. Mar. Environ. Res. 34: 215-219

Varanasi, U. (ed), 1989. Metabolism of polycyclic aromatic hydrocarbons in the aquatic environment. CRC Press, Boca Raton, pp. 341

Vindimian, E. & J. Garric, 1989. Freshwater fish cytochrome P-450-dependent enzymatic activities: a chemical pollution indicator. Ecotox. Environ. Safety, 18: 277-285

Williams, D.E. & D.R. Buhler, 1984. Benzo[*a*]pyrene-hydroxylase catalyzed by purified isoenzymes of cytochrome P-450 from ß-naphthoflavone-fed rainbow trout. Biochem. Pharmacol. 33: 3743-3753

Appendix 6

Measurement of the P450 1a system

Sampling and storage

Sacrifice the fish as quickly as possible. Dissect tissues carefully and avoid rupturing the gall bladder (bile may contain P450 inhibitors). All procedures must be performed at 1-4 °C (in a cold room or, alternatively, keep all buffers, homogenizer and samples on ice). Remove tissue immediately and transfer it into a beaker on ice. Remove excess blood from the tissues by washing once or twice in homogenizing buffer. If the fish are killed in the field without access to centrifuges, the tissues can be frozen in liquid nitrogen until further processing in the laboratory can take place. The tissues (1-2 gram pieces) should be frozen in liquid nitrogen (freezing on dry ice is not sufficient to retain the activity) and stored in liquid nitrogen until further processed. Addition of glycerol may give increased protection during freezing. The monooxygenase activity is not equally distributed in fish liver. Therefore, if the whole liver is not used for enzyme analysis, the liver should be minced and a subsample of the mix should be further processed. This sample can then represent the average activity in the liver. After homogenization and centrifugation the enzyme containing fractions (postmitochondrial supernatant (PMS) or microsomal fraction) can be stored frozen until analysis. Glycerol should then be added in order to increase protection during freezing. It is recommended that they be frozen in liquid nitrogen and stored in a -80 °C freezer.

Preparation of a postmitochondrial supernatant (PMS) and microsomes

There are many variations in methods for preparation of postmitochondrial supernatant (PMS) and microsomes and for measuring EROD activity and immunodetection of P450 1A. The methods briefly presented below have been used in our laboratories and have proven useful in field studies on fish (Förlin, 1980). The PMS, sometimes called the S-9 fraction, is the supernatant generated by centrifugation of a liver homogenate at 9-12.000 x g. The microsomes are precipitated by a 100-150.000 x g centrifugation of the PMS. When only PMS fractions are prepared from the sample, glycerol is recommended for addition to the homogenizing buffer.

Weigh the liver sample, add cold buffer and finely chop the tissue with scissors. If the tissue has been frozen it must be thawed before adding the buffer. Usually a phosphate buffer is used and, sometimes, antioxidants (*e.g.* DDT) and EDTA are added in order to retain enzyme activity. Homogenize gently in a tissue homogenizer (Potter-Elvehjem type) with a motor-driven teflon pestle. Centrifuge at 9-12.000 x g for 20 minutes at 1-4 °C (optimal speed and time must be tested for each new species). Separate supernatant (PMS) and pellet (and sometimes a floating lipid layer) carefully. The PMS produced in this step may be analyzed directly. The microsomal fraction is obtained by precipitation of PMS at 100-150.000 x g (ultracentrifugation) for 60 minutes (4 °C) (optimal speed and time should be tested for each new species). After centrifugation, carefully dispose cytosol fraction (supernatant). Suspend the pellet again in cold buffer to a protein concentration of 5-15 mg protein/ml (normally 1:1 buffer volume:tissue weight). Store the microsomes on ice and proceed to the enzyme assay which should be performed within a few hours. When assays are to be performed later the microsomes should be stored frozen. Freeze the microsomes (in appropriate aliquots) in liquid nitrogen and store in a -80 °C freezer.

EROD: ethoxyresorufin-O-deethylase - spectrofluorometric analysis

There are many variations in methods for measuring the O-deethylation of 7-ethoxyresorufin. The method presented briefly below is based on Burke & Mayer (1974) and on results and experiences obtained during the ICES-IOC EROD Intercalibration Workshop in Aberdeen, Scotland, September 1991.
The EROD activity analysis can be divided into three steps:
preparation of enzymes, incubation and detection. For further details on enzyme preparation the reader is referred to the previous section. The incubation mixture contains 7-ethoxyresorufin (25 nM), PMS or microsomes (approximately 50-500 $\mu g.ml^{-1}$), NADPH (50 μM) in an appropriate buffer (pH 7.0-8.5). Total incubation volume is 2 ml. pH optimum may vary considerably between fish species. pH 8.0 is suitable for *e.g.* rainbow trout (*Oncorhyncus mykiss*), perch (*Perca fluviatilis*) and cod (*Gadus morhua*). Ethoxyresorufin is dissolved in DMSO (dimethyl sulphoxide) and the stock solution (5 μM) can be stored at room temperature in lightproof bottles. Calculate the exact concentration of the stock solution by measuring the absorbance at 482 nm. The extinction coefficient for 7-ethoxyresorufin is 22.5 $cm^2.mmol^{-1}$. Stock solution of NADPH (10 mM) can be stored in a refrigerator for two weeks.
The reaction is started by the addition of NADPH. The rate of the reaction is measured as the rate of resorufin production and is recorded as changes in fluorescence using a spectrofluorimeter. Measure over 2-3 minutes. Leaving the recorder on, add 10 μl (2 pmol) resorufin standard to the incubation cuvette and record the increased fluorescence. Protein content of the tissue fraction must be determined. The Bradford or Lowry assay, using bovine serum albumin, is recommended (Bradford, 1976; Lowry *et al.*, 1951).
Calibration is achieved by either internal or external resorufin standards. Rhodamine B can be used as an external standard. Low purity of commercial resorufin makes it important to check the concentration of the standard. Dissolve resorufin in DMSO. This stock solution can be stored at room temperature in lightproof bottles. Calculate the exact concentration of stock resorufin solution (0.25 $nmol.ml^{-1}$) by measuring absorbance at 572 nm. The extinction coefficient for resorufin is 73.2 $mM^{-1}.cm^{-1}$. Observe that the Rhodamine B has a three times higher fluorescence than resorufin. The Rhodamine B standard is more stable than the resorufin standard, and can also be used as a secondary standard for the resorufin standard. The two standards should be compared a few times per year.
Use the linear change in fluorescence for calculation of EROD activity. It is important to include the resorufin standard in each sample because high protein concentration in the samples will quench the fluorescence.

Immunodetection of cytochrome P450: immunoblotting/Western blotting

Western blotting technique can be divided into six steps: preparation of enzymes, gel electrophoresis, protein blotting, blocking, antibody incubations and detection. The method briefly presented below follows the method of Towbin *et al.* (1979). For further details about enzyme preparations the reader is referred to the previous section.

In SDS-polyacrylamide gel electrophoresis (SDS-PAGE), protein samples are separated in non-gradient gels of 7-10% polyacrylamide gel. To solubilize protein aggregates and disulphide bonds, and to give proteins a similar charge/surface ratio, samples must be treated with SDS (sodium dodecyl sulphate) and 2-mercaptoethanol. After gel electrophoresis, the proteins in the gel are transferred to a nitrocellulose membrane, protein blotting. The gel is then packed into a sandwich system with close contact with nitrocellulose membrane for electrophoretic transfer of proteins to the membrane. After protein blotting, nitrocellulose membrane is incubated in blocking buffer (containing *e.g.* 5% non-fat dry milk) in order to block sites on the membrane not occupied by sample protein.

For immunodetection the nitrocellulose membranes are incubated first with primary antibody (*e.g.* rabbit anti-fish P450 1A1 IgG), washed, and then incubated with labelled secondary antibody (*e.g.* commercial Goat anti rabbit IgG conjugated with horseradish peroxidase, GAR-HRP). Cross reacting bands are stained using a commercial kit.

ELISA: enzyme-linked immunosorbent assay

There is a wide range of methods for detecting and quantifying antigens and antibodies. Different types of immunoassays are available. Indirect P450 1A1 ELISA, recently described by Goksøyr (1991) and Celander & Förlin (1991) is briefly presented below. ELISA techniques can be divided into five steps: preparation of enzymes, coating of wells, blocking, incubation with antibodies and detection. For further details on enzyme preparation the reader is referred to the previous section. Samples are diluted with buffer to a concentration of approximately 10-100 $\mu g.ml^{-1}$. Mix well and keep on ice.

Coating: make a diagram of a 96-well microplate, avoid the wells along the edges and mark 3 wells for each sample (triplicates). Fill each well with 100 µl. Put the buffer into all wells not containing samples. The plates are wrapped in aluminium foil and placed at 4 °C overnight (or at 37 °C for 1 hour). This will allow the proteins in the sample to adsorb to the well surface ("coating").

Blocking: in all wells, including one blank row along the edge, 200 µl of a blocking solution containing 5% non-fat powdered milk is added in order to block sites on well surfaces not occupied by sample proteins. Leave for 45-60 minutes at room temperature (or for 30 minutes at 37 °C).

Incubation with antibodies: for the primary antibody add 100 µl primary antibody solution (*e.g.* Rabbit anti-cod P450 1A1 IgG 1:300 diluted in blocking solution (Goksøyr, 1991) or Rabbit anti-perch P450 1A1 serum 1:1000 (Celander & Förlin, 1991)). Wrap the microplates in aluminium foil and place in an incubator at 37 °C for 1 hour (or overnight at 4 °C).

Labelled secondary antibody: add 100 µl labelled secondary reagent (GAR-HRP 1:2-3000 diluted in blocking solution). Wrap in aluminium foil and place in an incubator at 37 °C for 1 hour.

Between each step and after the last step the wells must be washed several times with buffer. Washing for 3 x 3 minutes with Tris or phosphate buffered saline solution with 0.05% Tween 20 is recommended.

Detection: prepare HRP colour reagent solution (commercial kit) just prior to use and then add 100 µl to each well. Let the colour reaction develop for 5-15 minutes (depending on the intensity of the reaction). Stop the reaction by adding 2 M H_2SO_4 or 2% oxalic acid. Read absorbance in the plate wells on a microplate reader at 405 nm. Subtract blanks (take the mean of the blank row) and the results can be expressed in absorbance units.

Immunohistochemical detection of cytochrome P450

Cell staining techniques can be divided into four steps: tissue preparation, fixation, addition of antibodies, and detection. There are many immuno-enzymatic staining methods which can be used to localize antigens. Fixation, antibody titre and dilutions, incubation time and temperature are critical factors for optimal immunohistochemical staining (see also Yevich & Yevich, this volume; Smolowitz *et al.*, 1991; Husøy *et al.*, 1993).

The tissue is sliced into thin sections and placed in a container containing fixative (10% buffered formalin has been found to be suitable for immunohistochemical localization of P450 1A1). During the excision of tissue samples, handle tissue with care and avoid mechanical damage such as squeezing or stretching.

Immediate fixation of tissue sections following dissection is imperative since autolysis begins rapidly following death. Fixation is best achieved in a fluid to tissue volume of 20:1. Tissue should be agitated in a container of fixative periodically (2-3 times) throughout the fixation period in order to ensure optimal penetration of the fixative. After fixation, the tissue is embedded in paraffin or in alternative embedding media. Different commercial staining kits are available. See also Yevich & Yevich (this Volume) for practical details on histological preparation techniques.

The general procedure for detection of cytochrome P450is as follows:

- deparaffinization of sections;
- blocking of endogenous peroxidase activity;
- incubation with normal (non-immune) goat serum;
- incubation with primary antibody (rabbit);
- incubation with labelled secondary antibodies (goat-anti rabbit);
- development;
- counterstaining and mounting.

Chapter 7

Metallothioneins as Indicators of Trace Metal Pollution

Stephen G. George[1] and **Per-Erik Olsson**[2]

[1] NERC Unit of Aquatic Biochemistry, University of Stirling, Sterling FK9 4LA, Scotland
[2] Department of Zoophysiology, University of Göteborg, 413 90 Göteborg, Sweden

ABSTRACT

From the data discussed we conclude that from both laboratory (calibration) experiments and field validation, induction of metallothionein synthesis in fish by Cd, Cu, Zn and Hg provides a suitable monitoring procedure to assess biological availability and impact of these metals in the aquatic environment and that it is capable of providing both an early warning of exposure and assessment of the area of global impact. The technique has been satisfactorily validated for a number of finfish species and so far preliminary studies indicate that *Mytilus* sp. is the only promising invertebrate candidate for monitoring use. A number of assay methods have been reviewed.

1. WHY USE A SPECIFIC BIOCHEMICAL INDICATOR FOR HEAVY METAL IMPACT ASSESSMENT ?

Ever since the Hg and Cd pollution tragedies of the 1950's and 1960's in Japan there has been considerable concern about contamination of the aquatic environment by heavy metals from industrial, mineral mining and processing sources and ensuing ecosystem damage or contamination of the food chain (Friberg *et al.*, 1974; Takeuchi, 1972). Great efforts have been made to determine fluxes of these metals in aquatic ecosystems, their speciation in natural waters and sediments (the ultimate sinks of most contaminants), to assess their bioavailability and to develop methods for determination of their environmental impact (Förstner & Wittmann, 1979).

These studies have clearly shown that gross analyses of water and sediments for toxic metals cannot be used to predict or assess environmental impact since the major proportion of these metals are present in the anoxic layer of sediments either in mineralised forms or strongly bound to insoluble organic residues (Förstner & Wittmann, 1979) and thus are not directly available to biota.

Biomonitoring of Coastal Waters and Estuaries. Edited by Kees J.M. Kramer.

Consequently assessments of the 'bioavailability' and 'biological impact' of these pollutant metals are of paramount importance and thus the concept of 'bioindicators' has arisen. It has been known for many years that biota, particularly the benthic invertebrates, are capable of accumulating concentrations of metals many thousand times greater than those present in their surrounding water and therefore many determinations of the accumulation of these metals in animals from supposedly polluted sites have been made. The concept of 'indicator organisms' such as filter feeding bivalves, particularly mussels, for suspended metals has received widespread acceptance (Goldberg, 1975).

Laboratory studies carried out over the past 20 years in our own and numerous other laboratories have established fluxes and accumulation rates of essential and non-essential metals in invertebrate species, as well as effects of environmental variables (metal speciation, salinity and temperature) which are essential for supporting such an approach to environmental monitoring (*e.g.* Bryan *et al.*, 1980; George, 1982). Through numerous field studies differences between species, effects of age (size) and life stages, including sexual cycles and moulting have been determined and analytical procedures standardised to take into account these factors and also to determine contamination, *e.g.* the purging of sediments from the gut (Boyden, 1977; Bryan *et al.*, 1980). These laboratory and field studies have also highlighted many difficulties in selecting suitable invertebrate organisms for use in biomonitoring. Many marine invertebrates, including commonly occurring species such as worms (both oligo- and polychaete), molluscs (including clams, mussels, oysters, scallops and winkles) and crustaceans (crabs and lobsters), possess specialised mechanisms for immobilising and accumulating excess metals in specific organs, cell types or in specific metalloproteins (George, 1982; George *et al.*, 1979; Mason & Nott, 1981; Thomson *et al.*, 1985). These accumulated metals may be mobilised by metabolic processes or insoluble deposits may be voided at certain times, such as reproduction, moulting or infection (George, 1982) and thus serious interpretative difficulties can arise if gross chemical analyses of organisms which measure all forms of metal are used to quantitatively evaluate environmental metal impact.

Alternative criteria for assessment of pollutant metal bioavailability have been sought and in recent years there has been considerable interest in developing physiological and biochemical methods of evaluating sub-lethal pollutant impact (Neff, 1985). For many years a 'false start' was made in this area by investigation of serum parameters reflective of cellular necrosis or protracted disturbances of ion or hormonal balance such as used in human studies. The fact that these are merely non-invasive methods which are used instead of gross pathology was overlooked. Where 'fatal' sampling can be used, alternative approaches are possible whereby primary responses are measured. Two classical parameters, which involve upregulation of gene expression in response to environmental pollutants, are induction of mixed function oxygenases by polyaromatic hydrocarbons and polyhalogenated biphenyls (see Förlin *et al.*, this Volume) and the induction of metallothionein by the heavy metals Cd, Cu, Hg and Zn (Hamilton & Mehrle, 1986; Haux & Förlin, 1988; Neff, 1985; Sulaiman *et al.*, 1991; Zafarullah *et al.*, 1989a). The remainder of this chapter will concentrate upon use of metallothionein as an indicator tool for assessing heavy metal impact and bioavailability.

2. METALLOTHIONEIN, AN INDUCIBLE METAL-BINDING PROTEIN

Metallothionein (MT) *i.e.* a metallo-derivate of the sulphur-rich protein, thionein, was first isolated by Vallee's group as a Cd-binding protein from equine renal cortex. It was suggested that the role of MT was to protect against the toxic effects of cadmium exposure and this was supported by the finding that large amounts of MT were formed in the liver of rabbits in response to repeated doses of Cd^{2+} which led to the proposal that once thionein synthesis had been initiated the animal would be resistant to Cd and consequently would tolerate higher doses of this toxic metal (Piscator, 1964). However, it soon became apparent that other metals, Cu and Zn in particular, could also induce MT synthesis. Although the exact function(s) of MT are still being debated, it is generally accepted that its major function is related to metabolism of the essential trace metals, Cu and Zn. Whilst most studies of MT have focused on liver and kidney, which are the primary sites for cadmium accumulation, MT has also been detected in many other organs and in various cultured cell types. MT has also been reported in other eukaryotes, such as invertebrates, plants and microorganisms (Hamer, 1986; Kägi & Nordberg, 1979; Karin, 1985; Kojima & Kägi, 1978; Webb, 1979; Zafarullah *et al.*, 1989a).

The structure and metal binding properties of MT have been extensively studied. MT's generally have a molecular weight of about 6000-7000 Daltons and consist of some 60 to 62 amino acids. The protein is unique in containing some 20 cysteine residues (*i.e.* 33%) which do not form disulphide bridges. Thus far, four different charge isoforms have been identified in mammals either at the protein or the genomic level. Isoforms I, II and IV consist of 60 to 62 amino acids. Isoform III differs from the other isoforms in that it consists of 68 amino acids, with a one amino acid insertion in the N-terminal and a six amino acid insertion in the C-terminal of the protein. While MTI & II are expressed in many cell types, MTIII is specifically expressed in astrocytes of the brain where it may be concerned with growth regulation (Uchida *et al.*, 1992); isoform IV only appears to be expressed in ectodermal tissues, tongue and skin (Palmiter, unpublished). In all MT isoforms the sequence position and number of cysteine residues are highly conserve within the protein. MT binds 6-7 g.atoms of Cd or Zn by tetrahedral co-ordination with the cysteine residues. MT is usually saturated with Zn and this can be displaced by varying amounts of Cd from a $CdZn_6MT$ to a Cd_7MT. The metals are organized in two clusters, designated cluster A and cluster B. Cluster A holds four metal ions, while cluster B holds three metal ions. By proteolytic cleavage of MT it has been shown that cluster B is contained in the amino terminal β-domain extending from amino acid 1 to 30, while cluster A is contained within the α-domain extending from amino acid 31 to 61 (Figure 1). Both domains are globular, with diameters of 1.5-2.0 nm and are linked by

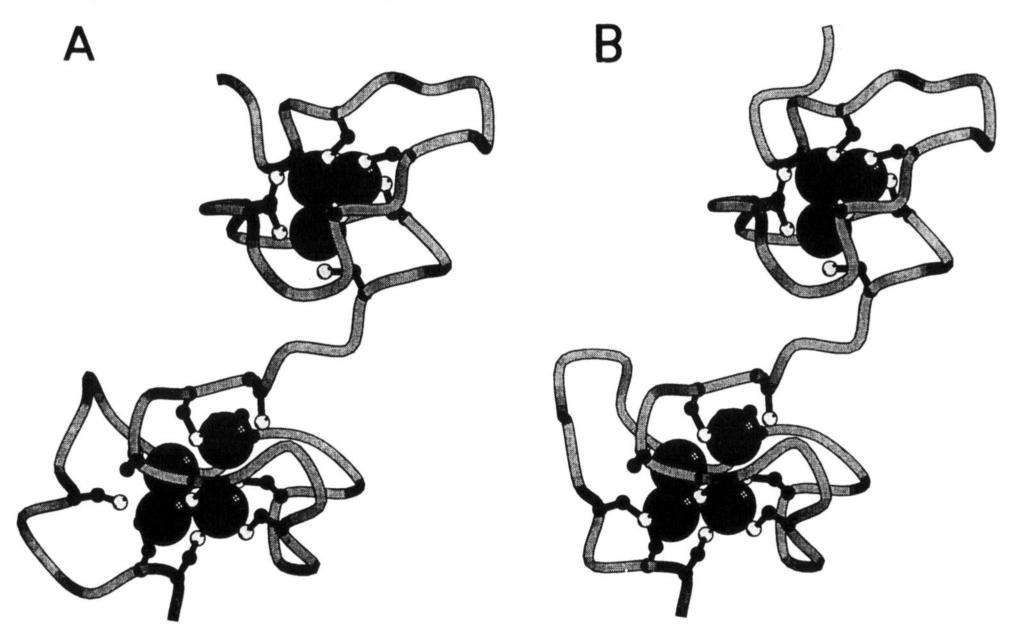

Figure 1. Structure of Cd/Zn-metallothioneins showing binding of metal atoms to cysteine residues and differences in structures of mammalian and fish proteins. a) Plaice MT modelled from rat MT crystal structure (Brookhaven Database data) by dr. P. Hodgson; b) Rat MT

residues 30 and 31 to form a prolate ellipsoid. Thus on size exclusion chromatography the protein behaves as though it has an apparent molecular weight of ca. 12-15,000 Daltons. Cu and Hg form tetragonal complexes and thus 11-12 g.atoms of these metals are bound by MT, the folding of CuMT and HgMT are not yet known.

2.1. Teleost metallothionein

2.1.1. Structure of fish MT's

The presence of MT in a teleost was first indicated in the goldfish (*Carassius auratus* L.) by Marafante *et al.* (1972) and then characterised in copper rock fish (*Sebastes caurinus*), eel (*Anguilla anguilla*) and plaice (*Pleuronectes platessa*) (Noel-Lambot *et al.*, 1978; Olafson & Thompson, 1974; Overnell & Coombs, 1979). Since these initial reports MT has been purified and characterised from many fish species (reviewed by George, 1990; Hogstrand & Haux, 1991a; Roesijadi, 1992; Zafarullah *et al.*, 1989a). Curiously the number of isoforms present in fish appears to vary between species. In the marine flatfish, dover sole (*Solea solea*), plaice (*P. platessa*), turbot (*Scopthalmus maximus*), winter flounder (*Pseudopleuronectes americanus*), and in several freshwater fish, stone loach (*Noemachelius barbatulus*) and pike (*Esox lucius*), only single isoforms of MT have been detected. In other teleost species two isoforms have been identified, including carp (*Cyprinus carpio*), eel, perch (*Perca fluviatilis*), rainbow trout (*Onchorynchus mykiss*), scorpionfish (*Scorpaena guttata*) and skipjack tuna (*Katsuwonus pelamis*).
Due to their high cysteine content, MT's are notoriously difficult to sequence by conventional protein sequencing methods and it was not until molecular biological techniques were applied that full sequence data for piscine MT's became available. To date nucleotide sequences have been obtained for MTA and MTB of *O. mykiss*, (Bonham *et al.*, 1987) and MTI's of *P. americanus* (Chan *et al.*, 1989), *P. platessa* (Leaver & George, 1989), *E. lucius* and *N. barbatulus* (Kille *et al.*, 1992). These have shown that the structure of fish metallothioneins is highly conserved (Kille *et al.*, 1991). Most of the observed amino acid changes are conservative replacements with the exception of an additional Ala^{31} located in the flexible 'hinge' between the two metal-binding domains of rainbow trout MTA and all the cysteine residues are in perfect alignment, including the ante-penultimate cysteine residue at residue 55, which is repositioned relative to that found in mammalian MT's where it is at position 57. The first 7 residues at the N-terminal end of the protein appear to form a 'hook' on the head or β-domain which determines antigenicity (Kikuchi *et al.*, 1988). The lack of cross reactivity between fish and mammalian MT's and antibodies (Norey *et al.*, 1990b) is probably explained by the difference in the structure of this 'hook' between fish and mammalian MT's; in the former there are 3 residues prior to the first cysteine residue, in the latter there are four. Nevertheless, MT's from a wide variety of fish species are recognised by antisera raised against another piscine MT which has permitted the development of quantitative immunoassays based upon enzyme linked immunoassays, ELISA (Chattergee & Maiti, 1990; Norey *et al.*, 1990b), and on radio immunoassays, RIA (Hogstrand & Haux, 1990b).

2.1.2. Inducibility of teleost MT, exogenous and endogenous influences

In most species basal levels of hepatic Zn are sequestered by MT, the Zn and MT concentrations in the liver displaying a linear relationship (Figure 2a), however, the exact proportionalities vary with age and sex *e.g.* in *P. platessa* from non-contaminated environments basal levels of 15-30 μg $MT.g^{-1}$ are present and proportionalities of [MT] = 4.2 [Zn] - 25 were found in juvenile fish, [MT] = 8.5 [Zn] - 170 in males and [MT] = 9.2 [Zn] - 160 in females (Overnell *et al.*, 1988; George, 1989). Moreover these proportionalities also appear to vary between seasons and populations (George, unpublished). In salmonid livers the basal levels of Cu are very high (150-350 $\mu g.g^{-1}$) which is reflected by high levels of MT. Thus in *O.*

mykiss liver values of 140-240 μg MT.g^{-1} are normal (Olsson *et al.*, 1987) whilst in Atlantic salmon, *Salmo salar*, returning to non-polluted rivers, values of ca. 100 μg MT.g^{-1} have been reported (George *et al.*, 1992b).

Noel-Lambot *et al.* (1978) first demonstrated that Cd was bound to MT and increased the synthesis of MT in gills and liver of *A. anguilla.* Laboratory experiments performed on several fish species and cell lines have shown that teleost MT's are inducible by Cd, Cu, Zn and Hg; that the MT and MT mRNA levels increase in correlation with the dose of administered heavy metals and that they are bound to MT (Chan *et al.*, 1989; Chatergee & Maiti, 1991; George, 1989; George & Young, 1986; George *et al.*, 1992a; Olsson *et al.*, 1989a,b; 1990a; Zafarullah *et al.*, 1990). In *P. platessa*, injected intraperitoneally with Zn (Overnell *et al.*, 1987) or Cd (George, 1989) good correlations between metal and MT levels were obtained *viz.* [MT] μg.g^{-1} liver = 6.6 [Zn, μg.g^{-1}] - 112 and [MT] = 36 + 9.3 [Cd] respectively (Figure 2b,c). For higher doses of Cd, when metal overload and hepatotoxicity becomes apparent, this relationship deviates from linearity (Figure 2d; George, 1989). In *O. mykiss,* exposed to Cd in the water (200 μg.l^{-1}) the relationship was [MT] = 160 + 24.2 [Cd] (Olsson *et al.*, 1989a). Similar linear relationships have also been observed in feral *P. fluviatilis* and *O. mykiss* from contaminated areas (Roch *et al.*, 1982; Olsson & Haux, 1986; Hogstrand & Haux, 1991b; Hogstrand *et al.*, 1991).

Experimental determination of these dose/response relationships are important for calibration purposes when considering MT as a tool in environmental monitoring and as an early warning system. However, it must be noted that the basal levels of MT have been shown to vary with time of year, reproductive state, water temperature and developmental state in both *O. mykiss* and *P. platessa* (Olsson *et al.*, 1987, 1989b, 1990b; Overnell *et al.*, 1988; George, unpublished). In juvenile fish, seasonal variations in MT of 2-3 fold are most probably related to altered metabolism during acclimation to lowered water temperature, although effects of altered feeding or photoperiod may also be influential (Olsson, 1985; George unpublished, Olsson unpublished). This temperature effect has also been demonstrated in studies with isolated hepatocytes of *O. mykiss* (Hyllner *et al.*, 1989). Hepatic MT levels are also elevated during sexual maturation and spawning. In male fish, MT mRNA and protein levels rise about 2-fold, whilst in female fish much larger increases in MT protein levels of up to 7-fold are observed (Figure 3) at the cessation of the period of exogenous vitellogenesis prior to spawning (Olsson *et al.*, 1987; Overnell *et al.*, 1988; George, unpublished). Studies with isolated trout hepatocytes have indicated that this may be due to oestradiol-induced mobilisation of Zn and Cu for metabolic purposes (Hyllner *et al.*, 1989). Other studies with primary hepatocyte cultures and established fish cell lines have shown that the MT levels are also influenced by hormones such as glucocorticoids and progesterone, as well as by noradrenaline (Burgess *et al.*, 1993; George *et al.*, 1992a; Hyllner *et al.*, 1989; Olsson *et al.*, 1990a) and in another study it has been shown that capture stress when taking fish from the field could cause a small elevation in MT levels in striped mullet (*Mugil cephalus* L.) (Baer & Thomas, 1991). Thus while MT displays a dose-dependent induction by heavy metals, there are a variety of other conditions that will ultimately result in a relatively small induction of MT levels which must be taken into account when sampling fish and in interpretation of data.

A few fish species appear to be metal-tolerant (*e.g.* the freshwater stone loach, *N. barbatulus*, Norey *et al.*, 1990a). The mechanism of this tolerance is not yet known and as far as their use in monitoring is concerned, it is wise to avoid such species. Similarly a few fish (*e.g.* the white perch, *Morone americana,* and possibly the squirrelfish, *Holocentrus rufis)* appear to have a metabolic defect resulting in hepatic accumulation of Cu (Bunton *et al.*, 1987; Hogstrand & Haux, 1990a) which would also rule out their use in monitoring programmes. With these exceptions other competing systems for metal sequestration are not present in fish, the protein and gene structures of MT's from several species have been elucidated, we have quite a thorough knowledge of endogenous and exogenous controls and variations in

MT expression and reasonably thorough validation has been performed to accept use in monitoring programmes. In juvenile fish the humoural and temporal influences noted above produce only small, 2 to 3-fold, variations in MT levels and these may be reduced or avoided if measurements are not carried out during periods of rapidly changing seasonal variations in water temperature. Relatively large variations are found in sexually maturing fish, especially females, where hormonally-mediated mobilisation of metals occurs and thus it is recommended that determinations are not carried out during the period of sexual maturation and only juveniles are utilised.

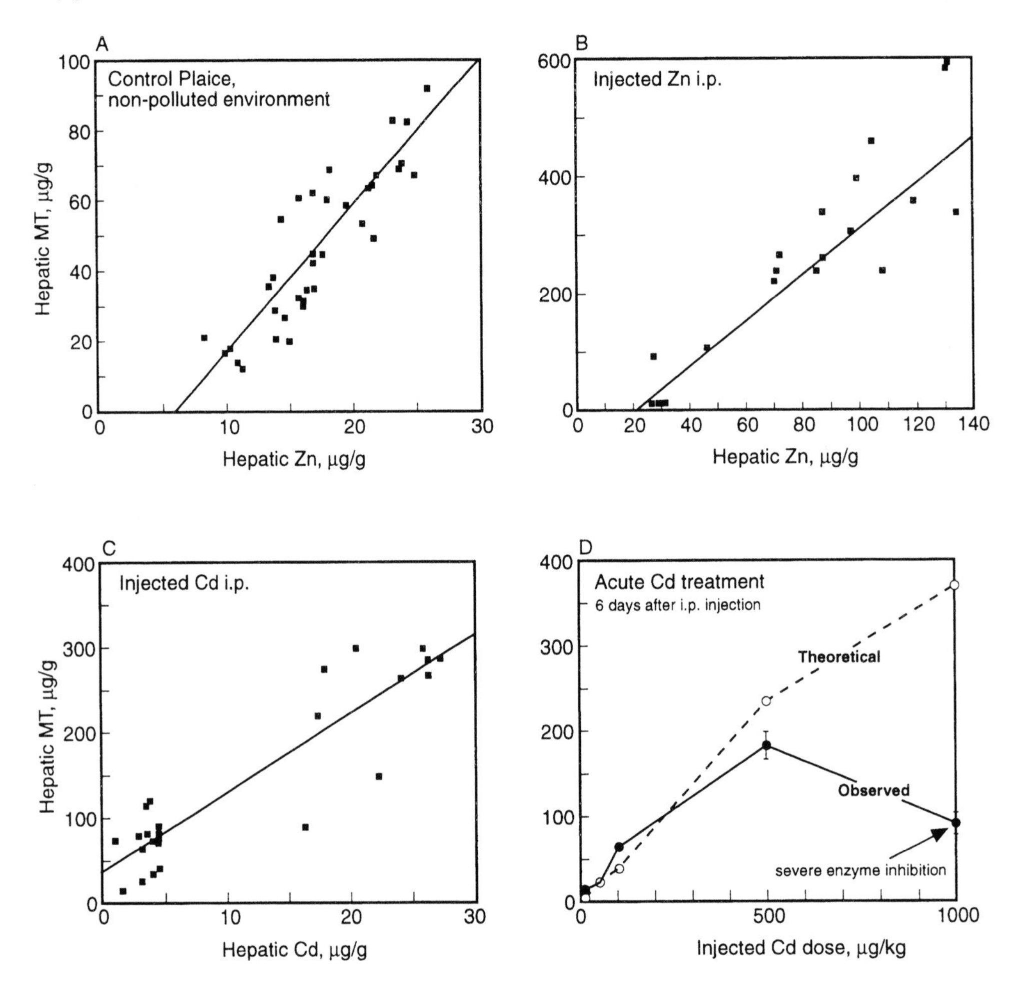

Figure 2. Proportionality between plaice (*Pleuronectes platessa*) liver MT content and metal concentrations. a) non metal-exposed fish; b) fish injected with Zn (data from: Overnell *et al.*, 1987); c) fish injected with subtoxic concentrations of Cd (data from: George *et al.*, 1992a); d) dose response to acute doses of Cd (data from: George *et al.*, 1992a)

2.2. Metal accumulation and invertebrate metallothioneins

As noted earlier, MT's are ubiquitous. Protein and gene cloning studies have demonstrated that the structure and metal-regulation of MT's are highly conserved throughout most Phyla. However, in many studies with invertebrates, less rigorous criteria have been applied to the identification and characterisation of MT's. The structure and mechanisms of regulation of invertebrate MT's are poorly understood and other systems may compete with MT for metal sequestration. Examples of the latter include:

- the occurrence of non MT Cd-binding proteins in many organisms;
- the presence of a Cu-containing respiratory pigment, haemocyanin in several genera;
- immobilisation of metals as inorganic precipitates in certain molluscs and in crustaceans, and
- accumulation of heavy metals as completely insoluble lipofuschin pigments in tertiary lysosomes of most Phyla (George, 1982; George & Pirie, 1980; Mason & Nott, 1981; Simkiss *et al.*, 1982; Stone & Overnell, 1985).

Thus where the proteins and induction response are inadequately characterised and competing pathways are present, the use of MT measurements in a particular species as a monitoring tool requires thorough experimental calibration before it can be validated.

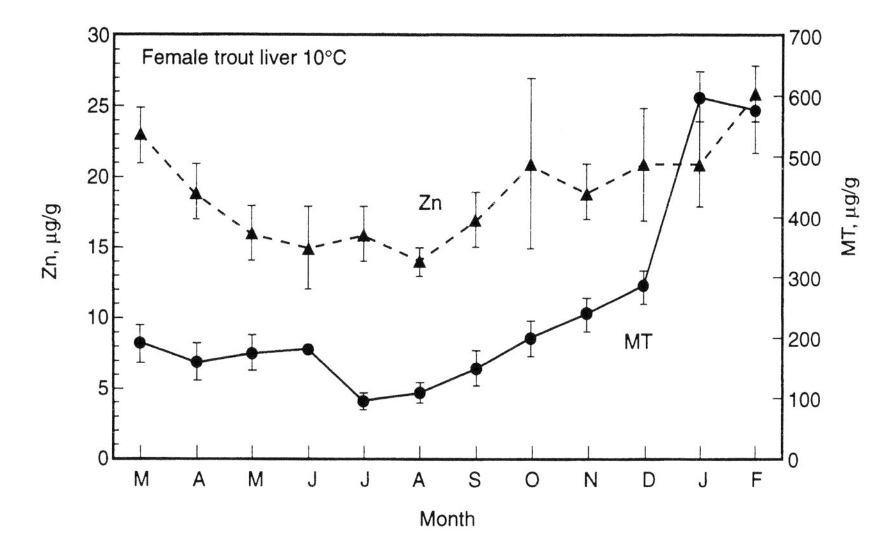

Figure 3. Annual variation in hepatic MT and Zn levels in adult female trout maintained at 10 °C (After: Olsson *et al.*, 1987)

2.2.1. Molluscs

In marine mussels (*Mytilus* sp.) there is a wealth of data for metal fluxes and detoxication mechanisms (see *e.g.* George, 1982; George & Viarengo, 1985). These fluxes vary between tissues. MT's are degraded in lysosomes and high concentrations of many metals can also accumulate in tertiary lysosomes which can lead to variations in half life and metal/MT content of tissues. *E.g.* Cu accumulates primarily in the digestive gland where lysosomes (containing Cu) are rapidly voided, resulting in a half life ($t_{1/2}$) for Cu of 6 days. On the other hand Cd and Zn are transported to the kidney and accumulate in this tissue; Cd has a $t_{1/2}$ of 60 days, whilst Zn has a $t_{1/2}$ of 300 days (George & Pirie, 1980; George & Viarengo, 1985). Metallothioneins have been purified from mussels (Frankenne *et al.*, 1980; George *et al.*, 1979), as many as 9 isoforms (both monomeric and dimeric) are present (Frazier *et al.*, 1985) and their amino acid sequences have been determined (Mackay *et al.*, 1990). The structures of the monomeric and dimeric isoforms differ, the monomeric MT's consist of 72 amino acids containing 21 cysteine residues whilst the dimeric subunits comprise 71 amino acids with a shift of the 6th cysteine from position 23 to 22 and an additional cysteine residue at position 55. Inducibility of mussel MT by Cd, Cu and Hg has been demonstrated (Kohler & Riisgård, 1982; Noel-Lambot, 1976; Pavicic *et al.*, 1993; Roesijadi, 1982; Viarengo *et al.*, 1980). Our data have also shown that these metals are probably transported to the lysosomes as metallothioneins and that the different tissue kinetics and metal accumulation can be explained by different rates of proteolysis of metal-thioneins, peroxidation to insoluble lipofuschin granules and lysosomal turnover (George & Viarengo, 1985). Roesijadi *et al.* (1988) raised an antiserum against the major dimeric charge isoform of HgMT from the mussel, *Mytilus edulis*, which displayed varying cross-reactivity against most of the isoforms and this was utilised in ELISA format to determine MT levels. In gills of control animals there was a basal level of 0.5 μg $MT.g^{-1}$ WW and on laboratory exposure to Cd (1-50 μg $Cd.l^{-1}$) or Hg (0.05-5 μg $Hg.l^{-1}$) for 28 days, levels increased in proportion to the metal concentrations in the tissue, although the sensitivities differed between metals. Exposure to 5 μg $Cu.l^{-1}$ increased MT concentrations but no increase upon exposure to 10 μg $Zn.l^{-1}$ was observed. These data are in accord with various chromatographic studies which have demonstrated inducibility of mussel MT by Cd, Cu and Hg but not Zn. Current data would therefore indicate that use of MT measurements in mussels as an indicator of Cd, Cu and Hg exposure but not Zn exposure is feasible and this species is a probable candidate for further investigation of the validity of MT as a monitoring tool.

Much work has also been carried out on metal uptakes in oysters, (*Crassostrea* or *Ostrea* spp.) particularly since oysters rather than mussels are found in most areas of the eastern coast of the USA. Oysters do possess a metallothionein of 75 amino acids containing 21 cysteine residues which is synthesised in response to Cd exposure (Frazier & George, 1983; Roesijadi *et al.*, 1991), however, they also accumulate very high concentrations (up to 5% DW) of Cu and Zn in their blood cells (Thomson *et al.*, 1985) and endogenous Cu can displace Cd from MT (Frazier & George, 1983). The types of metal-containing blood cells (Cu-, Cu/Zn- and Zn-containing) differs between oyster species (George *et al.*, 1984). Thus in our view, the occurrence of naturally high metal levels, metal displacement reactions and competition between sequestration mechanisms will probably rule out this species for biomonitoring use.

Investigation of metal-binding ligands in scallops, (*Pecten* sp.) which often contain elevated metal levels, have shown that they possess non-MT Cd-binding proteins (Stone *et al.*, 1986) and sequester Zn in phosphate granules in the kidneys (George *et al.*, 1980). Non-MT Cd-binding proteins which are not apparently Cd-inducible have also been identified in a gastropod mollusc, the whelk *Buccinum tenuissimum* (Dohi *et al.*, 1983). Clearly these animals are unsuitable for monitoring purposes.

Metal uptake rates and sequestration mechanisms in gastropod molluscs such as the littoral herbivorous periwinkle, *Littorina* sp, have been extensively studied (Mason & Nott, 1981; Mason *et al.*, 1984; Mason & Storms, 1993) and recently metallothionein-like metal-binding proteins have been identified (Langston *et al.*, 1989). The distribution of metals amongst various cytoplasmic ligands appears to be complex and these animals also contain haemocyanin. Thus in the absence of extensive studies of metal-interactions and analytical techniques, use of MT in *Littorina* sp. as an indicator of metal contamination is also probably premature.

*2.2.2. **Crustacea***

MT from the crab, *Scylla serrata,* was amongst one of the first non-mammalian MT's to be purified and characterised (Olafson *et al.*, 1979). The amino acid sequences of two isoforms from this species show that they comprise of 57-58 amino acids containing 17 cysteine residues. Sequence data and NMR studies indicate slightly different metal-binding properties from vertebrate thioneins and, as expected from the amino acid sequence, cross immunoreactivity of crab and fish or mammalian MT's is poor. The presence of MT in several other crab and lobster species has also been demonstrated (see Roesijadi 1992 for references). These studies and other independent studies on trace metal metabolism, accumulation and cellular localisation in Crustacea have demonstrated a complexity of inter-relationships between metals bound to MT or present in other pools such as the respiratory pigment haemocyanin, phosphate granules and tertiary lysosomes as well as the occurrence of very pronounced cellular and tissue effects with moult cycle (Figure 4). Thus in our opinion use of MT in Crustacea for environmental monitoring would be fraught with pitfalls.

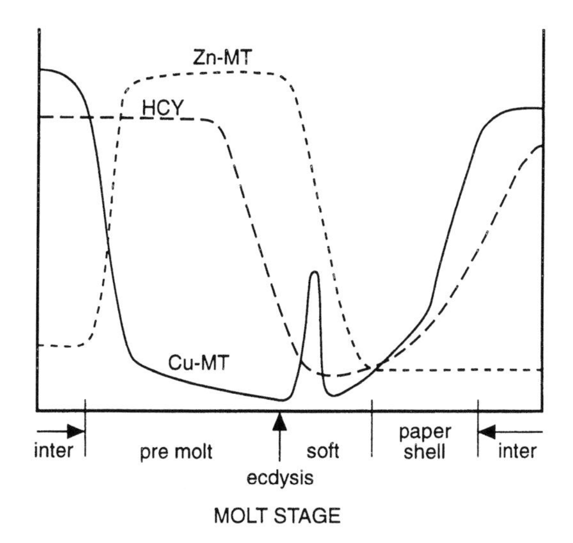

Figure 4. Dynamics of metals, haemocyanin and MT in crab digestive gland at moulting (After: Engel & Brouwer, 1993)

3. SELECTION OF 'SENTINEL' SPECIES FOR MONITORING USE AND METALLOTHIONEIN ASSAY METHODS

For teleosts where the biotic and abiotic influences on regulation of MT expression and effects of these factors on MT levels are quite well understood, measurement of MT may prove valuable for assessment of heavy metal bioavailability and as an indicator of heavy-metal pollution. In view of the difficulties noted above, the use of invertebrates must be viewed with great caution until sufficient validation has been carried out.

When designing field experiments a number of factors must be taken into account, the more important being:

- unless the inter-relationships are fully understood, the selected species should not have an alternative system which competes with MT for sequestration of the metals;
- the species should contain low 'natural' levels of metals and/or metallothionein; *e.g.* MT levels are naturally high in Atlantic salmon, *S. salar* eating a crustacean diet (George *et al.*, 1992b), in crabs and lobsters from seemingly unpolluted areas (Overnell & Trewhella, 1979) and in certain fish with atypical metal metabolism such as *Morone americana* (Bunton *et al.*, 1987) and *Holocentrus rufis* (Hogstrand & Haux, 1990a);
- the species should be sufficiently territorial (or sessile for invertebrates *e.g.* mussels) so that past pollutant exposure can be assigned to the geographic area of sampling. Furthermore, if sampling is carried out along a suspected pollution gradient then the point source can be identified and assay intercalibration problems are also avoided. Use of the 'gradient approach' is also useful in assessing potential damage to the ecosystem if a response over a wide geographic area is observed;
- defined sampling periods should be used which avoid periods of rapid change in environmental temperature (particularly spring when freshwater input from snow/ice melt can produce abrupt changes) and periods of sexual maturation (juveniles should be used whenever possible).

4. ASSAYS FOR ESTIMATION OF MT'S

In developing any monitoring system, including MT as a response parameter it is also important to evaluate and standardize the assay methods. The remainder of the chapter will discuss the use of several methods in the detection of MT or MT mRNA levels, illustrated by studies in fish. These include three methods which are not species dependent:

- chromatography, which is time consuming, but may be used to determine the relative induction of different MT isoforms;
- metal saturation methods, which utilise binding of radioactive metals such as ^{109}Cd and ^{203}Hg to measure MT levels and are applicable to measurements of Cd and Zn exposure, and
- polarography, which measures the thiol content of MT and can be used for effects of any metals.

Other methods are more highly specific for the species to which probes are prepared:

- the use of antibodies in radio immunoassays (RIA) or enzyme linked immunoassays (ELISA), and
- the use of complimentary nucleic acid probes for determination of MT messenger RNA (mRNA) levels.

An overview of the sensitivities of the various methods is presented in Table 1.
Detailed descriptions of the different methodologies are presented in the Appendix to this chapter.

Table 1. Sensitivities of different MT estimation procedures

method	detection limit	working range	practical detection limit
GPC (G-75)	50 µg MT	100 µg - 1 mg	100 µg MT.g^{-1}
HPLC AAS	5 µg MT	10 - 100 µg	5 µg MT.g^{-1}
Metal saturation	1 µg MT	5 - 65 µg	5 µg MT.g^{-1}
RIA	150 pg MT	0.15 - 7 ng	10 ng MT.g^{-1}
ELISA	1 pg MT	5 pg - 7 ng	1 ng MT.g^{-1}
Polarography	0.5 µg MT	0.5 - 150 µg	15 µg MT.g^{-1}

4.1. Chromatographic methods

Chromatographic techniques are useful for diagnosis of the presence of MT in an organism or identification of the particular pollutant metal in induced organisms, they also have the advantages that they can distinguish between different isoforms of MT and are technically simple, however, they require relatively large amounts of material unless sophisticated equipment is available and they are time consuming, which precludes their use for large numbers of samples. Metallothionein is generally identified by its behaviour and metal content on size exclusion chromatography, as a heat stable protein with an apparent molecular weight of ca. 12-15,000 Daltons. Different isoforms can subsequently be resolved by ion exchange or reverse phase chromatography.
The most commonly used procedure involves gel permeation chromatography (GPC) on columns of Sephadex G-75 (Figure 5) and calculation of the area of the MT peak (Chen & Ganther, 1975), however, semi-microscale analysis can be performed using size-exclusion HPLC. Since these columns work at low operating pressures the newer generation of semi-micro low pressure chromatography systems are suitable. Fractions can be collected and ana-

lyzed for metals or the instrument can be coupled directly to a flame-AAS, ET-AAS, ICP-AAS or ICP-MS instrument.

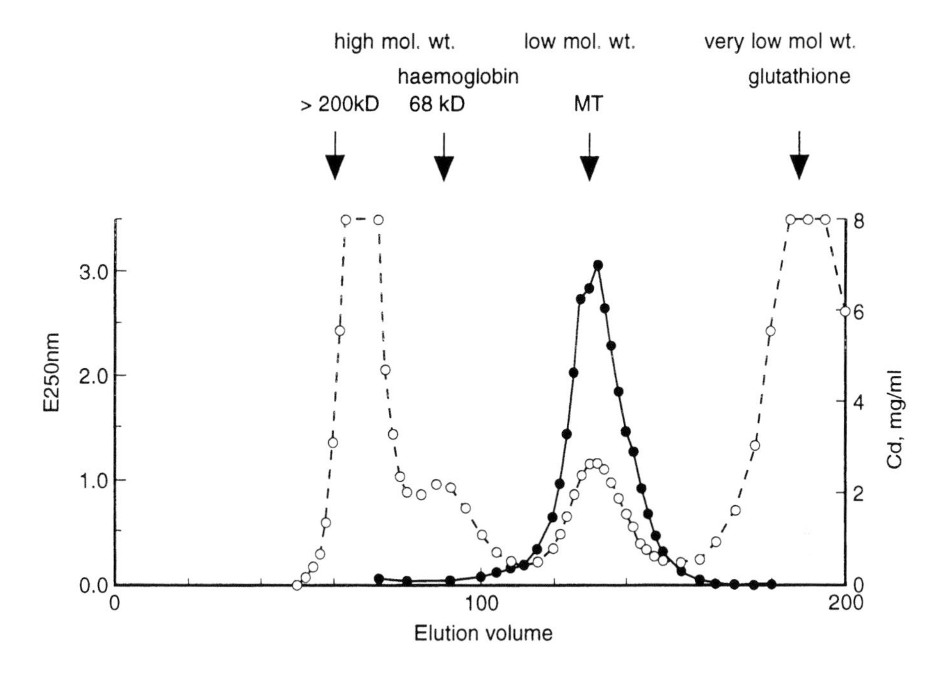

Figure 5. Elution profile on size exclusion chromatography of hepatic cytosol from Cd-exposed turbot, *Scopthalmus maximus*, on Sephadex G-75

4.2. Metal saturation methods

The binding affinities of metals with thionein are Hg>Cu>Cd>Zn, thus displacement and exchange with radioactive ^{203}Hg or ^{109}Cd can be used for estimation of MT. Whilst Hg can displace all metals including Cu, non-specific binding of Hg to other cytosolic components produces high backgrounds and lack of linearity. Assays are usually carried out with heat denatured cytosols or low speed supernatants.

These procedures for determination of MT are based upon:

- the high affinity of MT for Cd or Hg;
- the solubility of MT after heat treatment or in 5% trichloroacetic acid (TCA), and
- the ability of heat or TCA precipitation to remove Cd or Hg which is bound to other proteins.

The original ^{203}Hg-binding method of Piotrowski *et al.* (1973) utilising the TCA solubility of MT is not widely used due to its lack of sensitivity and interference by low molecular weight Hg-binding compounds such as glutathione, Se etc. A simple alternative method which is extensively used was described by Onasaka & Cherian (1982) and is based upon ^{109}Cd-binding to soluble fractions of heat denatured tissue homogenates followed by scavenging of excess ^{109}Cd by a red cell haemolysate. Eaton & Toal (1982) modified this procedure by use of pure haemoglobin for scavenging and we have successfully used their method for estimation of Cd- and Zn-thioneins in fish liver for quite a number of years.

4.3. Polarography

The use of the polarographic technique for determination of MT levels in field collected material is well suited for use on species where MT has not previously been characterized because it is based on the measurement of cysteines within the protein. This technique has been used for determination of piscine MT levels in several field studies (Roch *et al.*, 1982, Olsson & Haux 1986; Hogstrand & Haux, 1990a, 1991b; Hogstrand *et al.*, 1991b; Roch & McCarter, 1984) and has also been used to measure seasonal variations in MT in the mussel (*M. galloprovincialis*) collected from the Limski Kanal in the North Adriatic (Pavicic) *et al.*, 1993) and appears to be a good method in this species where there is a diversity of isoforms. Differential pulse polarography (DPP) is based on the detection of changes in current that occur when a compound is either oxidized or reduced and takes advantage of the specificity of the oxidation-reduction potential of the measured compound (Figure 6). Determination of cysteine-containing proteins, such as MT by differential pulse polarography is carried out by detection of cobalt-catalyzed, reduction of thiolic hydrogen (Olafson & Olsson, 1991). The use of Brdicka buffer, an ammonium buffer containing cobalt, makes it possible to quantify cysteine levels without any major interferences. Polarographic quantification of MT is dependent on the removal of other cysteine-containing proteins from the sample which is achieved by heat denaturation, leaving the heat stable MT in solution. The specificity of the procedure has been examined by chromatography of the polarographic active peak on size exclusion chromatography and anion exchange chromatography of *O. mykiss* liver samples (Olafson & Olsson, 1991) and it was confirmed that the two MT isoforms accounted for all

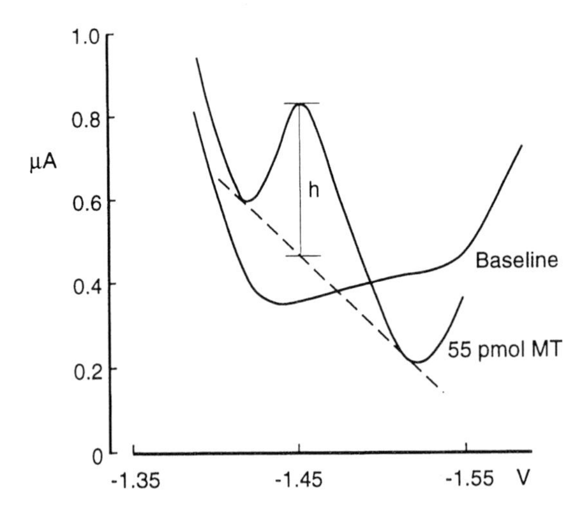

Figure 6. Polarographic profile of metallothionein estimation

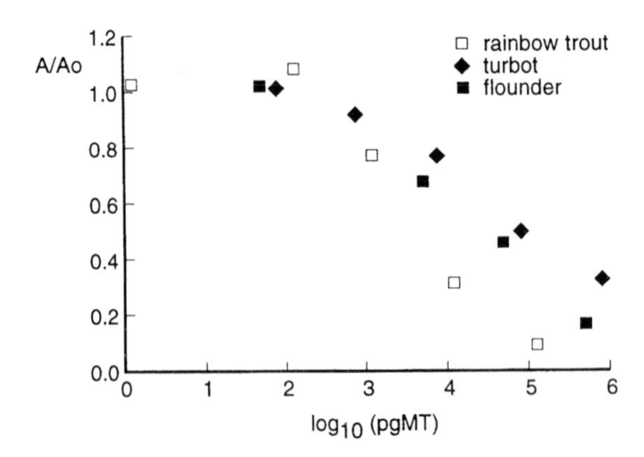

Figure 7. Standard curves for ELISA of Norey *et al.* (1990b) for MT's of different fish species (After: George *et al.* 1992a). MT's from rainbow trout (*O. mykiss*), flounder (*Platichthys flesus*) and turbot (*Scopthalmus maximus*) assayed with anti trout MT antibody

of the activity remaining after heat denaturation. The procedure for quantification of MT is based on the linear relationship between the cobalt reduction wave (Figure 7) and the protein concentration.

4.4. Immunoassays (ELISA and RIA)

The use of antibodies is the most sensitive method for quantization of MT protein (which may not be necessary for monitoring purposes where basal levels are naturally high) and they offer the advantages (over chromatography and polarography) that a high number of samples can be processed in a short time and (over metal saturation assays) that they are applicable to CuMT. The main drawback is the limited access to high quality MT antibodies and that in some instances there may be poor cross-species reactivity of the antibody.

Antibodies to MT's from aquatic species are not widely available; however, they can be raised by immunisation of mammals by standard or modified protocols (Chattergee & Maiti, 1990; Hogstrand & Haux, 1990b; Norey *et al.*, 1990b; Roesijadi *et al.*, 1988). As noted earlier, the antibodies must be raised against invertebrate or fish MT and not commercially available mammalian products since antibodies against mammalian MT's do not cross react with fish or invertebrate MT's. Extensive homology between different fish MT's does enable cross-species utilisation of antibodies (Chattergee & Maiti, 1990; Hogstrand & Haux, 1990a; Norey *et al.*, 1990b). Several protocols, originally developed for estimation of mammalian MT's have been adapted for aquatic species, heterogeneous ELISA's using antisera raised against *O. mykiss* MT (Norey *et al.*, 1990b) or catfish, *Heteropneustes fossilis,* MT (Chattergee & Maiti, 1990) and *Mytilus edulis* MT (Roesijadi *et al.*, 1988), and a radio-immunoassay using antiserum against *P. fluviatilis* MT (Hogstrand & Haux, 1990b).

The principal of the commonly used heterogenous, double antibody, competitive binding ELISA technique is competition between a known antigen (pure MT) in the solid phase (*i.e.* microplate) with a competitor antigen in the liquid phase (*i.e.* unknown fluid) for a primary antibody (anti-MT). A second antibody (anti-IgG for the species in which the primary antibody was raised) which has been conjugated with an enzyme, complexes with the primary antibody bound to the solid phase antigen. The amount of enzyme (*inter alia*, secondary antibody) bound is then determined from its activity when substrate is added in the fluid phase. In collaborative studies with Kay's group we have successfully used such an ELISA described by Norey *et al.* (1990b) for *O. mykiss* MT with cytosols from both liver and cultured cells of other species (Figure 7).

The principal of the radio immunoassay (RIA) described by Hogstrand & Haux (1990b) is competition between (^{125}I) radiolabelled MT and the unknown MT. Validation for HgMT has not been carried out. The RIA technique for *P. fluviatilis* MT has been evaluated against the polarographic technique and it has been shown that they are in good agreement with each other, $MT_{RIA} = MT_{DPP} \times 0.99 - 0.048$ ($r = 0.92$) (Hogstrand & Haux, 1992).

4.5. MT mRNA determinations

Control of MT synthesis is by transcription of DNA to produce MT mRNA which is then translated to produce the apoprotein. Thus the first measurable effect of heavy metal exposure is synthesis of MT mRNA and this can be quantified by use of a complimentary nucleic acid strand (which is usually radiolabelled) which binds (or hybridises) to form a double stranded nucleic acid molecule. This molecular biological approach using mRNA determinations by nucleic acid probes is an emerging technique and due to its very high sensitivity it enables detection of a cellular response to levels of metals which are about an order of magnitude lower than for other techniques. This enables determinations to be performed on very small tissue samples (including biopsies) and larvae.

Labelled probes are most commonly prepared from cloned complimentary mRNA strands (usually cDNAs) to the species in question. MT cDNA's from several fish have been cloned, including *O. mykiss* (Bonham *et al.*, 1987), *P. americanus* (Chan *et al.*, 1989), *P. platessa* (Leaver & George 1989), *N. barbatulus* and *E. lucius* (Kille *et al.*, 1991), and two invertebrates, the american oyster, *C. virginica* (Roesijadi *et al.*, 1991), and the sea urchin *Stongylocentrus purpuratus* (Nemer *et al.*, 1985). To date, field data are still being assessed and only data for mRNA estimations in experimentally treated fish and fish cells have been published.

5. FIELD APPLICATIONS

The use of MT levels for biomonitoring of environmental metal impact is now quite well validated and a few examples of these will be given which illustrate the different methods for MT determination. These methods involve gel permeation chromatography (GPC) on Sephadex G-75, the ^{109}Cd saturation assay, polarographic determination, and the radio-immuno assay (RIA).
The ELISA developed by Norey *et al.* (1990b) for trout MT has been used successfully in laboratory studies with cultured cell lines from this species and from turbot, *Scopthalmus maximus* (Burgess *et al.* 1993; George *et al.*, 1992a). The ELISA of Roesijadi *et al.* (1988) for *Mytilus* MT was used satisfactorily for determination of Cd, Cu and Hg-exposed animals, however, no data of field applications of these ELISA's have been reported to date. Neither have mRNA determinations in field studies been reported, however, mRNA analyses for MT mRNA in flounder (*Platichthys flesus*) from the Forth Estuary, are reported here.

5.1. Assessment of contaminant metal impact in lakes of the Campbell River system, Vancouver Island, Canada

Metallothionein levels in livers of rainbow trout, *O. mykiss,* obtained from lakes with different degrees of metal pollution were estimated by measurement of metals in different protein fractions resolved by gel permeation *chromatography* on Sephadex G-75 columns. In the initial study (Roch *et al.*, 1982) cadmium and copper in high molecular weight fractions, copper in low molecular weight fractions and MT, displayed an upward trend with increasing metal-contamination of the lakes. In a follow up study (Roch & McCarter, 1984) *O. mykiss* were caged in three lakes with different degrees of contamination having metal concentrations ranging from [< 1 μg Cu.l^{-1}, 44 μg Zn.l^{-1}] to [0.7 μg Cd.l^{-1}, 9 μg Cu.l^{-1}, 170 μg Zn.l^{-1}] and the livers analyzed after 2 and 4 weeks. Hepatic MT concentrations in fish from the most contaminated site were elevated after 2 weeks and from all sites after 4 weeks. There was a strong correlation between MT levels and the degree of Zn contamination measured by hepatic Zn concentrations ($r = 0.97$). In both these and caged fish over a two year period there was a significant correlation between hepatic MT and Zn contents which displayed the relationship $MT = 0.87 [Zn] + 56$.

5.2. Assessment of pollutant impact on the River Forth estuary, Scotland

MT levels in livers of flounder, *Platichthys flesus,* a benthic feeding Pleuronectid flatfish caught at various stations along the river and estuary were measured by *^{109}Cd saturation assay* (Sulaiman *et al.*, 1991). Fish caught from another estuary were used as controls. Liver samples (0.5 g) were taken immediately on sacrifice, frozen in liquid nitrogen and then stored at - 80 °C for up to 6 weeks before analysis. MT levels in fish from the River Forth estuary showed a clear gradient of induction from 34 ± 6 to 270 ± 57 μg.g^{-1}, with the highest levels at the turbidity maximum zone where heavy metals from the paint manufacturing plant at Grangemouth have accumulated in the sediments (Figure 8). Whilst the animals are not severely contaminated, the data localises the point source of contamination and demonstrates that whilst dissolved metal concentrations are low, a portion of the sediment/infaunal metals is biologically available and therefore passed up the food chain.

In the succeeding sampling year to the above study, we analyzed *MT mRNA levels* in livers of *P. flesus* caught at some of the same stations (Sawyer, Leaver & George, unpublished results). The relative mRNA levels (also shown in Figure 8) were elevated some 10-fold in fish caught at the station where hepatic MT protein levels were some 8-fold higher. These data therefore show that measurement of MT mRNA levels is also a valid monitoring procedure.

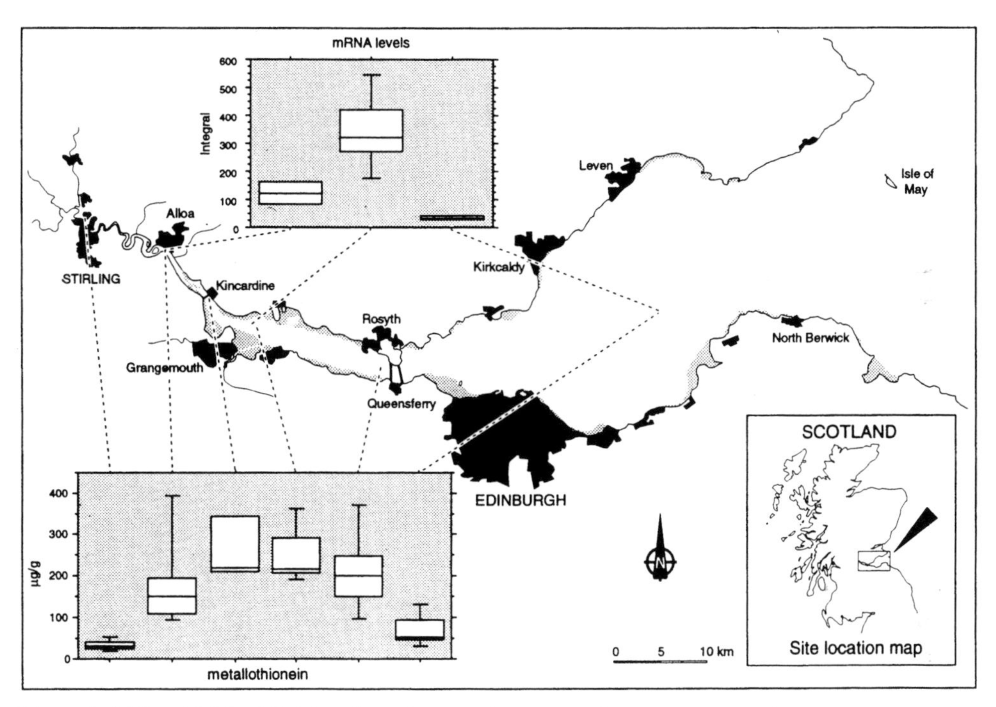

Figure 8. MT and MT mRNA levels in livers of flounder, *Platichthys flesus*, from the industrialised estuary of the River Forth. Fish were sampled and analyzed for protein in September 1989 (Sulaiman *et al.*, 1991) and mRNA in September 1990 (Sawyer *et al.* unpublished)

5.3. Assessment of cadmium pollution in the River Emån, Sweden

Using the *polarographic determination* it was possible to show good correlations between the cadmium burden and the hepatic MT levels in perch (*Perca fluviatilis*) caught in the cadmium-contaminated river Emån in the south east of Sweden (Olsson & Haux, 1986). The hepatic MT levels were found to increase in correlation to the cadmium content of the livers according to the relationship [MT] = 7.2 + 0.53 [Cd] ($r=0.84$, $n=20$).

5.4. Assessment of metal-contamination of subtropical fish species in Bermuda

Hogstrand & Haux (1990a) reported analysis of MT in fish from various stations in Bermuda by *radio-immunoassay (RIA)*. An antiserum raised against MT from *P. fluviatilis* was used for RIA and hepatic MT levels in grunt, *Haemulon sciurus* and tomtate, *H. aurolineatum* were measured using partially purified grunt CdMT as standard. Hepatic MT levels of ca. 800 $\mu g.g^{-1}$ were significantly (about 4 times) higher in *H. sciurus* from the inner Hamilton harbour, where sedimentary Cu and Zn levels were 5 to 10-fold elevated compared with a reference site. At another location, Castle Harbour, *H. aurolineatum* obtained 50 m from a dump site which contained elevated sedimentary Cu and Zn levels had MT levels which were about 2-fold higher than fish caught ca 4 km away. These data indicate that hepatic MT levels were an accurate reflection of sedimentary metal levels and that these metals were bioavailable.

5.5. Assessment of metal contamination in the North Sea

Metal (Cd, Cu and Zn) concentrations in sediments and tissues of adult dab (*Limanda limanda*) were determined at various stations along a cruise transect in the German Bight from the Elbe/Weser estuaries towards the Dogger Bank in the ICES/IOC Bremerhaven workshop (March 1990; collected papers in Stebbing *et al.*, 1992). Sediments at coastal stations closest to the estuaries contained generally elevated metal concentrations, a gradient of Zn contamination being evident. Hylland *et al.* (1992) determined liver metal concentrations in both male and female fish and hepatic MT levels by *radio-immunoassay (RIA)* using the anti-perch MT. In female livers MT and Zn concentrations appeared to be overidingly influenced by sexual maturation (tissue Zn mobilisation). Whilst MT levels in male fish livers were more closely correlated with total Cd and Cu levels, these did not reflect the sedimentary metal concentrations, high values were found at several offshore stations where metal concentrations in two benthic invertebrate species were also elevated. Thus there appeared to be a strong local dietary influence.

In an accompanying study (Stagg *et al.*, 1992) metal concentrations were measured in the gills of these male animals and branchial MT levels were measured by *polarography*. In this case there was a clear relationship between sediment and branchial metal concentrations which were elevated at the coastal stations, and also between the molar concentrations of Cd, Cu and Zn and branchial MT levels. Thus branchial MT levels appeared to be related to sedimentary metal concentrations and provide a longer term integration of metal bioavailability.

These two studies clearly demonstrate the futility of using sexually maturing female fish for monitoring purposes. They also indicate that in studies such as these over a large geographic area where sediments have very different characteristics (muddy vs. sandy) and the infauna differ in both species and metal content, that dietary effects might have a strong short-term influence on hepatic MT content which is not observed in the gills. In these situations measurements in gills are clearly advantageous.

CONCLUSIONS

We consider that the necessary criteria for species selection, knowledge of endogenous and exogenous influences, competitive metabolic systems and validation have only truly been met for the following fish species: *P. flesus*, *P. platessa*, *O. mykiss* and *S. maximus*, where MT has been characterised by purification to homogeneity or its cDNA cloned, field and/or laboratory validation of response have been carried out and assays for MT by Cd-saturation, polarographic, immunoassay or MT mRNA estimations, have all been satisfactorily calibrated. For other species including *H. aurolineatum*, *H. fluviatilis*, *H. sciurus*, *L. limanda* and *P. fluviatilis*, limited validation and calibration has been performed. Clearly the relationships between MT levels and tissue or environmental metal levels are complex and endobiotic influences must be ruled out in monitoring programs. Several studies have shown that sampling should not take place during periods of rapid temperature change or sexual maturation and that juvenile fish are preferable. If periods of sexual quiescence cannot be avoided then male fish must be used. In addition to the problems encountered with sexual maturation, the ICES/IOC study has also highlighted potential problems with differential rates of MT turnover in different organs which have probably highlighted the elevation of the hepatic MT levels (probably only short term) by dietary effects. This study also demonstrated the usefulness of using extrahepatic tissues (*e.g.* gill, kidney) for monitoring studies.

For the invertebrates, the response of mussels (*Mytilus* sp) to metal exposure is well characterised, numerous isoforms of MT have been isolated and sequenced, there is good knowledge of competing sequestration systems and current data indicate that MT measurements in mussel gills can be used as an indicator of Cd, Cu and Hg exposure but not Zn exposure. For other invertebrates, even crustaceans where MT has been characterised, too many interferences and competing sequestration systems are probably present for their definitive application as a monitoring tool.

A number of assay methods are available, each possessing advantages for particular applications, laboratories or cost efficiencies. All have been satisfactorily validated.

REFERENCES

Baer, K.N. & P. Thomas, 1991. Isolation of novel metal-binding proteins distinct from metallothionein from spotted seatrout (*Cynoscion nebulosus*) and Atlantic croaker (*Micropogonias undulatus*) ovaries. Mar. Biol. 108: 31-37

Bonham, K., M. Zafarullah & L. Gedamu, 1987. The rainbow trout metallothioneins: Molecular cloning and characterization of two distinct cDNA sequences. DNA, 6: 519-528

Boyden, C.R, 1977. Effect of size upon metal content of shellfish. J. mar. biol. Ass. U.K. 57: 675-714

Bryan, G.W., W.J. Langston & L.G. Hummerstone, 1980. The use of biological indicators of heavy metal contamination in estuaries. Occasional publication of the Mar. Biol. Ass. UK., Vol. I, Plymouth, pp. 73

Bunton, T., S.M. Baksi, S.G. George & J.M. Frazier, 1987. Abnormal copper storage in a teleost fish (*Morone americana*). Vet. Pathol. 24: 515-524

Burgess, D., N. Frerichs & S. George, 1993. Control of metallothionein expression by hormones and stressors in cultured fish cells. Mar. Environ. Res. 35: 25-28

Chan, K., W.S. Davidson, C. Hew & G.L. Fletcher, 1989. Molecular cloning of metallothionein cDNA and analysis of metallothionein gene expression in winter flounder tissues. Can. J. Zool. 67: 2520-2527

Chatterjee, A. & I.B. Maiti, 1990. Comparison of the immunological properties of mammalian (rodent), bird, fish, amphibian (toad), and invertebrate (crab) metallothioneins. Mol. Cell Biochem. 94: 175-181

Chatterjee, A. & I.B. Maiti, 1991. Induction and turnover of catfish (*Heteropneustes fossilis*) metallothionein. Mol. Cell. Biochem. 108: 29-38

Chen, R.W. & H.E. Ganther, 1975. Relative cadmium binding capacity of metallothionein and other cytosolic fractions in various tissues of the rat. Environ. Physiol. Biochem. 5: 378-388

Chomczynski, P. & N. Sacchi, 1978. Single-step method of RNA isolation by acid guanidinium thiocyanate-phenol-chloroform extraction. Anal. Biochem. 162: 156-159

Dohi, Y., K. Ohba & Y. Yoneyama, 1983. Purification and molecular properties of two cadmium binding proteins from the hepatopancreas of a whelk, *Buccinum tenuissimum*. Biochim. Biophys. Acta, 745: 50-60

Eaton, D.L. & B.F. Toal, 1982. Evaluation of the Cd/Hemoglobin affinity assay for the rapid determination of metallothionein in biological tissues. Toxicol. Appl. Pharmacol. 66: 134-142

Engel, D. & M. Brouwer, 1993. Crustaceans as models for metal metabolism: I. Effects of the molt cycle on blue crab metal metabolism and metallothionein. Mar. Environ. Res. 35: 1-5

Förstner, U. & G.T. Wittmann, 1979. Metal pollution in the aquatic environment. Springer Verlag, Berlin, pp. 486

Frankenne, F., F. Noel-Lambot & A. Disteche, 1980. Isolation and characterisation of metallothioneins from cadmium-loaded mussel, *Mytilus edulis*. Comp. Biochem. Physiol. 66C: 179-182

Frazier, J.M. & S.G. George, 1985. Cadmium kinetics in oysters - A comparative study of *Crassostrea gigas* and *Ostrea edulis*. Mar. Biol. 76: 55-61

Frazier, J.M., S.G. George, J. Overnell, T.L. Coombs & J.H.R. Kaji, 1985. Characterisation of two molecular weight classes of cadmium binding proteins of the common mussel, *Mytilus edulis*. Comp. Biochem. Physiol. 80C: 257- 262

Friberg, L., M. Piscator, G.F. Nordberg & T. Kjellstrom, 1974. Cadmium in the environment; 2nd Ed. CRC Press, Cleveland, Ohio, pp 93-100

Garvey, J.S., D.G. Thomas & H.J. Linton, 1987. Enzyme linked immunosorbent assay (ELISA) for metallothionein. Experientia Supplementum, 52 Metallothionein II. Birkhauser Verlag, Basel, pp. 335-342

George, S.G., 1982. Subcellular accumulation and detoxication of metals in aquatic animals. In: Physiological mechanisms of marine pollutant toxicity; W.B. Vernberg, A. Calabrese, F.P. Thurberg & F.J. Vernberg (eds). Academic Press, New York, pp. 1-52

George, S., 1989. Cadmium effects on plaice liver xenobiotic and metal detoxication systems: Dose-response. Aquat. Toxicol. 15: 303-310

George, S., 1990. Biochemical and cytological assessments of metal toxicity in marine animals. In: Heavy metals in the marine environment; R.W. Furness & P. Rainbow (eds). CRC Press, Boca Raton, pp. 123-142

George, S.G. & B.J.S. Pirie, 1980. Metabolism of zinc in the common mussel, *Mytilus edulis*. A combined ultrastructural and biochemical study. J. mar. biol. Ass. U.K. 60: 575-590

George, S.G. & A. Viarengo, 1985. A model for heavy metal homeostasis and detoxication in mussels. In: Marine pollution and physiology: Recent advances. F.J. Vernberg, F.P. Thurberg, A. Calabrese & W.B. Vernberg (eds). Belle Baruch Library in Marine Science, 13. Univ. South Carolina Press, Columbia, pp. 125-143

George, S.G. & P. Young, 1986. The time course of effects of cadmium and 3-methylcholanthrene on activities of enzymes of xenobiotic metabolism and metallothionein levels in the plaice, *Pleuronectes platessa.* Comp. Biochem. Physiol. 83C: 37-44

George, S.G., B.J.S. Pirie & T.L. Coombs, 1980. Isolation and elemental analysis of metal-rich granules from the kidney of the scallop, *Pecten maximus* (L.). J. exp. mar. Biol. Ecol. 42: 143-156

George, S.G., B.J. Pirie, J.M. Frazier & J.A. Thomson, 1984. Interspecies differences in heavy metal detoxication in oysters. Mar. Environ. Res. 14: 462-464

George, S.G., E. Carpene, T.L. Coombs, J. Overnell & A. Youngson, 1979. Characterisation of cadmium-binding proteins from the common mussel, *Mytilus edulis.* Biochim. Biophys. Acta, 580: 225-233

George, S.G., D. Burgess, M. Leaver & N. Frerichs, 1992a. Metallothionein induction in cultured fibroblasts and liver of a marine flatfish, the turbot *Scopthalmus maximus.* Fish Physiol. Biochem. 10: 43- 54

George, S.G., D. Groman, S. Brown & K. Holmes, 1992b. Studies of a fatal pollutant-induced hyperbilirubinaemia in spawning Atlantic salmon. Mar. Environ. Res. 34: 81-86

Goldberg, E.D., 1975. The mussel watch. A first step in global pollution monitoring. Mar. Pollut. Bull. 6: 111

Hamer, D., 1986. Metallothionein. Ann. Rev. Biochem. 55: 913-951

Hamilton, S.J. & P.M. Mehrle, 1986. Evaluation of metallothionein measurement as a biological indicator of stress from cadmium in brook trout. Trans. Am. Fish. Soc. 116: 551-560

Haux, C. & L. Förlin, 1988. Biochemical methods for detecting effects of contaminants on fish. Ambio, 17: 376-380

Hogstrand, C. & C. Haux, 1990a. Metallothionein as an indicator of heavy metal exposure in two subtropical fish species. J. exp. mar. Biol. Ecol. 138: 69-84

Hogstrand, C. & C. Haux, 1990b. A radioimmunoassay for perch (*Perca fluviatilis*) metallothionein. Toxicol. Appl. Pharmacol. 103: 56-65

Hogstrand, C. & C. Haux, 1991a. Binding and detoxification of heavy metals in lower vertebrates with reference to metallothionein. Comp. Biochem. Physiol. 100C: 137-141

Hogstrand, C. & C. Haux, 1991b. The importance of metallothionein for the accumulation of copper, zinc and cadmium in environmentally exposed perch, *Perca fluviatilis.* Pharmacol. Toxicol. 68: 492-501

Hogstrand, C. & C. Haux, 1992. Evaluation of differential pulse polarography for the quantification of metallothionein - A comparison with RIA. Anal. Biochem. 200: 388-392

Hogstrand, C., G. Lithner & C. Haux, 1991. The importance of metallothionein for the accumulation of copper, zinc and cadmium in environmentally exposed perch, *Perca fluviatilis.* Pharmacol. Toxicol. 68: 492-501

Hylland, K., C. Haux & C. Hogstrand, 1992. Hepatic metallothionein and heavy metals in dab *Limanda limanda* from the German Bight. Mar. Ecol. Progr. Ser. 91: 89-96

Hyllner, S.J., T. Andersson, C. Haux & P.E. Olsson, 1989. Cortisol induction of metallothionein in primary culture of rainbow trout hepatocytes. J. Cell. Physiol. 139: 24-28

Kägi, J.H.R. & M. Nordberg, 1979. Metallothionein II. Birkhauser Verlag, Basel, pp. 535

Karin, M., 1985. Metallothioneins: Proteins in search of function. Cell, 41: 9-10

Kikuchi, Y., N. Wada, M. Irie, H. Ikebuchi, J. Swada, T. Terao, S. Nakayama, S. Iguchi & Y. Okada, 1988. A murine monoclonal anti-metallothionein autoantibody recognises an N-terminal heptapeptide common to various metallothioneins. Mol. Immunol. 25: 1033-1036

Kille, P., P.E. Stephens & J. Kay, 1991. Elucidation of cDNA sequences for metallothioneins from rainbow trout, stone loach and pike liver using the polymerase chain reaction. Biochim. Biophys. Acta, 1089: 407-410

Kille, P., J. Kay, M. Leaver & S. George, 1992. Induction of piscine metallothionein as a primary response to heavy metal pollutants: applicability of new sensitive molecular probes. Aquat. Toxicol. 22: 279-286

Kohler, K. & H.U. Riisgård, 1982. Formation of metallothioneins in relation to accumulation of cadmium in the common mussel *Mytilus edulis*. Mar. Biol. 66: 53-58

Kojima, Y. & J.H.R. Kägi, 1978. Metallothionein. Trends Biochem. Sci. 3: 403-405

Langston, W.J., M.J. Bebianno & M. Zhou, 1989. A comparison of metal-binding proteins and cadmium metabolism in the marine molluscs *Littorina littorea* (Gastropoda), *Mytilus edulis* and *Macoma balthica* (Bivalvia). Mar. Environ. Res. 28: 195-200

Leaver, M.J. & S.G. George, 1989. Nucleotide and deduced amino acid sequence of a metallothionein cDNA clone from the marine fish, *Pleuronectes platessa*. EMBL accession #56743.

Lonnerdahl, B., A.G. Stranislowski & L.S. Hurley, 1980. Isolation of a low molecular weight zinc binding ligand from human milk. J. Inorg. Biochem. 12: 71-78

Mackay, E.A., J. Overnell, B. Dunbar, I. Davidson, P.E. Hunziker, J.H.R. Kagi & J.E. Fothergill, 1990. Polymorphism of cadmium induced mussel metallothionein. Experientia, 46: A36 (abstract No. 310)

Marafante, E., G. Pozzi & P. Scoppa, 1972. Detossicazione dei metalli pesanti nei pesci, isolamento della metallotioneina dal fagato di *Carrasius auratus*. Boll. Soc. Ital. Biol. Sper. 48: 109

Mason, A.Z. & J.A. Nott, 1981. The role of intracellular biomineralised granules in the regulation and detoxification of metals in gastropods with special reference to the marine prosobranch *Littorina littorea*. Aquat. Toxicol. 1: 239-256

Mason, A.Z. & S.D. Storms, 1993. Applications of directly coupled SE-HPLC/ICP-MS in environmental toxicology studies: A study of metal-ligand interactions in cytoplasmic samples. Mar. Environ. Res. 35: 19-23

Mason, A.Z., K. Simkiss & K. Ryan, 1984. The ultrastructural localization of metals in specimens of *Littorina littorea* collected from clean and polluted sites. J. mar. biol. Ass. U.K. 64: 699-720

Mazzucotelli, A., A. Viarengo, L. Canesi, E. Ponzano & P. Rivaro, 1991. Determination of trace amounts of metalloprotein species in marine mussel samples by high-performance liquid chromatography with inductively coupled plasma atomic emission spectrometric detection. Analyst, 116: 605-608

Neff, G., 1985. Use of biochemical measurements to detect pollutant-mediated damage to fish. ASTM, Special technical publication 854: 155-183

Nemer, M., D.G. Wilkinson, E.C. Travaglini, E.J. Sternberg & T.R. Butt, 1985. Sea urchin metallothionein sequence: Key to an evolutionary diversity. Proc. Natl Acad. Sci. USA. 82: 4992-4994

Noel-Lambot, F., 1976. Distribution of cadmium, zinc and copper in the mussel, *Mytilus edulis*. Existence of cadmium-binding proteins similar to metallothioneins. Experientia, 32: 423-326

Noel-Lambot, F., C. Gerday & A. Disteche, 1978. Distribution of Cd, Zn and Cu in liver and gills of the eel *Anguilla anguilla* with special reference to metallothionein. Comp. Biochem. Physiol. 61C: 177-187

Norey, C.G., A. Cryer & J. Kay, 1990a. A Comparison of Cadmium-induced metallothionein gene expression and Me^{2+} distribution in the tissues of cadmium-sensitive (rainbow trout, *Salmo gairdneri*) and tolerant (stone loach, *Noemacheilus barbatulus*) species of freshwater fish. Comp. Biochem. Physiol. 97C: 221-225

Norey, C.G., W.E. Lees, B.M. Darke, J.M. Stark, T.S. Baker, A. Cryer & J. Kay, 1990b. Immunological distinction between piscine and mammalian metallothioneins. Comp. Biochem. Physiol. 95B: 597-601

Olafson, R.W. & P.-E. Olsson, 1991. Electrochemical detection of metallothionein. Meth. Enzymol. 205: 205-213

Olafson, R.W., R.G. Sim & K.G.Boto, 1979. Isolation and chemical characterisation of the heavy metal binding protein metallothionein from marine invertebrates. Comp. Biochem. Physiol. 62B: 407- 416

Olafson, R.W. & J.A.J. Thompson, 1974. Isolation of heavy metal binding proteins from marine vertebrates. Mar. Biol. 28: 83-86

Olsson, P.-E., 1985. PhD. Thesis. Dept. Zoophysiology, Univ. Göteborg, Sweden, pp. 127

Olsson, P.-E., 1991. Purification of metallothionein by fast protein liquid chromatography. Meth. Enzymol. 205: 238-244

Olsson, P.-E. & C. Haux, 1985. Rainbow trout metallothionein. Inorg. Chim. Acta, 107: 67-71
Olsson, P.-E. & C. Haux, 1986. Increased hepatic metallothionein correlates to cadmium accumulation in environmentally exposed perch (*Perca fluviatilis*). Aquat. Toxicol. 9: 231-242
Olsson, P.-E., C. Haux & L. Förlin, 1987. Variations in hepatic metallothionein, zinc and copper levels during an annual reproductive cycle in rainbow trout, *Salmo gairdneri*. Fish Physiol. Biochem. 3: 39-47
Olsson, P.-E., A. Larsson, A. Maage, C. Haux, K. Bonham, M. Zafarullah & L. Gedamu, 1989a. Induction of metallothionein synthesis in rainbow trout, *Salmo gairdneri,* during long-term exposure to waterborne cadmium. Fish Physiol. Biochem. 6: 221-229
Olsson, P.E., M. Zafarullah & L. Gedamu, 1989b. A role of metallothionein in zinc regulation after oestradiol induction of vitellogenin synthesis in rainbow trout, *Salmo gairdneri*. Biochem. J. 257: 555-559
Olsson, P.E., S.J. Hyllner, M. Zafarullah, T. Andersson & L. Gedamu, 1990a. Differences in metallothionein gene expression in primary cultures of rainbow trout hepatocytes and the RTH-149 cell line. Biochim. Biophys. Acta, 1049: 78-82
Olsson, P.E., M. Zafarullah, R. Foster, T. Hamor & L. Gedamu, 1990b. Developmental regulation of metallothionein messenger RNA, zinc and copper levels in rainbow trout, *Salmo gairdneri*. Eur. J. Biochem. 193: 229-235
Onasaka, S. & M.G. Cherian, 1982. Comparison of metallothionein determination by polarographic and cadmium saturation methods. Toxicol. Appl. Pharmacol. 63: 270-274
Overnell, J. & T.L. Coombs, 1979. Purification and properties of plaice metallothionein, a cadmium binding protein from the liver of the plaice (*Pleuronectes platessa*). Biochem. J. 183: 277-284
Overnell, J. & E. Trewhella, 1979. Evidence for the natural occurrence of (cadmium-copper)-metallothionein in crab, *Cancer pagurus*. Comp. Biochem. Physiol. 64C: 69-76
Overnell, J., T.C. Fletcher & R. McIntosh, 1988. Factors affecting hepatic metallothionein levels in marine flatfish. Mar. Environ. Res. 24: 155- 158
Overnell, J., R. McIntosh & T.C. Fletcher, 1987. The enhanced induction of metallothionein by zinc, its half-life in the marine fish *Pleuronectes platessa*, and the influence of stress factors on metallothionein levels. Experientia, 43: 178-181
Pavicic, J., B. Raspor & D. Martincic, 1993. Quantitative determination of metallothionein-like proteins in mussels. Methodological approach and field evaluation. Mar. Biol. 115: 83-86.
Piotrowski, J.K., W. Bolanowska & A. Sapota, 1973. Evaluation of metallothionein content in animal tissues. Acta Biochim. Polonica, 20: 207-215
Piscator, M. 1964. Om kadmium i normala människonjurar samt redogörelse för isolering av metallothionein ur lever från kadmium exponerade kaniner. Nord. Hyg. Tidskr. 45: 76-82
Roch, M. & J.A. McCarter, 1984. Hepatic metallothionein production and resistance to heavy metals by rainbow trout (*Salmo gairdneri*). II. Held in a series of contaminated lakes. Comp. Biochem. Physiol. 77C: 77-82
Roch, M., J.A. McCarter, A.T. Matheson, M. Clark & R.W. Olafson, 1982. Hepatic metallothionein in rainbow trout (*Salmo gairdneri*) as an indicator of metal pollution in the Campbell River system. Can. J. Fish. Aquat. Sci. 39: 1596-1601
Roesijadi, G., 1982. Uptake and incorporation of mercury into mercury-binding proteins of gills of *Mytilus edulis* as a function of time. Mar. Biol. 62: 151-158
Roesijadi, G., 1992. Metallothioneins in metal regulation and toxicity in aquatic animals - Review. Aquat. Toxicol. 22: 81-114
Roesijadi, G., M.E. Unger & J.E. Morris, 1988. Immunochemical quantification of metallothioneins of a marine mollusc. Can. J. Fish. Aquat. Sci. 45: 1257-1263
Roesijadi, G., M.M. Vestling, C.M. Murphy, P.L. Klerks & C.C. Fenselau, 1991. Structure and time-dependent behaviour of acetylated and non-acetylated forms of a molluscan metallothionein. Biochim. Biophys. Acta, 1074: 230-236
Simkiss, K., M. Taylor & A.Z. Mason, 1982. Metal detoxification and bioaccumulation in molluscs. Mar. Biol. Lett. 3: 4-5
Stagg, R., A. Goksoyr, & G. Rodger, 1992. Changes in branchial $Na^+ K^+$ ATPase, metallothionein and P4501A1 in dab *Limanda limanda* in the German Bight: indicators of sediment contamination? Mar. Ecol. Progr. Ser. 91: 105-115
Stebbing, A.R.D., V. Dethelfsen & M. Carr (eds), 1992. Biological effects of contaminants in the North Sea. Results of the ICES/IOC Bremerhaven workshop. Mar. Ecol. Progr. Ser. 91: 1-361
Stone, H. & J. Overnell, 1985. Non-metallothionein cadmium binding proteins. Comp. Biochem. Physiol. 80C: 9-14
Stone, H.C., S.B. Wilson & J. Overnell, 1986. Cadmium-binding proteins in the scallop *Pecten maximus*. Environ. Health Perspect. 65: 189-191

Sulaiman, N., S. George & M.D. Burke, 1991. Assessment of sublethal pollutant impact on flounders in an industrialised estuary using hepatic biochemical indices. Mar. Ecol. Progr. Ser. 68: 207-212

Suzuki, K.T., 1980. Direct connection of high speed liquid chromatograph (equipped with gel permeation column) to atomic absorption spectrophotometer for metalloprotein analysis. Anal. Biochem. 102: 31-34

Takeuchi, T., 1972. Distribution of mercury in the environment of Minimata Bay and the inland Ariake Sea. In: Environmental mercury contamination; R. Hartnung & B.D. Dinman (ed). Ann Arbor Sci. Publ., Ann Arbor, pp. 79-81

Thomson, J.D., B.J.S. Pirie & S.G. George, 1985. Cellular metal distribution in the Pacific oyster, *Crassostrea gigas* (Thun.) determined by quantitative X-ray microprobe analysis. J. exp. mar. Biol. Ecol. 85: 37-45

Uchida, Y., K. Takio, K. Titani, Y. Ihara & M. Tomonaga, 1992. The growth inhibitory factor that is deficient in the Alzheimer's disease brain is a 68 amino acid metallothionein-like protein. Neuron. 7: 337-347

Viarengo, A., M. Pertica, G. Mancinelli, G. Zanicchi & M. Orunesu, 1980. Rapid induction of copper-binding proteins in the gills of metal exposed mussels. Comp. Biochem. Physiol. 67C: 215-218

Webb, M., 1979. The chemistry, biochemistry and biology of cadmium. Elsevier/North Holland, Amsterdam, pp. 465

Winge, D.R. & M. Brouwer, 1986. Discussion summary. Techniques and problems in metal-binding protein chemistry and implications for proteins in nonmammalian organisms. Environ. Health Perspect. 65: 211-214

Zafarullah, M., P.E. Olsson & L. Gedamu, 1989a. Rainbow trout metallothionein gene structure and regulation. In: Oxford survey on eucaryotic genes. Vol. 6. N. Maclean (ed). Oxford Univ. Press, Oxford, pp. 111-143

Zafarullah, M., P.E. Olsson & L. Gedamu, 1989b. Endogenous and heavy-metal-ion-induced metallothionein gene expression in salmonid tissues and cell lines. Gene, 83: 85-93

Zafarullah, M., P.E. Olsson & L. Gedamu, 1990. Differential regulation of metallothionein genes in rainbow trout fibroblasts, RTG-2. Biochim. Biophys. Acta, 1049: 318-323

Appendix 7

Metallothionein assay methods

SAMPLE COLLECTION, TRANSPORT AND STORAGE

Clearly assay of tissue from freshly killed animals is optimum, however, this is not practical for field sampling. For protein estimations, our method is to immediately excise livers (and kidneys), wrap them in aluminium foil and freeze in liquid nitrogen. On returning to the laboratory they are stored at -80 °C where they can be kept for a month or so before analysis, slight losses occur when stored at -20 °C. Relative losses on storage of tissues or tissue fractions at -80 °C and -20 °C are shown in Figure 9. Clearly storage of tissue cytosols is not acceptable. For mRNA estimations, tissue samples (30-150 mg) may be homogenised in the field in the extraction buffer and stored at -20 °C or lower for several weeks before extraction.

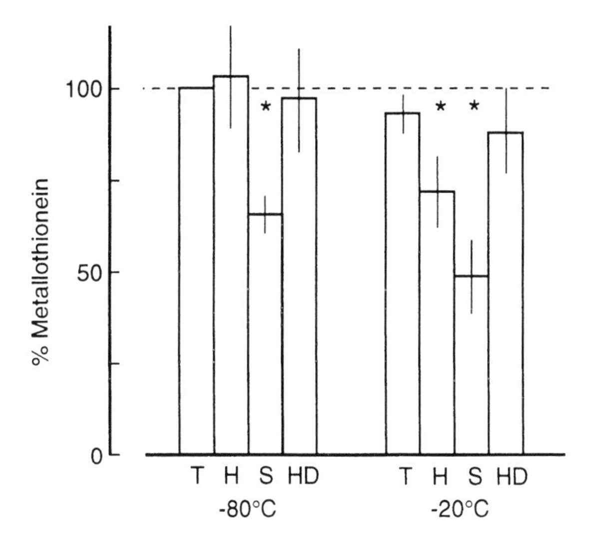

Figure 9. Effects of sample storage conditions on recovery of trout liver MT measured by polarography. Levels expressed relative to those in freeze clamped tissue stored at -80 °C (T). Storage at -80 °C or -20 °C as H: homogenate; S: cytosol; HD: heat denatured cytosol

ASSAY PROCEDURES

Sensitivities of different assays vary markedly (Table 1), however, with most techniques the ability to homogenise and fractionate a small tissue sample is the governing factor. In quoting practical detection limits we have assumed 100 mg tissue is fractionated in 1 ml medium and replicate assays are performed. It should be noted that basal MT levels in various fish species are generally between 10 and 250 $\mu g.g^{-1}$ liver.

Chromatographic methods

Preparation of cytosol

Degradation of MT by proteases released during homogenisation and oxidation of CuMT can cause problems during the lengthy procedures involved in chromatography therefore we have adopted a strategy of including protease inhibitors and an anti-oxidant in the homogenisation buffer. When it is suspected that high levels of CuMT may be present then procedures should be carried out in an inert atmosphere (Winge & Brouwer, 1987). Tissue is homogenised in 3 vols. 0.25 M sucrose, 50 mM phenylmethylsulphonyl fluoride, 2 mM dithiothreitol and 20 mM Tris-HCl, at pH 8.0 using a motor driven Potter type homogeniser and a cytosol fraction prepared by successive centrifugation at 10,000 g for 15 min and 100,000 g for 60 min to remove cellular organelles and microsomes. Alternatively, when determination of the partitioning of metals between high mol. wt. and MT fractions is not required, the sucrose can be omitted from the homogenisation buffer. The low speed supernatant is heat treated at 100 °C for 5 min and the denatured proteins are removed by recentrifugation at 10,000 g for 10 min. Heat treatment effectively overcomes problems of protease degradation and providing the column and buffers are sterile, also permits chromatography at room temperature.

Size exclusion chromatography on Sephadex G-75.

Routinely we apply a 1 ml sample of cytosol (or heat treated supernatant) to a 1.5 x 90 cm column of Sephadex G-75 (Pharmacia), swollen, packed and eluted with 10 mM Tris-HCl in 100 mM NaCl, pH 8.0. The column packing can chelate metals (Lonnerdahl *et al.*, 1980) and therefore we recommend pre-treatment with sodium borohydride before use. Fractions are collected and analyzed qualitatively for proteins (E280 nm), CdMT (E254 nm) and quantitatively for metals of interest (usually Cd, Cu and Zn) by AAS.

This procedure can be scaled up and used preparatively as an initial step in purification of MT for characterisation or preparation of antibodies.

Size exclusion chromatography by HPLC.

Semi-micro scale chromatography can be carried out using HPLC systems as follows. A heat treated cytosol fraction is filtered through a 0.22 µm syringe filter and up to ca 200-500 µl injected onto a 0.75x30 cm size exclusion column (TSK G2000SW, Toyo Soda Manufacturing Co. Tokyo or Waters Protein Pack 125, Millipore/Waters, Milford, USA) and eluted with 5 mM sodium phosphate and 50 mM NaCl, pH 7.0. Fractions can be collected and analyzed by flame or electro-thermal atomic absorption spectrometry (FAAS, ETAAS), or the liquid chromatograph (HPLC) can be coupled directly to a flame AAS for single element determination (George *et al.*, 1992; Suzuki, 1980). Fractions can be collected and metal concentrations determined by ETAAS. Alternatively for simultaneous multi-element determination other laboratories have used direct coupling to an inductively coupled plasma AAS (ICP-AAS; Mazzucotelli *et al.*, 1991) or mass spectrometer (ICP-MS; Mason & Storms, 1993). Correct maintenance of the column is essential to prolong life.

Calculation : The area of the low mol. wt. MT-containing peak eluted with an apparent mol. wt. of ca. 9-15 kDaltons is integrated and the amount of MT calculated using the stoichiometries of 7 g atoms Cd or Zn and 12 g atoms of Cu or Hg bound per mole MT (molecular weight of apoMT ca. 6000 Daltons).

Metal saturation methods

Cd saturation assay

Duplicate 0.25 g tissue samples are homogenised in 3 volumes of 20 mM Tris-HCl buffer, pH 8.6 using a Potter homogeniser or rotating blade homogeniser such as a Polytron® (Kinematica, Basel, Switzerland) and centrifuged in microfuge tubes for 10 min at 10,000 g. The supernatant is transferred to a fresh microfuge tube and heat treated for 2 min at 100 °C (using an aluminium heating block or boiling water bath). After cooling in ice the denatured proteins are removed by recentrifugation. The supernatant is diluted as necessary with Tris-HCl buffer and MT determinations are carried out in triplicate in microfuge tubes. Equal volumes of heat treated supernatant (diluted if necessary) or buffer (test blanks) and ^{109}Cd solution (1 μCi.ml^{-1}, 2 μg Cd.ml^{-1} prepared from carrier free ^{109}Cd and $CdCl_2$ in 10 mM Tris-HCl buffer) are equilibrated for 10 min at room temperature. Then excess ^{109}Cd is scavenged by addition of 0.5 vol. haemoglobin solution (2 % w/v crystalline haemoglobin in 10 mM Tris-HCl buffer) and after vortex mixing, the haemoglobin is heat denatured at 95 °C for 5 min and pelleted by centrifugation. After repetition of the haemoglobin scavenging step the radiolabelled MT present in the supernatant is determined by gamma counting.

Linearity: Up to about 65 % of Cd bound, if greater, dilute the heat treated supernatant.

Calculation: Assuming 7 g atoms Cd bound.mole^{-1} and mol. wt. of Cd MT is approximately 6820 Daltons (plaice) for a 20 % w/v homogenate and sample size of 200 μl.

$$\mu g\ MT.g^{-1}\ tissue\ WW = \frac{(cpm_{test} - cpm_{blank}) \times 86.5\ (\times\ dilution\ of\ supernatant)}{cpm_{total}} \qquad (A1)$$

Notes: For highly induced samples dilution is necessary so that less than 65 % of the Cd is bound. An increase in sensitivity of up to 10 times can be obtained by equilibration with 0.1 μg.ml^{-1} Cd but accuracy is reduced to ±10 % and endogenous GSH may lead to an overestimation of MT. This method can be used for Cd- and Zn-induced animals only as Cd does not displace Cu or Hg. The linearity of the assay does not appear to be good with kidney samples.

Immunoassays

It is necessary to purify MT from the animal of interest for use as standards in the immunoassay procedures. MT is fairly easy to purify from vertebrates and sufficiently pure MT can be obtained by a three stage chromatographic procedure, using a Sephadex G-75 column to achieve separation from high and low molecular weight material, a DEAE Sephadex A-25 or a Mono Q column to separate the MT's based on ionic charge and finally a Sephadex G-50 chromatography step to change the buffer to a volatile one prior to lyophilization and weighing (Olsson, 1991; Olsson & Haux, 1985).

Enzyme linked immunoassays (ELISA)

The ELISA developed by Norey *et al.* (1990b) for rainbow trout MT is performed as follows. Individual wells in alternate lanes of 96 well ELISA plates, (Falcon 'Pro-bind', Becton Dickinson Labware, Lincoln, NJ) are coated overnight with pure rainbow trout MT (140 ng.ml^{-1}) and then washed in PBS/Tween (0.05 % Tween-20, 0.1 M sodium phosphate, 0.9 % NaCl, pH 7.2). MT- containing unknowns are diluted 1:1 with competitor antigen (0, 2 x 10^2, 2 x 10^3, 2 x 10^4 or 2 x 10^5 pg MT) in PBS/Tween and mixed with primary antisera, (usually about 1/2000 dilution of primary sera). The mixtures are then added (in duplicate) to coated and uncoated wells of the plates and incubated for 1 h at room temperature, washed, and an optimised concentration of diluted enzyme-labelled secondary antibody (alkaline phosphatase or peroxidase) is added and incubated for 1 hour. After washing the amount of enzyme-linked secondary antibody bound to the immobilised antigen is determined by colorimetric assay.

Calculation: Absorbance values from wells incubated with samples, corrected for non-specific binding (corresponding uncoated wells) {A-Abl} are expressed as a fraction of the absorbance obtained when no antigen is present {Ao}. {A-Abl}/{Ao} for the standard MT can be plotted against $\log_{10}$ pg MT to produce a sigmoidal standard curve against which the unknowns are read. A more accurate method is to convert the data to a straight line by means of a Logit Log plot.

Applicability: Differing sensitivities or deviation from linearity has been observed in cross-species assays (Norey *et al.*, 1990b), thus it is important to determine this from a calibration curve using pure antigen from the species in question (see Figure 7).

The ELISA described by Roesijadi *et al.* (1988) for mussel MT is essentially similar and has a working range of 1-20 ng MT for the major dimeric isoform. Cross-reactivity of one of the minor isoforms (from gills of induced animals) was not equal and possible differential induction of the isoforms requires thorough investigation.

Radioimmunoassay (RIA)

The principal of the assay described by Hogstrand & Haux (1990b) is competition between radiolabelled MT and the unknown MT. MT is labelled with [^{125}I]-labelled Bolton-Hunter reagent, 3-(4-hydroxy,5-[^{125}I]iodophenyl)proprionate (Amersham).

The RIA is performed as follows. Aliquots of either:

i) 50 µl of a post mitochondrial supernatant prepared from a 5% w/v tissue homogenate in 50 mM Tris/HCl buffer, pH 8.0 (diluted in RIA buffer: 0.5 M Tris/HCl, pH 8.0, 0.1% w/v gelatin, 0.1% NaN_3),
ii) RIA buffer for non-specific binding (NSB) controls, or
iii) MT standards,

are reacted with primary antibody in RIA buffer and 3 x 10^5 cpm [^{125}I] MT. After reaction for 24 h at 4 °C, the immune complexes are then immunoprecipitated with a secondary antibody (anti-IgG to the species used for raising the primary antibody in). The precipitated ^{125}I is pelleted by centrifugation and determined by gamma counting.

Calculation: Specific binding of ^{125}I, corrected for non-specific binding (by subtraction of counts in NSB tubes) (Bo) is expressed as the fraction of the total radioactivity (T) added (Bo/T%). A standard curve for pure MT is plotted, against which unknowns are read.

Polarography

The sample preparation for polarographic determinations (differential pulse polarographic mode, DPP) is similar to that used in metal saturation assays (homogenisation, centrifugation and heat treatment). The final heat-treated supernatant generally being filtered through a 0.45 µm membrane filter. Samples from chromatography can be introduced directly into the polarographic cell without pretreatment. Determination conditions vary according to the instrument used. We use a PARC model 174A polarographic analyser and a 303A SMDE mercury electrode (Princeton Applied Research, EG & G, Princeton, USA) and with this electrode a modified Brdička supporting electrolyte of 1 M NH_4Cl, 1 M NH_4OH and 1.2 mM $\{Co(NH_3)_6\}Cl_3$ is used with an aliquot (100 µl) of Triton X-100 (working solution 0.125 % v/v) added to 10 ml of electrolyte to suppress a maximum that appears close to the cobalt peak. Since this is a temperature-dependent reaction the electrolyte must be kept at constant temperature for analysis. Operational conditions for this equipment are given in Olafson & Olsson (1991).

MT mRNA determinations

Experimental protocols based upon hybridisation of MT mRNA with ^{32}P-labelled cDNA or cRNA probes are outlined below. Development of non-radioactive methods is under way. It is not necessary to isolate mRNA, the assays may be performed using total nucleic acids (tNA).

RNA extraction

We routinely use the acid guanidinium thiocyanate procedure of Chomczynski & Sacchi (1987) for isolation of RNA.

Probe preparation

We use full length plaice cDNA (EMBL #X 56743, George *et al.*, 1992a) or a partial length 216bp *Alu*I fragmant of tMT-B cDNA (Zafarullah *et al.*, 1989b) as probes, although it is possible to use a synthetic 30-40 mer oligo-nucleotide as probe. Whilst we have used high specific activity [^{32}P]-labelled cRNA probes derived from these, cDNA probes are conveniently labelled by random priming using a commercially available kit which uses ca. 50 µCi a [^{32}P] dCTP / 50 ng DNA template. The plaice probe is available from the authors for non-commercial uses.

mRNA assay procedures

Solution hybridization (or RNAase protection assay) can be used although we generally use northern blotting with typically 10 µg total RNA/lane (Figure 10a). This procedure has the advantage that the integrity of each RNA sample can be assessed, however, once hybridisation conditions have been determined so that non-specific binding is eliminated, mRNA levels in a large number of very small samples can be determined more easily by slot blotting (Figure 10b).

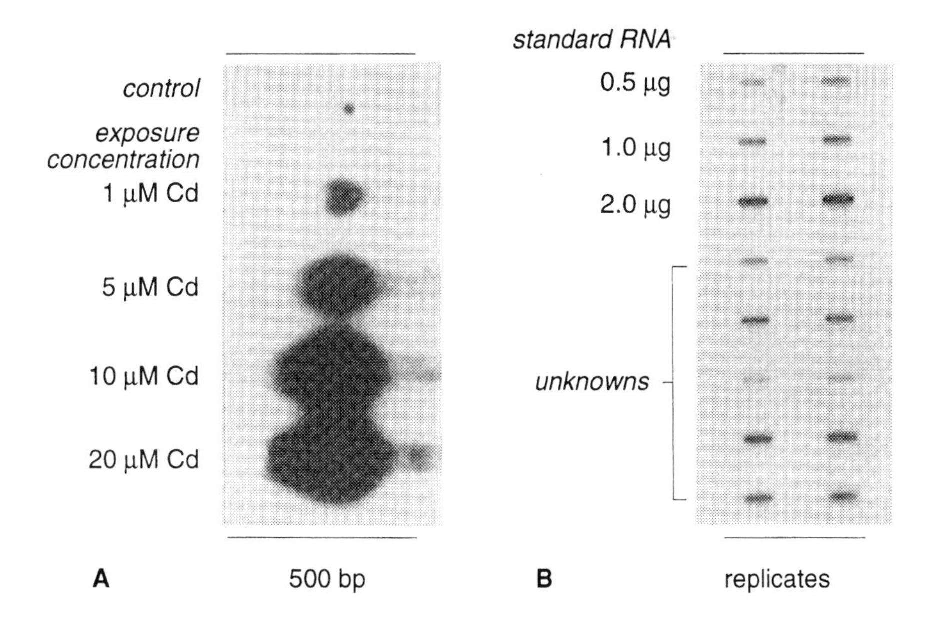

Figure 10. Detection of metallothionein mRNA by hybridisation to plaice cDNA probes.
a) Northern blot of total RNA from Cd exposed turbot (*S. maximus*) fin cell line (After: George *et al.*, 1992a); b) Slot blot of total RNA from Cd-exposed turbot liver

For *Northern blotting* tNA samples are separated according to molecular size by electrophoresis on agarose gels and hybridising bands of the correct size detected by hybridisation and autoradiography. Specific mRNA can be quantified by scanning densitometry of these autoradiographs. This procedure is essential for determination of hybridisation and washing conditions when using a probe with RNA from another species and is also useful for determining if degradation has occurred, especially in field sampled material. We use glyoxal gels since they are relatively non-toxic and can be run without undue precautions which are necessary with formamide or methyl mercury gels. Our procedure is described and referenced in George *et al.* (1992a). tNA samples are glyoxalated at 50 °C for 1 h, then separated on 1 % agarose gels run in 40 mM MOPS/acetic acid, pH 7.0, 10 mM sodium acetate, 1 mM EDTA, containing 0.5 $\mu g.ml^{-1}$ ethidium bromide at ca 70 V for 1.5-2 h using buffer recirculation. After electrophoresis the gel is photographed under UV illumination to determine if degradation of the sample has occurred (smearing or loss of 18 and 23S rRNA bands) then the RNA is transferred from the agarose gel to a nitrocellulose or nylon filter either by capillary blotting using 20X SSC (3 M NaCl and 0.3 M trisodium citrate) as blotting buffer when transfer is complete within 24 hours, or by vacuum blotting using 10X SSC for about 40-60 min. If nitrocellulose filters are used they should be dried and subsequently baked for 1 h at 80 °C in a vacuum oven. If nylon filters are being used, the RNA may be UV-crosslinked (5 min) to the filter. Filters are prehybridised on a rotisserie at 60-65 °C for 1 h with 20 ml of freshly prepared hybridisation solution (4 ml 5 M NaCl, 2 ml 0.5 M sodium phosphate pH 7.6, 12 ml water, then 2 ml 50X Denhardts solution, 200 μl 10 % SDS, 200 μl 0.5 M EDTA, 100 μl herring sperm DNA (20 $mg.ml^{-1}$, boiled before use)). 10 ml of the prehybridisation solution is then discarded and the remaining 10 ml is spiked with 10^7 cpm of probe (it is essential to boil the probe for ca 2 min, before addition). Hybridisation is then carried out overnight (16-20 h) at the same temperature. The radioactive solution is carefully discarded (although it can be reused) and the filter washed at room temperature with two 100 ml portions of 2X SSC/0.1 % SDS for 15 min each. This is repeated with 1X SSC/0.1 % SDS and then the counts on the filter checked by a hand monitor. If there are very few counts further washing should not be carried out. The filter is then put down to x-ray film and an autoradiograph prepared. An example is shown in Figure 10a. The filter can always be rewashed at a higher stringency if there is non specific binding or just re-exposed for shorter or longer periods.

Chapter 8

Use of Histopathology in Biomonitoring Marine Invertebrates

Paul P. Yevich and **Carolyn A. Yevich**

Environmental Research Laboratory
US Environmental Protection Agency, Narragansett, Rhode Island 02882, USA, retired

ABSTRACT

Histopathology is a valuable discipline for biomonitoring the effects of anthropogenic contaminants in the marine environment on marine invertebrates. Several histopathological biomonitoring studies show how histopathology can determine if physical, chemical or biological agents are having detrimental effects on the cells and tissues of marine organisms. It shows if there is interference with the health of the animals, especially with reproduction. This knowledge is valuable to the other disciplines involved with biomonitoring.

Examples of the successful use of histopathological techniques are given for the monitoring of several dump sites and dredged spoils, of power plant effluent, oil spills, and as part of the U.S. Mussel Watch Program. Several forms of abnormal and diseased tissues were recorded, as well as increased numbers of parasites.

A summary of the various aspects practical to the histopathological techniques is presented.

1. INTRODUCTION

Effects! Effects! Effects! That is what biomonitoring is all about. It seeks to answer the question: are there any detrimental effects on organisms at the individual, population or ecological level? This chapter is concerned with the effects on marine invertebrates found in estuarine or coastal areas that have been contaminated by anthropogenic sources.

Almost all present day biomonitoring programs involve chemical analysis of marine organisms, water or sediment. This tells us what chemicals are present and at what concentration for that given time period. However, these analyses do not tell us about the health of the animal populations. One of the essential tools for determining the effects of physical, chemical and biological agents on plants and animals is histopathology. It is an integral part of toxicity testing within the food and drug industry and in cancer and other medical research.

Biomonitoring of Coastal Waters and Estuaries. Edited by Kees J.M. Kramer.

Histopathology involves the microscopic examination of cells and tissues of an organism. First the normal microscopic structure of the organism as it goes through various physiological changes is determined. Then deviations from normal are looked for to determine the histopathology. A determination is made of any interference by the observed histopathology with the behaviour, physiology, metabolism, growth, nutrition and reproduction of the organism.

Histopathological studies can tell a great deal about an organism. The following is a list of some of the information that histopathologic examination can provide.

- First and foremost, it can tell if there are any histopathologic changes due to anthropogenic contaminants.
- Physiology as related to anatomical structure.
- What histopathologic changes are normally found in the animals, such as the atypical cell hyperplasia found in the soft shell clams, *Mya arenaria*, collected from the east coast of the United States (Barry *et al.*, 1971).
- Do the observed histopathological changes interfere with physiological functions such as growth, reproduction, respiration and nutrition?
- The sex of the organism.
- The stage of the reproductive cycle, how many times a year does it spawn and what stage of gametogenesis is it in. This is of importance to chemists doing lipid analysis for chlorinated hydrocarbons.
- Are the histopathologic changes noted in the cells and tissues reversible or irreversible?
- If there is interference with reproduction, is it caused by parasitic castration, nutrition or pollutants?
- Are parasites present? What type? What are their cycles? Do they leave reversible or irreversible changes such as parasitic granulomas?
- Degree of parasitism. Has the exposure to a pollutant caused the parasitic load to increase or decrease? The authors' studies of *Geukensia demissa*, and lobsters, *Homaris americanus*, collected from a highly polluted PCB site contained very few parasites in comparison to those collected from less polluted areas.
- Is there interference with growth? Is it due to pollutants, parasitism or lack of nutrition?
- What are the defense mechanisms of the animal? If the blood cells are involved, which type of cell? What role do the various types of mucous cells play? What detoxifying mechanisms are present?
- What cycle does the digestive tract, particularly the digestive diverticula, go through?
- Are there known target organs for which the contaminants have a predilection?

Examination of the literature on biomonitoring of marine organisms or of bioindicators of marine contaminants finds little mention of histopathology. Hinton *et al.* (1992) include histopathology as a tool for bioindicators, but the greater portion of the chapter is devoted to fish, not invertebrates. A study of the literature survey produced by Cantillo (1991) on Mussel Watch programs throughout the world shows that only 34 of the 1134 references contained the words histology or histopathology.

Out of these 34 references, only 6 did histopathologic studies on animals collected from the field. Studies were conducted at a GEEP workshop by Moore (1986). He showed lysomal membranes were affected in mussels collected from an area polluted by PAHs, PCBs and metals near Langesundfjord, Norway. Lowe (1988) doing studies on mussels, *Mytilus edulis*, collected from the same area, found an effect on the reproductive tissue, while Auffret (1988) referred to a greater incidence of granulomas in the mussels from the same polluted sites, which could be a consequence of chronic pollutant exposure.

Berthou *et al.* (1987) did a 7 year histopathologic study on the oysters, *Crassostrea gigas*, and *Ostrea edulis* collected from the Amoco Cadiz oil spill sites. Histopathologic lesions were noted in the digestive tract and gonads. Neff & Haensley (1982) also studied the long term effect of the Amoco Cadiz oil spill on the oyster, *Crassostrea gigas*. They showed necrosis of the digestive and reproductive tracts.

Only two extensive multiple site studies have been conducted to date. The animals were collected from each site once a year. One of these studies was the EPA 1976 Mussel Watch Program (Yevich & Barszcz, 1983) and the other is the ongoing NOAA National Status and Trends Mussel Watch Program. In the NOAA Mussel Watch Program, mussels, *Mytilus edulis* and *Mytilus californianus* are collected from 128 sites on the east and west coasts of the US. Oysters, *Ostrea sandvicensis*, are collected from 2 sites in Hawaii. The mussels have been collected for 6 years and the oysters for only 1 year. All the animals were fixed in Dietrich's fixative, cross sectioned and examined for gonadal index and neoplasia. The histopathologic findings are at present only available for the year 6 (Anon, 1992). The incidence of abnormal gonadal formation, trematodes and neoplasia is discussed.

Histopathologic studies have been conducted by Bowmer (1987, 1989) and Bowmer & Van der Meer (1991) on the histopathological condition of the fresh water zebra mussel, *Dreissena polymorpha*, and the marine bivalve *Macoma baltica* collected from polluted sites in the Netherlands.

2. CONDUCTING HISTOPATHOLOGIC STUDIES

The techniques presented in the appendices are based on the authors 27 years of experience with the histopathologic examination of over 350 different species of marine organisms. Using these techniques, histopathologic studies can be done on any marine plant or animal.

The cost of setting up a histopathologic technique laboratory is small in comparison to many other analytical disciplines. The time required for preparation of organisms for microscopic examination can be very short. Most tissue can be prepared for histopathologic examination within 48 hours; while zooplankton, copepods, amphipods and larvae can be prepared for examination within one day.

The histopathologist has many tools available to aid in the study of the effects of contaminants on the cells and tissues of marine animals. Traditional is the light microscope; distinguishing between different tissues is supported by various staining techniques. This light microscope can now be attached to a video camera; a picture of the histopathologic changes may be stored on tape or on CD-ROM along with any other data required. With the touch of a few buttons, this computer data can be recalled and more information added to the file. Colour photographs can be made of the lesions in a matter of a few minutes. With image analyzers, the lesions and cells can be quantified by size, volume and area. The fluorescent microscope allows various cellular components involved with immune responses to be studied after the cells have been exposed to different immuno-fluorescent compounds. Histochemistry is available to help determine the presence of and/or changes in mineral deposits, immune responses and cellular changes.

3. TECHNIQUES

Acceptable slides cannot be produced unless the tissues are properly prepared. It is important, therefore, that care is exercised in each step of the preparation process.

A typical sequence of handling samples for histopathological examination involves:

- a general description of the sample;
- fixation;
- decalcification;

- final trimming;
- dehydration and embedding;
- sectioning;
- staining;
- cover slipping and recording;
- storage.

Table 1 gives a general idea of the times involved in the various steps of the processing of tissues for histological examination. Details of these technical methods are presented in the Appendixes A and B and the references therein. For more details the reader is referred to some standard works on histological techniques *(e.g.* Lison, 1953; Graumann & Neumann, 1958; Davenport, 1964; Jones, 1964; Lillie & Fullmer, 1976; Sheehan & Hrapchak, 1980; Bancroft & Stevens, 1990; Carson, 1990). General works on marine pathology include Perkins & Cheng (1990), while Leake (1975) describes comparative histology; histopathological atlasses are presented by Murchelano & MacLean (1990).

When animals are received for tissue preparation, detailed observations of the appearance of the organisms should be made and recorded along with information such as the date, species (see *e.g.* Turgeon *et al.,* 1988; Williams *et al.,* 1988), area of collection and physical parameters such as salinity and temperature. This information should be recorded in a log book and each animal should be given a unique identification number.

After the gross observations are made and recorded, the animals are placed in fixative. Adequate fixation with the proper fixative is critical for the preservation of the cellular inclusions and secretions as well as the normal cell structure of aquatic animals (Appendix A). Helly's fixative made with zinc chloride instead of mercuric chloride is the fixative of choice for most invertebrates (Russel, 1947; Barszcz & Yevich, 1975). However, Dietrich's fixative is used for fish, copepods and other small zooplankton. Table 2 summarizes the suggested times required for proper fixation in the two fixatives of various groups of marine organisms.

Table 1. General procedures for processing tissues and estimated process time

Procedure	Process time necessary
Initial fixation	20 - 30 minutes
Initial trimming	as required
Final fixation	2 - 24 hours
Decalcification of bone or exoskeleton	determine empirically minimum amount necessary
Final trimming	as required
Washing in water	16 - 24 hours
Storage of tissues	until processed
Dehydration, clearing and infiltration	4 - 16 hours
Embedding	as necessary
Sectioning	as required
Staining	as required

After a short fixation period, the animals are removed and cut into small pieces and returned to fixative.

After fixation, the tissues containing bone or chitin are decalcified to allow cutting by the microtome (Brian, 1966). Most decalcification of both crustaceans and fish is done with commercially prepared products containing dilute hydrochloric acid (HCl) and a chelating agent. After this process the animals are given a final trimming, then washed to remove excess fixative.

Table 2. Suggested times in fixatives

Species	Helly's[1]	Dietrich's[2]
Zooplankton	not recommended	2+ h
Sponges	8-16 h	8+ h
Hydroids	2 h	2+ h
Anemones	2-16 h	2+ h
Corals	1-16 h	8+ h
Worms	15 min - 2 h	4+ h
Barnacles	2-4 h	4+ h
Bivalves	2-16 h	8+ h
Gastropods	4-16 h	8+ h
Cephalopods	16 h	16+ h
Crabs and Lobsters	16 h	16+ h
Shrimp	4-16 h	8+ h
Echinoderms	16 h	16+ h

[1] The amount of time in Helly's fixative is dependent on the size and density of the tissue
[2] Time given is minimum amount; tissues may be left in Dietrich's fixative indefinitely

To prepare the tissues for penetration by the paraffin, which makes them firm enough to be cut into very thin sections, the tissue must be dehydrated and cleared. After infiltration with paraffin, the tissues are embedded into paraffin blocks. Tissue sections are cut on a rotary microtome (Steedman, 1960), mounted on slides and stained *(e.g.* Gurr, 1960; Conn *et al,* 1965; Gray, 1973, 1975; Green, 1990).

For routine microscopic examination, slides are stained with Harris' hematoxylin and eosin. The hematoxylin stains the nucleus and some mucous secretory cells. The eosin stains the cytoplasm elements of the tissues. The time spent in each of the stains is determined by the type of fixative used and whether or not the tissue has been decalcified. Tissues fixed in Helly's stain more quickly than those fixed in Dietrich's fixative or have been in decalcifying fluid.

The only way to learn what is normal and what is histopathologic is to look at a large number of animals from "clean" areas and compare them histologically with animals that have been stressed in some way. It must be kept in mind that the invertebrates may go through extensive natural cyclic morphological and histological changes. These must be taken into account when determining histopathologic effects.

4. CASE STUDIES IN HISTOPATHOLOGICAL MONITORING

The authors were involved with several biomonitoring studies while at the Environmental Research Laboratory, Narragansett (Environmental Protection Agency, EPA). Examples of the successful use of histopathological techniques are given here for several aspects of anthropogenic sources, including the monitoring of several dump sites and dredged spoils, power plant effluent, oil spills, and U.S. Mussel Watch Program.

4.1. Monitoring Prudence Island

From 1979-1982 twenty-five mussels, *Mytilus edulis,* were collected periodically from a station on Prudence Island, Narragansett Bay. During 1979, a few cases of myodegeneration of the byssus muscle, inflammation and cyst formation in the thread forming area (septa) and abnormal thread formation were observed.

This continued through 1980, involving 8 to 10 animals per study. In March, 1981, there was a massive kill of mussels. These animals also showed byssus organ involvement, but it was more extensive than noted in 1979. It involved the necrosis of the entire byssus organ and a portion of the digestive diverticula. Bacterial studies were negative. This condition continued to be observed in 1982, but it was not as severe. The disease was localized at Prudence Island for mussels collected from other stations did not show such histopathologic changes. We were unable to find the causative factors of the disease.

4.2. Monitoring of dump sites

The Histopathology group at EPA was involved with the biomonitoring of several dump sites along the east coast of the United States. These included the Black Rock Harbor dump site, the Brenton Reef dump site, and the Philadelphia and DuPont dump sites. At the Black Rock Harbor dump site, the mussels were kept in cages (Active Biomonitoring, ABM) at various stations around the dump site. Histopathologically, nothing of significance was noted despite the fact that lesions were found in mussels exposed in the laboratory to the Black Rock Harbor sediment.
Histopathologic examination of *Artica islandica,* the black quahog, collected from various stations around the Brenton Reef dump site, showed tumors of the heart and fusion of gill filaments, as well as swelling of the interlamellar connective tissue of the gills. Silver was found in the basement membrane of the kidneys of animals collected from the Philadelphia dump site. Nothing of significance was noted in the animals from the DuPont dump site.

4.3. Monitoring of dredged spoils

Another example of the value of histopathology in biomonitoring was the study of the effects of dredge spoils on the sea-anemone *Cerianthopsis americanus* (Peters & Yevich, 1989). Over five hundred specimens were collected from central Long Island Sound, USA, between January 1982 and October 1984. The collection stations included a control area and three areas located two hundred (station 200mE), four hundred (station 400mE) and one thousand meters (station 1000mE) east of the primary discharge of the dredge spoils removed from Black Rock Harbor (Figure 1). *C. americanus* collected prior to the dumping of the dredge spoils were in good to excellent condition. Specimens collected from the control, 400mE and 1000mE sites continued to show good to excellent health after the completion of the dumping. Animals from the 200mE station, however, were in poor health in June of 1983 because of accumulations of cellular debris and necrosis evident in all areas of the body. The epidermis was vacuolated with erosion and loss of mucous secretory cells and ptychocysts interfered with tube formation by the anemones. Animals examined one year later from the 200mE site showed improvement in the health of the animals.

4.4. Monitoring of oil spills

Several oil spill sites along the coast of Maine were monitored histopathologically over a period of years. The most significant findings were tumors of the reproductive tract and the hemopoietic system (Yevich & Barszcz, 1977). Numerous other studies were conducted on oil spill sites, including the Amoco Cadiz spill. Histopathologic studies showed that when animals were exposed to high concentrations of petroleum products, the first response was necrosis and sloughing of the mucosa and acute inflammation. As the exposure continued, the digestive diverticula became involved and, as found at the Amoco Cadiz spill, the reproductive tract was affected (Berthou *et al,* 1987).

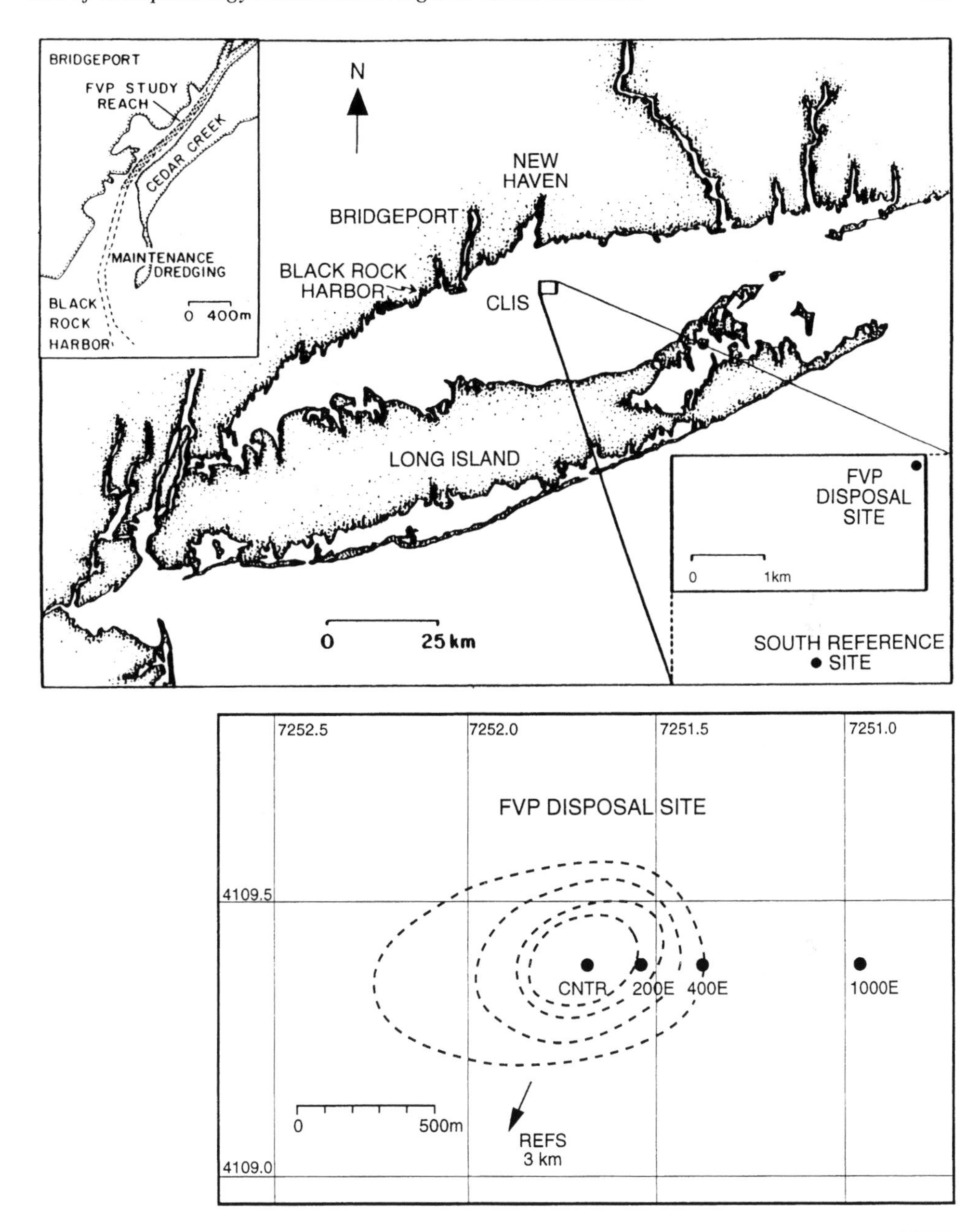

Figure 1. Central Long Island Sound disposal site and Black Rock Harbor (BRH) dredge site and FVP sampling stations.

4.5. Monitoring cooling water of a power plant

A biomonitoring study of mussels collected from a power plant effluent in New Bedford, (Gonzalez & Yevich, 1976) showed that as the heat of the effluent increased, the mussels lost cilia from the gills, followed by necrosis of the digestive diverticula. When mussels were exposed to heated sea water in the laboratory, the same histopathologic changes were observed, thus showing a cause and effect relationship.

4.6. The mussel watch program

During the U.S. Mussel Watch Program (see also O'Connor *et al.*, this volume) of 1976 (Yevich & Barszcz, 1983) our histopathologic studies showed that mussels, *M. edulis,* collected from the east coast of the United States were in poorer health than those collected from the west coast. This was especially true of the mussels collected from Boston to Cape Henlopen. These areas have high population density and high industrial activity. The animals were in poor health because of extensive parasitism (Figures 2, 3 and 4). They often contained three to five different species of parasites and the number of parasites per animal was greater than seen in animals from other areas.

The west coast mussels, *M. californianus,* had swelling of the interlamellar connective tissue of the gill filaments (Figures 5 and 6). This was considered a significant effect because it could interfere with respiration and food collecting. The swelling also caused entrapment of micro-organisms, debris and sediment.

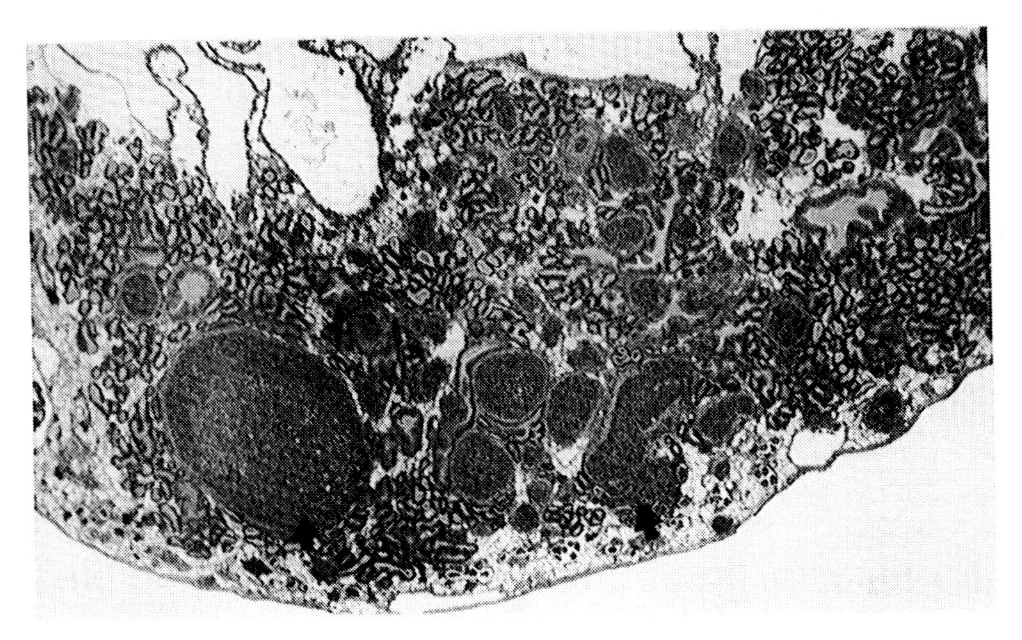

Figure 2. *Mytilus edulis*: Section through the digestive diverticula. Arrows point to very large granulomas which consist of inflammatory amebocytes, cellular debris and parasites in various stages of degeneration. There are numerous smaller granulomas scattered throughout the section (40x)

Hematopoietic tumors were noted in mussels, *M. edulis,* from San Francisco Bay and a tumor was found in the plicate gills of an oyster from Matagordo Bay.

The most prevalent parasite in the east coast mussels was the larval form, *Cercaria milfordensis.* Sporocyst development coincided with the period of gonadal development. In most animals, the sporocysts completely took over the reproductive tract. The parasite did not kill its host, but the interference with reproduction would be detrimental to the population over time. One animal from the east coast had a tumor (mesothelioma) on the pericardial wall of the heart (Figure 7).

4.7. Comparison of laboratory and field situations

Although the above studies worked out well, showing field and lab correlation, histopathologic changes that are seen in animals exposed in toxicity studies in the laboratory are rarely seen in animals collected from the field. We have exposed over 26 different species of marine organisms in the laboratory to various concentrations of cadmium chloride. The target organ in almost all species was the kidney, which showed necrosis of the epithelial cells with subsequent replacement of the columnar epithelium by squamous epithelial cells. In all of our field biomonitoring studies, we have seen this same kidney condition in only two areas.

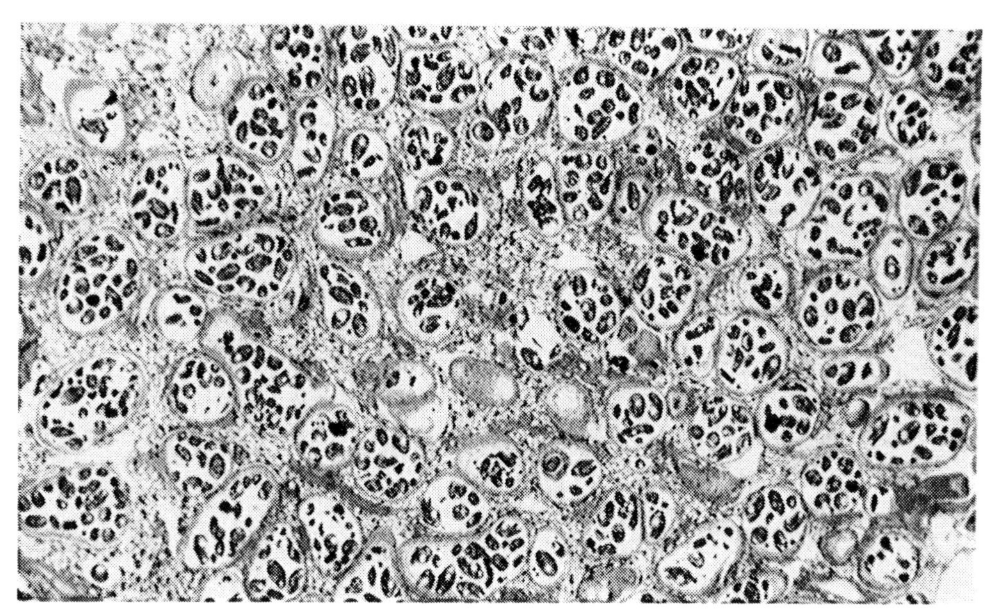

Figure 3. *Mytilus edulis*: Section of the mantle showing the reproductive tract completely taken over by the sporocysts and various stages of cercariae development of one of the *Buchephalidae trematodes* (100x)

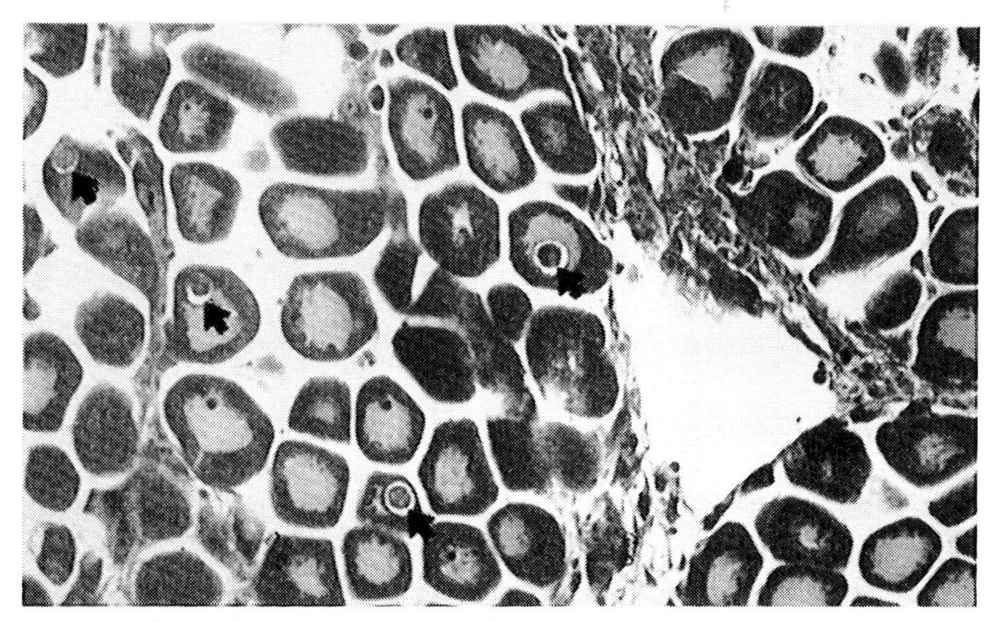

Figure 4. *Mytilus edulis*: Arrows point to spores of microsporidia located in the ova (400x)

During Mussel Watch 1976 (Yevich & Barszcz, 1983) 15 of 25 oysters examined from Caleasieu Lake, had kidney lesions that were the same as those found in the kidneys of oysters exposed to 15 $\mu g.l^{-1}$ of cadmium in the laboratory. The same type of kidney lesions were noted in oysters collected from a site in the Pawcatuck River, Rhode Island, which is highly polluted with cadmium and other contaminants.

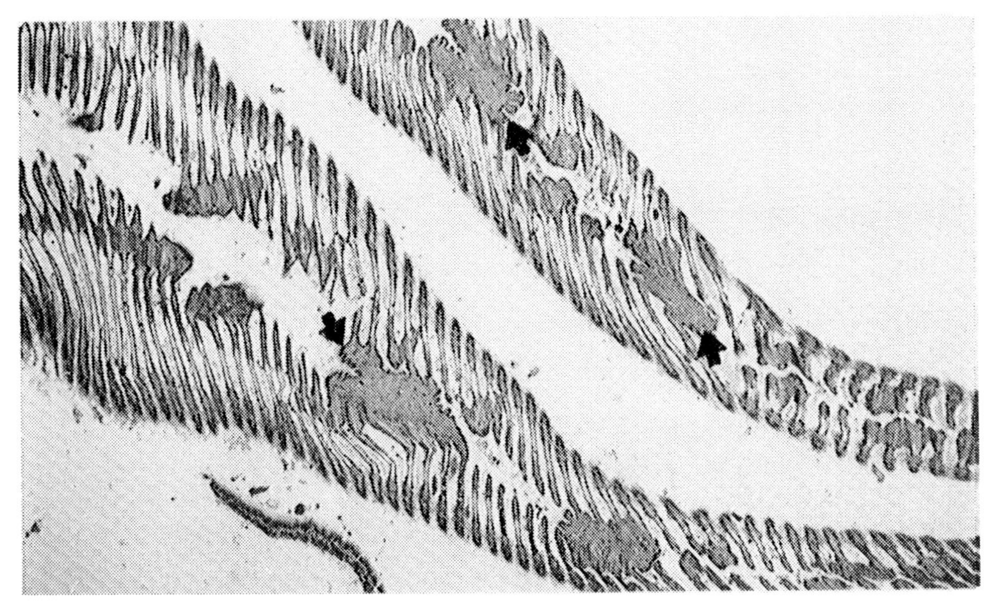

Figure 5. *Mytilus californianus*: Section through the gills. Arrows point to normal interlamellar connective tissues (40x)

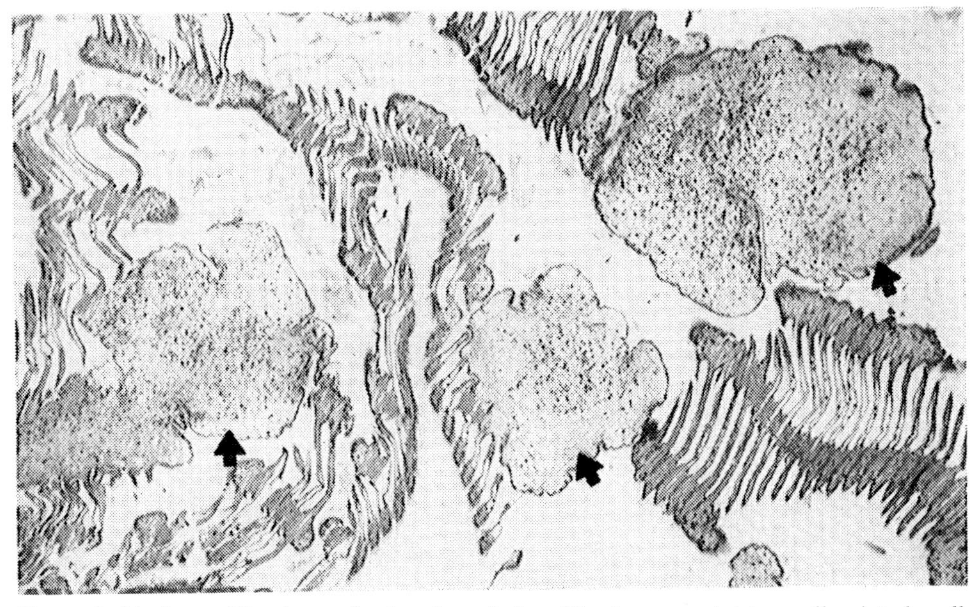

Figure 6. *Mytilus californianus*: Section through the gills. Arrows point to swollen interlamellar connective tissue with infiltration of the swollen tissue by amebocytes (40x)

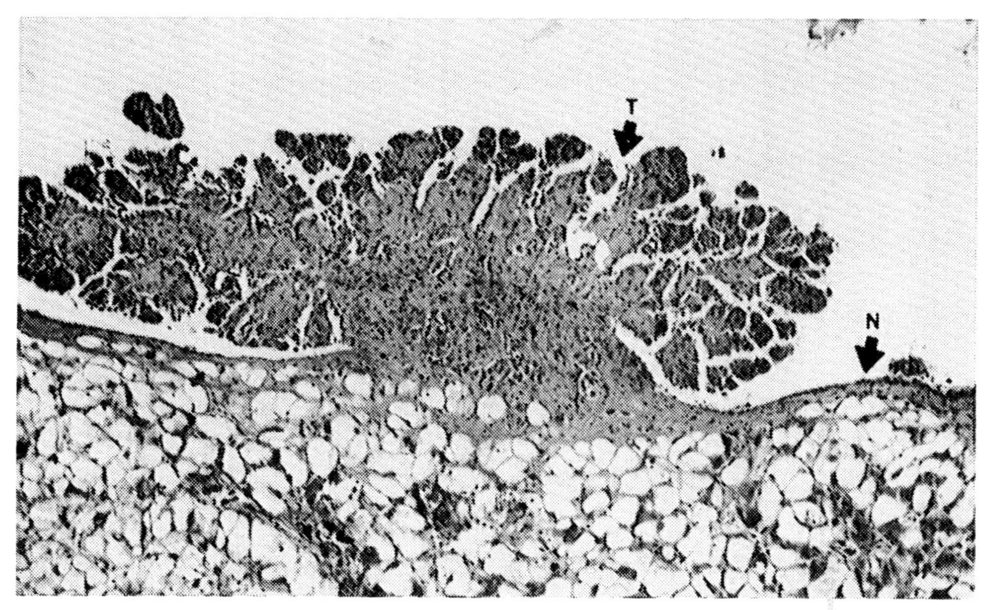

Figure 7. *Mytilus edulis*: Section through a tumor growing from the wall of the pericardial cavity. Arrow N points to the normal pericardial wall. Arrow T points to the tumor, which is a mesothelioma (250x)

5. AN IDEALISTIC HISTOPATHOLOGIC BIOMONITORING PROGRAM

Most biomonitoring histopathologic studies have been conducted on marine invertebrates, mainly molluscs (mussels and oysters) which were collected only once a year. To do histopathologic biomonitoring of marine invertebrates, several phyla (worms, crustaceans, mollusc) should be collected at least once a month or bi-weekly for a year. Invertebrates with short life spans should be examined at least every 2 to 3 days. From this the various histopathologic cyclic changes can be learned. Animal collections should be limited to only a few polluted sites plus control or reference sites. More can be learned about animals histopathology and defense mechanisms from studying a few polluted sites than from doing extensive coastal surveys.

At least 25 animals, if available, should be collected each time. Histopathologic studies have shown that if only 12 or fewer animals are collected, they may be all one sex, but with 25, a more even ratio is found.

When trimming, take as much tissue as possible. Cross sections of such animals as oysters, clams and mussels will not do as they do not contain enough representative tissue types. Sagittal sections give a better view of the histopathology and health of the animal. A careful examination of the slides of animals from clean and polluted areas involves many comparisons of equivalent tissues before conclusions can be made.

If it is at all possible, collection of animals should be coordinated with the collection by analytical chemists and other scientists. It is important to get the data on the physical parameters of the collecting sites in order to draw conclusions about observed histopathology.

CONCLUSION

Histopathologic biomonitoring studies by the authors have shown that histopathology is a valuable discipline for determining effects on cells and tissues.

The findings displayed in the above case studies are examples of how histopathology is an important part of any biomonitoring program. There is no discipline that can tell as much about the effects of anthropogenic pollution on a marine organism. Histopathologic studies should be conducted in conjunction with scientists such as chemists, physiologists, microbiologists and ecologists and should be part of any team studying the effects of contaminants on marine organisms found in estuaries and coastal waters.
The cost of carrying out histopathologic studies is small in comparison to that of chemistry and other disciplines. The tools are now available for quantifying the histopathologic changes. Most important, there are people trained in marine invertebrate histopathology who are able to interpret the changes seen in the tissues of marine invertebrates exposed to many different environments.

Note: The opinions expressed in this chapter are solely those of the authors. This chapter has not been subjected to US EPA review and therefore does not reflect the views of the agency and no official endorsement should be inferred.

REFERENCES

Anon, 1992. Year 6 final report on National Status and Trends Mussel Watch Project. Collection of bivalves and surficial sediments from coastal US Atlantic and Pacific locations. Prepared for US Dept. of Commerce, NOAA, Rockville, MD, pp. 167

Auffret, M., 1988. Histopathological changes related to chemical contamination in *Mytilus edulis* from field and experimental conditions. Mar. Ecol. Prog. Ser. 46: 101-107

Bancroft, J.D., 1975. Histochemical techniques, 2nd Ed. Buttersworth, London, pp. 726

Bancroft, J.D. & A. Stevens, 1990. Theory and practice of histological techniques. Churchill Livingstone, New York

Barry, M.M., P.P. Yevich & N.H. Thayer, 1971. Atypical hyperplasia in the soft shell clam. *Mya arenaria.* J. Invert. Path. 17: 17-27

Barszcz, C.A. & P.P. Yevich, 1975. The use of Helly's fixative for marine invertebrate histopathology. Comp. Path. Bull. 12: 4

Barszcz, C.A. & P.P. Yevich, 1976. Preparation of copepods for histopathological examination. Trans. Amer. Micros. Soc., 95: 104-108

Bayne, B.L., 1967. Marine mussels; their ecology and physiology. Cambridge Univ. Press, Cambridge, pp. 506

Bell, T.A. & D.V. Lightner, 1988. A handbook of normal penaeid shrimp histology. World Aquaculture Society, Baton Rouge, Louisiana

Berthou, F., G. Balouet, G. Bodennec & M. Marchand, 1987. The occurrence of hydrocarbons and histopathological abnormalities in oysters for seven years following the wreck of the Amoco Cadiz in Brittany (France). Mar. Environ. Res. 23: 103-133

Binyon, J., 1972. Physiology of Echinoderms. Pergamon, Oxford

Bird, A.F., 1971. The structure of nematodes. Academic Press, New York, pp. 318

Boolootian, R.A., 1966. Physiology of echinodermata. Interscience Publ., New York

Boon, M.E. & L.P. Kok, 1987. Microwave cookbook of pathology. Coulomb Press, Leiden, pp. 224

Bowmer, T., 1987. Histopathology and condition assessment of the freshwater mussel *Dreissena polymorpha* as a measure of contaminant effect: a preliminary investigation. Netherlands Organization for Applied Scientific Research, TNO report R87/157, Delft, pp. 28

Bowmer, C.T., 1989. Pathological screening and condition assessment of *Macoma balthica* from the Delfzijl Harbour canal. Netherlands Organization for Applied Scientific Research, TNO Report R89/393, Delft, pp. 24

Bowmer, C.T. & M. van der Meer, 1991. Reproduction and histopathological conditions in first year zebra mussels, *Dreissena polymorpha*, from Haringvliet, Volkerakmeer and Hollands Diep Basins. Netherlands Organization for Applied Scientific Research, TNO Report R91/132, Delft, pp. 36

Brian, E.B., 1966. The preparation of decalcified sections. Charles C. Thomas, Springfield

Cantillo, A.Y., 1991. Mussel watch worldwide literature survey 1991. NOAA Technical Memorandum NOS ORCA 63. National Status and Trends Program for Marine Environmental Quality. National

Oceanic and Atmospheric Administration, National Ocean Service. US Dept. of Commerce, Washington, pp. 129
Carson, F.L., 1990. Histotechnology. ASCP Press, Chicago, pp 294
Conn, H.J., M.A. Darrow & V.M. Emmel, 1965. Staining procedures, 2nd Ed. The Williams & Wilkins Co., Baltimore, pp. 692
Culling, C.F.A., R.T. Allison & W.T. Barr, 1985. Cellular pathology technique. Buttersworths, London, pp. 642
Davenport, H.A., 1964. Histological and histochemical technics. B. Saunders, Philadelphia, pp. 401
Freeman, W.H. & B. Bracegirdle, 1971. An atlas of invertebrate structure. Heinemann Educational Books, London, pp. 128
Fretter, V. & A. Graham, 1976. A functional anatomy of invertebrates. Academic Press, New York, pp. 589
Galtsoff, P.S., 1964. The American oyster, *Crassostrea virginica*, (Gmelin). Fishery Bulletin, Vol. 64, Fish and Wildlife Service, U.S. Government Printing Office, Washington, pp. 246
Giese, A.C. & J.S. Pease, 1979. Reproduction of marine invertebrates, Vols. 1-5. Academic Press, New York
Gonzales, J.G. & P.P. Yevich, 1976. Responses of an estuarine population of the blue mussel, *Mytilus edulis,* to heated water from a steam generating plant. Mar. Biol. 34: 177-189
Graumann, W. & K. Neumann, 1958. Handbuch der Histochemie, Vol. 1-7. Gustav Fisher Verlag, Stuttgart
Gray, P., 1973. The encyclopedia of microscopy and microtechnique, Van Nostrand Reinhold, New York, pp. 638
Gray, P., 1975. The microtomist's formulary and guide. Robert E. Krieger Publ., Huntington, NY, pp. 794
Green, F.J., 1990. The Sigma-Aldrich handbook of stains, dyes and indicators. Aldrich Chemical Co. Inc., Milwaukee, pp. 776
Gurr, E., 1960. Encyclopedia of microscopic stains. Leonard Hill Books, London, pp. 498
Guyer, M.F., 1953. Animal micrology, 5th Ed. Univ. of Chicago Press, Chicago, pp. 327
Harrison, F.W. & R.R. Cowden, 1976. Aspects of sponge biology. Academic Press, New York, pp. 354
Harrison, F.W. & J.O. Corliss, 1991. Microscopic anatomy of invertebrates, Vols. 1-15. Wiley-Liss, New York
Hinton, D.E., P.C. Baumann, G.R. Gardner, W.E. Hawkins, J.D. Hendricks, R.A. Marchelano & M.S. Okihiro, 1992. Histopathologic biomarkers. In: Biomarkers; R.J. Huggett, R.A. Kimer, P.M. Mehrle & H.L. Bergman (eds). Lewis Publ., Michigan, pp. 155-209
Lowe, D.M., 1988. Alterations in cellular structure of *Mytilus edulis* resulting from exposure to environmental contaminants under field and experimental conditions. Mar. Ecol. Prog. Ser. 46: 91-100
Hobart, W.L. (Ed), 1975. Diseases of crustaceans. Marine Fisheries Review, Vol. 37, Nos. 5-6, National Marine Fisheries Service, Seattle, Washington
Hochachka, P.W., 1983. The mollusca, Vol. 1. Academic Press, New York
Howard, D.W. & C.S. Smith, 1983. Histological techniques for marine bivalve mollusks. NOAA, National Marine Fisheries Service, Woods Hole, pp. 97
Johnson, P.T., 1968. An annotated bioliography of pathology in invertebrates other than insects. Burgess, Minneapolis, pp. 322
Johnson, P.T., 1980. Histology of the blue crab, *Callinectes sapidus*. Praeger, New York, pp. 440
Jones, R.M., 1964. McClung's handbook of microscopical technique, 3rd Ed. Hafner Publ., New York, pp. 790
Lauchner, G., 1983. Diseases of mollusca: Bivalvia, Chapter 13. In: Diseases of marine animals, Vol. II; O. Kinne (ed). Biologische Anstalt Helgoland, Hamburg, pp. 467-1038
Leake, L.D., 1975. Comparative histology. Academic Press, New York, pp. 738
Lillie, R.D. & H.M. Fullmer, 1976. Histopathologic technic and practical histochemistry. McGraw-Hill, New York, pp. 501
Lillie, R.D., E.H. Stotz & V.M. Emmel, 1990. H.J. Conn's biological stains, 9th Ed. Sigma Chemical Co., St. Lewis, Missouri
Lison, L., 1953. Histochimie et cytochimie animale: Principles et methodes. Gauthier-Villars, Paris
Luna, L.G., 1968. Manual of histologic staining methods of the Armed Forces Institute of Pathology, 3rd Ed. McGraw-Hill, New York, pp. 258
Moore, M.N., 1988. Cytochemical responses of the lysosomal system and NADPH-ferrihemoprotein reductase in molluscan digestive cells to environmental and experimental exposure to xenobiotics. Mar. Ecol. Prog. Ser. 46: 81-89

Murchelano, R.A. & S.A. MacLean, 1990. Histopathology atlas of the registry of marine pathology. NOAA, National Marine Fisheries Service, Oxford, MD, pp. 77

Neff, J.M. & W.E. Haensley, 1982. Long-term impact of the Amoco Cadiz crude oil spill on oysters *Crassostrea gigas* and plaice *Pleuronectes platessa* from Aber Benoit and Aber Wrac'h, Brittany, France. I. Oyster histopathology, II Petroleum contamination and biochemical indices in stress in oysters and plaice. NOAA-CNEXO, Rockville, pp. 269-327

Pearse, A.G.E., 1980. Histochemistry, 3rd Ed. Williams & Wilkins, Baltimore

Perkins, F.O. & T.C. Cheng, 1990. Pathology in marine science. Academic Press, San Diego. pp. 538

Peters, E.C. & P.P. Yevich, 1989. Histopathology of *Ceriantheopsis americanus* (Cnidaria: Ceriantharia) exposed to Black Rock Harbor dredge spoils in Long Island Sound. Dis. Aquat. Org. 7: 137-148

Preece, A., 1972. A manual for histologic technicians, 3rd Ed. Little, Brown & Co., Boston

Russell, W.O., 1947. The substitution of zinc chloride for mercuric chloride in Zenker's fluid. Tech. Methods Bull. Int. Assoc. Med. Museums, 21: 47

Sheehan, D.C. & B.B. Hrapchak, 1980. Theory and practice of histotechnology, 2nd Ed. Battell Press, Columbus

Shumway, S.A. (Ed), 1991. Scallops: Biology, ecology and aquaculture, Elsevier, New York, pp. 1095

Sindermann, C.J., 1990. Principal diseases of marine fish and shellfish, 2nd Ed. Vol. 1, Diseases of marine fish; Vol. 2, Diseases of marine shell fish. Academic Press, New York, pp. 521; pp. 516

Sparks, A.K., 1972. Invertebrate pathology. Academic Press, New York, pp. 387

Sparks, A.K., 1985. Synopsis of invertebrate pathology, exclusive of insects. Elsevier, Amsterdam, pp. 423

Steedman, H.F., 1960. Section cutting in microscopy. Blackwell, Oxford, pp. 172

Thompson, S.W. & R.D. Hunt, 1966. Selected histochemical and histopathological methods. Charles C. Thomas, Springfield, pp. 1639

Turgeon, D.D., A.E. Bogan, E.V. Coan, W.K. Emerson, W.G. Lyons, W.L. Pratt, C.F.E. Roper, A. Scheltema, F.G. Thompson & J.D. Williams, 1988. Common and scientific names of aquatic invertebrates from the United States and Canada: Mollusks. American Fisheries Society Special Publication 16, Bethesda, MD

Vacca, L.L., 1985. Laboratory manual of histochemistry. Raven Press, New York

White, K.M., 1937. *Mytilus*. L.M.B.C. Memoirs, Vol. 31. Liverpool Univ. Press, Liverpool, pp. 117

Wilbur, K.M. & C.M. Yonge, 1966. Physiology of mollusca, Vol. 2. Academic Press, New York

Williams, A.B., L.B. Abele, D.L. Felder, H.H. Hobbs, R.B. Manning, P.A. McLaughlin & I.P. Farfamte, 1988. Common and scientific names of aquatic invertebrates from the United States and Canada: Decapod crustaceans. American Fisheries Society Special Publication 17, Bethesda, MD

Yevich, P.P. & C.A. Barszcz, 1977. Neoplasia in soft-shell clams, *Mya arenaria,* collected from oil impacted sites. Ann. N.Y. Acad. Sci. 298: 409-426

Yevich, P.P. & C.A. Barszcz, 1983. Histopathology as a monitor for marine pollution. Results of the histopathologic examination of the animals collected for the U.S. 1976 Mussel Watch program. Rapp. P.-v. Réun. Cons. int. Explor. Mer, 182: 96-102

Appendix 8A

General Procedures for Tissue Preparation

A typical sequence of handling samples for histopathological examination involves: a general description of the sample, fixation and decalcification, final trimming, dehydration and embedding, sectioning, staining, cover slipping and recording, and finally storage. Table 1 gives a general idea of the times involved for the various steps involved in processing tissues. Some recent general works on histochemical techniques include Pearce (1980), Culling *et al.* (1985), Vacca (1985), Boon & Kok (1987), Bancroft & Stevens (1990). Many helpful and practical hints on histotechniques can particularly be found in Jones (1964), Guyer (1953), Preece (1972) and Gray (1973, 1975).

FIXATION

The most important step in the preparation of marine organisms for histopathologic examination is the process of fixation. Fixation is necessary to prevent autolysis (tissue self-destruction) and to preserve the tissue in as life-like condition as possible. For detailed information the reader is referred to various handbooks, such as: Davenport (1964), Thompson & Hunt (1966), Bancroft (1975), Lillie & Fulmer (1976), Pearce (1980) and Bancroft & Stevens (1990).
It is essential that animals are fixed as soon as possible after removal from water. Tissues must be immersed in at least twenty times their volume of fixative. The proper size container must be chosen taking into account the size of the specimen and the volume of fixative required. Fixation does not have to be done in any special containers. Any type of baby food jar, tin can, plastic bag, food container or plastic bottle with the top cut off will serve the purpose.

Fixatives

Helly's fixative, made with zinc chloride instead of mercuric chloride (Russel, 1947), is the fixative of choice for most invertebrates (Barszcz & Yevich, 1975). However, Dietrich's fixative is used for fish, copepods and other small zooplankton.
Helly's Fixative
Helly's fixative is an excellent cytological fixative for aquatic invertebrates, preserving the nucleoplasm of the nuclei of ova and the granules of the secretory cells and amoebocytes better than other fixatives tested (Barszcz & Yevich, 1975). Tissues fixed in Helly's stain intensely with hematoxylin and eosin and many special stains. However, the amount of time that the tissue remains in this fixative must be carefully controlled because prolonged fixation will cause excessive hardening and brittleness, making sectioning difficult or impossible.

The length of time the tissue can be left in Helly's is dependent on the size of the pieces and the density of the material. Twenty-four hours is usually considered the maximum allowable time (see Table 2).

Formula for Helly's Fixative

1,000 g zinc chloride ($ZnCl_2$)
500 g potassium dichromate ($K_2Cr_2O_7$)
20 l distilled water

Add 5 ml of 37-40% formaldehyde per 100 ml of fixative at time of use.

Zinc chloride is used in Helly's fixative instead of mercuric chloride because it is less toxic and does not form a precipitate that must be removed during staining. It has the same fixing and mordanting qualities as mercuric chloride.

Dietrich's Fixative

Dietrich's fixative gives good cellular detail and there is little shrinkage or distortion of the cells. Animals can be stored in Dietrich's for 2-3 months if necessary. The major drawback to using Dietrich's or any other fixative that contains glacial acetic acid for invertebrates, is that it causes the loss of secretions and granules and loss of nucleoplasm of ova. However, it is useful for very small invertebrates such as copepods (Barszcz & Yevich, 1976).

Formula for Dietrich's Fixative

9,000 ml distilled water
4,500 ml 95% ethanol
1,500 ml 40% formaldehyde
300 ml glacial acetic acid

DECALCIFICATION

After fixation is complete, tissues containing bone or chitin must be decalcified to allow cutting by the microtome. Most decalcification of both crustaceans and fish is done with commercially prepared products containing dilute hydrochloric acid (HCl) and a chelating agent. The amount of time necessary to decalcify varies with the type of tissue and the amount of area involved (Brian, 1966). Experience will indicate the length of time necessary for each species. There are many methods of determining the end point of decalcification mentioned in the histotechnique books. We do so by attempting to trim the tissue. As soon as it will cut easily, the tissue is removed from the decalcifying fluid, trimmed and washed. It is important not to leave the tissue in a decalcifying fluid any longer than necessary because it has an adverse effect on the staining reaction of the tissues.

The use of Dietrich's fixative shortens the time necessary in decalcifying fluid because the glacial acetic acid acts as a decalcifying agent. Small animals such as amphipods and copepods that are left in Dietrich's fixative for 2-3 days need no further decalcification. Fish larvae and very small fish that have been in Dietrich's for 3-4 days will need little if any decalcification, but it will be necessary to place larger fish in decalcifying fluid for several hours (Table 2).

It is important that all tissue that has been in a decalcifying fluid must be washed for 24 hours to allow proper staining reactions.

WASHING AND FINAL TRIMMING

All tissue is washed overnight (16 hours) in a running water bath to remove all excess fixative. If the tissue has been decalcified, it must be washed for at least 24 hours to remove the acid. On the authors' field collecting trips, the animals collected from each station were put in a mesh bag, together with an identifying tag, and placed in the water storage tank of a toilet. Periodic flushing of the toilet washed the tissues. Along the ocean or estuaries, a stake was pounded into the soil. The mesh bags were tied to the stake by a rope and allowed to hang in the water which rinsed out the excess fixative.

After fixation and decalcification are completed, large pieces of tissue are cut into sections no more than 2-3 mm thick and small enough in diameter to fit into the cassettes and eventually into the embedding molds. The surface of tissue to be examined is placed face down in the cassette to enable the embedder to orient the tissue correctly in the paraffin block. The animal's identification number is included in each cassette on a small piece of paper.

Any form of knife or cutting edge can be used to trim the fixed tissues. We have used razor blades for years. In place of regular metal or plastic cassettes, cheese cloth bags can be used to carry the tissues through the dehydrating solutions.

DEHYDRATION, CLEARING AND EMBEDDING

After washing, the cassettes are stored in a dehydration solution until processed. Tissues are prepared for embedding in paraffin by taking them through a series of dehydration and clearing steps.

Graded ethanol solutions are generally used for dehydration, and xylene or chloroform are used for clearing the tissues and preparing them for infiltration by the paraffin. The handbooks mentioned list the various alcohols, xylenes and cedar wood oil, among other reagents for the dehydration and clearing process.

If the tissue is to be processed within two weeks, it can go directly into the dehydrating reagent used in that procedure, but if the tissues are to be stored longer than two weeks, the use of 70% ethanol is recommended as a storage reagent. If the alcohol becomes discolored, it should be changed.

After dehydration and clearing, the tissues are carried through at least three changes of melted paraffin. It is recommended that tissues be embedded in paraffin with a melting point of 56-57 °C. This can be done in metal containers in a slightly warm oven. The temperature of the oven should not exceed the melting temperature of the paraffin as the heat will harden the tissues. A hot plate, bunsen burner or stove can also be used, but the temperature must be carefully controlled. The senior author used in the field a five gallon paint can, heated with a 100 W bulb to keep the paraffin melted.

The 16 hour or night run is convenient to use for machine processing of most tissues. For hand processing on the 16 hour schedule, the cassettes can be left in fresh dehydrant overnight and the schedule completed the next day. Many small containers such as watch glasses, match boxes, and aluminum foil can be used as embedding molds as well as paper boats. Glycerine applied to the inside of the glass and aluminum molds will aid the removal of the cooled paraffin block.

Tissues that have been stored in a dehydrating solution for several days can be processed on the day run. This schedule can also be used for soft tissues that are easily penetrated by the reagents. Penetration by the paraffin is enhanced by placing the tissue under vacuum infiltration for at least one half hour. This also helps to remove air bubbles trapped in the tissue. Tissues should be embedded as rapidly as possible to minimize further exposure to the hot paraffin. The short run (10 minutes in each reagent) is used for zooplankton, small fish larvae and eggs, nerve fibers and ganglions, and any other delicate tissues. These tissues should not be put into a vacuum infiltrator.

SECTIONING

A rotary microtome is necessary for cutting the blocks. For most slides, it is set at 6 microns. Because of the presence of sand and shell particles found in many molluscs and other invertebrates, disposable knives are used (Steedman, 1960). The tissue ribbons are floated on a warm water bath to allow the sections to flatten and expand to their original shape. One, two or more sections, depending on the size of the block, are mounted on 2.5 x 75 mm etched slides that have been marked with the animal's identification number. The slides are drained, put into a staining rack and held in a 57 °C oven for at least 3-4 hours until the protein in the tissue adheres to the slide. With this treatment, no adhesives such as gelatin or albumen are usually necessary. When removed from the oven, the slides can be stained or stored in slide boxes at room temperature until needed. After sectioning, the block is sealed by dipping the cut surface in paraffin to prevent dehydration and it is filed for future reference. If upon microscopic examination, something of particular interest is found in the tissues, the block can be recut and special stains done as necessary.

STAINING

Various handbooks deal with staining and dying techniques (*e.g.* Gurr, 1960; Conn *et al.*, 1965; Luna, 1968; Gray, 1975; Green, 1990; Lillie *et al.*, 1990).
For routine microscopic examination, slides are stained with Harris' hematoxylin, the nuclear stain, and eosin, the cytoplasmic stain. They can be purchased ready-made or made up from the formulations given in the various reference books. The time spent in each of the stains is determined by the type of fixative used and whether or not the tissue has been decalcified. Helly's fixed tissue stains more quickly and intensely than Dietrich's fixed tissue stain. If the tissue has been decalcified, the staining times must be increased.
If special histological features are to be reviewed after the initial reading of the slides, the blocks can be recut and special stains applied.

COVERSLIPPING AND RECORDING FINISHED SLIDES

After the slides are stained, they are coverslipped using a synthetic resin and 24 x 60 mm coverslips and labelled with a permanent tag bearing the name of the laboratory and the animal's identification number. The number of slides prepared from each animal and the date the slides are completed are recorded in the log book. After the slides have been examined by the histopathologist, they are filed for future reference.

Appendix 8B

Guidelines for Fixing and Decalcifying Specific Groups of Animals

In the following sections groups of invertebrates are discussed for their optimal treatment. In general, all organisms are fixed in Helly's fixative, except zooplankton.
Apart from the following general references on invertebrate histology (*e.g.* Johnson, 1968; Freeman & Bracegirdle, 1971; Sparks, 1972, 1985; Leake, 1975; Fretter & Graham, 1976; Giese & Pease, 1979; Harrison & Corliss, 1991) specific references are included that focus on the histology and histological techniques of specific groups of organisms.

ZOOPLANKTON

Dietrich's fixative is added directly to the collecting medium. As the animals die, they settle to the bottom of the container at which time more of the collecting medium can be decanted and fixative added. A few drops of eosin added to the Dietrich's fixative in the container will make the zooplankton more visible, aiding in the embedding process. The minimum fixation time for small copepods is 2 hours. They can be left in Dietrich's fixative indefinitely (Barszcz & Yevich, 1976).

SPONGES

After initial fixation, sections are taken from various areas of the sponge representative of surface and interior. These should be no more than 2-3 mm thick and of a diameter to fit into a cassette. If it is desirable to examine boring sponge, the entire shell of the bivalve containing the sponge can be put into fixative for 8-16 hours and then decalcified to allow the sponge to be sectioned.

HYDROIDS AND ANEMONES

To keep them extended during fixation, they can be gradually narcotized by adding a 1-2% solution of magnesium sulphate to the water that the animals are in and then gradually increasing the concentration up to 30% before adding fixative. Anemones usually are trimmed in such a way as to allow examination of the tentacles, body cavity and base of the animal. Hydroids should be fixed for 2-4 hours and anemones for 2-16 hours depending on size of tissue.

CORALS

True corals or stony corals must be carefully broken into sections small enough to fit into the fixative container. If they are small, they can be left in seawater until they extend and are narcotized like anemones and hydroids before fixation. After fixation, the coral must be decalcified before it can be trimmed to the proper size for processing. Cross and longitudinal sections are taken. All tissue must be well washed after decalcification. Horny corals containing gorgonin do not need to be decalcified. The fixation time varies from 15 minutes to 2 hours depending on the size of the colonies. Longer fixation will result in poor tissue sections.

WORMS

Worms less than 3 cm long, can be fixed whole. They can be dropped directly into fixative or narcotized first and then put into fixative. If they are put on paper towel and the fixative added with an eye dropper they will remain flat and can be put into a larger container of fixative to complete fixation. Larger worms, after initial fixation, are cut into sections about 3 cm long. When trimming, cross, sagittal and longitudinal sections are taken from each animal. Species such as the sea mouse are trimmed into cross and sagittal sections small enough to fit into the cassettes. Fixation times varies from one hour for small worms to 16 hours for the larger, thicker sections.

BARNACLES

Barnacles are put directly into fixative. If the shells are very large, they can be cracked with shears. After fixation the shell is decalcified and the tissues trimmed as desired. Fixation should last from 2-4 hours depending on the size of the barnacles.

BIVALVES

Hard-shell clams, such as the black clam *(Artica islandica)* and the quahog *(Mercenaria mercenaria)* are opened by cutting the ligament and wrenching the shell apart to tear the muscle. Using a quahog knife, the animal is shucked free of both valves and dropped into fixative. Care should be taken not to tear the mantle when removing the animals. Animals such as the soft-shell clam *(Mya arenaria)* and the surf clam *(Spisula solidissima)* can be opened by inserting the quahog knife into any available opening between the mantle and the valve. The knife is moved with a sweeping motion to loosen the mantle and muscle from the shell and wrenching the shells apart with a twist of the wrist. Scallops are opened by inserting a spatula between the mantle and valve and with a sweeping motion, cutting the muscle from the valve. Mussels are opened by forcing a sharp quahog knife between the two valves slightly above the byssus threads so that the tip of the knife is between the mantle and the shell. With a sweeping motion, the muscle and mantle are cut loose from the shell.

As the above animals are removed from their shells, they are put into individual bottles of fixative for 15-30 minutes, then removed and trimmed. Large clams are usually cross-sectioned into 3-4 mm sections beginning at the siphon. Oysters are cross-sectioned through the body mass just to the side of the pericardial cavity opposite the adductor muscle; then each half is cut sagittally. All scallops are cut sagittally into 2, 3, or 4 sections depending on the size of the animal. Mussels are trimmed by separating the mantel from the body mass and

cutting the body mass sagittally into 2 or 4 sections. Fixation time for most bivalve tissues is from 16 to 24 hours.
Small bivalves (up to 1.5 cm) are dropped into fixative after their shells have been cracked slightly, but not crushed. This allows the fixative to penetrate, and the animals will be easier to remove after fixation. They are cut sagittally, or if very small, processed whole. If the spat are too small to open and remove from their shells, after the appropriate fixation time (6-12 hours), the fixative is removed and decalcifying fluid is added. Decalcification is complete when no hard areas of shell remain. Be sure to wash tissues adequately after decalcification.
Shells of aquatic animals can be prepared for study by the grinding techniques used for preparing teeth for microscopic examination.
For a treatise of the histology for marine bivalve molluscs one is referred to Howard & Smith (1983), principal diseases are treated by Sindermann (1990), while scallops are reviewed by Shumway (1991)

GASTROPODS

Shelled gastropods must be removed from their shells to allow fixative penetration. This is done by cracking the shell and slowly chipping it away from the animal with ronguers or bone forceps. Care must be taken not to injure the body mass during the chipping process. Animals like slipper shells *(Crepidula fornicata)* and limpets *(e.g. Patella* sp.) are dropped into fixative, shell and all. After a short period of fixation, they can be easily pulled from their shell and trimmed as desired. If the snails are large, they must be trimmed into 3 mm thick sections to allow penetration by the fixative. Fixation time ranges from 4 hours for small animals to 16 hours for large ones.

CEPHALOPODS

The animals are put whole into fixative until dead, then trimmed as desired.

CRABS & LOBSTERS

Crabs and lobsters are put live into containers with enough fixative to cover the animals. The containers should be covered to prevent the fixative from being splashed out by the dying animal. After death, the legs and claws are cut from the body. The claws should be opened with scissors or cut in half sagittally, if they are too large to permit penetration by the fixative, and returned to fixative in a large jar.
The body of a crab is prepared by cutting the abdomen from the carapace and pulling the body apart, before returning it to the fixative. After overnight fixation, the muscle and internal organs are cut from the body and put into cassettes. All chitinous tissue, including the stomach, is put in a decalcifying fluid until soft, then trimmed according to interest.
Lobsters are prepared by loosening the carapace from the thorax and cutting it off behind the eyes. The intestine, reproductive tract, heart, and hepatopancreas are cut free and the body of the animal is cut through the midline from the anterior to the posterior end. If the animal is very large, pieces of muscle and exoskeleton are cut off the tail. All tissue is returned to fixative overnight. After overnight fixation, soft tissues, trimmed to 2-3 mm, are ready to be put into cassettes. At this time, the ventral nerve fibers and their ganglions are easily removed from between the muscle bundles. The nerve fiber is trimmed to suitable lengths, washed, and then processed on a short run. All hard tissue, including the stomach, is

decalcified and trimmed as desired. These tissues must be washed for 24 hours before processing. See *e.g.* Hobart (1975), Johnson (1980).

Hermit Crabs

Hermit crabs are pulled from their shells and put into fixative. Larger crabs must be cut into several sections. After fixation, the chitinous portions of the animals are decalcified and the animals trimmed sagittally or cross-sectioned.

Shrimps

Shrimp are put live into fixative, and when dead, are removed and trimmed. Small shrimps such as sand shrimp (*Crangon septemspinosa*) are cut sagittally and large shrimp are either cross-sectioned or cut sagittally, then returned to fixative for 4-16 hours. All shrimps are decalcified before processing (Bell & Lightner, 1988).

ECHINODERMS

Starfish, if small, are put directly into fixative. Large animals, after initial fixation, are cut to permit better penetration of the fixative. Fragile starfish disintegrate when placed directly into fixative. These animals should be carefully placed in a solution of 1-2% magnesium sulphate ($MgSO_4$). This concentration is gradually increased until the animals are narcotized and then the fixative can be added.

Sea cucumbers are fixed briefly and then cut with a razor blade both dorsally and ventrally to allow penetration by the fixative for up to 16 hours. After fixation the animals are decalcified and trimmed to the proper size. See *e.g.* Boolootian (1966) and Binyon (1972).

Appendix **8C**

Comparative Histology of Selected Molluscan Tissues

The determination of the histopathologic effects of stress on marine invertebrates is very difficult unless there is knowledge of the histological variations that the organs go through during cyclic and physiological changes. Many of the histological changes seen in the tissues of invertebrate would be classified as pathologic by a mammalian or human pathologist. The marine histopathologist soon learns to regard these changes as normal.

During the past 27 years, we have observed many histological changes in marine animals which are considered part of the normal cycle or physiological response of these animals. This section will discuss, without going into great detail, some of the tissue changes seen in commercially important species of bivalve molluscs.

Useful handbooks on the biology and physiology of mollusca include *e.g.* White, 1937; Wilbur & Yonge, 1966; Bayne, 1967; Hochachka, 1983; Howard & Smith, 1983; Lauchner, 1983; Sindermann, 1990).

***Mercenaria mercenaria,* quahog**

Kidney: consists of folds, will contain black concretions between the folds (Figure 8).

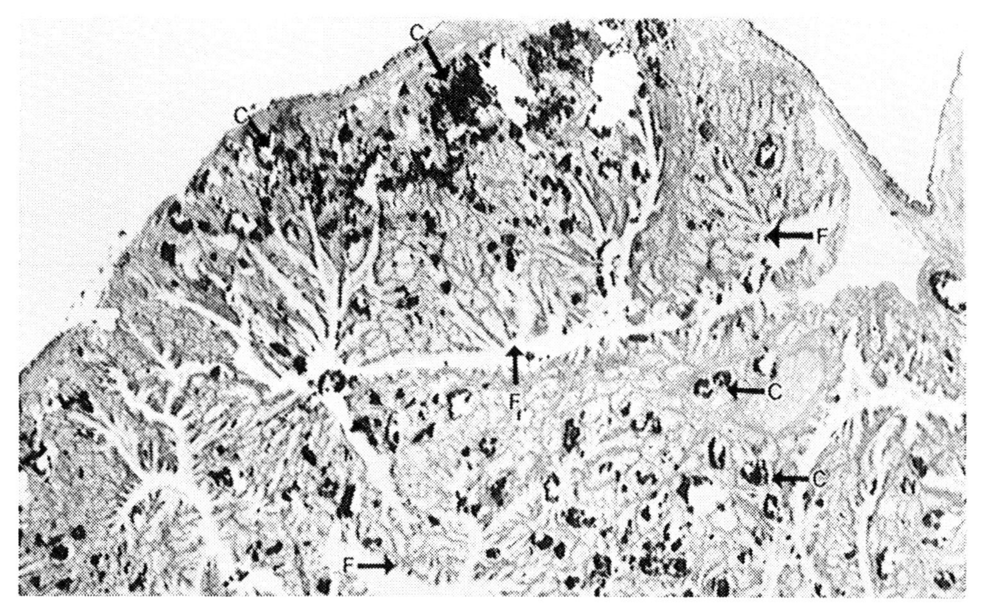

Figure 8. *Mercenaria mercenaria*: Section through the kidney. Arrow C points to numerous black concretions. Kidney consists of folds, Arrows F, rather than tubules (40x)

Reproduction: synchronous. Has genital pore as well as excretory pore. Gonadal ducts have one side lined with an epithelium containing red granules.
Heart: has a truncus arteriosus.
Brown cell system: red gland only containing large vesicles.
Gills: water tubes contain blue mucous secretory cells. In a highly polluted area, the mucous cells may not be present.

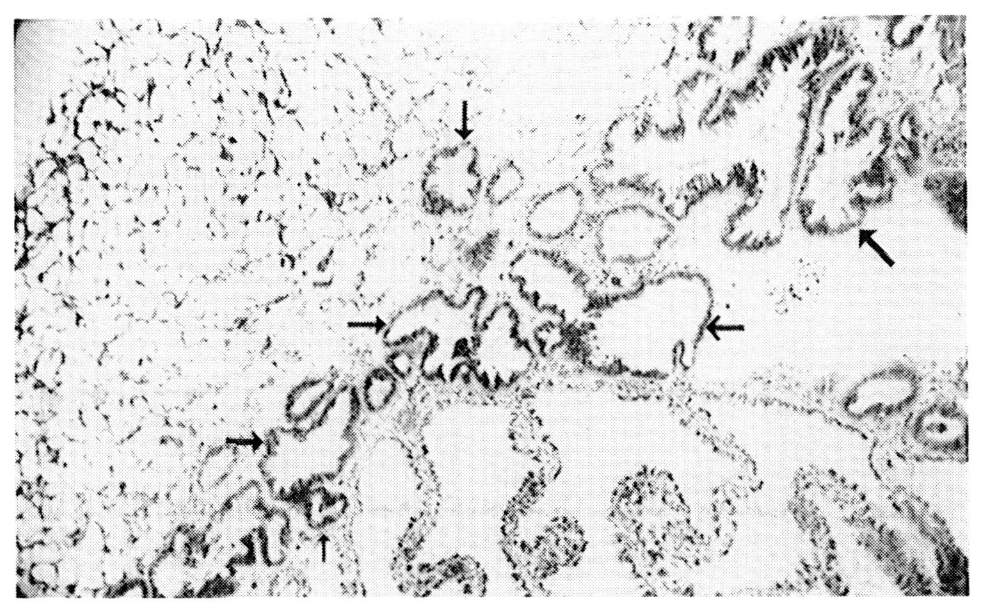

Figure 9. *Mytilus edulis*: Section through the kidney. Arrows point to the kidney tubules whose walls are lined with cuboidal to columnar cells. The cytoplasm of the cells is jammed with fine black particulates (40x)

***Mytilus edulis*, common or blue mussel**
Kidney: made up of tubules in which the epithelium may be jammed with grayish black granules. Will form calcium concretions in the tubules when under stress (Figure 9).
Reproduction: most of the reproductive tract is in the mantle. Orange mantle - female, white mantle-male. Reproductive follicles may also be found among the digestive diverticula tubules and in the visceral mass.
Brown cell system: found on the auricle wall and consist of fine vesicles in the cells.
Gills: filamentous and held together by ciliary junctions and interlamellar connective tissue. Blue staining secretory cells usually are found on the gill filaments. Numerous red granular secretory cells indicates stress.

***Crassostrea virginica*, oyster**
Kidney: consists of tubules. Concretions found in tubules when under stress. Pink, brownish or purplish granules may be found in tubular epithelium.
Reproduction: one side of gonadal duct is lined with ciliated cuboidal to columnar epithelium.
Brown cell system: found around the muscle bundles of the auricle and in the visceral connective tissue. Under stress, the brown cells in the auricle will decrease while increasing in the visceral connective tissue.
Gills: water tubes are lined with red granular secretory cells. In long term, low level exposure to contaminants, these cells will be lost.

Digestive system: the stomach duct epithelium will contain areas in which the columnar epithelium is jammed with red enzymatic granules. Tubules of the digestive diverticula may dilate to a squamous epithelium which may be due to a stage in the digestive cycle. See also Galtsoff (1964).

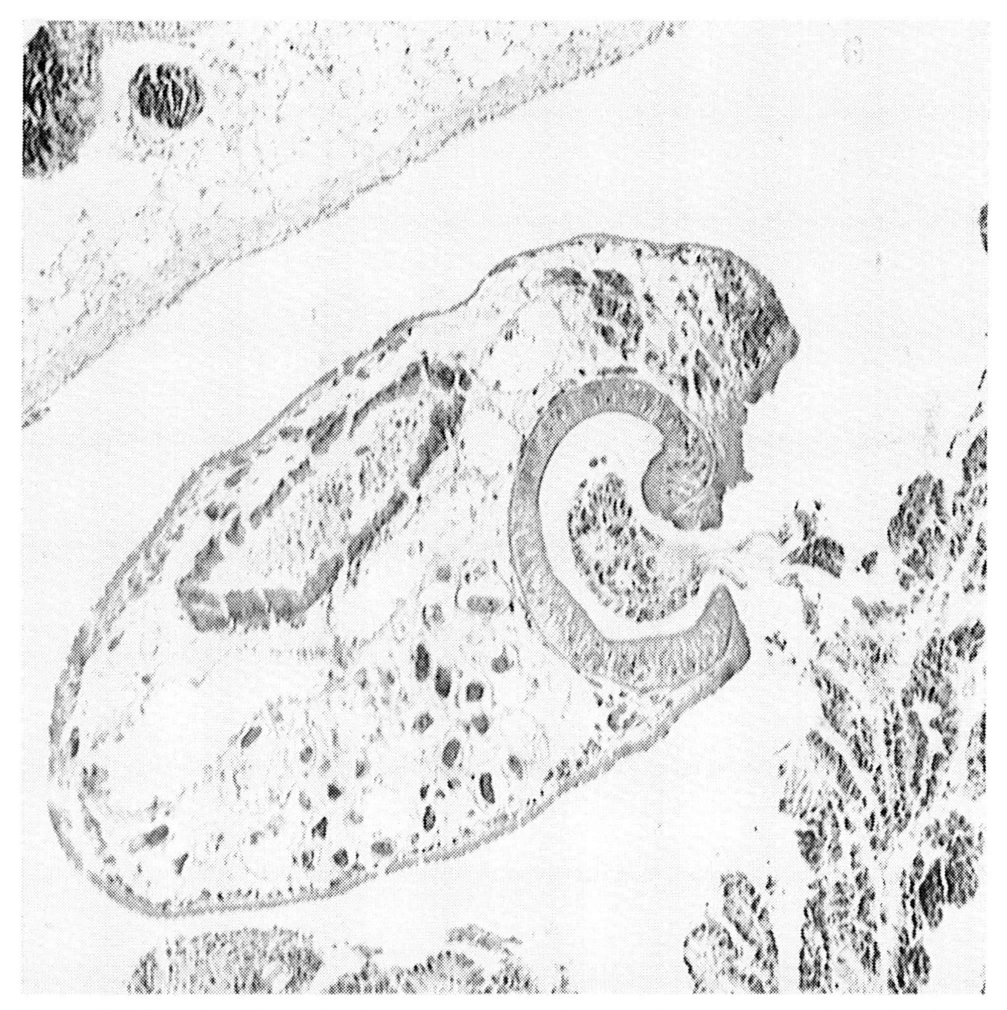

Figure 10. The authors' favourite: 'eat your heart out'. A trematode consumes the heart of a *Mytilus edulis*. (40x)

***Argopecten irradians*, bay scallop**

Kidney: located outside the visceral mass. Consists of folds. During spawning season, the clear columnar cells fill up with calcium concretions which are sloughed into the lumen, increasing the size of the kidney 2 or 3 times.

Reproduction: hermaphroditic. Male portion - white, female - orange. Gonadal ducts lined with blue secretory cells.

Brown cell system: in the auricle muscle bundles.

Digestive system: digestive tubules contain yellowish to green inclusions. Some say they are waste debris, some say dinoflagellates.

***Mya arenaria*, soft shell clams**

Kidney: consists of folds, may have concretions when under stress.

Reproduction: after spawning, the male follicles may contain atypical cell inclusions and the female follicles contain follicle cell inclusions.

Brown cell system: found on the wall of the auricle and in red gland. Brown cells contain fine vesicles.

***Spisula solidissima*, surf clam**

Kidney: consists of folds, concretions formed under stress.

Brown cell system: auricle walls and red gland. Cells made up of fine vesicles.

Secretory cells: numerous blue staining mucous secretory cells are found in the epithelium throughout the whole animal.

Chapter 9

Biomonitoring of Tributyltin (TBT) Pollution using the Imposex Response of Neogastropod Molluscs

Peter E. Gibbs and **Geoffrey W. Bryan**†
Plymouth Marine Laboratory
Citadel Hill, Plymouth, Devon PL 1 2PB, UK

ABSTRACT

Tributyltin (TBT) compounds have been used as biocides in marine antifouling paints since the mid-1960s: high levels of TBT were recorded in coastal waters during the 1980s. The leachates of these paints are now known to have deleterious effects on many non-target organisms: the most sensitive group appears to be neogastropod molluscs, the females of which signal exposure to TBT by becoming masculinised, *i.e.* develop male sex characters, notably a penis and sperm duct. This phenomenon - 'imposex' -is induced at concentrations less than 1 $ng.l^{-1}$, below the level of TBT detection by chemical methods. The degree of development of imposex is dose-dependent and thus it can be used as a sensitive bioindicator of TBT pollution. This review outlines the stages in its development in relation to TBT exposure and the various indices that have been used as comparative measures for biomonitoring purposes. Sampling techniques, application of indices and recognition of effects, leading in certain species to sterilization and population extermination, are discussed.

1. INTRODUCTION

Marine antifouling paints containing tributyltin (TBT) compounds as biocides were introduced in the mid-1960s and because of their effectiveness and ease of application they rapidly gained popularity, particularly in the pleasure boat industry. In the 1980s, usage of these paints had become so widespread that high levels of TBT leachate were recorded in coastal waters, notably near marinas (see Waldock *et al.*, 1987). Concomitant studies demonstrated the toxicity of TBT compounds to non-target organisms (see Rexrode, 1987; Waldock *et al.*, 1987): these compounds are now known to produce a variety of pathological condi-

Biomonitoring of Coastal Waters and Estuaries. Edited by Kees J.M. Kramer.

tions in marine species at relatively low concentrations (see Bryan & Gibbs, 1991) but, in terms of sensitivity, none rivals that of the 'imposex' response induced in neogastropods (Gibbs *et al.*, 1991c). Legislation to restrict the use of TBT paints on small boats in French waters was introduced in 1982; similar regulations for UK waters followed in 1987 and latterly for many worldwide regions. Consequently, TBT concentrations in seawater have declined in many coastal areas, but with the continued use of TBT paints on large vessels (more than 25 m length) and the persistent release of TBT trapped in sediments (see Langston *et al.*, 1990) the levels of pollution still remain a cause for concern. Even if a total ban on TBT paint usage could be implemented now, the toxic effects of leachates would be detectable for many years.

Prosobranch gastropods exhibit all types of sexuality but in the majority the sexes are separate and unchanged throughout the life of the individual (Fretter & Graham, 1962). Neogastropods are strictly gonochoristic and fertilisation is internal, sperm being transferred by copulation. Typically, eggs are encapsulated before laying and larval development usually includes a planktonic phase of variable duration (few weeks to several days) but in some species this phase is omitted and development is direct, as in the chief subject of this account, the dogwhelk *Nucella lapillus* (L.). Recognition of the sexes was, until the late 1960s, a routine matter, involving purely a determination of the presence or absence of a penis. This situation changed around 1970 when penis-bearing females were noted in widespread populations of at least four neogastropods, namely, *Nucella (or Thais) emarginata* (Deshayes) [California: Houston, 1971], *Ilyanassa (= Nassarius) obsoleta* (Say) [Connecticut: Smith, 1971], *Nucella lapillus* [south-west England: Blaber, 1970] and *Ocenebra erinacea* (L.) [south-west France: Poli *et al.*, 1971]. These observations coincided with the introduction of TBT antifouling paints and the link was confirmed by subsequent studies that demonstrated the degree of masculinisation in all four species increased massively in the 1980s (refs in Gibbs *et al.*, 1990).

The term 'imposex' was coined by Smith (1971) to describe 'a superimposition of male characters on to unparasitised and parasitised females'. The term is now generally used exclusively to describe the masculinizing effect on gastropods of exposure to tributyltin compounds. Recognition of the syndrome is not straightforward. The definition implies the active intervention of an external agent causing an unnatural degree of masculinisation; thus the presence of a vestigial penis on a female should not be identified as imposex *per se* unless corroborative evidence links the condition to TBT pollution. As an example, the case of masculinized females of the American oyster-drill *Urosalpinx cinerea* (Say) may be cited. British populations of this species (introduced around 1900) are now in decline, almost certainly because of the sterilising effect of advanced imposex induced by high TBT pollution along the Essex and Kent coastlines (see Gibbs *et al.*, 1991d). Sex reversal in *U. cinerea* was extensively searched for in the Essex populations in the 1940s and again in the 1950s but no evidence of this was found. Yet contemporary studies of U.S. east coast populations indicate that some females possessed small penes in the pre-TBT paint era: Griffith & Castagna (1962) report Chincoteague Bay females with vestigial penes collected in 1958-59, and one such female is illustrated by Carriker (1943, Figure 1) from New Jersey. Clearly, these, along with other old records, should be viewed as having a separate status to those originating after the advent of TBT paints.

Proof that imposex is a modern phenomenon is largely based on comparison of historical and present-day samples from specific sites. However, preserved samples of a common species such as the dog-whelk, *N. lapillus,* can prove surprisingly difficult to locate, but where achieved (Bryan *et al.*, 1986; Bailey & Davies, 1988a) the evidence is conclusive.

2. IMPOSEX: A BRIEF HISTORY

The first record of female masculinisation was that of Blaber (1970) who described penis-like outgrowths on *N. lapillus* females collected in Plymouth Sound; although lacking any ducts, he regarded the structures as incipient penes. This observation prompted Smith (1971) to examine Connecticut *I. obsoleta*; he found comparable, but more advanced, masculinisation (development of penis, vas deferens and oviducal convolution) occurring in this species and this led him to invent the term 'imposex'. Subsequent studies showed that the phenomenon was not a consequence of parasitism (Smith, 1980), but could be induced in the laboratory by exposure to tributyltin compounds at concentrations calculated to be in the range of parts per billion ($\mu g.l^{-1}$)(Smith, 1981a-d). This finding led Miller & Pondick (1984) to analyse New England *N. lapillus* for metals associated with boating activity; they found that Sn levels in whole body tissues were below the level of detection ($<0.5\ \mu g.g^{-1}$) for their analytical method and, whilst not ruling out the possibility of organotin compounds being responsible for imposex, concluded that elevated Cu levels were indicative of the syndrome. Concurrently in Europe, studies into the effects of TBT were largely focused on the Pacific oyster, *Crassostrea gigas* (Thunberg), notably its reduced spatfall and abnormal shell growth (summarised in Alzieu, 1991). However, the widespread nature of imposex in several neogastropod species in French waters had been demonstrated (Féral, 1980a,b & refs therein) and experimental evidence linked the condition in *O. erinacea* to TBT exposure (Féral & Le Gall, 1982, 1983). Up to this time no evidence had been found to suggest that imposex reduced the reproductive capacity of even highly affected populations. But interest in the syndrome was heightened following the discovery that, around the south-west peninsula of England, *N. lapillus* populations were declining in abundance in all of the major inlets used for boating activities, particularly close to marinas. Declining populations were characterised by

- a general scarcity of individuals;
- no indication of breeding activity, either recent (capsules) or in the previous few years (juveniles);
- survivors all appeared old, predominantly male.

Extensive field surveys revealed that the degree of imposex increased with proximity to boating centres and correlated with body tissue concentrations of Sn in a hexane-extractable form that included tributyltin and its degradation product dibutyltin; all of the evidence pointed towards imposex being initiated at an ambient TBT concentration of less than 1 ng $Sn.l^{-1}$ (Bryan *et al.*, 1986). Females from heavily contaminated areas all exhibited a characteristic symptom, namely, blockage of the oviduct by an overgrowth of sperm duct (=vas deferens) tissue; this blockage prevented the release of egg capsules which then accumulated within the capsule gland. The masses of such aborted capsules sometimes grew to the extent that the capsule gland wall was breached. Such gross malformation of the female tract caused sterilisation, thus explaining population decline, and was considered probably responsible for the higher mortality rate observed in females. Hyperplasia of genital duct tissues was noted as a feature of both sexes (Gibbs & Bryan, 1986). Experiments demonstrated that bioaccumulation of tin was associated with increased development of imposex; no evidence was (or has) been gained to indicate that a loss of tin leads to a remission of imposex (Bryan *et al.*, 1987). In natural populations it was thought likely that the diet contributed less than half of the body burden of TBT resulting from life-long exposure (Bryan *et al.*, 1989). Laboratory rearing from hatching to maturity (>2 years) showed that even at water TBT concentrations of only 1-2 ng $Sn.l^{-1}$ some females were sterilised and at 3-5 ng $Sn.l^{-1}$ all were sterilised; when TBT levels were maintained at around 20 ng $Sn.l^{-1}$, the process of masculinisation advanced to the stage where oogenesis was suppressed and spermatogenesis promoted (Gibbs *et al.*, 1988). These data accorded with field observations [Note that concentrations expressing TBT as Sn can be converted to TBT by multiplying by 2.44].

The metabolic disturbance underlying the imposex phenomenon remains to be elucidated in detail but some insight into the probable mechanisms has been gained. Spooner *et al.* (1991) found that exposure of *N. lapillus* females to TBT in water increased the testosterone titre; artificially increasing the latter by injection, in the absence of TBT, induced imposex. These observations, along with others, give grounds to believe that testosterone accumulation may result from the inhibition of the cytochrome P-450 dependent aromatase responsible for the conversion of testosterone to estradiol-17ß. Whether this inhibition is a direct result of the action of TBT is unclear. Féral & Le Gall (1983) found that penis growth in *O. erinacea* is controlled by neurohormonal factors released by the cerebral ganglia. These neurohormones may mediate steroid metabolism; if this is the case, then any imbalance in their production will be reflected in abnormal steroid titres.

Since 1990, imposex has been studied in many species worldwide. A complete appraisal is beyond the scope of this chapter but certain features of imposex appear to be applicable to all neogastropods:

- it is induced by TBT exposure at or below 1 ng $Sn.l^{-1}$;
- its degree of development is dose-controlled;
- advanced development causes sterility in those species where there is encroachment of the anterior oviduct by vas deferens tissue;
- in sterilised species the effect at the population level depends largely on the dispersive capacity of the larvae, *i.e.* length of the planktonic existence.

Overall, there is now little room for doubt of the connection between TBT exposure and masculinization of female neogastropods; in fact, this link has to be viewed as one of those exceptional cases where a biological effect has been proven to be caused by an identified compound.

3. ANATOMY OF NEOGASTROPOD REPRODUCTIVE SYSTEM

The basic form of the male and female reproductive systems of neogastropods appears to be similar throughout the group although detailed descriptions are wanting for many genera; for example, little is known in a comparative sense about *Conus* gonoducts (A.J. Kohn, 1992, personal communication). Fretter (1941) produced the first account of the functional anatomy of the genital ducts of four European species, comparing *N. lapillus, O. erinacea, Nassarius (Hinia) reticulatus* (L.) and *Buccinum undatum* (L.): this oft-cited paper is a classic. In terms of the use of imposex in neogastropods as a bioindicator of TBT pollution, the structure and layout of the male and female gonoducts are of most relevance; these aspects are best known for *N. lapillus*, here used as an example.

The external characters of animals with the shell removed are shown in Figure 1. The most convenient character by which to distinguish the sexes is the presence of the sperm-ingesting gland in the female; this gland is often dark brown but may have a reddish hue or be black. In the male a simple duct, the vas deferens, carries sperm from the testis to the penis located on the head behind the right tentacle. The proximal section of this sperm duct is highly convoluted, forming the vesicula seminalis in which sperm are stored. The middle section lies within the mantle cavity and forms the prostate, recognisable by its swollen, highly glandular structure (Figure 1c). From the prostate the distal portion of the duct, the pallial vas deferens, crosses the floor of the mantle cavity to the base of the penis and thence to its tip as the penial duct. The pallial vas deferens is a subsurface tube formed by overgrowth and fusion of the edges of a shallow groove, the line of closure being marked by a thin surface streak throughout its length; subdermally the fused epithelia become highly convoluted (Figure 2). Ontogenetically, the pallial vas deferens does not appear *in toto*; instead, infoldings of the epithelia appear adjacent to the penis and to the prostate and these two sections migrate towards each other to join up about mid-way. This pattern of development is duplicated as a feature of imposex.

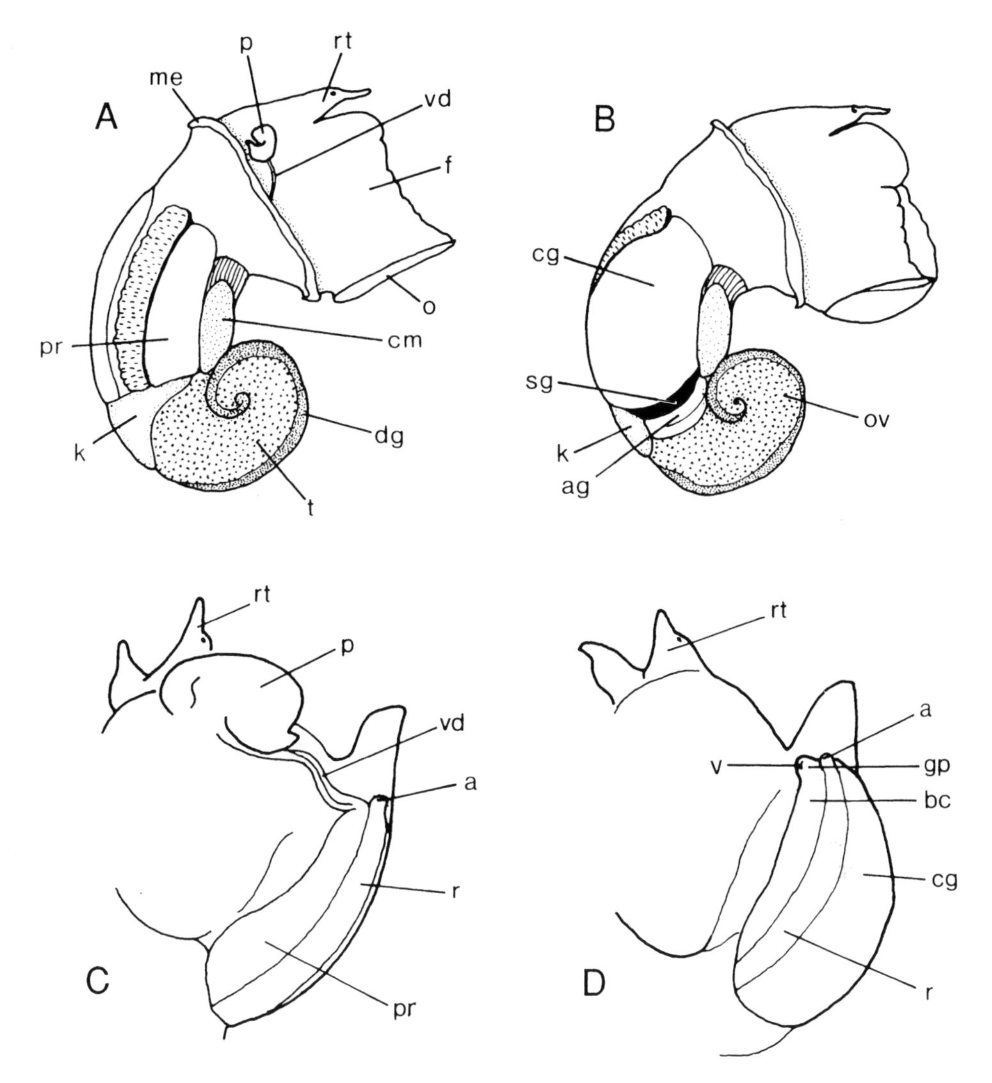

Figure 1. *Nucella lapillus*. Male (A) and female (B) removed from shell; (C,D) same, with mantle cavities opened mid-dorsally to display anterior portions of gonoducts.
Abbreviations: a, anus; ag, albumen gland; bc, bursa copulatrix (site of); cg, capsule gland; cm, columella muscle; dg, digestive gland; f, foot; gp, genital papilla; k, kidney; me, mantle edge; o, operculum; ov, ovary; p, penis; pr, prostate; r, rectum; rt, right tentacle; sg, sperm-ingesting gland; t, testis; v, vulva; vd, vas deferens. (A and B from Gibbs *et al.*, 1987, with permission)

By contrast, the structure of the female gonoduct is more complex as a result of modifications that have evolved to permit encapsulation of the fertilised ova. Internal fertilisation is achieved by copulation, the penis entering the oviduct through the vulva situated at the tip of the genital papilla (Figure 1d). Sperm are deposited in the pouches of the bursa copulatrix and are then passed into the ventral channel of the capsule gland. Eggs are carried forward from the ovary along the oviduct to the albumen gland where they are mixed with secretions and probably fertilized before passing into the capsule gland. In manufacturing the capsule this gland produces first protein matter and then the mucoid material forming the outer fibrous coat of the capsule. The capsules exit from oviduct via the vulva to the mantle cavity and are transported to the ventral pedal gland situated on the sole of the foot where they are moulded into their typical vase-like shape before being fixed to a rock substratum (see Fretter, 1941). The presence of the pedal gland on females is a useful character when a non-sacrificial method of separating the sexes is desired (see below).

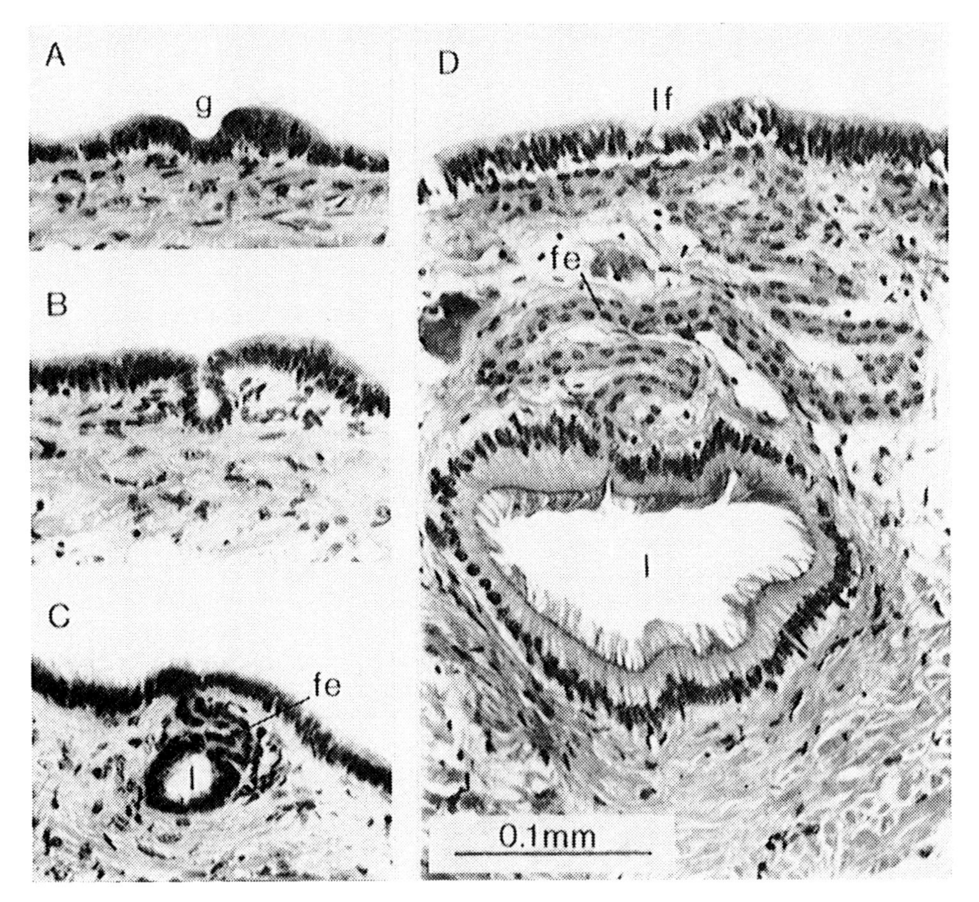

Figure 2. *Nucella lapillus*. Successive stages in the formation of the male vas deferens, from open groove (A, B) to closed, sunken duct (C, D), by infolding of the mantle floor epithelium. Note that line of fusion remains after development is complete, *viz.* convoluted epithelial connection from duct to surface in D.
Abbreviations: fe, fused epithelium; g, groove; l, lumen of duct; lf, line of fusion on surface of mantle cavity floor

4. DEVELOPMENT AND CHARACTERISATION OF IMPOSEX

Experience has shown that the degree of masculinisation exhibited by female *N. lapillus* is dose-dependent. The state of each female represents that individual's integrated response to the TBT exposure experienced during its life-time, particularly that prevailing when juvenile. It is during the first year of life that the reproductive tract develops and the evidence suggests that the extent of steroid imbalance induced by TBT exposure during this phase is the major factor in determining the degree of masculinisation exhibited by the mature female. If the level of TBT pollution increases, imposex is promoted even in adults; but if it decreases the degree of imposex remains unchanged since all the evidence points to the process being irreversible. *N. lapillus* may survive for in excess of ten years, especially in sheltered habitats; thus the imposex levels exhibited by a population can reflect levels of pollution that have long since ameliorated.

On examining *N. lapillus* females from different populations experiencing different levels of TBT exposure, a complete spectrum of imposex expression is encountered. Imposex being irreversible, all of these stages can be ordered in a series of increasing masculinisation to illustrate the probable sequence of events leading from initial induction to sterilisation and beyond. This process is a continuum, and for descriptive purposes, it is necessary to identify the major steps which can be used to characterise its progression. Six morphological stages are readily recognisable in *N. lapillus* (Figure 3); these are as follows:

- *Stage 1* is marked by the appearance of an infolding of the mantle cavity epithelium adjacent to the genital papilla - this is the developing proximal section of the pallial vas deferens;
- *Stage 2* is signalled by the initial phase of penis formation, heralded by a second epithelial infolding (developing penial duct/vas deferens) which is subsequently raised on a protruding ridge;
- *Stage 3* is the continuation of penis development from a low ridge to a protusion of recognisable shape and the extension of the distal section of the vas deferens from the base of the penis;
- *Stage 4* follows the fusion of the distal and proximal sections of the vas deferens and further penis enlargement may occur;
- *Stage 5* is marked by a proliferation of the proximal vas deferens tissue overgrowing the genital papilla to the extent that the vulva is displaced, constricted or no longer visible;
- *Stage 6* results from the blockage of the vulva, an event manifested by the accumulation of aborted capsules in the lumen of the capsule gland; a single capsule may be present or several to many compressed together into a translucent or dark-coloured mass.

Normal breeding activity continues through stages 1 to 4. The critical phase in the sequence is stage 5 since any capsules formed subsequently cannot be expelled from the oviduct. The presence of a high percentage of such sterilised females in a population signals its eventual demise. Recognition of stage 5 can be somewhat subjective because, in some circumstances, it is likely that occlusion of the vulva occurs slowly over a period of time, perhaps years. Thus, there may be a transition phase when the breeding capacity is reduced because restriction of the vulval diameter limits the size of capsule that can be extruded. In the most advanced cases, histological sections show vas deferens tissues invading the oviduct to the extent that the bursa copulatrix is replaced by a prostate.

Further characters of stage 6 give an indication of the long-term survival of at least some affected females, *i.e.*

- masses are translucent when formed (see Gibbs *et al.*, 1988, Figure 1) and subsequently darken with age and
- capsules continue to accumulate to the extent that the distended oviduct wall ruptures and the liberated mass then becomes fixed to the interior of the shell by secreted material, probably conchiolin (see Gibbs & Bryan, 1986, plate 1).

The amount of secreted material is often considerable and this suggests the 'post-rupture' survival time may be quite lengthy but the inequality of the sex ratio in declining populations indicates this injury must hasten the death of such severely-affected females.

The above arrangement of stages is largely based on the development of the vas deferens and thus was designated the Vas Deferens Sequence (VDS) (see Gibbs *et al.*, 1987). It was designed to be a rapid and simple method of assessment using a stereomicroscope only. It was recognised that the imposex syndrome in *N. lapillus* was complex and the response of the individual and of populations was subject to considerable variation. For example, it was appreciated that in some individuals the proximal vas deferens was not detectable even though penis formation had commenced; the absence of a detectable stage 1 in some individuals was considered to be relatively unimportant; rarely, penis formation was entirely lacking and in these cases the stages 3 and 4 were scored according to vas deferens development. Although penis development is important in the context of providing a conspicuous biomarker, it is unimportant in terms of the biological consequence of imposex, *i.e.* sterilisation and population extinction. The latter is an effect of vas deferens development, hence the emphasis needs to be on the formation of this duct; the absence of a penis introduces an interesting variation but one which can be readily accommodated in a VDS scheme. The main requirement of any scoring system is that it is kept simple, because not only does it have to be used by non-specialists, but also because the time and resources available for widescale surveys are often limited. In modern idiom, it has to be 'user-friendly'.

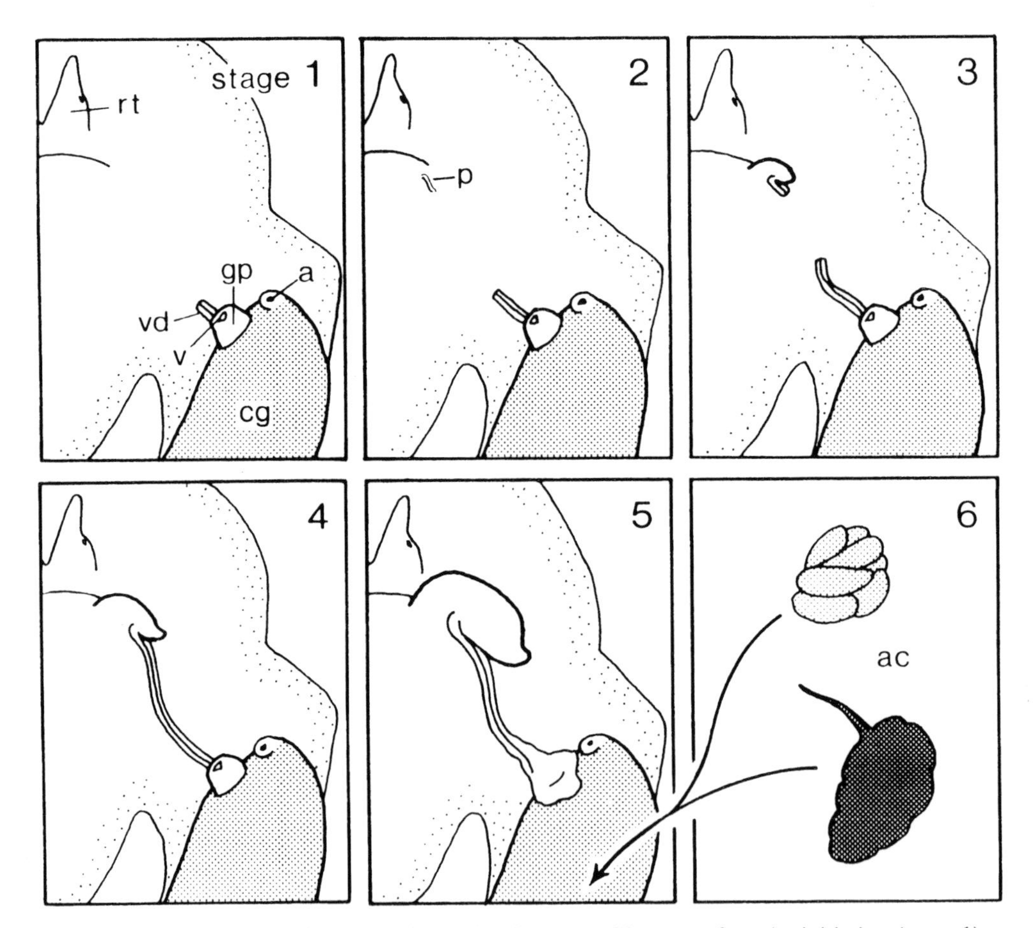

Figure 3. *Nucella lapillus*. Six stages in the development of imposex from its initiation (stage 1) to sterilisation (stage 5) and subsequent accumulation of aborted capsules in the capsule gland (stage 6). Stages are numbered according to the Vas Deferens Sequence (VDS) scheme (see text).
Abbreviations: a, anus; ac, accumulations of aborted capsules; cg, capsule gland; gp, genital papilla; v, vulva; vd, vas deferens (From: Gibbs, 1993; with permission)

The principle of a VDS scheme has found universal application although definition of the stages requires modification to suit specific variations. Thus, modified schemes have been applied to *Nucella lima* (Gmelin) (Short *et al.*, 1989; Stickle *et al.*, 1990) and to *Lepsiella scobina* (Quoy & Gaimard) (Stewart *et al.*, 1992). The scheme of Oehlmann *et al.* (1991) includes a number of variations which can be arranged in temporal sequences of development of penis and vas deferens. For descriptive purposes, their scheme may offer certain advantages, notably in being applicable to a wide variety of species. However, these authors offer a cautionary statement noting that an unambiguous assignment to the stage/type of their classification requires specimen fixation, and that even with the use of the scanning electron microscope, a clear decision is sometimes impossible without histological sections. It might also be added that their scheme overlooks the possibility of sterilisation in females lacking penes, a feature recently discovered in Brittany *N. lapillus* populations (M. Huet, 1992, personal communication).

Experimentally, the imposex response of *N. lapillus* to TBT exposure is remarkably predictable. In the field, the timing, duration and level of exposure must all affect its expression; the form and intensity of imposex observed in the adult represent an integration of the total dosage throughout its life, especially during the juvenile period when the ontogeny of the genital tract is proceeding. The contribution of genetic factors remains to be investigated but clearly may contribute to individual subtleties in the morphological expression of the syndrome.

5. INDICATORS OF IMPOSEX INTENSITY

5.1. Female penis size

In areas of light TBT contamination, masculinisation may be exhibited by only a fraction of the female population and thus the degree of imposex in a population can be expressed in simple terms, such as the percentage of penis-bearing females or more generally as the incidence of occurrence. For the latter it is essential to decide whether early characters have been considered in addition to penis formation, *i.e.* the presence of precursors of the vas deferens (proximal and distal infoldings). Once the level of TBT contamination in water reaches the level of chemical detection (ca 0.2 ng $Sn.l^{-1}$) such gauges become redundant because typically all females exhibit masculinisation. Some means of measuring the degree of masculinisation then becomes necessary. The most obvious and easily measured is penis length; this parameter is most sensitive for imposex at its intermediate levels of expression (stages 3 & 4) when the major part of development of the penis proper occurs. Thereafter attention focuses on the genital papilla region and disruption of the anterior oviduct by vas deferens tissue.

Penis length is a valid and convenient parameter in experiments designed to test the effects of various agents (see below) and also to determine temporal trends in a single population. However, use of female penis length as a comparative measure of the intensity of imposex between populations requires caution because, firstly, the maximum penis size attainable in the female will be comparable to that of the male in the same population and, secondly, in *N. lapillus* males, penis size is related to body size which is subject to considerable variation between populations; for example, populations in situations exposed to high wave action are often markedly smaller than those living in shelter. A further consideration is that comparisons based simply on length do not convey the contrast in mass between, say, a penis of 2 mm and one of 4 mm. Such factors need to be accommodated for any geographical assessment of imposex based on female penis development and the Relative Penis Size Index (RPSI) was designed for this purpose.

The basis for the RPSI is the biometric data presented in Bryan *et al* (1986, Figure 2): these show that the weight (or volume) the penis is related to the cube of its length and therefore the ratio:

$$\text{RPSI} = \frac{(\text{mean length of female penis})^3}{(\text{mean length of male penis})^3} \times 100 \qquad (1)$$

can be used as a measure of the relative size of the female penis in any population. A level of 50% indicates that the average female penis has half the bulk of that of the male. The relationship between RPSI and TBT concentration in ambient water and body tissues is shown in Figure 4. It has to be stressed that this index is based on the lengths of penes of *non-narcotised* animals. The relationship of mass to length will not apply, obviously, to penis lengths measured on animals that have been narcotised; failure to appreciate this fact has misled some authors (Oehlmann *et al.*, 1991; Stroben *et al.*, 1992a,b) to discredit the value of cubing the data. In some *N. lapillus* females, penis development appears to be inhibited; the incidence of this feature shows great variability between populations from being very rare around southwest England to quite common in parts of Brittany (M. Huet, 1992, personal communication). Clearly, where the incidence of non-penis-bearing females is high, the application of the RPSI must be qualified. In devising an RPS scheme to species possessing penes which do not approximate a cylindrical form, other biometric methods need to be considered, such as length or area (see below).

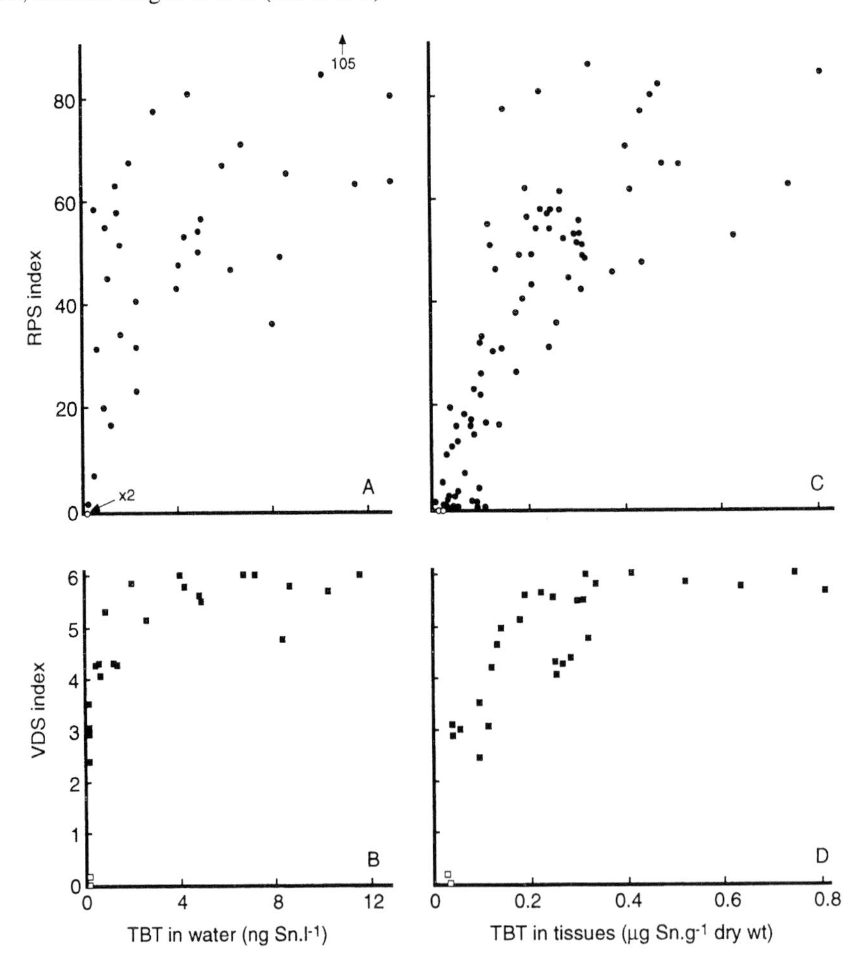

Figure 4. *Nucella lapillus*. Relationships between tin as TBT in (A, B) ambient seawater, (C, D) body tissues of females, and Relative Penis Size (RPS) and Vas Deferens Sequence (VDS) indices. Water and animal samples taken at same time at sites around south-west England 1984-1986. Values for two clean-water sites (Isle of Mull, west Scotland; open symbols) are shown for comparison (adapted from Gibbs *et al.*, 1987).

5.2. Vas deferens development

Calculation of the mean stage of the Vas Deferens Sequence (described above) provides an index by which to compare the intensity of imposex in different populations and relationship to TBT in water and tissues (Figure 4). Any population with a VDSI above 4 contains sterilised females and therefore has a reduced breeding capacity. The VDSI does not, however, give an indication of the proportion of females that are sterilised. In areas of high TBT contamination and advanced imposex, expression of this proportion as a percentage of the total provides a meaningful indicator of the reduction in the reproductive potential of a population (see, for example, Gibbs *et al.*, 1991b). Sterilisation may result from one of at least two types of malformation: (i) a simple blockage of the oviduct, as in *N. lapillus*, or (ii) incomplete fusion of the oviduct during its ontogeny so that it remains open to the mantle cavity, an abnormality that can interpreted as a mimic of the male tract (Gibbs *et al.*, 1991c). The latter is found in *O. erinacea* and *U. cinerea*: the nature of this abnormalty appears to preclude accumulation of capsules within the oviduct and in such species no stage 6 is identifiable.

5.3. Convolution of proximal oviduct

In most species the proximal section of the oviduct appears as a relatively straight tube extending from the ovarian tissue to the albumen gland and its structure appears to undergo little modification with imposex. However, in *I. obsoleta* this part of the oviduct becomes highly convoluted so as to resemble its male counterpart, the vesicula seminalis. Smith (1980) used the degree of oviducal convolution to complement penis and vas deferens development in calculating his total measure [*Z*] of imposex expression. The usefulness of this character as an index in other species similarly affected has yet to be tested.

6. EXAMPLES OF APPLICATION

6.1. Geographical surveys

The early work of Smith (1981a-d) provided field evidence indicating TBT to be the cause of imposex in *I. obsoleta*. Soon after, refinements in chemical methods made TBT detection to <1 ng.l^{-1} possible and facilitated the demonstration of the extreme sensitivity of the imposex response in *N. lapillus* (Bryan *et al.*, 1986). When it was realised the degree of development was dose-related, the potential of the syndrome as a biomonitoring technique became readily apparent (Gibbs *et al.*, 1987, 1988). Subsequently, *N. lapillus* has been examined throughout its European distribution (Norway to Portugal) to determine the extent of TBT contamination. Elsewhere, imposex in other neogastropod species has proved to be an equally valuable TBT bioindicator.

RPS and VDS indices for numerous *N. lapillus* populations are now available. These pinpoint the sources of TBT contamination, chiefly boat and ship facilities (marinas, mooring and maintenance) and mariculture units (chiefly salmon farms) but also give a broad indication of the dissemination of this biocide by currents and ship movements. The technique is well-exemplified by the investigations of the effects of TBT from large commercial vessels, chiefly gas and oil tankers, visiting Sullom Voe in the Shetland Islands (see Bailey & Davies, 1988b; Davies & Bailey, 1991). Inside the Voe, all *N. lapillus* populations exhibited well-developed imposex with RPSIs between 34 and 81% (Figure 5a) and VDSIs exceeding 4.0 at all stations indicating widespread sterilisation (Figure 5b). The latter was most evident close to the loading jetties of the oil terminal where up to 90% of females were stages 5 and 6 and juveniles were scarce. Outside of the Voe, in the open waters of Yell Sound, RPSI values were low; no sterilised females were found but stage 4 predominated in some populations which were considered to be threatened. Overall, the survey demonstrated that the operation of large vessels and supporting facilities has led to significant TBT contamination in Shetland waters. Similar surveys of imposex intensity in *N. lapillus* populations inhabiting European waters variously used by pleasure craft, fishing boats and larger vessels, and also for fish farming, are described in Bryan *et al.* (1986), Bailey & Davies (1989), Spence *et al.* (1990a), Gibbs *et al.* (1991a,b) and Harding *et al.* (1992).

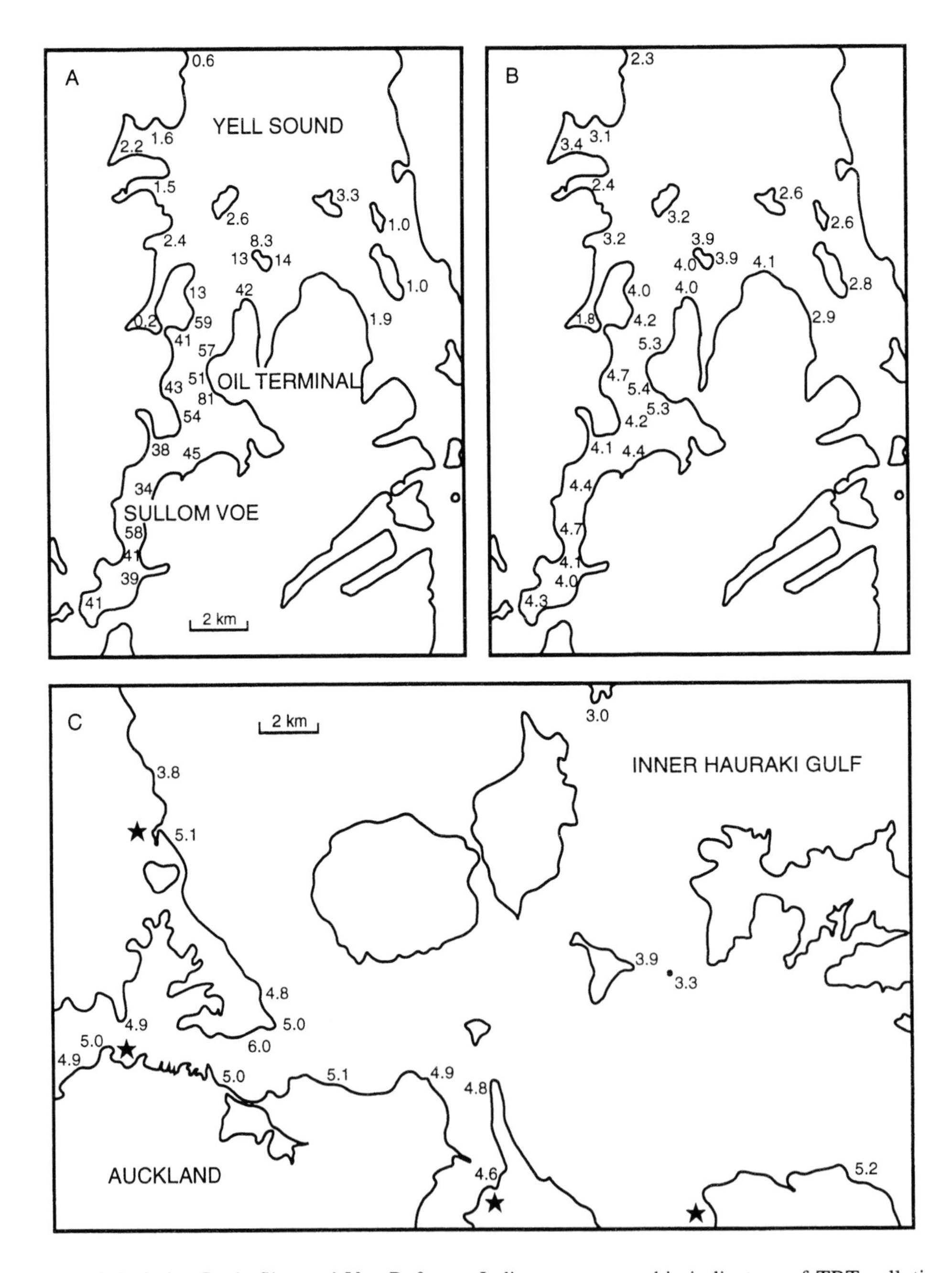

Figure 5. Relative Penis Size and Vas Deferens Indices as geographic indicators of TBT pollution levels. (A) RPSIs and (B) VDSIs of *N. lapillus* populations on shores close to the oil terminal in Sullom Voe, Shetland in 1987 (modified from Bailey & Davies, 1988b). (C) VDSIs of *L. scobina* populations in the environs of Auckland and inner Hauraki Gulf, northern New Zealand; stars mark sites of marinas (adapted from Stewart *et al.*, 1992)

Comparable field surveys of imposex in neogastropods have been carried out worldwide, including the north-east Pacific (Bright & Ellis, 1990; Short *et al.*, 1989; Stickle *et al.*, 1990; Ellis & Pattisina, 1990), south-east Asia (Saveedra Alvarez & Ellis, 1990) and New Zealand (Smith & McVeagh, 1991; Stewart *et al.*, 1992). All of these studies have demonstrated not only the efficiency of imposex as an indicator of TBT pollution but also the intrinsic variability of its development, both in its form and in its net effect on the female tract. Bright & Ellis (1990), for example, were able to recognise three different imposex types within one genus (*Nucella*). It would seem that the first appearance of the vas deferens close to the genital papilla found in *N. lapillus* (VDS stage 1) is not typical for many neogastropods *(e.g. N. lima, L. scobina)* and consequently some modification of the original VDS scheme is necessary to accommodate specific variations. Thus, Stewart *et al.* (1992) devised a scheme of seven stages to describe the development of imposex in *Lepsiella scobina*; this sequence takes into account minor differences but retains the principle of sterilisation at stage 5 and thus the basis for interpreting VDSI values, illustrated in Figure 5c, is the same as that for the *N. lapillus* index.

There is general agreement amongst workers that a VDS index has a wider application as a measure of the intensity of imposex than an RPS index. Enlargement of the female penis is a feature of intermediate imposex (stages 3 & 4) and thus the RPSI as a measure of intensity is limited to these stages. Other factors also have to be considered when relating female penis size to that of the male. Firstly, the male penis may show some variation in size according to breeding condition or even the effects of TBT exposure: in *I. obsoleta*, for example, healthy males, and also males parasitised by trematodes, normally lose the penis after breeding, but both groups tend to retain it where imposex-inducing conditions are strong (Curtis & Barse, 1990); and secondly, the form of the male penis varies according to species and thus the relationship between the parameter measured, usually length, and volume ('bulk') has to be considered for each individual species; for example, the penis of *L. scobina* is flattened and recurved with a long apical flagellum and thus Stewart *et al.* (1992) considered its development in imposex to be more realistically expressed as the Relative Penis Length Index (RPLI):

$$\text{RPLI} = \frac{\text{(mean length of female penis)}}{\text{(mean length of male penis)}} \times 100 \qquad (2)$$

rather than by the RPSI. Similarly, the RPSIs for *Thais orbita* (Gmelin) and *Morula marginalba* (Blainville) given in Wilson *et al.* (1993) are based on penis area. Some workers have employed penes lengths of narcotised animals (see above) but variability in the reaction of specimens to anaesthetics would seem to introduce a further source of error.

6.2. Monitoring of long-term trends

Laboratory and field experiments using adult *N. lapillus* have shown that enlargement of a small female penis can be rapidly achieved by exposure to TBT; by contrast, vas deferens development advances at a much slower rate (see for *e.g.* Bryan *et al.*, 1987). Female penis length has proved to be an effective measure of imposex promotion in a variety of studies of the syndrome, including effects of

- different paint types applied to the shell surface (Figure 6a);
- different organotin compounds, both dissolved and injected (see Bryan *et al.*, 1988);
- transplantation from areas of low contamination to sites of high contamination (Figure 6b).

Early investigations used unsexed samples and relied on an equal sex ratio to recover an adequate subsample of females. However, the efficiency of data gathering was much improved in later studies using narcotisation to select suitable females for experimentation (see Appendix).

For long-term monitoring of trends in imposex, the most efficient technique is reckoned to be routine sampling of pre-adult females aged about one year (recognisable by the sharp-edged margin of the body whorl and a small, immature ovary); the developmental state of the penis in this group reflects a general decrease, as well as an increase, in the water TBT concentration over the previous year whereas the penis state of adults will reflect only an increase because of the irreversible nature of imposex (recruitment of less-affected juveniles to the adult population will, of course, eventually lead to a decline in imposex intensity but, given

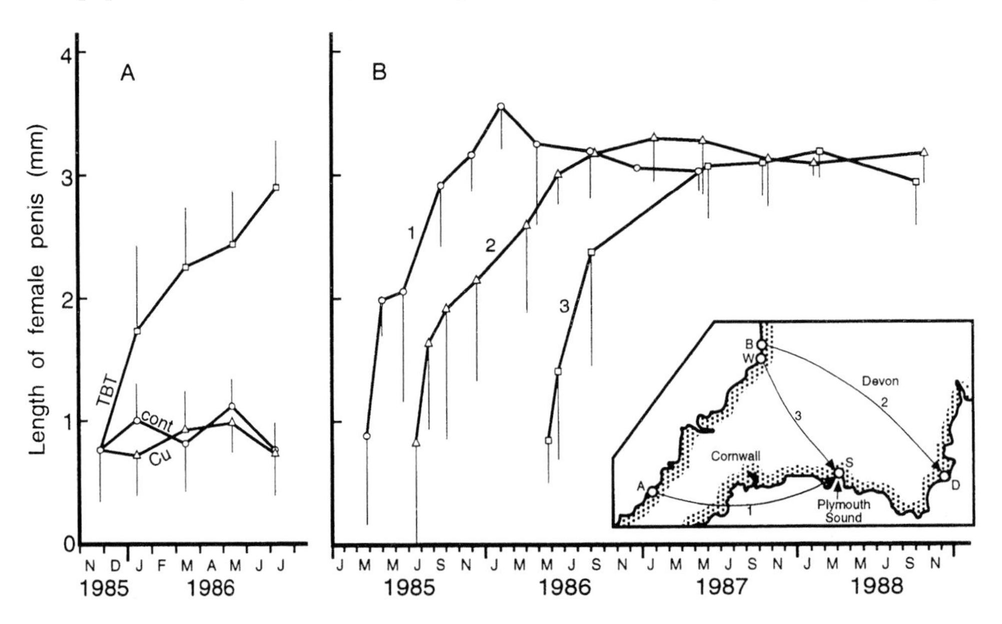

Figure 6. *Nucella lapillus*. Female penis length as an indicator of TBT pollution. (A) Effect of TBT, copper and enamel (control) paints applied to shell surface (modified from Bryan *et al.*, 1987); (B) Effect of tranplantation of animals from areas of low TBT contamination (north Cornwall: A, St Agnes; B, Bude; W, Widemouth Bay) to sites of high TBT pollution (south Devon: S, Sutton Harbour; D, Dart Estuary). In both figures values shown are means; vertical bars indicate standard deviations

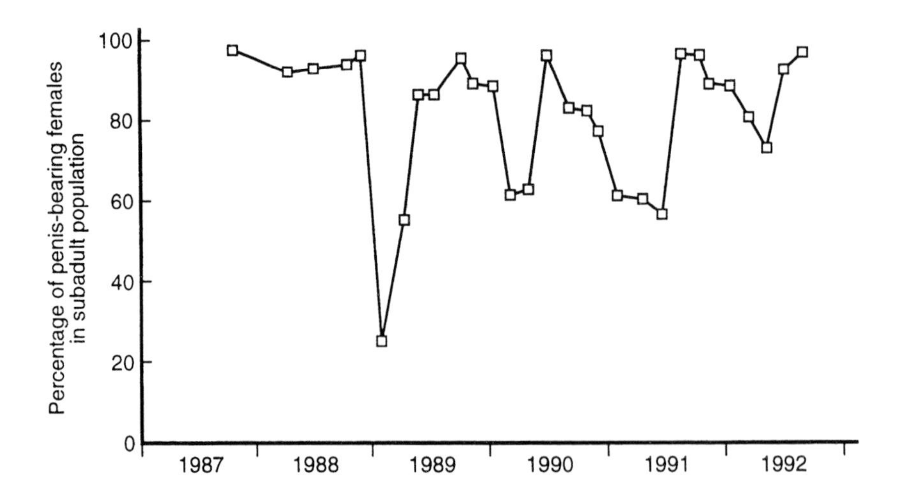

Figure 7. *Nucella lapillus*. Penis-bearing in subadult females of a population at a 'clean' site -Widemouth Bay, north Cornwall - where ambient water TBT concentrations are at, or below, the level of chemical detection (ca 0.2 ng $Sn.l^{-1}$)

that adult survival may exceed ten years, it must take several years or more for a significant decline to be apparent: this may be adequate for long-term surveillance but not for the purpose of monitoring the effectiveness of any recent legislation restricting usage of antifouling paints).

In areas of trace TBT concentration (<0.5 ng $Sn.l^{-1}$), rapid assessment can be made using the incidence of penis-bearing in pre-adult females: as illustrated in Figure 7, seasonal fluctuations become apparent, probably as a result of subtle changes in ambient TBT levels differentially affecting successive cohorts during their juvenile life when sensitivity is highest.

Following the 1987 legislation for U.K waters restricting the usage of TBT paints, attention is now focused on the rate of amelioration of TBT pollution. Of particular interest is the repopulation of those areas denuded of *N. lapillus*. This process depends on the success of breeding by immigrants carried from less-affected areas. Recent pilot experiments have shown that animals transplanted from breeding populations to areas formerly highly contaminated (south-west England inlets) will reproduce and the progenies will survive. However, at maturity (1.5 - 2 years old) these progeny prove to be sterile (VDS stage 5), indicating ambient water TBT contamination have not yet declined to below the critical concentration of 2 ng $Sn.l^{-1}$ required to allow breeding and long-term recolonisation. Successive transplantations will probably be necessary to demonstrate if and when natural repopulation becomes feasible.

We are indebted to Dr. R. Abel (U.K. Department of the Environment) and to Messrs L.G. Hummerstone, P.L. Pascoe and G.R. Burt (all of the Plymouth Marine Laboratory) for their assistance with the investigations of imposex in British species; this work was supported by the U.K. Department of the Environment under Contracts DGR 480/51, PECD 7/8/103 & PECD 7/8/175 and by the EEC under Contract ENV-686-UK(H).

REFERENCES

Alzieu, C., 1991. Environmental problems caused by TBT in France: assessment, regulations, prospects. Mar. Environ. Res. 32: 7-17

Bailey, S.K. & I.M. Davies, 1988a. Tributyltin contamination in the Firth of Forth (1975-87). Sci. Total Environ. 76: 185-192

Bailey, S.K. & I.M. Davies, 1988b. Tributyltin contamination around an oil terminal in Sullom Voe (Shetland). Environ. Pollut. 55: 161-172

Bailey, S.K. & I.M. Davies, 1989. The effects of tributyltin on dogwhelks (*Nucella lapillus*) from Scottish coastal waters. J. mar. biol. Ass. U.K. 69: 335-354

Balls, P.W., 1987. Tributyltin (TBT) in the waters of a Scottish Sea Loch arising from the use of antifoulant treated netting by salmon farms. Aquaculture, 65: 227-237

Batley, G.E. & M.S. Scammell, 1991. Research on tributyltin in Australian estuaries. Appl. organomet. Chem. 5: 99-105

Blaber, S.J.M., 1970. The occurrence of a penis-like outgrowth behind the right tentacle in spent females of *Nucella lapillus* (L.). Proc. malac. Soc. Lond. 39: 231-233

Bright, D.A. & D.V. Ellis, 1990. A comparative survey of imposex in northeast Pacific neogastropods (Prosobranchia) related to tributyltin contamination, and choice of a suitable bioindicator. Can. J. Zool. 68: 1915-1924

Bryan, G.W. & P.E. Gibbs, 1991. Impact of low concentrations of tributyltin (TBT) on marine organisms. In: Ecotoxicology of metals: current concepts and application; M.C. Newman & A.W. McIntosh (eds). Lewis Publ., Ann Arbor, pp. 323-361

Bryan, G.W., P.E. Gibbs., & G.R. Burt. 1988. A comparison of the effectiveness of tri-n-butyltin chloride and five other organotin compounds in promoting the development of imposex in the dog-whelk, *Nucella lapillus*. J. mar. biol. Ass. U.K. 68: 733-744

Bryan, G.W., P.E. Gibbs, G.R. Burt, & L.G. Hummerstone. 1987. The effects of tributyltin (TBT) accumulation on adult dog-whelks, *Nucella lapillus*: long-term field and laboratory experiments. J. mar. biol. Ass. U.K. 67: 525-544

Bryan, G.W., P.E. Gibbs, R.J. Huggett, L.A. Curtis, D.S. Bailey, & D.M. Dauer, 1989. Effects of tributyltin pollution on the mud snail, *Ilyanassa obsoleta*, from the York River and Sarah Creek, Chesapeake Bay. Mar. Pollut. Bull. 20: 458-462

Bryan, G.W., P.E. Gibbs, L.G. Hummerstone, & G.R. Burt, 1986. The decline of the gastropod *Nucella lapillus* around south-west England: evidence for the effect of tributyltin from antifouling paints. J. mar. biol. Ass. U.K. 66: 611-640

Bryan, G.W., P.E. Gibbs, L.G. Hummerstone, & G.R. Burt, 1989. Uptake and transformation of ^{14}C-labelled tributyltin chloride by the dog-whelk, *Nucella lapillus*: importance of absorption from the diet. Mar. Environ. Res. 28: 241-245

Carriker, M.R., 1943. On the structure and function of the proboscis in the common oyster drill, *Urosalpinx cinerea* (Say). J. Morph. 73: 441-506

Curtis, L.A. & A.M. Barse, 1990. Sexual anomalies in the estuarine snail *Ilyanassa obsoleta*: imposex in females and associated phenomena in males. Oecologia, 84: 371-375

Davies, I.M. & S.K. Bailey, 1991. The impact of tributyltin from large vessels on dogwhelk (*Nucella lapillus*) populations around Scottish oil ports. Mar. Environ. Res. 32: 201-211

Donard, O.F.X., S. Rapsomanikis & J.H. Weber, 1986. Speciation of organic tin and alkyltin compounds by atomic absorption spectrometry using electrothermal quartz furnace after hydride generation. Analyt. Chem. 58: 772-777

Ellis, D.V. & L.A. Pattisina, 1990. Widespread neogastropod imposex: a biological indicator of global TBT contamination. Mar. Pollut. Bull. 21: 248-253

Féral, C., 1980a. Variations dans l'évolution du tractus génital mâle externe des femelles de trois Gastéropodes Prosobranches gonochoriques de stations Atlantiques. Cah. biol. mar. 21: 479-491

Féral, C., 1980b. Influence de la qualité de l'eau de mer sur la différenciation d'un tractus génital mâle externe chez les femelle d'un Mollusque Gastéropode gonochorique: *Ocenebra erinacea* (L). C. r. hebd. séanc. Acad. Sci. 291: 775-778

Féral, C., & S. Le Gall. 1982. Induction expérimentale par un polluant marin (le tributylétain), de l'activité neuroendocrine contrôlant la morphogénèse du pénis chez les femelles d'*Ocenebra erinacea* (Mollusque Prosobranche gonochorique). C. r. hebd. séanc. Acad. Sci. 295: 627-630

Féral, C., & S. Le Gall. 1983. The influence of a pollutant factor (tributyltin) on the neuroendocrine mechanism responsible for the occurrence of a penis in the females of *Ocenebra erinacea*. In: Molluscan neuro-endocrinology. Proc. Int. Minisymp. Molluscan Endocrinology, 1982; J. Lever & H.H. Boer (eds). North Holland Publ. Amsterdam, pp. 173-175

Fretter, V., 1941. The genital ducts of some British stenoglossan prosobranchs. J. mar. biol. Ass. U.K. 25: 173-211

Fretter, V., 1946. The pedal sucker and anal gland of some British Stenoglossa. Proc. malac. Soc. Lond. 27: 126-130

Fretter, V. & A. Graham, 1962. British prosobranch molluscs: their functional anatomy and ecology. London, Ray Society

Gibbs, P.E., & G.W. Bryan. 1986. Reproductive failure in populations of the dog-whelk, *Nucella lapillus*, caused by imposex induced by tributyltin from antifouling paints. J. mar. biol. Ass. U.K. 66: 767-777

Gibbs, P.E., & G.W. Bryan. 1987. TBT paints and the demise of the dog-whelk, *Nucella lapillus* (Gastropoda). In: Proceedings Oceans '87 Conference, Halifax, Nova Scotia, September 28-October 1, 1987. Vol. 4. International Organotin Symposium, Institute of Electrical and Electronics Engineers, New Jersey, pp. 1482-1487

Gibbs, P.E., G.W. Bryan, P.L. Pascoe, & G.R. Burt. 1987. The use of the dog-whelk, *Nucella lapillus*, as an indicator of tributyltin (TBT) contamination. J. mar. biol. Ass. U.K., 67: 507-523

Gibbs, P.E., P.L. Pascoe, & G.R. Burt, 1988. Sex change in the female dog-whelk, *Nucella lapillus*, induced by tributyltin from antifouling paints. J. mar. biol. Ass. U.K, 68: 715-731

Gibbs, P.E., G.W. Bryan, P.L. Pascoe, & G.R. Burt. 1990. Reproductive abnormalities in female *Ocenebra erinacea* (Gastropoda) resulting from tributyltin-induced imposex. J. mar. biol. Ass. U.K. 70: 639-656

Gibbs, P.E., G.W. Bryan, & P.L. Pascoe, 1991a. TBT-induced imposex in the dog-whelk, *Nucella lapillus*: geographical uniformity of the response and effects. Mar. Environ. Res. 32: 79-87

Gibbs, P.E., G.W. Bryan, & S.K. Spence, 1991b. An assessment of the impact of TBT pollution on the populations of the dog-whelk, *Nucella lapillus*, around the coastline of south-east England (Sussex and Kent). Oceanol. Acta, Spec. Vol. 11: 257-261

Gibbs, P.E., P.L. Pascoe, & G.W. Bryan, 1991c. Tributyltin-induced imposex in stenoglossan gastropods: pathological effects on the female reproductive system. Comp. Biochem. Physiol. 100C: 231-235

Gibbs, P.E., B.E. Spencer, & P.L. Pascoe, 1991d. The American oyster drill, *Urosalpinx cinerea* (Gastropoda): evidence of decline in an imposex-affected population (River Blackwater, Essex). J. mar. biol. Ass. U.K. 71: 827-838

Griffith, G.W. & M. Castagna, 1962. Sexual dimorphism in oyster drills of Chincoteague Bay, Maryland-Virginia. Chesapeake Sci. 3: 215-217

Hall, J.G. & S.Y. Feng, 1976. Genital variation among Connecticut populations of the oyster drill, *Urosalpinx cinerea* Say. (Prosobranchia: Muricidae). The Veliger 18: 318-321

Harding, M.J.C., S.K. Bailey & I.M. Davies, 1992. UK Department of the Environment TBT imposex survey of the North Sea. Scottish Office Agriculture and Fisheries, Working Paper 9/92 & Annexes 10/92-17/92

Hodge, V.F., S.L. Seidel, & E.D. Goldberg, 1979. Determination of tin(IV) and organotin compounds in natural waters, coastal sediments and macroalgae by atomic absorption spectrometry. Anal. Chem. 51: 1256-1259

Houston, R.S., 1971. Reproductive biology of *Thais emarginata* (Deshayes,1839) and *Thais canaliculata* (Duclos, 1832). Veliger, 13: 348-357

King, N., M. Miller, & S. de Mora, 1989. Tributyltin levels for sea water, sediment, and selected marine species in coastal Northland and Auckland, New Zealand. New Zealand J. Mar. Freshw. Res. 23: 287-294

Langston, W.J., G.W. Bryan, G.R. Burt & P.E. Gibbs, 1990. Assessing the impact of tin and TBT in estuaries and coastal regions. Funct. Ecol. 4: 433-443

McKie, J.C., 1987. Determination of total tin and tributyltin in marine biological materials by electrothermal atomic absorption spectrometry. Anal. Chim. Acta, 197: 303-308

Miller, E.R. & J.S. Pondick, 1984. Heavy metals in *Nucella lapillus* (Gastropoda: Prosobranchia) from sites with normal and penis-bearing females from New England. Bull. Environ. Contam. Toxicol. 33: 612-620

Oehlmann J., E. Stroben, & P. Fioroni, 1991. The morphological expression of imposex in *Nucella lapillus* (Linnaeus)(Gastropoda: Muricidae). J. mollusc. Stud. 57: 375-390

Poli, G., B. Salvat, & W. Strieff, 1971. Aspect particulier de la sexualité chez *Ocenebra erinacea.* Haliotis, 1: 29-30

Rexrode, M., 1987. Ecotoxicology of tributyltin. In: Proceedings Oceans '87 Conference, Halifax, Nova Scotia. September 28-October 1, 1987. Vol. 4. International Organotin Symposium. Institute of Electrical and Electronics Engineers, New Jersey, pp. 1443-1455

Ritzema, R., R.W.P.M. Laane & O.F.X. Donard, 1991. Butyltins in marine waters of the Netherlands in 1988 and 1989; concentrations and effects. Mar. Environ. Res. 32: 243-260

Saavedra Alvarez, M.M. & D.V. Ellis, 1990. Widespread neogastropod imposex in the northeast Pacific: implications for TBT contamination surveys. Mar. Pollut. Bull. 21: 244-247

Short, J.W., S.D. Rice, C.C. Brodersen, & W.B. Stickle, 1989. Occurrence of tri-n-butyltin-caused imposex in the North Pacific marine snail *Nucella lima* in Auke Bay, Alaska. Mar. Biol. 102: 291-297

Smith, B.S., 1971. Sexuality in the American mud-snail *Nassarius obsoletus* (Say). Proc. malac. Soc. Lond. 39: 377-378

Smith, B.S., 1980. The estuarine mud snail, *Nassarius obsoletus:* abnormalities in the reproductive system. J. mollusc. Stud. 46: 247-256

Smith, B.S., 1981a. Reproductive anomalies in stenoglossan snails related to pollution from marinas. J. Appl. Toxicol. 1: 15-21

Smith, B.S., 1981b. Male characteristics on female mud snails caused by antifouling bottom paints. J. Appl. Toxicol. 1: 22-25

Smith, B.S., 1981c. Tributyltin compounds induce male characteristics on female mud snails *Nassarius obsoletus* = *Ilyanassa obsoleta*. J. Appl. Toxicol. 1: 141-144

Smith, B.S., 1981d. Male characteristics in female *Nassarius obsoletus*: variations related to locality, season and year. Veliger, 23: 212-216

Smith, P.J., & M. McVeagh, 1991. Widespread organotin pollution in New Zealand coastal waters as indicated by imposex in dogwhelks. Mar. Pollut. Bull. 22: 409-413

Spence, S.K., S.J. Hawkins & R.S. Santos, 1990a. The mollusc *Thais haemastoma* - an exhibitor of 'imposex' and potential indicator of tributyltin pollution. Mar. Ecol. 11: 147-156

Spence, S.K., G.W. Bryan, P.E. Gibbs, D. Masters, L. Morris & S.J. Hawkins. 1990b. Effects of TBT contamination on *Nucella* populations. Funct. Ecol. 4: 425-432

Spooner, N., P.E. Gibbs, G.W. Bryan, & L.J. Goad, 1991. The effect of tributyltin upon steroid titres in the female dogwhelk, *Nucella lapillus*, and the development of imposex. Mar. Environ. Res. 23: 37-49

Stewart, C., S.J. de Mora, M.R.L. Jones & M.C. Miller, 1992. Imposex in New Zealand neogastropods. Mar. Pollut. Bull. 24: 204-209

Stickle, W.B., J.L. Sharp-Dahl, S.D. Rice & J.W. Short, 1990. Imposex induction in *Nucella lima* (Gmelin) via mode of exposure to tributyltin. J. exp. mar. Biol. Ecol. 143: 165-180

Stroben, E., J. Oehlmann & P. Fioroni, 1992a. The morphological expression of imposex in *Hinia reticulata* (Gastropoda: Buccinidae): a potential indicator of tributyltin pollution. Mar. Biol. 113: 625-636

Stroben, E., J. Oehlmann & P. Fioroni, 1992b. *Hinia reticulata* and *Nucella lapillus*. Comparison of two gastropod tributyltin bioindicators. Mar. Biol. 114: 289-296

Ten Hallers-Tjabbes, C.C., J.F. Kemp & J.P. Boon, 1993. Imposex in Whelks (Buccinum undatum) from the open North Sea: Relation to shipping traffic intensities. Mar. Pollut. Bull. (in press)

Unger, M.A., W.G. MacIntyre, J. Greaves & R.J. Huggett, 1986. GC determination of butyltins in natural waters by flame photometric detection of hexyl derivatives with mass spectrometric confirmation. Chemosphere, 15: 461-470

Waldock, M.J., J.E. Thain & M.E. Waite, 1987. The distribution and potential toxic effects of TBT in UK estuaries during 1986. Appl. organomet. Chem. 1: 287-301

Ward, G.S., G.C. Cramm, P.R. Parrish, H. Trachman & A. Slesinger, 1981. Bioaccumulation and chronic toxicity of bis (tributyltin) oxide (TBTO): tests with a saltwater fish. In: Aquatic toxicity and hazard assessment; D.R. Branson & K.L Dickson (eds). Associate Committee on Scientific Criteria for Environmental Quality, Philadelphia, Pennsylvania, pp. 183-200

Wilson, S.P., M. Ahsanallah & G.B. Thompson, 1993. Imposex in neogastropods: an indicator of tributyltin in eastern Australia. Mar. Pollut. Bull. 26: 44-48

Appendix 9

Biomonitoring of Tributyltin (TBT) Pollution

DETERMINATION OF TBT IN NEOGASTROPODS AND SEA WATER

Most methods for measuring TBT concentrations in neogastropods include an initial step involving solvent extraction from a homogenate dispersed in concentrated HCl. Various techniques have been used to analyse the extracts. The simplest methods involve chemical separation followed by atomic absorption analysis (AAS) (Ward *et al.*, 1981). A technique of this type was used by Bryan *et al.* (1986) and involved extraction of the HCl-acidified homogenate with hexane followed by back-extraction of di- and monobutyltins with sodium hydroxide solution: Sn in the organic extract was determined by graphite-furnace atomic absorption (ETAAS), the detection limit being about 10 ng $Sn.g^{-1}$ dry tissue. Extensive tests with various organotin compounds showed the method to be fairly specific for TBT although it could not be separated from tripropyl- or tetrabutyltin. However, these compounds are rare in the marine environment. A similar technique was used by Stroben *et al.* (1992a). The method of McKie (1987) is based on similar principles and was used by Bailey & Davies (1988a,b) and Stewart *et al.* (1992): the latter reported a detection limit of about 5 ng $Sn.g^{-1}$ dry wt. In their work on *N. lima* in Auke Bay, Alaska, Short *et al.* (1989) extracted TBT from an HCl digest with a solution of tropolone in benzene. After pentylation by Grignard reaction, the final separation and measurement was carried out with a gas chromatograph coupled to the graphite furnace of an atomic absorption system. The detection limit was about 10 ng $Sn.g^{-1}$ dry tissue.

Several different methods have been employed to measure TBT in sea water associated with neogastropods. Bryan *et al.* (1986) and Stroben *et al.* (1992a) determined TBT in 1 litre acidified samples by a procedure similar to that used for tissues, the detection limits being of the order of 1 ng $Sn.l^{-1}$ (see above). King *et al.* (1989) employed a method developed by Hodge *et al.* (1979) in which organotins were reacted with sodium borohydride to produce their volatile hydrides. After being trapped in a liquid-nitrogen-cooled tube the hydrides were separated on the basis of their different boiling points and determined as Sn by atomic absorption in a quartz furnace. A comparable method having a detection limit of about 2 ng $Sn.l^{-1}$ was developed by Balls (1987) and used by Bailey & Davies (1988) in their work on *N. lapillus*. Both Ritsema *et al.* (1991) and Batley & Scammell (1991) used the method of Donard *et al.* (1986). Solvent-extracted tin species are converted to their hydrides, trapped in liquid nitrogen and separated on a chromatographic column. By progressively raising the temperature of the column, organotins are driven off and measured as Sn by atomic absorption in a heated quartz cell: the detection limit was about 1 ng $Sn.l^{-1}$. In a study of imposex in *I. obsoleta* (Bryan *et al.*, 1989), seawater TBT concentrations were determined by the method of Unger *et al.* (1986). Organotins are solvent-extracted from sea water and derivitized with n-hexyl magnesium bromide to form tetra-alkyltins that are measured by gas chromatography with flame photometric detection. A detection limit of about 0.5 ng $Sn.l^{-1}$ was reported.

Although a range of techniques has been employed to determine TBT in neogastropod tissues and in water, virtually all authors have reached the same conclusion; namely that imposex in neogastropods is generally induced by TBT concentrations of no more than a few ng $Sn.l^{-1}$.

SAMPLING AND APPLICATION OF INDICES

General points concerning sample numbers, sex ratios and sampling errors in studies of *N. lapillus* imposex are discussed in Bryan *et al.* (1986) and Gibbs *et al.* (1987). Although gender determination is normally straightforward in species of the widespread genera *Nucella* and *Thais* (which usually possess a distinctive sperm-ingesting gland), difficulties have been encountered with other genera, such as *Cronia* and *Morula* (Ellis & Pattisina, 1990). In such species gonadal sectioning may be necessary, but, in cases where the females are severely affected the possibility of masculinisation of the ovary (see Gibbs *et al.*, 1988; Stewart *et al.*, 1992) should be borne in mind, otherwise an erroneous gender identification may result (*cf.* Oehlmann *et al.*, 1991).

Measurement of penis size is usually based on length, determined under a stereo-microscope using either 1 mm graduated graph paper or eyepiece graticule. Penes may be recurved, or possess an apical flagellum, or both; the system of measurement must be adapted to suit specific characters, *i.e.* curves measured in one-third steps, flagellum measured separately. The sample number required to produce consistent RPS values for example will probably vary according to the complexity of penis structure; in the case of *N. lapillus,* which has a simple penis shape, 10-15 penis length measurements for both sexes yield RPS values varying by only a few percent even when performed by different workers.

A general protocol for imposex determination and recording is given in Ellis & Pattisina (1990).

SELECTION OF FEMALES FOR EXPERIMENTATION

As mentioned above, early experiments on *N. lapillus* used large, unsexed samples for experiments and subsequent subsampling relied on an equal sex ratio to recover adequate numbers of females. Efficiency was greatly improved in later studies by screening the initial sample using narcotisation to select only those females suitable for experimentation. The following technique is recommended. Anaesthetise all specimens in $MgCl_2.6H_2O$ (75 $g.l^{-1}$ distilled water) for two hours to obtain full relaxation. By holding the shell and gripping the operculum with forceps the animal can then be gently eased from the shell sufficiently to be able to examine both the sole of the foot and the anterior region of the mantle cavity. Under a stereomicroscope, and securing the operculum against the shell with a thumbnail, females can be readily distinguished by the presence of a ventral pedal gland situated on the sole of the foot behind the accessory boring organ (see Fretter, 1946). For some experiments, *i.e.* those designed to test the promotability of imposex using penis length as a measure, it is desirable to minimise the range of penis length in the initial sample; whilst under narcosis it is a simple operation to select those females with penes of similar size. After this treatment, animals should be returned to sea water for a period of two to three days to permit full recovery; providing manipulation has been gentle, mortalities are rare.

KEY NEOGASTROPODS IDENTIFIED AS EXHIBITING IMPOSEX WITH POTENTIAL AS A TBT BIOINDICATOR

Recent studies have identified the masculinisation of neogastropod females as being imposex in various worldwide regions. The following is a selection of species considered to have potential as TBT indicators; three categories are recognised: species known, or thought, to be (a) sterilised; (b) not sterilised and (c) species for which data on the effect of advanced imposex are not yet available.

North-east Atlantic:

(a)	*Nucella lapillus*	refs in text
	Ocenebra erinacea (L.)	Gibbs *et al.*, 1990
	Thais haemastoma (L.)	Spence *et al.*, 1990
	Urosalpinx cinerea	Gibbs *et al.*, 1991d
(b)	*Nassarius reticulatus*	Bryan & Gibbs, 1991; Stroben *et al.*, 1992a,b
	Buccinum undatum (L.)	Ten Hallers-Tjabbes *et al.*, 1993

North-west Atlantic:

(a)	*Nucella lapillus,*	Miller & Pondick, 1984
	Urosalpinx cinerea	Hall & Feng, 1976
(b)	*Ilyanassa obsoleta*	refs in text Bryan *et al.*, 1989

North-east Pacific:

(a)	*Nucella lima*	Stickle *et al.*, 1990
	Nucella lamellosa (Gmelin),	Bright & Ellis, 1990
	Neptunia phoenicia (Dall)	ditto
(b)	*Nucella canaliculata* (Duclos),	Bright & Ellis, 1990
	Nucella emarginata (Deshayes),	ditto
	Ocenebra lurida (Middendorff),	ditto
	Searlesia dira (Reeve),	ditto
	Colus halli (Dall)	ditto

South-east Asia:

(c)	*Cronia margariticola* (Broderip),	(Ellis & Pattisina, 1990)
	Thais luteostoma (Holten),	ditto
	Morula musiva (Kiener)	ditto

South-east Australia:

(c)	*Thais orbita* (Gmelin),	Wilson *et al.*, 1993
	Morula marginalba (Blainville)	ditto

New Zealand:

(a)	*Lepsiella scobina,*	Stewart *et al.*, 1992
	Thais orbita (Gmelin),	ditto
	Haustrum haustorium (Gmelin)	ditto
(c)	*Xymene ambiguus* (Philippi),	Stewart *et al.*, 1992
	Taron dubius (Hutton),	ditto
	Cominella virgata (H. & A. Adams),	ditto
	Amalda (Baryspira) australis (Sowerby)	ditto

Parallel studies of more than one species are frequently necessary. Thus, around Europe rocky habitats can be monitored using *N. lapillus* and *O. erinacea*, and soft-bottom areas using *N. reticulatus*. Some knowledge of the breeding biology of species selected is necessary to interpret population effects, particularly recruitment through larval dispersal. Extinction of local populations is unlikely in species with even a short planktonic larval phase, for example *O. erinacea*, but is probable in those having direct development, such as *N. lapillus*. The rate of population decline will depend on longevity; many neogastropods appear to be long-lived (>10 years), particularly in sheltered habitats, and thus a lack of recruitment may not be obvious unless breeding activity is investigated.

Chapter 10

Biomonitoring of Genotoxicity in Coastal Waters

Sergey V. Kotelevtsev, Ludmila I. Stepanova and **Vadim M. Glaser**

Department of Biology, Moscow State University, 119899 Moscow, Russia

ABSTRACT

Increasing amounts of mutagenic compounds are being detected in the tissues of marine organisms. Test-systems for biological indication and biological monitoring of carcinogenic and mutagenic compounds in coastal marine ecosystems are reviewed here. Their practical applicability is shown in several case studies. The Ames test was used for assays of tissue extracts from organisms obtained from various parts of the Black Sea. The experiments showed that mutagens could be detected in fish and other aquatic animals originating from different regions of the Black Sea.

1. INTRODUCTION

It is customary to divide anthropogenic effects on ecosystems into two groups:

- *acute stress,* which is characterized by a sudden start, a rapid increase of intensity and a short duration of the disturbances;
- *chronic stress*, where disturbances of low intensity are prolonged or repeated frequently; these are "constantly disturbing" effects (Odum, 1954; Phiodorov & Gilmanov, 1980).

As a rule, ecosystems possess resistance and resilience that help them to bear periodic severe or acute effects. Moreover, a number of populations need severe stochastic disturbing effects for their development, such as fires or sharp changes in climatic conditions. Marine coastal ecosystems can restore themselves successfully after effects which occur on one occasion only. These effects may not be limited to natural effects such as storms, volcanic eruptions or sudden drastic changes in water temperature. Exposure, to even a considerable amount of toxic substances once only may not be especially dangerous, if these compounds are not stable and are quickly removed from the ecosystem. However, chronic disturbances can result in marked, stable and irreversible consequences, especially in the case of pollutants that are not found in nature (Lyakhovich & Tsyrlov, 1981). As a rule these compounds,

Biomonitoring of Coastal Waters and Estuaries. Edited by Kees J.M. Kramer.

which are alien to organisms, xenobiotics, take part neither in energetic nor in structural metabolism. They may, however, be taken up by tissues and may undergo chemical transformation while reacting with vitally important biological molecules such as DNA, RNA or proteins.

Carcinogenic and mutagenic compounds are, theoretically, the most dangerous that occur in an ecosystem, as their action may exert an effect beyond that of the individual and may be active through several generations. Accumulation of genotoxic compounds in ecosystems could modify biocoenoses irreversibly (Phiodorov & Gilmanov, 1980). In addition, these mutagenic compounds are often carcinogenic and cause tumour formation both in aquatic organisms (including marine fish) and in humans (Anderson, 1993). According to some estimates they make up no less than 5% of the total amount of anthropogenic pollutants of ecosystems (Dubinin, 1977). They may not be active as acute toxicants, but when accumulated in the organism they may display chronic action.

Mutations are hereditary changes resulting in an increase or decrease of the amount of genetic material or in changes of chromosome structure or DNA nucleotide sequence. Mutations can arise spontaneously or through the action of various factors: chemical substances and physical factors ultraviolet (UV) or ionizing radiation.

The different mutation types are:

- changes of ploidy, *i.e.* chromosome number in the nucleus;
- chromosome mutations;
- point mutations - changes of sequence in DNA at the level of single nucleotides.

Chromosome mutations form a large group of chromosome reorganization mechanisms. They include changes in linkage groups (translocation), chanches in the sequence of their location within a chromosome (inversion), fragmentation of chromosomes, loss of chromosome parts (deletion) as well as doubling of chromosome parts (duplication) or inclusion of other genetic elements into chromosomes (insertion).

Among the genotoxic substances a distinction should be made between chemicals that are intrinsically reactive and, thus, may form DNA adducts directly, and those compounds which require a metabolic transformation or activation before they become genetically active. Examples of directly acting chemicals include alkylsulphonic esters, epoxides, aromatic N-oxides, aromatic nitro-compounds, lactones, alkylnitrosoureas and alkylnitrosamides. Indirectly acting compounds include polycyclic aromatic hydrocarbons (PAHs), aromatic amines, alkyl and arylnitrosamines, aromatic azo-compounds and aliphatic vinyl compounds (Wrisberg, 1991). Apart from the xenobiotics, natural products may also interfere genotoxically, *e.g.* naturally formed PAHs, mycotoxins and polybromomethanes and other alkyl mono- and di-halides which are compounds released by macroalgae, such as *Ulva lactuca, Enteromorpha linza, Ascophyllum nodosum* and *Fucus vesicolosus* (Gschwend *et al.*, 1985).

Xenobiotics are distributed world-wide. The increase in the frequency of some diseases, disturbances of the immune system and some hereditary pathologies among the human population, may be attributed to the presence of xenobiotics in the environment. Significantly, it is noted that between 5 to 25 % of human cancers are connected with the pollution of air, food products and drinking water by carcinogenic and mutagenic substances (Anderson, 1993; Poroschenko & Abilev, 1988).

Mutagenic and carcinogenic compounds produced as a result of industrial and agricultural activities will, sooner or later reach the marine ecosystem, either transported by rivers or by atmospheric deposition. A considerable part of these substances is introduced into the seas directly as spills from ships carrying oil, oil products or other toxic chemicals, but sometimes as a result of calamities. Substantial amounts of mutagenic compounds reach the marine ecosystem during the course of oil and gas extraction in continental shelf areas. During the last ten years there has been a sharp increase in the number of tumours recorded in the tissues of marine fish, including fish for human consumption (De Flora *et al.*, 1991; Moore *et al.*,

1989; Metcalfe, 1990; Vethaak, 1993). Some of these diseases have a virus type nature but there is increasing evidence that they may be connected with the pollution of seas and oceans by mutagenic and carcinogenic xenobiotics (Mix, 1986; Moore *et al.*, 1989).
One of the features of mutagenic and carcinogenic compounds is that they can display biological effects even at very low concentrations, which hampers their detection, particularly in biological tissues, through chemical assays (Khelnitsky & Brodsky, 1990). On the other hand, it is not possible to determine whether a given substance has potential carcinogenic and/or mutagenic properties by means of chemical methods. That is the reason why biological testing and biological indication of carcinogenic and mutagenic compounds is attaining a growing significance. The observation that the concentration of mutagenic compounds is increasing in marine products used for human nutrition accounts for the necessity to monitor for genotoxic effects in coastal areas and especially in those areas where commercial fisheries are located.

2. TEST-SYSTEMS FOR THE ANALYSIS OF MUTAGENIC AND CARCINOGENIC COMPOUNDS

Various test systems are employed for investigating whether environmental factors alter genetic material, ranging from chemical detection of adduct formation to assays for DNA effects for the determination of different mutations. The complexity and duration of a test varies highly from one test system to another. Most genotoxic tests, *e.g.* for aquatic organisms, are time consuming and, thus, expensive. Hence, they are infrequently carried out and then usually only while performing model experiments.
Genotoxicity tests can be classified according to the information they provide:

- assays for gene (point) mutations
  on micro-organisms or cultured animal cells. Assays for both direct and reverse mutations are used. The most widespread among these is the Ames test, based on reversion of bacteria to histidine prototrophy (Ames *et al.*, 1975; Maron & Ames, 1983);
- assays for chromosomal aberrations (CA);
  eucaryotic *in vivo* and *in vitro* tests are known, which can often be detected using a light microscope. It can be primarily defined as a cytogenetic analysis of structural disturbances of chromosomes in metaphase and anaphase of mitosis (Pesch *et al.*, 1987) as well as a micronucleus test (*e.g.* Schmid, 1975; Mackey & MacGregor, 1979);
- recombination assays
  are, primarily, cytogenetic assays for *in vivo* and *in vitro* sister chromatid exchange (SCE) in eucaryotic systems (Alink *et al.*, 1980; Brunetti *et al.* 1986; Dixon & Clarke, 1982) as well as for reciprocal mitotic crossing over and mitotic gene conversion in yeast;
- assays for dominant lethals and recessive lethals
  (performed on mice and on *drosophila,* respectively). These tests are not yet adapted for use in the analysis of aquatic ecosystems and they are, therefore, not examined in detail; they are widely used, however, for analysis of genotoxicity of various chemical compounds, including the testing of mutagens found in water ecosystems (Abilev & Poroshenko, 1986);
- assay for morphological defects
  The investigation of mutations in mammals, fish or other animals by observation of their origin in a series of generations by means of morphological characteristics, is a long and expensive process. Nevertheless, studies on mice, rats and fish are carried out when analyzing carcinogenic features of a number of chemical compounds. This analysis not only detects mutagenesis but carcinogenic and teratogenic effects are also taken into account.

A cause-effect relationship will often be problematic, but in some cases these test-systems are sensitive and informative enough (Zakharov, 1993).

The most widely used among them are biotests on *daphnia* during which the effect of xenobiotics is studied in several generations of these crustacea. In this assay not only survival, fertility and mobility, but the presence of anomalies in development, abnormalities etc. are registered as well (Isacova & Kolosova, 1988). Such test systems could include patho-morphological, histological studies which are carried out frequently enough and include studies on tissues from marine organisms (see also Yevich & Yevich, this Volume).

- assays for DNA effects

  This group of assays includes the registration of DNA adducts with the help of different methods. They include *e.g.* the alkaline elution for the determination of DNA single-strand breakages, the DNA unscheduled (repair) synthesis (Kohn, 1983) or the SOS chromotest (umu-test; Oda *et al.*, 1985; von der Hude *et al.*, 1988). These test systems have been used successfully for analysis of genotoxicity in marine and freshwater ecosystems (Kurelec *et al.*, 1989; Zahn *et al.*, 1983; Wrisberg & Rhemrev, 1992).

  Mutagen induced lesions in the DNA structure change the nature of fluorescence and allow possible gene mutations to be judged (Jenner *et al.*, 1990). New methods apply flow laser fluorometer to investigate direct lesions of the DNA structure. Alternatively, the analysis of DNA adducts can be performed by chromatographic and other chemical analytical techniques (McCarthy *et al.*, 1989; Peakal, 1992; Stein *et al.*, 1989).

2.1. Cytogenetic methods

In the previous section, the most important of a large number of possible test systems available for the detection of genotoxicity or genetic instability were presented. At present, only a limited number of these tests have been adapted for the analysis of genotoxicity in marine ecosystems. In this section those which are most often used or could be adopted for bio-indication and bio-monitoring of aquatic ecosystems will be discussed.

2.1.1. Registration of chromosome aberrations (CA)

Accumulation of CA in somatic cells is one of the factors resulting in malignant proliferation, as well as aging and degeneration of tissues. There is a high degree of correlation (about 90%) between the ability of chemical substances to cause CA and gene mutations (Kirsch-Volders *et al.*, 1984). In spite of the differences in the mechanisms of these two types of mutations, the cytogenetic activity of the tested substance can indicate its ability to induce gene mutations. CA can be induced in cell cultures *in vitro* and by actively dividing somatic animal cells *in vivo*. Usually, the breaks of a chromosome with the formation of free or bound fragments and intra- and inter-chromosomal exchanges by chromosome parts is registered. For the investigation of different types of CA induced by various mutagenic agents, the analysis of the mitotic metaphase stage is the most informative. The division of cells could be delayed at this stage with the help of mitotic poisons (*e.g.* colchicine). However, the metaphase analysis requires that the caryotype of the tested organism should be well known and be suitable for cytogenetic analysis: *i.e.* big, but not numerous chromosomes which are quite distinct morphologically after differential staining. For the quantitative estimation of CA induction it is easier to analyze cells at the anaphase stage. In this case it is possible to register the increase of chromosome fragments and dicentrics, which become apparent in the anaphase, as so called bridges.

For study of CA induction *in vitro*, the lymphocytes of peripheral blood of animals stimulated to mitotic division by phytohaem glutinine are used. The advantages of the method are:

simplicity, accessibility of the material, synchronism of cell cycles and a low level of spontaneous mutations. The acentric fragments make up about 90% of the aberrations discovered. Animal bone marrow cells represent a convenient system for the investigation of CA induction *in vivo*. The main qualities of this method are a high proliferation activity of cells and the simplicity of making preparations. The morphology of chromosomes of laboratory animals (mice) is well studied. The number of metaphases with CA as well as the total amount of structural disturbances are registered.

These tests are also widely used for the investigation of marine ecosystems, *e.g.* using molluscs (Al-Sabti & Kurelec, 1985). The application of this test is limited by the necessity of chosing marine animals with a small quantity of large chromosomes.

2.1.2. Micronuclear test

Micronuclei (MN) are cytoplasmic chromatins with the appearance of small nuclei. They arise from acentric chromosomal fragments that are not included in one of the daughter nuclei at cell division. Other MN arise from chromosomes lagging at anaphase, due to dysfunction of the spindle apparatus (Wrisberg, 1991). The micronuclear test does not require a specific karyotype and, therefore, it is much easier to use than the CA analysis at the mitotic metaphase stage (Mackey & MacGregor, 1979; Lasne *et al.*, 1984). This test system was initially proposed for testing the erythroid cells of bone marrow (Schmid, 1975), but later it was shown that it is a suitable test for all dividing cells of animals (Heddle *et al.*, 1983). Micronuclei are stained like nuclei (by Giemza, accridine orange and other DNA stains) and are usually quite distinct at cytological preparations (Hayashi *et al.*, 1983; Iliinskich *et al.*, 1992).

This test has been used for the analysis of genetic defects in marine biota. However, the registered increase of micronuclei number in the cell did not always indicate the presence of mutagens in the environment, nor did they correlate with the appearance of tumours in fish tissues (Scarpato *et al.*, 1990; Smith, 1990). Nevertheless, an *in vivo* genotoxicity test, focussing on the production of micronuclei in the blue mussel, *Mytilus edulis,* has been developed and applied successfully (Majone *et al.*, 1987; Wrisberg & Rhemrev, 1992; Wrisberg *et al.*, 1992).

2.1.3. Sister chromatid exchange (SCE)

Although the exchange of chromatid parts (crossing over) between sister chromatids occurs spontaneously, it could be considerably induced in response to effects of mutagenic factors and, therefore, it could be used for estimating their activity (Alink *et al.*, 1980; Brunetti *et al.*, 1986; Dixon & Clarke, 1982). Essentially it is the assay for the recombinogenic effect of DNA damaging agents, but the difference between SCE and CA is rather relative because CA arise by the recombination mechanism also. The assay for SCE is one to two thousand times more sensitive compared to the assays for CA, but their execution is more difficult. It is necessary that two sister chromatids should be stained differently. This is achieved by treating the growing cells with 5-bromodeoxyuridine (BUdR) in the course of 2 cell cycles with subsequent histological staining by Giemza. Because of the semiconservative mechanism of DNA replication, one of the two sister chromatids will contain BUdR in one of DNA strands and the other - in both strands. The first one will be dark coloured and the second one will be light coloured. The mosaically coloured chromatids are formed as a result of their exchanges. The shortcoming of the method consists of the fact that BUdR can raise the SCE level by itself. This technique is used successfully for analysis of the genotoxicity of marine and freshwater ecosystems, *e.g.* using molluscs and some species of marine fish (Harrison & Jones, 1982; Jones & Harrison, 1987).

One of the considerable limits to the use of this technique is the fact that it is rather complicated, the results may vary considerably and only animals with the special karyotype are available for use for this analysis (Lasne *et al.*, 1984; Pesch *et al.*, 1987).

2.2. Alkaline elution method

Cultivated rodent or human cells (with or without metabolic activation) or even animals treated by mutagenic agents are used as test objects. In the last case, cell suspensions are prepared from animal organs. The cells are gathered on filters (ca 2 μm pore size). The cells are exposed to gentle lysis by a detergent directly on the filter which results in the release of high molecular weight DNA. Subsequently, an alkaline eluting solution (pH 12) is passed through the filter slowly over several hours. DNA will be denaturated and is eluted from the filter at a speed proportional to the number of breaks. One can judge the quantity of single-strand lesions of DNA by the inclination of the curves of elution kinetics in the experiment and in the untreated control (Kohn, 1983). The method has been applied *e.g.* to the haemolymph of mussels, *M. provincialis* (Bihari *et al.*, 1990).

2.3. Microbial tests: the Ames test

Micro-organisms are used for a quick estimation of the mutagenic effects of chemical substances. The advantages of using bacteria (or yeasts) as test objects are evident. Firstly, they produce a great number of generations in only a few hours. The advantages of micro-organisms as test objects are: a high speed of reproduction; simplicity of cultivation; the possibility for the use of selective media for the selection of rare mutants and haploidy, due to which all arising recessive mutations are displayed phenotypically. The shortcomings are the in principal differences in organization of the genetic apparatus and in the system of reparation of pre-mutational disturbances of DNA of procaryotes and eucaryotes, as well as the lack of the system of metabolic activation of xenobiotics which hampers extrapolation to higher animals of the results obtained. But the latter disadvantage may be eliminated by using the system of metabolic activation obtained from tissues of higher animals (mammals, fish).

The Ames test *Salmonella* microsomes, is based on the use of the set of tester strains specially constructed by Ames and his collaborators (Ames, 1971; Ames *et al., 1973*, 1975; McCann *et al.* 1975; Levin *et al.*, 1982). The strains bear mutations of histidine deficiency of different types. Various mutagenic agents cause reversions to histidine prototrophy (His+-phenotype) in tester strains, of which the frequency can be estimated by a number of colonies grown on the plates with minimal selective medium without histidine. The use of different tester *Salmonella* strains in the assay makes it possible to distinguish between a mutagen causing a mutation of the frame-shift or of the base-substitution type. To improve sensitivity, in addition to one of the his-mutations, all tester strains bear rfa and uvr B mutations resulting in the lack of cell wall lypopolysaccharide and the defect of excision repair, respectively, which enhances their sensitivity to mutagenic agents (strains TA 1535, TA 1537, TA 1538)(Ames *et al.*, 1973). Some *Salmonella* strains thus constructed (*e.g.* TA 97, TA 98 or TA 100) contain the plasmid pKM 101 which further increases their sensitivity (McCann *et al.*, 1975; Levin *et al.*, 1982). Strains TA 1537, TA 1538, TA 97 and TA 98 register frame-shift mutations and strains TA 1535 and TA 100 register replacement of bases.

The following diagram shows the two routes of the induction of genotoxic effects:

compound —— direct action ——→ genotoxic effect
compound —— metabolic activation ——→ genotoxic effect

For this reason it is possible to register the effect of direct mutagens only in microorganism cultures, *i.e.* only the effect of compounds interacting directly with the genetic material and displaying mutagenic effect themselves. However, the great number of xenobiotics show their mutagenic properties only after metabolic activation by the monooxygenase system (Abilev, 1986; Anderson, 1993).

Ames *et al.* (1975) proposed a test where enzymes from the rat liver cytochrome P-450 system were used for *in vitro* activation of the investigated substances. For a special analysis of the effects of mutagenic and carcinogenic compounds on aquatic organisms it is possible to use the microsomal fraction from the liver of fish (including marine fish) which have been exposed to injection of 20-methylcholanthrene (Kotelevtsev *et al.* 1986a; Kotelevtsev *et al.*, 1987a) and even the mono-oxygenase system from mollusc tissue (Marsh *et al.*, 1991).

This test is in use in a great number of laboratories and is convenient for the analysis not only of chemical compounds but of *e.g.* waste waters, food products etc. as well as marine samples (Blevina *et al.* 1987; Pittinger *et al.* 1987). By statistical analysis one may distinguish between the various degrees of induced effect. In studies of several hundreds of chemical compounds, it was shown that about 90% of substances causing carcinogenesis in mammals proved to be mutagenic in the Ames test. On the other hand, more than 90% of compounds that did not cause mutations do not display carcinogenic effect in this test, although some such 'non-carcinogens' were active in the Ames test (so-called 'false-positive results'). So if the substance displays mutagenic properties in the Ames test there is a high probability that it will have both a mutagenic and a carcinogenic effect upon man and animals (Ames *et al.*, 1975; Abilev & Poroshenko, 1986).

3. PRINCIPLES OF THE BIOLOGICAL TESTING OF MUTAGENIC AND CARCINOGENIC COMPOUNDS IN MARINE ECOSYSTEMS

For the biological (toxicological) testing of a chemical substance or a mixture of substances, the marine organisms are exposed under controlled laboratory conditions to defined concentrations of the chemicals and the possible effects are recorded as a function of time. That concentration of the tested substance which does not cause animals or plants to die, or to show any other biological effects such as marked changes in growth rate or development and functional state, even after chronic exposure, is considered to be harmless. This principle is used for the analysis of the potential toxicity of individual chemical compounds or their mixtures and of industrial waste materials that are or may be discharged into the aquatic ecosystem. The method is, however, of limited use for biological (genotoxic) testing of mutagenic and carcinogenic compounds. Nevertheless, some methods have been worked out for the detection of mutagenic effects in higher organisms, *e.g.* with daphnia or with small marine pelagic crustacea (Kainukova, 1988). They are able to produce several generations in 1-2 months allowing the appearance of abnormalities and some other hereditary changes over the generations to be registered. As they do not have a sufficiently developed system of metabolic activation of xenobiotics and are only accumulators of promutagenic compounds, one will observe the action of direct mutagenic substances only (Kotelevtsev *et al.*, 1987a).

For the biological testing of mutagenic compounds it is necessary to use special genetic test systems: Firstly, it is possible to study chromosome aberrations, or DNA damage, with aquatic animals exposed to injections of the tested substances (laboratory tests) or organisms caught in the areas polluted by mutagenic and carcinogenic xenobiotics (fish, molluscs) (Smith, 1990; Iliinskich *et al.*, 1992). Secondly, it is possible to place selected test organisms (*e.g.* molluscs, fish) in special cages directly into the aquatic system to be tested (active biomonitoring, ABM; see also De Kock & Kramer, this Volume). After a certain period of time the organisms may be used for the determination of chromosome aberrations, conducting micronuclear analysis or studying damage in DNA by the techniques mentioned above (Scarpato *et al.*, 1989; Waranazi *et al.*, 1989). Thirdly, it is possible to study (by means of a genetic test system) the accumulation of mutagenic and carcinogenic substances in tissues of marine animals, algae and sediments. In this case it is necessary to extract and isolate the mutagenic compounds from aquatic animal tissues. The latter method seems to be the most valuable for carrying out biological monitoring in the field, since it allows, on the one hand, the use such as the SOS chromotest (a test system that lends itself to automation), of various genetic test systems for the analysis of extracts (Kotelevtsev *et al.*, 1986b; Reifferscheid *et al.*, 1990). Although some genotoxins may escape detection, it allows the extent of metabolic activation of xenobiotics in tissues of various organisms to be judged, using the mono-oxygenase systems from the livers of warm-blooded animals and fish for metabolic activation.

4. EXTRACTION OF MUTAGENIC COMPOUNDS FROM TISSUES OF MARINE ORGANISMS, WATER AND SEDIMENTS

Separation and concentration of xenobiotics is accomplished by various chromatographic techniques. Two types of extraction are performed for preparing samples for analysis with the use of biological tissues: aqueous extraction and extraction by organic solvents.

Xenobiotics are usually extracted from the water by using one or more hydrophobic solvents, *e.g.* acetone, hexane or chloroform. The different methods of xenobiotic extraction from tissues of marine and fresh water organisms and from water have no fundamental differences. The choice of the method depends upon the nature of the compounds one intends to isolate from the sample to be tested. Various organic solvents are mostly used for the extraction of mutagenic compounds that are the most widespread in coastal marine ecosystems (including oil products, pesticides and herbicides) (Parkinson, 1974; Parry *et al.*, 1976). The preparation of tissue extracts for mutagen activity assay is begun by homogenization of the tissue, then followed by either one of the next two procedures: I. Aqueous (acidic or alkaline) extraction, followed by sedimentation of proteins and pH adjustment; concentration of mutagens; II. Extraction by organic solvents; concentration of mutagens by either removal of solvent by a rotor evaporator or concentration on resins of a XAD type.

After extraction, mutagens are dissolved in the medium of the particular assay and analyzed in one or more biological test systems. One should realize, however, that genotoxins may get lost during the process of extraction and concentration.

More technical information is presented in the technical Appendix to this chapter.

5. CASE STUDIES

Quite a large number of studies for genotoxicity in marine waters have been reported, carried out either in the field or in laboratory experiments. Often they included the genotoxic effects upon *Mytilus* sp. (*e.g.* Dixon & Clarke, 1982; Harrison & Jones, 1982; Al-Sabti & Kurelec, 1985; Dixon *et al.*, 1985; Brunetti *et al.*, 1986; Scarpato *et al.*, 1990; Wrisberg, 1991). Other organisms which have been tested for the genotoxic effects include fish (*e.g.* Kurelec *et al.*, 1981; Alink *et al.*, 1980), sponges (Zahn *et al.*, 1983) or polychaetes (Pesch *et al.*, 1987). Several well identified compounds or mixtures were tested, such as PAHs (Kurelec *et al.*, 1981; Zahn *et al.*, 1983) and oil compounds (Kurelec *et al.*, 1981), paper pulp industrial effluents (Wrisberg & van der Gaag, 1992) or dredged materials from harbours and docks (Dixon & Clarke, 1982; Allen *et al.*, 1983; Pesch *et al.*, 1987) and general surveys of polluted seawater (*e.g.* Frezza et al., 1982; Scarpato et al., 1990).

Several case studies, depicting genotoxic studies in coastal waters follow here, exemplifying different analytical approaches. We consider it important to carry out studies of genotoxicity which result not only in the detection of effects caused by the mutagens themselves. Mechanisms of the activity xenobiotics should also be taken into account and the metabolism of these compounds, the character of their accumulation in tissues and the manner of their transfer in food chains should be investigated.

It is also important to identify species which could serve well as monitoring organisms for the pollution of marine ecosystems by mutagenic compounds.

In our laboratory, the biological monitoring of genotoxicity of the aquatic ecosystems of the southern part of Lake Baikal, the Rybinsk Lake and the Black and the Caspian Seas have been carried out for many years by means of the Ames test and the SOS chromotest (Kotelevtsev *et al.*, 1986a,b; Kotelevtsev & Hänninen, 1992). In some cases we used the metabolic activation system from not only rat liver but from the liver of fish inhabiting the investigated areas as well (Kotelevtsev *et al.*, 1986a, 1987a).

It turned out that the activity of the metabolic activation system of standard carcinogens from the fish liver in the Ames test depended considerably upon both the species involved and on the method of induction of mono-oxygenases (Kotelevtsev *et al.*, 1987a). For instance, benzo[*a*]pyrene and 2-aminoanthracene are metabolized more actively in the liver of the Black Sea scad (*Trachurus* sp.) compared to the liver of bullheads (*Gobius* sp.) (Kotelevtsev *et al.*, 1986b, 1987b). Therefore, in scad tissues, xenobiotics would be oxidized and released from the tissues, but the population of bullheads pays for it by an elevated risk of the rise of malignant tumours. In the bullheads' tissues, the metabolism of xenobiotics is slower, therefore, they may accumulate in this fish to significantly greater amounts (Kotelevtsev *et al.*, 1987b).

Furthermore, it is necessary to take the possibility of concentration of xenobiotics in food chains into account as, for instance: molluscs - bullheads - scad - man.

In the case of the Caspian Sea, the mutagenic compounds in the extracts of tissues from aquatic organisms were detected even in relatively clean areas which were remote from human settlements. Mutagenic compounds were found to be present in tissues of almost all organisms inhabiting the Black Sea, especially in the more polluted areas.

The character of the accumulation of mutagenic compounds in marine tissues depends not only upon the type of pollutant but also upon the activity of metabolizing enzymes in the tissues.

5.1. Genotoxicity of tissue extracts from Black Sea organisms

In this section we will examine the results of the analysis of mutagenic compounds in tissues of the Black Sea organisms. The survey work was carried out in the period June-July 1989-1991 respectively, in the Black Sea, on board Moscow University's scientific research ship "Experiment". The material was collected by trawling (using a beam-trawl) the coastal areas at a depth of 7-12 meters, while individual species were collected by divers. Immediately after collection, the species was identified and the material was fixed in a two fold volume of acetone and stored at about 4 °C. In total, six groups of species, which were collected at 32 Black Sea sampling stations were studied. Unfortunately, at some stations it was not possible to catch representatives of the main classes of these organisms. Nevertheless, the species described in this work can be considered typical for each of the explored areas. The species collected were bivalves (*Chamella gallina*) and sea snails (*Rapana thomasiana*), polychaetes (*Nereis* sp), crustacea (*Crangon crangon, Clibanarius erythropus*) and fish (*Gobius* sp).

Isolation of xenobiotics from the organisms' tissues and from the bottom sediments was carried out by means of repeated extractions by a mixture of organic solvents (acetone:hexane:chloroform = 1:1:1). The obtained xenobiotics were separated from the extraction mixture in a rotor evaporator, dried by lyophilization, dissolved and diluted in dimethylsulphoxide (DMSO) proportionally to the dry weight of a sample at a rate of 1 ml of DMSO per 100 mg of tissue dry weight. The samples were tested for genotoxicity.

Analysis of the mutagenic properties of each sample was performed in the Ames test in four variants. The first two of them identify the presence of direct mutagens (identified as "-MA") in the extracts of xenobiotics as causing both types of point gene mutations in the test organism (*Salmonella typhimurium*): a frameshift of the genetic code (TA 98 strain) and a base substitution (TA 100 strain). In the two other variants the presence of indirect mutagens (called "+MA") were determined as causing a drastic increase of the frequency of point gene mutations in the conditions of the PCB mixture Arochlor 1254-induced metabolic activation system from rat liver (S9 fraction).

The results of typical experiments on assay of extracts from animal tissues in the Ames test are shown in Table 1. They are represented as a number of colonies of His+ revertants of *Salmonella* per plate. The number of spontaneous mutations in the control experiment was determined in the presence of DMSO in the incubation medium. As a control promutagen causing mutagenic effects after metabolic activation in the microsomal monooxygenase system, 2-aminoanthracene and benzo[*a*]pyrene were used.

Table 1. Mutagenic activity of different tissue extracts (Black Sea, Pomorie) in tester strains TA 98, TA 100 of *S. typhimurium* (MA = metabolic activation)

Sample	**Dose ml**	**TA 98**		**TA 100**	
	per plate	**+MA**	**-MA**	**+MA**	**-MA**
Dimethylsulphoxide	0.1	38	20	112	108
2-Aminoanthracene	0.5[a]	4420	18	2150	104
2-Nitrofluorene	10.0[a]	843	34	682	107
Gobius sp. (liver)	0.1	40	21	154	115
Rapana thomasiana (digest. gland)	0.1	58	23	141	103
Chamella gallina	0.1	60	114	110	101
Clibanarius erythropus	0.1	61	54	190	165
Polychaete sp.	0.1	76	18	121	111

[a] in μg

Table 2. Selected species for the monitoring of genotoxicity in the Black Sea

Species	**% samples content mutagenic compound**
Polychaetes (*Nereis* sp.)	77 ± 6.0
Bullhead (muscles) (*Gobius* sp.)	53 ± 7.0
Shrimps (*Crangon crangon*)	51 ± 3.2
Hermit crab (*Clibanarius erythropus*)	46 ± 4.2
Bivalves (*Chamella gallina*)	27 ± 4.5
Sea snail (digest. gland) (*Rapana thomasiana*)	27 ± 3.1

The main results of the investigation of extracts of tissues of the Black Sea organisms are represented in Figure 1. The analysis of the data showed that in the majority of the investigated areas, including those relatively remote from direct sources of pollution, a significant amount of mutagenic compounds could be detected. They displayed both direct mutagenic and promutagenic effects. The TA 98 strain is the most informative of the two test strains.

The results indicate that the investigated Black Sea coastal organisms mainly accumulate direct mutagens, causing mutations of a frame-shift type. It is possible to arrange the investigated organisms in a decreasing order by their ability to accumulate direct mutagens. For the TA 98 strain: polychaetes = shrimps > bullheads > sea snails = bivalves = hermit crab. For the TA 100 strain: shrimps > bullheads > polychaetes = hermit crab = sea snails > bivalves.

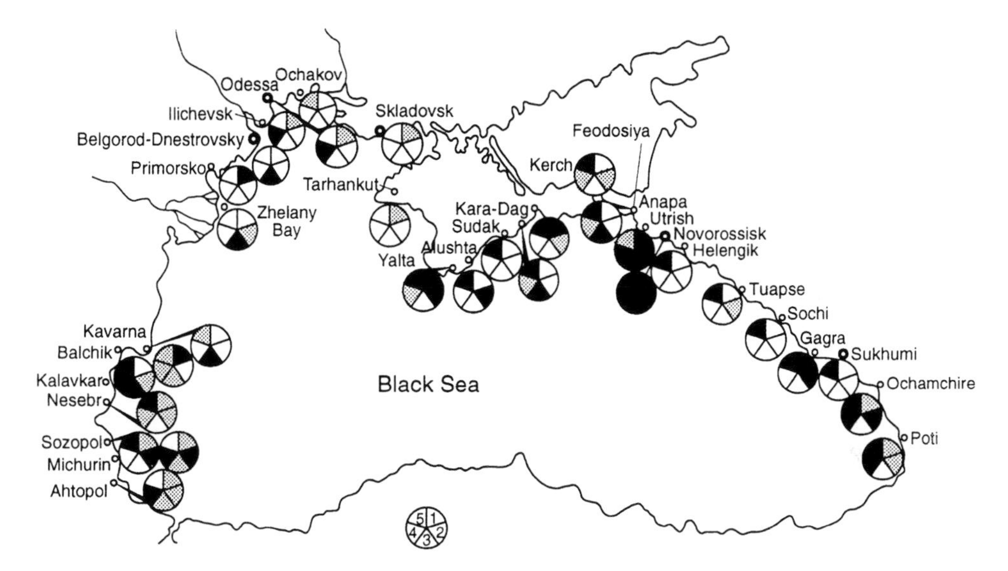

Figure 1. Genotoxicity of tissue extracts from Black Sea organisms identified by the Ames test. In each segment the genotoxic effect (white = no mutagenic effect found; shaded = mutagenic effect found; black = no organisms collected), and the five species investigated (1: Bullhead (muscles) (*Gobius*) sp.; 2: Hermit crab (*Clibanarius erythropus*); 3: Bivalves (*Chamella gallina*); 4: Sea snail (digestion gland) (*Rapana thomasiana*); 5: Polychaete (*Nereis*) sp.)) are indicated

This approach allows the most active accumulators of the mutagens of the Black Sea such as polychaetes, shrimps or bullheads to be chosen (Table 2). These organisms can thus be used as biomonitors for genotoxicity in Black Sea coastal waters.

Compounds were also found in the tissue extracts that have promutagenic action which is revealed in the Ames test after the metabolic activation by the mono-oxygenase system from the liver of Arochlor 1254-induced rats. The Black Sea water organisms accumulate these compounds to a different extent: polychaetes *C. erythropus* and *R. thomasiana* should be selected as typical accumulators of these xenobiotics. It indicates that the system of metabolic activation of xenobiotics in tissues of these animals is not sufficiently active in comparison with other species and warm-blooded animals and cannot effectively hydroxylize mutagenic substances that enter the organism. Promutagens are rarely accumulated in organisms of the Black Sea bivalve molluscs (*C. gallina*).

It is necessary to mention that *R. thomasiana* and *C. erythrophus* display the same tendencies in the frequency of accumulation of direct mutagenic and of promutagenic substances. The Black Sea polychaetes and bullheads are the most active accumulators of both types of mutagenic compounds. We should realize, however, that it is natural that in each specific case the type of compounds accumulating in the tissues depends upon the character of pollutants entering the ecosystem, the features of nutrition and the activity of the organism's detoxication system, which could be influenced by seasonal variations.

CONCLUSION

Mutagenic and carcinogenic compounds are now registered in the tissues of marine organisms more often. The increase in the concentration of these compounds, as well as mutations caused by a chronic exposure to radioactive irradiation connected with the pollution of seas by radionuclides, may result in possible changes in species composition and even ultimately change some aquatic ecosystems.

On the other hand, accumulation of mutagenic and carcinogenic compounds in tissues may represent a direct danger for man, as a consumer of sea products.

It is necessary to carry out regular monitoring programmes for genotoxicity of marine ecosystems, to determine and, if possible, to eliminate the sources of pollution of seas and oceans by genotoxic compounds. These problems can only be solved through international cooperation.

REFERENCES

Abilev, S.K., 1986. Metabolic activation of mutagens. In: Sum of science and technique. Genetic 9; G.G. Poroshenko (ed). VINT, Moscow, pp. 5-96, (in Russian)

Abilev, S.K., G.G. Poroshenko, 1986. Fastened techniques of prognosing of mutagenic and blastomere features of chemical compounds. In: Sum of science and technique. Toxicology, 14; G. Stepanenco & L.M. Fonschtein (eds). VINT, Moscow, pp. 9-171 (in Russian)

Alink, G.M., E.M.H. Frederix-Wolters, M.A. van der Gaag, J.F.J. van de Kerkhof & C.L.M. Poels, 1980. Induction of sister-chromatid exchanges in fish exposed to Rhine water. Mutation Res. 78: 369-374

Allen, H.E., K.E. Noll & R.E. Nelson, 1983. Methodology for assessment of potential mutagenicity of dredged materials. Environ. Technol. Lett. 4: 101-106

Al-Sabti, K. & B. Kurelec, 1985. Induction of chromosomal abberations in the mussel *Mytilus galloprovincialis* watch. Bull. Environ. Contam. Toxicol. 35: 660-665

Ames, B.N., 1971. The detection of chemical mutagens with enteric bacteria. In: Chemical mutagens: Principles and methods for their detection, Vol. 1; A. Holaender (ed). Plenum Press, New York, pp 267-282

Ames, B.N., F.D. Lee, W.E. Durston, 1973. An improved bacterial test system for the detection and classification of mutagens and carcinogens. Proc. Nat. Acad. Sci. USA, 70: 782-786

Ames, B.N., J. McCann & E. Yamasaki, 1975. Method for detecting carcinogens and mutagens with the *Salmonella*/mammalian microsomes mutagenicity test. Mutat. Res. 31: 347-364

Anderson, D., 1993. Mechanisms of mutagenicity and tumor formation. In: Molecular aspects of oxidative drag metabolizing enzymes: their significance in environmental toxicology, chemical carcinogenesis and health. Book of abstracts and short lecture notes. F. Arinc (ed). Middle East Technical University, Turkey, pp 19-26.

Anthony, T., 1979. Methylene chloride. In: Kirk-Othmer encyclopedia of chemical technology, Vol. 5; G.J. Bushey, L. Campbell, C. Eastman, A. Klingsberg & L. van Nes (eds). Wiley, New York, pp. 686-693

Belisario, M.A., V. Buonocore, E. De Marinis & F. De Lorenco, 1984. Biological availability of mutagenic compounds absorbed into diesel exhaust particulate. Mutat. Res. 135: 1-9

Bihari, N., R. Batel & R.K. Zahn, 1990. DNA damage determination by the alkaline elution technique in the haemolymph of the mussel *Mytilus galloprovincialis* treated with benzo[*a*]pyrene and 4-nitroquinoline-N-oxide. Aquat. Toxicol. 18: 13-22

Bjeldanes, L.F., K.R. Grose, P.H. Davis, D.H. Stuermer, S.K. Healy & J.S. Felton, 1982. An XAD-2 resin method for the efficient extraction of mutagens from fried ground beef. Mutat. Res. 105: 43-49

Blevina, R.D., Mohr C.D., Pancorbo O.C., 1987. Mutagenicity activity of fish muscle extracts. Environ. Mutagens (Suppl), 8: 17-19

Brunetti, R., I. Gola & F. Majone, 1986. Sister-chromatid exchange in developing eggs of *Mytilus galloprovincialis* Lmk. (Bivalvia). Mutation Res. 174: 207-211

Commoner, B., A. Vithayathil, P. Dolara, S. Nair, E. Madyastha & G.C. Cuca, 1978. Formation of mutagens in beef and beef extract during cooking. Science, 201: 913-916

De Flora, S., M. Bagnasco & P. Lanacchi, 1991. Genotoxic, carcinogenic and teratogenic hazards in the marine environment with special reference to the Mediterranean Sea. Mutat. Res. Rev. Genet. Toxicol. 258: 285-320

Dixon, D.R. & K.R. Clarke, 1982. Sister-chromatid exchange: A sensitive method for detecting damage caused by exposure to environmental mutagens in the chromosomes of adult *Mytilus edulis*. Mar. Biol. Lett. 3: 163-172

Dixon, D.R., I.M. Jones & F.L. Harrison, 1985. Cytogenetic evidence of inducible processes linked with metabolism of a xenobiotic chemical in adult and larval *Mytilus edulis*. Sci. Total Environ. 46: 1-8

Dubinin, N.P., 1977. The genetic results of environmental pollution. Nauka, Moscow, pp. 3-20, (in Russian)

Felton, J.S., S. Healy, D. Stuermer, C. Berry, H. Timourian & F.T. Hatch, 1981. Mutagens from the cooking of food, improved extraction and characterisation of mutagenic fraction from cooked ground beef. Mutat. Res. 88: 33-34

Frezza, D., B. Pegoraro & S. Presciuttini, 1982. A marine host-mediated assay for the detection of mutagenic compounds in polluted sea waters. Mutat. Res. 104: 215-223

Gennari, G. & G. Jori, 1970. Acetone-sensitized anaerobic photo-oxidation of methionine. FEBS Lett. 10: 129-131

Grabow, W.O.K., J.S. Burger & C.A. Hilner, 1981. Comparison of liquid extraction and resin adsorption for concentrating mutagens in Ames *Salmonella*/microsomal assays on water. Bull. Environ. Contam. Toxicol. 27: 442-449

Harrison, F.L. & I.M. Jones, 1982. An *in vivo* sister-chromatid exchange assay in the larvae of the mussel *Mytilus edulis:* response to 3 mutagens. Mutat. Res. 105: 235-242

Hayashi, M., T. Sofuni & M. Ishidate, 1983. An application of acridine orange fluorescent staining to the micronucleus test. Mutat. Res. 120: 241-247

Heddle, J.A., M. Hite, B. Kirkhart, K. Mavournin, J.T. MacGregor, G.W. Newell & M.F. Salamone, 1983. The induction of micronuclei as a measure of genotoxicity, a report of the U.S. Environmental Protection Agency gene tox program. Mutat. Res. 123: 61-118

Holt, L.A., B. Milligan, D.F. Rivett & F.H.C. Stewart, 1977. The photo-decomposition of tryptophan peptides. Biochem. Biophys. Acta, 499: 131-138

Iliinskich, N.N., V.V. Novitskyi, N.N. Vanchugova & I.N. Iliinskich, 1992. Micronucleus analysis and the cytogenetic instability, Tomsk. University Press, Tomsk, pp. 272, (in Russian)

Isacova, E.F. & L.B. Kolosova, 1988. Biotest methods with *Daphnia magna*. In: Biotest methods; A.N. Krainukova & A.P. Braginski (eds). Acad. Sc. USSR Press, Chernogolovka, pp 50-57 (In Russian)

Iwaoka, W.T., Ch.A. Krone, J.J. Sullivan, E.H. Meaker, C.A. Johnson & L.S. Miyasato, 1981. A source of error in mutagen testing of foods. Cancer Lett. 11: 225-230

Jamasaki, E. & B.N. Ames, 1977. Concentrations of mutagens from urine by adsorption with the non polar resin XAD-2: Cigarette smokers have mutagenic urine. Proc. Natl Acad. Sci. USA, 74: 3555-3559

Jenner, N.K., G.K. Ostrander, T.J. Kavanagh, I.S. Livesey, M.W. Shen, S.C. Kim & E.H. Holmes, 1990. A flow-cytometric comparison of DNA content and glutathione levels in hepatocytes of English sole (*Parophrys vetulus*) from areas of differing water quality. Environ. Contam. Toxicol. 19: 807-815

Jones, I.M. & F.L. Harrison, 1987. Variability in the frequency of sister-chromatid exchange in larvae of *Mytilus edulis:* implications for field monitoring. J. exp. mar. Biol. Ecol. 113: 283-288

Jori, G., G. Galiazzo & E. Scoffone, 1969. Photodynamic action of porphyrins on amino acids and proteins. 1. Selective photooxidation of methionine in aqueous solution. Biochem. 8: 2868-2875

Kadhim, M. & J.M. Parry, 1984. The detection of mutagenic chemicals in the tissues of shellfish exposed to oil pollution. Mutat. Res. 136: 93-105

Kainukova, A.N., 1988. Biotests and the health of water systems. In: Biotest methods; A.N. Krainukova & A.P. Braginski (eds). Acad. Sci. USSR Press, Chernogolovka, pp 4-14, (In Russian)

Khelnitsky, R.A. & E.S. Brodsky, 1990. Mass spectrometry of environmental pollution. Chimia, Moscow, pp. 184, (in Russian)

Kirsch-Volders, M., M. Radman, P. Jeggo & P. Verschare, 1984. Chromosome aberrations in somatic cells. In: Mutagenicity, carcinogenicity and teratogenicity of pollutants; M. Kirsch-Volders (ed). Plenum Press, London, pp. 6-58

Kohn, H.W., 1983. The significance of DNA-damaging assays in toxicity and carcinogenicity assessment. Ann. N.Y. Acad. Sci. 407: 106-118

Kotelevtsev, S.V., S.L. Stvolinskyi & A.M. Beim, 1986a. Ecotoxicological analysis on the basis of the biological membrane. Moscow State University, Moscow, pp. 120 (in Russian)

Kotelevtsev, S.V., Yu.P. Kozlov & L.I. Stepanova, 1986b. Eco-toxicological biomonitoring by the methods of physico-chemical biology. Biol. Sci. 1: 19-30, (in Russian)

Kotelevtsev, S.V., L.I. Stepanova, L.V. Ponomariova, G.V. Buevitch, I.M. Limarenko, A.M. Beim, V.M. Glaser & Yu.P. Kozlov, 1987a. Polycyclic aromatic hydrocarbon induction of the mono-oxygenase activity in fish tissues for use in the biomonitoring of the pollution. Exper. Oncology, 9: 46-49, (in Russian)

Kotelevtsev, S.V., L.I. Stepanova, I.M. Limarenco, Yu.P. Kozlov, 1987b. Different metabolic pathways of polycyclic xenobiotics in fish tissues. In: Cytochrome P-450 and environmental health; V.V. Lyakhovich & I.B. Tsyrlov (eds), Nauka, Novosibirsk, pp. 21-22

Kotelevtsev, S.V. & O. Hänninen, 1992. Biotechnology in analysis water ecosystems: Hydrobionts tissue isoforms of cytochrome P-450. In: Proc. 7th Int. Conf. Cytochrome P-450: Biochemistry and biophysics; A.I. Archakov & G.V. Bachmanova (eds). INCO-TNC, Joint Stock Company, Moscow, pp. 475-480

Krone, Ch.A. & W.T. Iwaoka, 1983. Differences in observed mutagenicity associated with the extraction of mutagens from cooked fish. J. Agr. Food Chem. 31: 428-431

Kurelec, B., M. Protic, S. Britvic, N. Kesic, M. Rijavec & R.K. Zahn, 1981. Toxic effects in fish and the mutagenic capacity of water from the Sava river in Jugoslavia. Bull. Environ. Contam. Toxicol. 26: 179-187

Kurelec, B., A. Garg, S. Krcva & R.C. Gupta, 1989. DNA adducts as biomarkers in genotoxic risk assessment in the aquatic environment. Mar. Environ. Res. 28: 317-321

Lasne, C., Z.W. Gu, W. Venegas & I. Chouroulinkov, 1984. The *in vitro* micronucleus assay for detection of cytogenetic effects induced by mutagen-carcinogens: comparison with the *in vitro* sister-chromatid assay. Mutat. Res. 130: 237-282

Levin, D.E., E. Yamasaki & B.N. Ames, 1982. A new Salmonella taster strain, TA 97 for the detection of frameshift mutagens. A run of cytosines as a mutational hot-spot. Mutat. Res. 94: 315-330

Loper, J.C., M.W. Tabor & L. Distlerath, 1981. A new microsomal activation dependent mutagen isolated from old residues of drinking water. Environ. Mutagens, 3: 306-310

Lyakhovich, V.V. & I.B. Tsyrlov, 1981. The induction of the metabolic xenobiotics enzyme. Nauka, Novosibirsk, pp. 242, (in Russian)

Mackey, B.E. & J.T. MacGregor, 1979. The micronucleus test: Statistical design and analysis. Mutat. Res. 64: 195-204

Majone, F., R. Brunetti, I. Gola & A.G. Lewis, 1987. Persistence of micronuclei in the marine mussel, *Mytilus galloprovincialis,* after treatment with mitomycin C. Mutat. Res. 191: 157-161

Maron, D.M. & B.N. Ames, 1983. Revised methods for the *Salmonella* mutagenicity test. Mutat. Res. 113: 173-215

Marsh, S., J.K. Chipman & D. Livingstone, 1991. Activation of carcinogens to mutagenic products by the mussel, *Mytilus edulis*. Mutat. Res. Environ. Mutagens and Related Subj. 252: 191

McCann, J., N.E. Spingarn, J. Kobori, B.N. Ames, 1975. Detection of carcinogens as mutagens: bacterial tester strains with R factor plasmids. Proc. Nat. Acad. Sci. USA, 72: 979-983

McCarthy, J.F., D.N. Jacobson, L.R. Shugart & B.D. Jimenez, 1989. Pre-exposure to 3-methylcholanthrene increases benzo(a)pyrene adducts on DNA of bluegill sunfish. Mar. Environ. Res. 28: 323-328

Metcalfe, C.D., 1990. Chemical contaminants and fish tumors. Sci. Total Environ. (Special issue) 94: 1-168

Mix, M.C., 1986. Cancerous diseases in aquatic animals and their association with environmental pollutants: a critical literature review. Mar. Environ. Res. 20: 1-141

Moore, M.S., R. Smolowitz & J.J. Stegeman, 1989. Cellular alternations preceding neoplasia in *Pseudopleuronectes americanus* from Boston Harbor. Mar. Environ. Res. 28: 425-429

Oda, J., S.I. Nakamura, I. Oki, T. Kato & H. Shinagawa, 1985. Evaluation of the new system (umu-test) for the detection of environmental mutagens and mutagens. Mutat. Res. 147: 219-229

Odum, E.P., 1954. Fundamentals of ecology. Saunders, Philadelphia, pp. 384

Osborne, L.L., R.W. Davies, K.R. Dixon & R.L. Moore, 1982. Mutagenic activity of fish and sediments in the Sheep river Alberta. Wat. Res. 16: 899-902

Parkinson, J.A., 1974. Organic pesticides. In: Chemical analysis of ecological materials; S.F. Allen (ed). Wiley, New York, pp. 332-356

Parry, J.M., D.J. Tweats & M.A.J. Al-Mossawi, 1976. Monitoring of the marine environment for mutagens. Nature, 264: 538-540

Peakal, D.B., 1992. Studies on genetic material. In: Animal biomarkers as pollution indicators; D.B. Peakall, M.H. Depledge & B. Sanders (eds). Chapman & Hall, London, pp. 290

Pesch, G.G., C. Müller, C.E. Pesch, A.R. Malcolm, P.F. Rogerson, W.R. Munns, G.R. Gardner, J. Heltshe, T.C. Lee & A. Senecal, 1987. Sister-chromatid exchange in a marine polychaete exposed to a contaminated harbor sediment. In: Short-term bioassay in the analysis of complex environmental mixtures, Vol. V; M.D. Waters *et al.* (eds), Plenum, New York, pp. 237-253

Phiodorov, V.D. & T.G. Gilmanov, 1980. Ecology. Moscow State University, Moscow, Leninskie Gori, pp. 464, (in Russian)

Pittinger, C.A., A.L. Buikema & J.O. Falkinham, 1987. *In situ* variations in aromatic hydrocarbons. Environ. Toxicol. Chem. 6: 51-60

Poroschenko, G.G. & S.K. Abilev, 1988. Anthropogenic mutagens and natural anti-mutagens, In: Sum of science and technique. Genetic. 12; V.A. Shevchenko (ed). VINT, Moscow, pp. 207 (in Russian)

Reifferscheid, G., J. Heil & R.K. Zahn, 1990. An automated microplate version of the umu-test for rapid detection of genotoxins in environmental samples. (In prep)

Scarpato, R., A.G. Cegnetti, F. DiMarino & L. Migliore, 1989. Mutagenic monitoring of marine environments. Mutat. Res. Environ. Mutagenes Related Subj. 216: 315

Scarpato, R., L. Migliore, G. Alfinito-Cognetti & R. Barale, 1990. Induction of micronuclei in gill tissue of *Mytilus galloprovincialis* exposed to polluted marine waters. Mar. Pollut. Bull. 21: 74-80

Schmid, W., 1975. The micronucleus test. Mutat. Res. 31: 9-15

Smith, J.W., 1982. Mutagenicity of extracts from agricultural soil in *Salmonella*/microsome test. Environ. Mutagens, 4: 369-370

Smith, I.R., 1990. Erythrocytic micronuclei in wild fish from lakes Superior and Ontario that have pollution - associated neoplasia. J. Great Lakes Res. 16: 139- 142

Stein, J.E., W.L. Reichert & U. Varanasi, 1989. Covalent binding of environmental contaminants to hepatic DNA of marine flatfish: laboratory and field studies with English sole and winter flounder. Mar. Environ. Res. 28: 345-346

Vethaak, A.D, 1993. Fish disease and marine pollution. A case study of the flounder (Platichthys flesus) in Dutch coastal and estuarine waters. PhD thesis, University of Amsterdam, pp. 155

Vian, C.J., S.M. Sherman & P.S. Sabharwal, 1982. Comparative extraction of genotoxic components of air particulates with several solvent systems. Mutat. Res. 105: 133-137

von der Hude, W., C. Behm, R. Gurtler & A. Basler, 1988. Evaluation of the SOS-chromotest. Mutat. Res. Environ. Mutagenes and Related Subj., 203: 81-94

Waranazi, U., W.L. Reichert, B.T. Le Eberhart & J.E. Stein, 1989. Formation and persistence of benzo(a)pyrene diolepoxide-DNA adducts in liver of English sole *(Parophrys vetulus)*. Chem. Biol. Interact. 69: 203-216

Wrisberg, M.N., 1991. Aquatic genotoxicity. The use of the blue mussel for detection of geotoxins in the marine environment. PhD thesis, Technical University of Denmark, Copenhagen, pp. 131

Wrisberg, M.N. & R. Rhemrev, 1992. Detection of genotoxins in the aquatic environment with the mussel *Mytilus edulis*. Wat. Sci. Technol. 25: 317-324

Wrisberg, M.N. & M.A. van der Gaag, 1992. *In vivo* detection of genotoxicity in waste water from a wheat-rye straw paper pulp factory. Sci. Total Environ. 121: 95-108

Wrisberg, M.N., C.M. Bilbo & H. Spliid, 1992. Indiction of micronuclei in haemocytes of *Mytilus edulis* and statistical analysis. Ecotox. Environ. Safety, 23: 191-205

Zahn, R.K., G. Zahn-Daimler, W.E.G. Müller, M. Michaelis, B. Kurelec, M. Rijavec, R. Batel & N. Bihari, 1983. DNA damage by PAH and repair in a marine sponge. Sci. Total Environ. 26: 137-156

Zakharov, V.M., 1993. Analysis of variation of structure as a method of bio-monitoring pollution. In: Bio-indicators of chemical and radioactive pollution; D.A. Krivolutsky (ed). MIR, Moscow, pp 187-198

Appendix 10

Sampling of marine animals for analysis of genotoxicity

It is possible to catch organisms both from a ship, by trawling (beam-trawl or other similar techniques) and/or by using light diving equipment in coastal areas. As a rule, the highest amounts of xenobiotics are accumulated by filter feeding and (detrivorous) deposit feeding species. Those species placed higher in the food chain but which do not possess systems that allow them to metabolize and actively remove xenobiotics are of special interest. Examples are crustacea and polychaetes.
While selecting and catching animals it is important to identify the species precisely and to determine their habitat. One should take possible ways of migration into account, even the possibility of active transportation in bilge water.

Fixation and transport

In the analytical procedure for the content of mutagenic and carcinogenic compounds in tissues, the methods for material fixation, storage and extraction are important as genotoxic compounds could be introduced into the material with contaminated solvents or as the result of improper storage and delivery of the samples to the laboratory. Therefore, we recommend that genotoxic control steps be carried out at all stages of fixation and extraction of the xenobiotics. The extraction step is especially sensitive to contamination.

Extraction of genotoxic compounds

The mechanical destruction of tissues is performed with the help of homogenizers of a Potter or Poltroon type after addition of water or organic solvent. Aqueous extracts of animal tissues contain a relatively large amount of proteins which can hamper isolation of mutagenic components in the subsequent procedures. Therefore coagulants such as water free sodium sulphate (Parkinson, 1974; Osborne *et al.*, 1982; Krone & Iwaoka, 1983), ammonium sulphate (Commoner *et al.*, 1978) or sulphosalycilic acid (Belisario *et al.*, 1984) are added to the homogenate. The method of hydrolysis of proteins by hydrochloric and nitric acids is widespread (Commoner *et al.*,1978; Bjeldanes *et al.*, 1982; Kadhim & Parry, 1984), but the use of polar organic solvents (*e.g.* acetone, methanol, ethanol) for the extraction of tissues of aquatic organisms is more common (Felton *et al.*, 1981; Krone & Iwaoka, 1983). These solvents allow not just the proteins to coagulate but they will also transfer (as much as possible) to a soluble phase, especially hydrophobic mutagenic and promutagenic compounds. The method of the acetone extraction is the easiest and the most convenient for isolation of mutagens from organic samples (Felton *et al.*, 1981). The yield of mutagens from the investigated material is about twice as high by this method compared to the aqueous extraction method (Commoner *et al.*, 1978). Mixtures of acetone (1:1) with other solvents, such as benzene, hexane or water (Vian *et al.*, 1982), methanol or ethanol (Krone & Iwaoka, 1983) are widely used. Parallel to these methods, non-polar solvents are also widely used for extraction of mutagens from the various compartments: hexane (Parkinson, 1974; Kurelec *et al.*, 1981), chloroform (Loper *et al.*, 1981), dichloromethane (Grabow *et al.*, 1981), benzene (Smith, 1982), cyclohexane, benzene (Osborne *et al.*, 1982). They are often used together with protein coagulants (Krone & Iwaoka, 1983). Sometimes, mixtures of polar and non-polar solvents are used for their extraction, while taking into account the different chemical nature of the various mutagens and promutagens (Smith, 1982). In each case, it is necessary to select the best method of extraction while taking into account the chemical features of mutagens and the nature of the studied material.

It should be realized that some extraction procedures *i.e.* where ammonium ions are present in solvents, can induce the formation of artificial mutagens since they have the potential ability to form amides and amines in the course of reactions of ammonium ions with organic solvents (Felton *et al.*, 1981; Iwaoka *et al.*, 1981). It was also shown that in the course of extraction many solvents, or the contaminants, can alter the tested substances through oxidation (Jori *et al.*, 1969; Gennari & Jori, 1970; Holt *et al.*, 1977). The use of chlorinated solvents can also result in artifacts in testing of mutagens since the latter contain stabilizers. Besides, chlorinated solvents are not stable and can form hydrochloric acid with the subsequent formation of radicals that easily polymerize non-volatile toxic substances in solution (Anthony, 1979). A wide range of organic solvents (alcohols, ethers, aliphatic hydrocarbons) may change the tested material through the reaction of etherification, without affecting the mutagenic potential. From this point of view, they are inert and are the most suitable for extraction.

When considering the different chemical nature of mutagens and promutagens, different solubilities in water and aqueous solutions may be assumed. For this reason, the most complete extractions of the tested compounds use a method of sequential extraction, changing the pH of the investigated samples during each step (Commoner *et al.*, 1978; Grabow *et al.*, 1981; Bjeldanes *et al.*, 1982). However, this method also has shortcomings since the possibility exists of obtaining artifact results while studying the extracted mutagens in the Ames test (Grabow *et al.*, 1981; Krone & Iwaoka, 1983). The use of protein coagulants in this method leads to the binding of mutagens by precipitated proteins owing to electrostatic, ionic and hydrophobic interactions (Krone & Iwaoka, 1983).

When considering the hydrophobic nature of most mutagens it may be assumed that the use of organic solvents for the isolation of mutagens is preferable to aqueous extraction. In comparison with other methods of isolation of mutagenic compounds the use of synthetic resins, *e.g.* of the XAD type, does not cause the appearance of artifact mutagens. It is a separation method with a high yield of mutagen recovery, while not changing their chemical structure and, as a result, it often increases the mutagenic response in the *Salmonella* microsomes Ames assay (Jamasaki & Ames, 1977; Krone & Iwaoka, 1983).

Despite the relative universality of the enumerated methods, the aims of the investigation should be taken into account in each separate case. The selection of a proper technique, or combination of methods, should be directed by the nature of the compounds to be tested and the compartment to be examined.

Ames test

The test consists of the addition to half of the probes of a suspension of the tester bacteria, the studied substance (pollutant), rat liver postmitochondrial fraction S9 and cofactors of mono-oxygenases into 2 ml of melted 0.6% soft agar (45 °C) ("+MA"). The samples are quickly mixed and covered onto selective minimal agar. The other half of the probes

("-MA") does not contain the metabolic activation components. Each substance is assayed in triplicate. The plates are incubated at 37 °C for 48 h. Then, the quantity of His+ revertant colonies on the plates is calculated. If the average quantity of colonies in the experiment is 2-10 times higher than in the control the mutagenic activity of the substance is considered to be slight, if it is 10-100 times higher the mutagenic activity is considered to be medium and if it is 100 or more times higher then the mutagenic activity is considered to be strong.

Chapter 11

The Scope for Growth of Bivalves as an Integrated Response Parameter in Biological Monitoring

Aad C. Smaal[1] and **John Widdows**[2]

[1] National Institute for Coastal and Marine Management, RIKZ
P.O. Box 8039, 4330 EA Middelburg, The Netherlands
[2] Plymouth Marine Laboratory, Prospect Place
Plymouth PL1 3DH, United Kingdom

ABSTRACT

Scope for growth (SFG) is an integrated physiological parameter reflecting the energetic balance between processes of energy acquisition (feeding and absorption) and energy expenditure (metabolism and excretion). The scope for growth can range from maximum positive values under optimum conditions declining to negative values when the animal is severely stressed and utilizing body reserves. It has been widely used to evaluate aquatic environmental quality, and to test the toxic effects of environmental contaminants. In many cases bivalves such as the blue mussel, *Mytilus edulis*, have been used as an indicator species, because of their world-wide occurrence and their ability to bioaccumulate contaminants. Mussels have shown a quantitative and predictable reduction in scope for growth in response to a wide range of contaminants. Therefore scope for growth can now be used to assess the impact of environmental contamination and provide a toxicological interpretation of tissue residue chemistry data.

1. REVIEW OF THE FIELD

1.1. Energy budget and scope for growth

Survival of a population in a certain habitat is a fundamental criterion in the evaluation of environmental quality; the ability of the population to grow and reproduce, *i.e.* to maintain a positive energy budget, can be seen as a primary prerequisite of survival. However, direct estimates of growth and reproduction of test organisms in biomonitoring programmes are not only time consuming because growth can be slow, but are also difficult to measure and interpret: reproductive output (*i.e.* gamete loss) may occur during the period of measurement,

Biomonitoring of Coastal Waters and Estuaries. Edited by Kees J.M. Kramer.

and the causes of changes in growth are difficult to identify, particularly the separation of 'nutrient effects' from toxicant effects in the field (see Kimball & McElroy, 1993).
Instead, scope for growth provides an instantaneous measure of the potential growth of an animal under specific conditions. However, it does not mean that this amount of energy is actually converted into somatic or gonad tissue at a specific site; as misinterpreted by Jørgensen (1990). Accurate prediction of this is more difficult and requires detailed time-integrated information on food availability at each specific site. In spite of this, a number of studies have shown good agreement between growth and scope for growth (Gilfillan & Vandermeulen, 1978; Bayne & Worrall, 1980; Grant & Cranford, 1991).
The energy budget of an animal represents an integration of the basic physiological responses such as feeding, food absorption, respiration, excretion and production. Each component is converted into energy equivalents ($J.h^{-1}$) and alterations in the amount of energy incorporated into growth and production can be described by the balanced energy equation of Winberg (1960):

$$C = P + R + U + F \quad (1)$$

where C = total consumption of food energy, P = total production of shell, somatic tissue and gametes, R = respiratory energy expenditure (*i.e.* costs of maintenance, feeding, digestion and growth), U = energy loss in excreta, and F = faecal energy loss.
The absorbed ration (A) is the product of energy consumed (C) and the absorption efficiency (ae) of energy from the food (A = ae x C). Production may then be expressed as:

$$P = A - (R + U) \quad (2)$$

When production (P) is estimated from the difference between the energy gains (energy absorbed) and the energy losses (energy expenditure via respiration and excretion), it is referred to as 'scope for growth' (Warren & Davis, 1967). Scope for growth (SFG) can range from maximum positive values under optimum conditions, but declining to negative values when the animal is severely stressed and utilizing its body reserves for maintenance. The scope for growth is defined as "an index of the energy available for growth and reproduction" (Bayne *et al.*, 1985). It not only provides an integration of the physiological parameters involved in the energy budget, but also an understanding of the impact of environmental stress which can discriminate between natural and anthropogenic stresses (Figure 1). This is achieved by estimating scope for growth under well defined and standardized conditions.

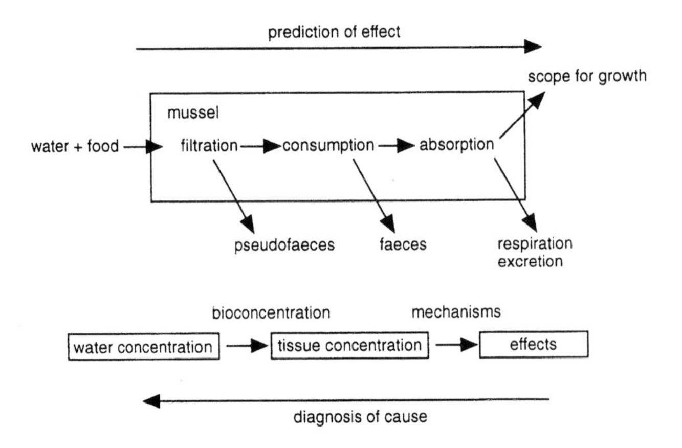

Figure 1. Scope for growth as part of the individual energy budget, in an ecotoxicology framework (After: Widdows & Donkin, 1992)

The advantages of using scope for growth in environmental biomonitoring are:

- SFG provides a sensitive, quantitative and integrated stress response over a wide range of conditions from optimal to lethal, and is therefore responsive to environmental pollutants;
- A decline in growth has ecological relevance and is readily interpretable as a deleterious effect;
- SFG is an early warning indicator of cellular and metabolic responses at the individual level, which allows detection of responses to environmental change prior to measurable effects on the growth, reproduction or survival of the individual;
- SFG represents an instantaneous estimate of growth potential which can be applied rapidly and cost-effectively in both the laboratory and field situations.

The main disadvantages are:

- the initial costs of equipment, and
- the measurement of individual physiological responses requires a higher level of training and care compared to lethal toxicity testing methods.

1.2. The use of mussels

Although the term 'scope for growth' was first applied to fish (Warren & Davis, 1967), its more recent application to ecotoxicology and biological monitoring has mainly involved marine invertebrates, such as bivalves (Bayne *et al.*, 1985; Widdows & Johnson, 1988; Widdows *et al.*, 1990), gastropods and crustaceans (Stickle *et al.*, 1987; Wang & Stickle, 1987).

In this chapter we will focus on the use of mussels in monitoring environmental quality and pollution impact. They are particularly suitable for combined chemical and biological monitoring and the application of scope for growth as a response parameter, because mussels are:

- sedentary and therefore cannot escape from pollution;
- abundant and widely distributed for regional mapping and global pollution;
- commercially and ecologically important;
- suspension feeders that pump and sample large volumes of water;
- bio-concentrators of chemical contaminants by factors of 10 to 100,000 and so facilitate analytical detection of contamination;
- sensitive to a wide range of contaminants at environmentally realistic concentrations;
- responsive over a wide range of conditions from sublethal to lethal levels of pollution;
- providing a time-integrated bioaccumulation of environmental contaminants and an integrated biological response to 'total pollution load' by reflecting different mechanisms of toxicity;
- unaffected by handling stress, so can be readily transplanted and measured;
- amenable to laboratory and field studies;
- cost-effective measurements of environmental pollution.

In the case of 'active' biological monitoring (ABM) programmes, test animals are selected at random from a stock population at a clean reference site prior to field acclimatization (de Kock & Kramer., this Volume); in case of 'passive' biological monitoring (PBM) mussels are sampled from wild populations (O'Connor *et al.*, this Volume). After selection of animals, preferably of one size class, they are transported and subsequently held in the laboratory under standard conditions, *i.e.* clean seawater, constant or ambient temperature, salinity and food level. The feeding conditions should preferably be close to natural seawater with seston concentrations below the pseudofaeces threshold (minimum seston concentration at which pseudofaeces rejection occurs). Adjustment to the standard laboratory conditions takes 24 hours, primarily to flush gut contents and adjust the absorption efficiency. This period should not be longer than 24 hours to avoid significant metabolic conversion or elimination of contaminants, which can occur with hydrocarbons (Widdows *et al.*, 1987a).

The procedures for the scope for growth measurements of mussels include selection and pre-treatment of the animals, measurement of food consumption and absorption efficiency, excretion rate of ammonia, respiration rate, and calculation of the scope for growth (see Appendix). Although results normally are presented as the integrated scope for growth response, information on feeding rate, respiration rate and absorption efficiency are useful to identify the cause and the mechanism of toxicity of contaminants (Figure 1). Feeding rate can be reduced by nonspecific narcotic agents such as hydrocarbons or by neurotoxic substances such as copper and tributyltin (TBT); respiration rate can be enhanced by uncoupling of oxidative phosphorylation by phenols or TBT; absorption efficiency can decrease due to membrane damage by hydrocarbons (Widdows & Donkin, 1992). It should be emphasized that measurements can be executed in clean seawater under standard laboratory conditions, thus removing any possible confounding effects of natural environmental stresses, and simply quantifying the underlying effects of pollution induced stress.

2. DISCUSSION OF THE TECHNIQUE

2.1. Sampling of mussels

In monitoring programmes mussels can be sampled from wild populations and brought to the laboratory for measurement. In many cases mussels can be transplanted from a 'clean' reference stock, to cages (either attached to the substratum or buoys) at specific field sites and acclimatized for a number of weeks, before transfer to the laboratory for measurement. The use of mussels from a reference population will minimize variation due to different previous experiences in various locations (*e.g.* differences in body size of wild mussels, even with the same shell length), and possible genetic or phenotypic adaptation to prevailing contaminant levels. Moreover, transplantation allows monitoring at sites without indigenous mussel populations.

When sampling/exposure sites are too far from a laboratory, and rapid transportation of samples is difficult, measurements can be executed using a mobile laboratory or on board ship. It has been found that mussels can be transported air exposed in a 'cool box' for 24 hours, and then recovered for measurement within 12 - 24 hours of reimmersion in the laboratory (Widdows & Salkeld, unpublished data).

During the spawning period the variability of the individual measurements can increase, due to differences in reproductive maturation of the test animals. This is illustrated in Table 1, showing SFG estimations of wild mussels from an unpolluted area before (February) and during the spawning period (April). In April the variance about the mean was 3-4 fold higher than in February, in spite of the larger number of animals. Moreover, the SFG was significantly lower in April (Smaal & Korporaal, unpublished results).

2.2. Scope for growth measurement

Estimation of the scope for growth of mussels requires measurement of components of the energy budget (feeding rate, food absorption, respiration and excretion) under standardized laboratory conditions (in terms of food quantity and quality, temperature and salinity, and most importantly 'clean' seawater). Results are expressed per gram dry weight of mussel tissue, by using the allometric relation between the measured process rate and the body weight of each animal (see Appendix).

2.2.1. Feeding rate

Food consumption is measured in terms of clearance rate which is the volume of water cleared of particles per unit time. The clearance rate can be measured in a static or a flow-

Table 1. Energy budget and scope for growth (in $J.h^{-1}.g^{-1}$ DW) of wild mussels from a clean reference site in the Oosterschelde (NL) in February (n = 5) and April 1987 (n = 12). Mean values ± S.E. in parentheses.
CR = clearance rate in $l.h^{-1}.g^{-1}$ DW; C = consumption; A = absorption; R = respiration; U = excretion; SFG = scope for growth. Food concentration was 0.18 $mg.l^{-1}$ POM

Date	CR	C	A	R	U	SFG
16/2	2.92 (0.29)	12.35 (1.16)	11.73 (1.10)	2.34 (0.36)	<.082	9.31 (0.59)
18/2	3.61 (0.78)	15.27 (0.93)	14.60 (0.49)	2.66 (0.17)	<.082	11.92 (0.17)
19/2	3.34 (0.22)	14.13 (0.82)	13.14 (0.82)	2.62 (0.11)	<.082	10.44 (0.77)
20/2	3.59 (0.27)	15.19 (1.10)	14.57 (1.02)	2.60 (0.19)	<.082	11.90 (0.85)
6/4	2.35 (0.66)	10.16 (1.62)	8.31 (1.29)	5.28 (0.36)	.94 (.13)	1.59 (1.45)
8/4	3.22 (0.40)	13.62 (2.79)	11.58 (2.75)	6.70 (0.49)	.32 (.03)	4.56 (2.44)
13/4	2.90 (0.40)	12.22 (1.72)	10.41 (1.38)	5.61 (0.65)	.23 (.03)	4.57 (1.15)
15/4	2.75 (0.27)	11.63 (1.04)	7.21 (1.19)	8.04 (0.64)	.51 (.03)	-1.34 (1.12)

through system, preferably with separate individuals so that closed, non-feeding individuals can be detected and excluded from the calculations. In a static system the animals are held in a known volume of water containing food particles; filtration results in an exponential decrease in food concentration over time. Clearance rate in static systems is then calculated using the following equation (see Coughlan, 1969):

$$CR = V(\log_e C_1 - \log_e C_2)/t \quad (3)$$

where CR = clearance rate, V = volume of water, C_1 and C_2 = concentration at beginning and end of time interval t.

In flow-through systems, the feeding rate is estimated by measuring the in-and out-flow concentrations of particles and recording the water flow rate through each chamber. Flow-through systems are preferred because steady state conditions can be maintained: food supply can be kept constant over a longer period of time and the depletion of oxygen and accumulation of ammonia is avoided.

The clearance rates of individual animals are measured in separate chambers. In practice 15 individual replicates are required in order to reduce the coefficient of variation and increase the ability to detect population differences. Measurements have also been executed with groups of individuals in raceway systems. This allows the use of larger quantities of individuals. The use of individuals, however, gives a better estimate of the variance, but is more time consuming. At least three measurements are made over a period of 2 - 3 hours. This allows the calculation of the mean and the maximum clearance rate for each individual.

Phytoplankton culture or natural seston can be used as a food source, but the composition must remain consistent over time to allow comparison of feeding rate and absorption efficiency.

In flow-through systems, using small volume chambers and high flow rates, there is no significant recirculation of water within the chamber by the mussel and the clearance rate can be calculated as:

$$CR = ((C_{in}-C_{out})/C_{in}) \times Q \quad (4)$$

where C_{in} = concentration inflow, C_{out} = concentration outflow and Q = flow rate. The inflow concentration is based on the outflow concentration from a control chamber, without an animal. This allows correction for possible sedimentation when using natural seston. The concentration of particles > 3 μm, which is the minimum size retained by bivalve gills with 100 % efficiency, is measured using an electronic particle counter.
When high flow rates cannot be achieved, a lower flow rate in combination with a larger chamber and thorough mixing can be used. In this case:

$$CR = ((C_{in} - C_{out})/C_{out}) \times Q \quad (5)$$

where the concentration surrounding the animal is equal to the outflow concentration. Both methods have been evaluated by Widdows (1985) and did not show significant differences.
Food consumption is estimated by multiplying the clearance rate by the food concentration, expressed in terms of mg particulate organic matter (POM) per litre. This is measured by taking samples of the inflow (or the outflow of the control) and analyzing the POM (or POC) content according to standard methods (Strickland & Parsons, 1972). The POM concentration can be converted into energy equivalents using a standard conversion factor (see Appendix)

*2.2.2. **Food absorption***

The amount of food that is digested and absorbed by the animal can be estimated either directly or indirectly. In both cases, the production of pseudofaeces (excess material rejected by palps and gills) should be avoided, because the estimate of absorption is based on the comparison of food and faeces. Consequently, food concentrations below the pseudofaeces threshold (3-5 $mg.l^{-1}$; Bayne, 1976) are used. Also in both methods, the net absorption is estimated: metabolic faecal loss is not taken into account (see Hawkins & Bayne, 1992).
The direct method requires quantitative collection of all the faeces produced during the period of the consumption measurement. It also requires an exact estimate of the total food uptake during the measurement, which can be achieved by taking a continuous subsample of in- and out-flow: $C = (C_{in}-C_{out}) \times Q$. These integrated samples and the total amount of faeces have to be analyzed for total dry weight. The balance between food uptake and defaecation is by definition the amount of absorbed material: C - F = A
The indirect, indicator method only requires representative subsamples of faeces and food. The inorganic content of food and faeces is then used as an inert non-absorbed tracer, passing through the gut. The fraction of organic matter in the faeces is related to the organic fraction in the food. The absorption efficiency (ae) can then be calculated according to Conover (1966):

$$ae = (F-E)/(1-E)F \quad (6)$$

where F = ash-free dry weight:dry weight ratio (AFDW/DW) of the food and E = ash-free dry weight:dry weight ratio of faeces.
This method requires a measurable proportion of inorganic matter in the food, which is normally the case when natural seawater, possibly enriched with cultured algae, is used as food. Other methods to estimate food absorption include the use of radio-tracers, but this is not cost-effective for use in routine monitoring programmes. The indirect (Conover) method is based on the assumption that only organic matter is digested. The method is rather sensitive to small changes in the organic matter content of faeces at low absorption efficiencies, *i.e.* at high food concentrations, when the proportion of organic matter in the faeces is close to that of the food.

In recent literature, the direct budget method was used in preference to the indirect method when measuring food absorption in field studies (Iglesias *et al.*, 1992). For practical reasons, however, the Conover method is recommended. Absorption rate is then calculated as food consumption rate multiplied by absorption efficiency.

2.2.3. Respiration

The total energy costs associated with maintenance, activity, digestion and growth can be estimated by the rate of oxygen consumption, carbon dioxide liberation or heat dissipation. Heat production measurements by direct calorimetry are, however, not suitable for rapid and routine measurements. Under immersed normoxic conditions, mussels are fully aerobic and therefore oxygen uptake provides an accurate estimate of metabolic expenditure following conversion to energy equivalents (see Appendix). The oxygen consumption rate can be estimated in a static or in a flow-through set up. The static system involves monitoring the decline in oxygen concentration in a sealed chamber over a period of time. Conditions are not constant, because the oxygen and food concentration will decrease, and carbon dioxide and excretion products will increase. By using short measurement periods these changes do not have a significant effect on the rate of oxygen uptake. Another type of static system is the manometric method, measuring the reduction in gas volume associated with oxygen uptake, while carbon dioxide is absorbed by potassium hydroxide. This is generally used for smaller organisms (Lawton & Richards, 1970). In a flow-through system, oxygen uptake is measured by the difference in oxygen concentration of the in- and out-flow at a known constant flow rate. The major limitation is the slower throughput of samples and the limited number of animals that can be run simultaneously.

2.2.4. Excretion

Excretion of metabolic waste products represents a relatively small component of the energy budget. In mussels the main excretion product is ammonia, with only 10-20 % of total N-loss through amino acids or urea. For scope for growth estimations N excretion is of minor importance, with only a small fraction (typically < 5% of metabolic energy expenditure) of the absorbed energy lost through excretion. Organisms under stress, however, can show higher excretion rates. The estimation of excretion consists of measuring the ammonia concentration before and after incubation of individuals in a known volume of filtered seawater. The ammonia concentration is measured by standard analytical techniques (Solorzano, 1969) and then converted into energy equivalents.
The scope for growth is simply calculated from the above mentioned components:

$$C = CR \times POM \quad (7)$$
$$A = C \times ae \quad (8)$$
$$SFG = A - R - U \quad (9)$$

3. FACTORS AFFECTING THE SCOPE FOR GROWTH

3.1. Natural factors

Various factors can have an effect on the scope for growth of mussels. Intrinsic factors include body size of the animal, reproductive condition, genetic characteristics and parasitic infestations.
Feeding, respiration and excretion rates increase with increasing body size; the slope of the allometric relationship between body size and a particular rate ranges from b = 0.4 - 0.8 (Bayne *et al.*, 1985).

The reproductive condition has impact on the scope for growth because mussels with ripe gonads show relatively high respiration and ammonia excretion rates, which are only partially compensated by increased feeding rates (Table 1). Moreover, populations of mussels sampled during the spawning period show a larger variation in scope for growth, due to a mixture of ripe and fully spawned mussels experiencing post spawning stress. Consequently, the spawning period should be avoided in scope for growth measurements.
Genetic differences between individuals and populations may have a slight effect on the scope for growth. Multiple locus heterozygotes show a higher growth rate compared to homozygotes and this is attributed to a lower maintenance metabolism, measured as oxygen consumption. However, heterozygosity explained only 8 % of the differences in energy requirements (Hawkins & Bayne, 1991). The correlation between heterozygosity and enhanced growth is extensively discussed in Gosling (1992). From transplantation experiments it is shown that stock differences can account for up to 27 % of the differences in shell length, but it is not clear to what extent stocks represent different genotypes (Gosling, 1992). Kautsky *et al.* (1990) have shown that physiological responses and growth rate in mussels transplanted between the North Sea and the Baltic reflected environmental rather than genetic differences.
Parasitic infestations may affect the scope for growth. For example, mussels infested with *Pinotheres pisum* (pea crabs) exhibit reduced filtration rates, reduced growth rates and shell shape distortions (Seed & Suchanek, 1992).
Laboratory conditions can affect the scope for growth of mussels when they exceed the adaptive capacity of field acclimatized animals. Mussels are able to adapt to extrinsic factors such as food quantity and quality, salinity, oxygen concentration, temperature, and aerial exposure, and therefore maintain SFG independent of these factors over a wide range. These factors are reviewed by Bayne *et al.* (1985), Seed & Suchanek (1992) and Hawkins & Bayne (1992). The adaptation time depends on the factor and exact field and laboratory conditions. It is therefore recommended that laboratory conditions are close to ambient field conditions.

3.2. Anthropogenic factors: pollution

Contamination is a physico-chemical phenomenon resulting from the discharge to the environment of chemicals in excess of normal concentrations. When contaminants are sufficiently elevated to cause deleterious biological effects, this is termed pollution.
The scope for growth can be influenced directly, as an acute response to chemical contaminants in water (Manley, 1983; Redpath & Davenport, 1988) and sediment (Eertman *et al.*, 1993a), or after chronic exposure and long-term accumulation of pollutants (see Widdows & Donkin, 1992 for review). Monitoring of acute responses in field exposed animals would require field measurements of SFG using a mobile laboratory or research vessel. However, when measurements of field exposed animals are performed under standard laboratory conditions, only chronic responses are monitored.
Various field studies have shown significant differences in the SFG of mussels collected from transects along estuaries or pollution gradients (Table 2). SFG correlated significantly with the tissue concentrations of a number of chemical contaminants. However, mussels from the field can be influenced by a large number of contaminants. Chemical analyses include only a small fraction of the large number of potentially toxic substances entering coastal waters.
The relationship between scope for growth and contaminant concentrations in the tissues can be evaluated on the basis of laboratory exposures (Table 3). Scope for growth typically shows a linear decline with a log increase in the toxicant concentration in the tissues, reaching zero and negative SFG values at the highest (lethal) concentration. The impact of contaminants is also reflected in the reduction of clearance rates relative to the control (Table 3).

Table 2. Scope for growth and tissue contaminant concentrations of bivalves (M.a. = *Mya arenaria*; M.e. = *Mytilus edulis*; A.z. = *Arca zebra*) from various field sites, from wild and transplanted populations. Tissue concentrations and SFG expressed per g dry weight of flesh (DW); conversion wet weight to dry weight: DW = 0.2 x WW. Main pollutants represent the contaminants which show significant negative correlation with SFG, except for Dutch coastal waters. Minimum tissue concentrations and maximum scope for growth in most cases as measured in animals from 'clean' reference sites. n.d. = not determined.

location	species	main pollutant [d]	tissue conc. range ($mg.kg^{-1}$ DW)	exposure time (days)	SFG range ($J.h^{-1}.g^{-1}$ DW)	references
Casco Bay (Maine, USA)	M.a.	AHs	5.5-13.0	168	18.5 to 130.2[a]	Gilfillan *et al.* (1977)
Narragansett Bay (RI, USA)	M.e.	PAHs Ni	40-560 3.6-16	28	8 to 30	Widdows *et al.* (1981)
San Francisco Bay (USA)	M.e.	Cr chlordane dieldrin PCBs	7.5-18.0 0.02-0.42 0.02-0.25 0.14-9.0	field samples	3.8 to 43.8	Martin *et al.* (1984)
Sullom Voe (Shetlands)	M.e.	AHs	0.5-8.2	field samples	7.0 to 23.6	Widdows *et al.* (1987b)
Langesundfjord (Norway)	M.e.	Cd PAHs PCBs	1.6-2.7 2.2-15.4 0.08-0.28	field samples	-1.1 to 8.0	Widdows & Johnson (1988)
Hamilton Harbour (Bermuda)	A.z.	Pb TBT PCBs AHs	3.18-7.34 0.18-1.11 0.001-0.032 5-40[b]	field samples	3.4 to 9.8	Widdows *et al.* (1990)
Greenwich Cove (RI, USA)	M.e.	Sewage	n.d.	30	-4.2 to 3.4	Nelson (1990)
Dutch coast (NL)	M.e.	Cd PCBs PAHs	0.22-1.44 0.03-0.44 0.01-0.26	42	0.6 to 2.4[c]	Smaal *et al.* (1991)

[a] SFG expressed as mg C per 100 mg animal;
[b] in MDU = marine diesel units from GC;
[c] no SFG measured, clearance rate in $l.h^{-1}.g^{-1}$ DW, no significant correlation with tissue concentrations;
[d] AH: aromatic hydrocarbons, PAHs: polycyclic aromatic hydrocarbons, PCBs: polychlorinated biphenyls, TBT: tributyltin

Despite some difficulties in comparing these studies, due to differences in methodology of SFG and chemical analyses, Table 3 shows remarkable similarities in the physiological responses. Hydrocarbons show a less dramatic effect on SFG and clearance rate over a wide concentration range, compared to the other contaminants (*e.g.* Cu and TBT). The total number of contaminants studied with respect to effects on the scope for growth is still limited and more studies are needed to establish toxicant concentration - effect relationships.

Table 3. Scope for growth (SFG) and clearance rate (CR) of mussels after chronic exposure to various contaminants in the laboratory or in mesocosms. Tissue concentrations and SFG expressed in g^{-1} DW; conversion wet weight to dry weight: DW = 0.2 x WW. Minimum tissue concentration and maximum SFG as measured in the control animals. nd = not determined.

contaminant[f)]	water conc. ($\mu g.l^{-1}$)	exposure time (d)	tissue conc. range ($mg.kg^{-1}$ DW)	SFG range ($J.h^{-1}.g^{-1}$ DW)	Minimum CR in % of control	references
AHs (oil-WAF[a)])	30	14-140	45.0-98.0[b)]	1 to 8.5	80	Widdows *et al.* (1982)
AHs (oil-WAF[a)])	28-125	240	3.8-342.6	-7.0 to 9.8	30	Widdows *et al.* (1985,1987a)
PAHs (selected) + Cu	3-124 0.5-20	95 95	1.1-22.8 7.3-59	-1.5 to 2.9	19	Widdows & Johnson (1988)
Fluoranthene Benzo[*a*]pyrene	0.5-6[c)] 0.5-6[c)]	28 28	0.05-7.3 0.003-15.8	nd nd	55 50	Eertman *et al.* (1993a)
Cu	1-100	7	nd	-4.2 to 3.4	5	Sanders *et al.* (1991)
PCB_{1254}	0-8[d)]	42	0.03-2.6	-0.3 to 44.8	nd	Martin *et al.* (1984, 1985)
TBT DBT	0.01-5 1-300	4-8 4	0.15-18.9 0.4-127	-8.0 to 13.85 nd	2 0	Widdows & Page (1993)
PCP	0.5-786	3	2.5-210.5	-5.44 to 15.2	2.9	Widdows & Donkin (1991)
Sewage sludge	0-0.42%	33	-[e)]	-5.52 to 25.16	10	Butler *et al.* (1990)

a) WAF = water accommodated fraction;
b) concentration in tissue excluding the digestive gland;
c) nominal concentration after addition to phytoplankton;
d) after correction of dimension errors in Table 5 of Martin (1985);
e) no significant correlation with measured trace metal tissue concentrations;
f) see Table 2, and DBT: dibutyltin, PCP: pentachorophenol

4. APPLICATION OF SCOPE FOR GROWTH

4.1. SFG as a response parameter

The use of a response parameter, such as scope for growth, in biological monitoring has two functions:

- as a measure of deleterious biological effects, ranging from optimal, sublethal to lethal effects, and
- when coupled to tissue residue chemistry, as an indicator of the agents causing the adverse effects.

4.1.1. Sensitivity

Pollution monitoring requires a parameter that is sensitive to a wide variety of toxic substances but is maintained relatively independent of variations in natural stressors. Field studies clearly indicate that scope for growth shows a significant response to relatively low levels of pollution (Table 2). For example, tissue concentrations in the North Atlantic area range from 0.1-300 $mg.kg^{-1}$ DW for PAHs, from 0.004-10 $mg.kg^{-1}$ DW for PCBs, from 0.04-10.2 $mg.kg^{-1}$ DW for TBT, and from 0.1-36.2 $mg.kg^{-1}$ DW for Cd (see Widdows & Donkin, 1992 for review). These values are in the same range as values shown in Table 2.

Scope for growth has been compared with other lethal and sublethal physiological response parameters (Widdows & Donkin, 1991). They concluded that:

- physiological components of the energy budget are responsive to environmental levels of pollutants;
- physiological energetic and growth responses are considerably more sensitive than lethal responses, and
- larval stages can be > 10 fold less sensitive to some contaminants than adult mussels (Cu, petroleum hydrocarbons, sewage sludge).

The latter is attributed to the absence of a developed nervous system and their reliance on energy reserves rather than on direct feeding (see also Widdows & Donkin, 1992). With respect to TBT, the adult mussel is also more sensitive than larval stages, but less sensitive than the toxicant-specific 'imposex' response of the gastropod *Nucella lapillus* (see Gibbs & Bryan, this Volume). In this evaluation, no distinction was made between the SFG as an integrated energy budget measurement, and the response of the components of the SFG, *i.e.* feeding rate, absorption efficiency and respiration/excretion rate. In many cases the SFG response appears to depend mainly on the feeding rate response, which seems to respond more sensitively than other components of the energy budget (Widdows & Johnson, 1988; Butler *et al.*, 1990). However, this is related to the type of contaminant; for example, low concentrations of TBT and pentachlorophenol (PCP) reduce the feeding rate while enhancing the respiration rate, and consequently the response is best integrated and expressed as SFG (Widdows & Donkin, 1991).

Donkin *et al.* (1989) made a comparison of the effects of hydrocarbons on clearance rate of *Mytilus edulis*, and mortality of *Daphnia magna* and *Artemia*. Clearance rate appears as a sensitive parameter of the effects of low-molecular weight hydrocarbons, which is ascribed to the narcotic effect of these compounds. As part of a routine monitoring programme, Smaal *et al.* (1991) reported stress response measurements of transplanted mussels exposed along a pollution gradient in the Dutch coastal zone and the river Scheldt Estuary. In this case, estimated clearance rates did not correlate with tissue contaminant concentrations (see Table 2). In comparison to other field studies of SFG, the range and maximum tissue concentrations are relatively low, which might explain the absence of significant correlations. A more sensitive response in this study was shown by measuring 'survival in air'; this parameter showed good correlation with tissue concentrations, particularly of PCBs (Smaal *et al.*, 1991; see also: Eertman *et al.*, 1993b, Eertman & de Zwaan, this Volume).

In conclusion, scope for growth generally provides a good indicator of adverse biological effects, caused by levels of contamination that occur at many field sites.

4.1.2. Causality

The relationship between environmental contamination and biological response consists of two steps: (i) from concentration of the agent in the environment to tissue concentration, and (ii) from tissue concentration to biological effect (Figure 1).

Knowledge of these cause-effect relationships allows diagnosis of the cause of biological effects and prediction of effects of known contaminants in the environment or accumulated in the tissues of mussels. However, it is only possibly to establish cause-effect relationships experimentally for a limited number of environmental contaminants. For prediction of toxic effects of the many thousands of organic contaminants which enter the marine environment, it is necessary to apply a Quantitative Structure-Activity Relationship (QSAR) approach (Könemann, 1980; Widdows & Donkin, 1992). This enables the fate and effect of structurally related organic compounds to be predicted from their physico-chemical and structural properties. This approach has been applied for the feeding activity of mussels by Donkin *et al.* (1989, 1991) and Donkin & Widdows (1990). These studies clearly demonstrate for the mussel that there is an inverse relationship between hydrophobicity (expressed as the partition coefficient between octanol and water: log K_{ow}), and the concentration of a compound in the water inducing 50 % reduction in clearance rate. At the same time the bioconcentration factor increases with increasing hydrophobicity. As a result, organic compounds with widely differing hydrophobicity have equal toxicity when expressed on a tissue concentration basis. This relationship holds for aromatic and aliphatic hydrocarbons with a log $K_{ow} < 5$ or 6 respectively. According to Donkin *et al.* (1989) compounds with higher partition coefficients accumulate in tissues without showing effects on feeding rate.

In contrast, recent results of Eertman *et al.* (1993a) show a sensitive response of clearance rate to fluoranthene and benzo-[*a*]-pyrene: TEC_{50} values for clearance rate were 9.5 and 16.1 $mg.kg^{-1}$ DW respectively (see Table 3). This is much lower than reported for fluoranthene by Donkin *et al.* (1989). Except for differences in methods, the dissimilarity is as yet unexplained.

The single QSAR line for the less hydrophobic compounds implies that there is a common mechanism of toxicity that acts at similar tissue concentrations for a range of compounds. In complex mixtures there is a simple concentration additive effect (Widdows & Donkin, 1991). For hydrocarbons, the mechanism is termed a nonspecific narcosis effect on the ciliary feeding activity.

In most field situations, there are complex mixtures of structurally unrelated contaminants. Effects on different components of the energy budget can then provide insight into the modes of action of different contaminants. Simple additive effects on clearance rate were reported by Widdows & Johnson (1988) for copper and aromatic hydrocarbons. A re-interpretation of data of Strömgren (1986) showed a similar additive effect. Both contaminants affect the ciliary feeding activity, although apparently through different mechanisms: copper shows neurotoxic action on the neural control of the gill cilia, while hydrocarbons have a narcotic effect on the gill cilia (Widdows & Donkin, 1992). An example of an antagonistic effect is reported for TBT and hydrocarbons (Widdows & Donkin, 1991). Exposure of mussels to a mixture of these compounds, resulting in tissue concentrations of 65-150 mg naphthalene kg^{-1} DW and 0.1-1.31 mg TBT kg^{-1} DW showed up to 50% higher clearance rates than predicted from separate exposures on the basis of proportional additivity. In harbours, contaminated with potentially lethal TBT concentrations (5-10 mg TBT kg^{-1} DW), the combined input of hydrocarbons apparently may thus enable mussels to survive (Widdows & Donkin, 1991).

From these studies it is apparent, that lower molecular weight hydrocarbons (*i.e.* with a log $K_{ow} < 5$) are ecotoxicologically important, although in many chemical environmental monitoring programmes, few of these compounds are included in the routine analysis of abiotic and biotic compartments.

4.2. Case studies: field and laboratory applications

Field studies using scope for growth have been reported since 1977 (see Table 2 for examples from open literature).

An important conclusion of field and mesocosm studies using SFG, has been the need for well defined 'clean' reference stations, and the ability to place the findings of each study into a broader context. The SFG of mussels in the vicinity of the Sullom Voe oil terminal

(Shetland islands) and from the Solbergstrand mesocosms (Norway) are summarized in Figure 2. It shows that there is an inverse linear relationship between SFG and the log tissue concentration of 2 and 3 ring hydrocarbons over more than three orders of magnitude. The mesocosm control mussels had tissue contaminant concentrations in the same range as mussels from the oil loading jetties in Sullom Voe and the Shetland reference site is significantly cleaner than the Solbergstrand area (Widdows & Donkin, 1992). The same holds for the pollution gradient study of the Langesundfjord, which was also related to the Shetland refer

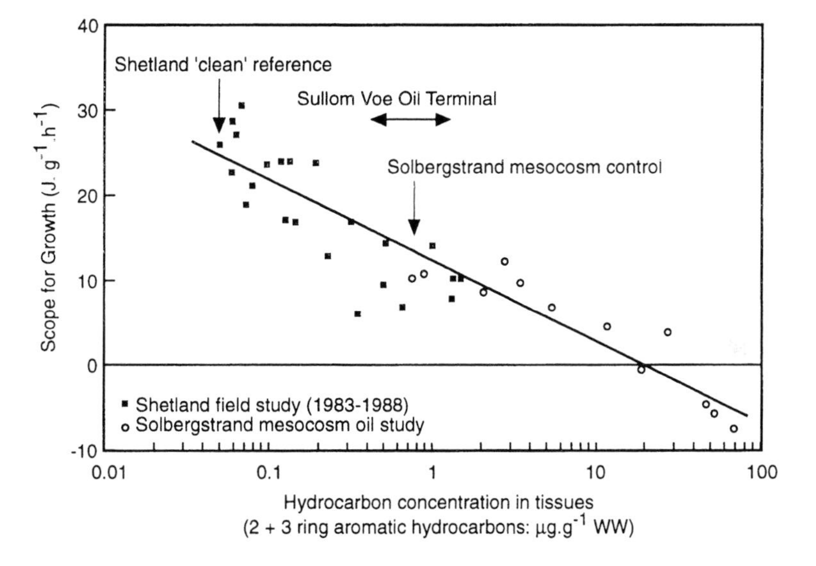

Figure 2. Relationship between scope for growth and the log tissue concentration of 2 and 3 ring hydrocarbons in mussels, *Mytilus edulis*, from experiments at the Sullom Voe oil terminal and the Solbergstrand mesocosms (From: Widdows & Donkin, 1992; with permission)

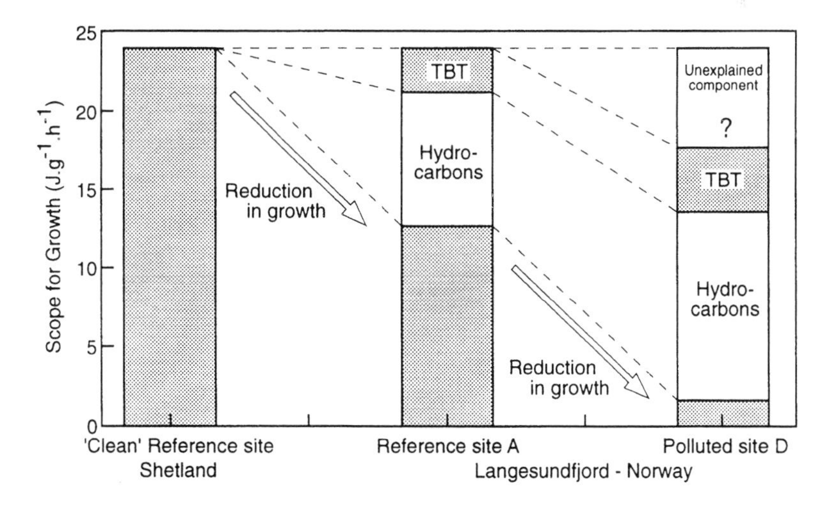

Figure 3. The effect of pollution on the scope for growth of mussels from a gradient in the Langesundfjord (Norway) in comparison to a clean reference site in the Shetland islands. The observed reduction in SFG can partly be attributed to effects of TBT and hydrocarbons. The remaining part is an indication of effects of unidentified pollutants and/or of PCBs, dioxins and Hg, for which there are no established tissue concentration-response relationships at present (From: Widdows & Donkin, 1992; with permission)

ence site, as shown in Figure 3 (Widdows & Donkin, 1991, 1992). The Langesundfjord 'reference site' was found to be significantly contaminated with TBT and hydrocarbons at concentrations that reduced the SFG of mussels, relative to the SFG at the reference site in the Shetlands. The Langesund Fjord site D, nearest to the pollution source, showed reduced scope for growth to a greater extent than could be explained by TBT and hydrocarbon concentrations in the mussels. Metal concentrations were below the effects threshold, so this additional reduction probably reflected the toxic action of other contaminants (*e.g.* dioxins, PCBs, Hg and other unidentified contaminants).
SFG or feeding rate of mussels has also been applied in the laboratory as a bio-assay of unknown complex mixtures of contaminants, such as sewage sludge (Butler *et al.*, 1990) or contaminated harbour sediment (Nelson *et al.*, 1986; Eertman *et al.*, 1993a). In all cases a significant relationship between clearance rate and sludge or sediment concentration was observed. These results show the potential of physiological energetics of mussels as a sensitive bio-assay technique.

CONCLUSIONS

In conclusion, field application of the scope for growth can be used to quantify changes in environmental quality, particularly in comparison to a well defined 'clean' reference site. Chemical analysis of tissue concentrations, in combination with laboratory and mesocosm derived tissue concentration - response relationships, can be used to predict and interpret the causes of the observed effects. Moreover, feeding rate and SFG of mussels have the potential for a sensitive toxicity bio-assay.
However, further research is needed to establish tissue concentration - effect relationships of environmentally relevant concentrations for a wider range of contaminants and mixtures of contaminants.

The authors are grateful to Drs R. Eertman and J. Everts for the stimulating discussions and for critically reading the manuscript.

REFERENCES

Bayne, B.L. (ed), 1976. Marine mussels, their ecology and physiology. Cambridge Univ. Press, Cambridge, pp. 506

Bayne, B.L. & C.M. Worrall, 1980. Growth and production of mussels *Mytilus edulis* from two populations. Mar. Ecol. Prog. Ser. 3: 317-328

Bayne, B.L., D.A. Brown, K. Burns, D.R. Dixon, A. Ivanovici, D.R. Livingstone, D.M. Lowe, M.N. Moore, A.R.D. Stebbing & J. Widdows, 1985. The effects of stress and pollution on marine animals. Praeger Press, New York, pp. 384

Butler, R., B.D. Roddie & C.P. Mainstone, 1990. The effects of sewage sludge on two life history stages of *Mytilus edulis*. Chem. Ecol. 4: 211-219

Conover, R.J., 1966. Assimilation of organic matter by zooplankton. Limnol. Oceanogr. 11: 338-354

Coughlan, J., 1969. The estimation of filtering rate from the clearance of suspensions. Mar. Biol. 2: 356-358

Donkin, P., J. Widdows, S.V. Evans, C.M. Worrall & M. Carr, 1989. Quantitative structure-activity relationships for the effect of hydrophobic organic chemicals on rate of feeding by mussels (*Mytilus edulis*). Aquat. Toxicol. 14: 277-294

Donkin, P. & J. Widdows, 1990. Quantitative structure-activity relationships in aquatic invertebrate toxicology. Rev. Aquat. Sci. 2: 375-398

Donkin, P., J. Widdows, S.V. Evans & M.D. Brinsley, 1991. QSARs for the sublethal responses of marine mussels (*Mytilus edulis*). Sci. Total Environ. 109/110: 461-476

Eertman, R.H.M., M.L.F.M.G. Groenink, B. Sandee, H. Hummel & A.C. Smaal, 1993a. Response of the blue mussel *Mytilus edulis* following exposure to PAHs or contaminated sediments. Mar. Environ. Res. (in press)

Eertman, R.H.M., A.J. Wagenvoort, H. Hummel & A.C. Smaal, 1993b. 'Survival in air' of the blue mussel *Mytilus edulis* L. as a sensitive response to pollution induced environmental stress. J. Exp. Mar. Biol. Ecol. 170: 179-195

Gilfillan, E.S., D.W. Mayo, D.S. Page, D. Donovan & S.A. Hanson, 1977. Effects of varying concentrations of hydrocarbons in sediments on carbon flux in *Mya arenaria*. In: Physiological responses of marine biota to pollutants; F.J. Vernberg, A. Calabrese, F.P. Thurberg & W.B. Vernberg (eds). Academic Press, New York, pp. 299-314

Gilfillan, E.S. & J.H. Vandermeulen, 1978. Alterations in growth and physiology of soft-shell clams *Mya arenaria*, chronically oiled with Bunker C from Chedabucto Bay, Nova Scotia, 1970-76. J. Fish Res. Bd. Can. 35: 630-636

Gnaiger, E. & H. Forstner, 1983. Polarographic oxygen sensors. Aquatic and physiological applications. Springer, Berlin, pp. 360

Gosling, E., 1992. Genetics of *Mytilus edulis*. In: The mussel *Mytilus*: ecology, physiology, genetics and culture; E. Gosling (ed). Elsevier, Amsterdam, pp. 309-382

Grant, J. & P.J. Cranford, 1991. Carbon and nitrogen scope for growth as a function of diet in the sea scallop *Placopecten magellanicus*. J. mar. biol. Ass. U.K. 71: 437-450

Hawkins, A.J.S. & B.L. Bayne, 1991. Nutrition of marine mussels: factors influencing the relative utilizations of protein and energy. Aquaculture, 94: 177-196

Hawkins, A.J.S. & B.L. Bayne, 1992. Physiological interrelations, and the regulation of production. In: The mussel *Mytilus*: ecology, physiology, genetics and culture; E. Gosling (ed). Elsevier, Amsterdam, pp. 171-212

Iglesias, J.I.P., E. Navarro, P. Alvarez Jorna & I. Armentia, 1992. Feeding, particle selection and absorption in cockles *Cerastoderma edule* (L.) exposed to variable conditions of food concentration and quality. J. Exp. Mar. Biol. Ecol. 162: 177-198

Jørgensen, C.B., 1990. Bivalve filter feeding: hydrodynamics, bioenergetics, physiology and ecology. Olsen & Olsen, Fredensborg, pp. 140

Kautsky, N., K. Johanneson & M. Tedengren, 1990. Genotypic and phenotypic differences between Baltic and North Sea populations of *Mytilus edulis* evaluated through reciprocal transplantations. I. Growth and morphology. Mar. Ecol. Prog. Ser. 59: 203-210

Kimball, D.M. & A.E. McElroy, 1993. Characterizing the annual reproductive cycle of *Mytilus edulis* from Boston Harbor and Cape Cod Bay - a comparison by means of stereology and condition indices. Mar. Environ. Res. 35: 189-196

Könemann, H., 1980. Quantitative structure-activity relationships and additivity in fish toxicity studies of environmental pollutants. Ecotoxicol. Environ. Safety, 4: 415-421

Lawton, J.H. & J. Richards, 1970. Comparability of Cartesian diver, Gilson, Warburg and Winkler methods of measuring the respiratory rates of aquatic invertebrates in ecological studies. Oecologia, 4: 319-324

Manley, A.R., 1983. The effects of copper on the behaviour, respiration, filtration and ventilation activity of *Mytilus edulis*. J. Mar. Biol. Ass. U.K. 63: 205-222

Martin, M., G. Ichikawa, J. Goetzl, M. de los Reyes & M.D. Stephenson, 1984. Relationships between physiological stress and trace toxic substances in the bay mussel *Mytilus edulis*, from San Francisco Bay, California. Mar. Environ. Res. 11: 91-110

Martin, M., 1985. State mussel watch: toxics surveillance in California. Mar. Pollut. Bull. 6: 140-146

Nelson, W.G., D.K. Phelps, W.B. Galloway, P.F. Rogerson & R.J. Pruell, 1986. Effects of Black Rock Harbour dredged material on the scope for growth of the blue mussel *Mytilus edulis*, after laboratory and field exposure. Technical report D-86 US EPA, Narragansett, R.I. for US Army engineer waterways experiment station, Vicksburg, Miss. pp. 125

Nelson, W.G., 1990. Use of the blue mussel, *Mytilus edulis*, in water quality toxicity testing and *in situ* marine biological monitoring. Aquatic Toxicology and Risk Assessment: thirteenth volume, ASTM STP 1096; W.G. Landis & W.H. van der Schalie (eds). ASTM, Philadelphia, pp 167-175

Redpath, K.J. & J. Davenport, 1988. The effect of copper, zinc and cadmium on the pumping rate of *Mytilus edulis*. Aquat. Toxicol. 13: 217-226

Sanders, B.M., L.S. Martin, W.G. Nelson, D.K. Phelps & W. Welch, 1991. Relationships between accumulation of a 60 kDa stress protein and scope-for-growth in *Mytilus edulis* exposed to a range of copper concentrations. Mar. Environ. Res. 31: 81-97

Seed, R. & T.H. Suchanek, 1992. Populations and community ecology of *Mytilus*. In: The mussel *Mytilus*: Ecology, physiology, genetics and culture; E. Gosling (ed). Elsevier, Amsterdam, pp. 87-170

Smaal, A.C., A. Wagenvoort, J. Hemelraad & I. Akkerman, 1991. Response to stress of mussels (*Mytilus edulis*) exposed in Dutch tidal waters. Comp. Biochem. Physiol. 100C: 197-200

Solorzano, L., 1969. Determination of ammonia in natural waters by the phenolhypochlorite method. Limnol. Oceanogr. 14: 779-801

Stickle, W.B., M.A. Kapper, T.C. Shirley, M.G. Carb & S.D. Rice, 1987. Bioenergetics and tolerance of the pink shrimp (*Pandalus borealis*) during long-term exposure to the water-soluble fraction and oiled sediment from Cook Inlet crude oil. In: Pollution physiology of estuarine organisms; W.B. Vernberg, A. Calabrese, F.P. Thurberg & F.J. Vernberg (eds). Univ. of S. Carolina Press, pp. 87-106

Strickland, J.D.H. & T.R. Parsons, 1972. A practical handbook of seawater analysis. Bull. Fish. Res. Bd Can. 167: 1-311

Strömgren, T., 1986. The combined effect of copper and hydrocarbons on the length growth of *Mytilus edulis*. Mar. Environ. Res. 19: 251-258

Wang, S.Y. & W.B. Stickle, 1987. Bioenergetics, growth and molting of the blue crab *Callinectes sapidus*, exposed to the water-soluble fraction of South Louisiana crude oil. In: Pollution physiology of estuarine organisms; W.B. Vernberg, A. Calabrese, F.P. Thurberg & F.J. Vernberg (eds). Univ. of S. Carolina Press, pp. 107-126

Warren, C.E & G.E. Davis, 1967. Laboratory studies on the feeding bioenergetics and growth in fish. In: The biological basis of freshwater fish production; S.D. Gerking (ed). Blackwell Scientific, Oxford, pp. 175-214

Widdows, J., 1985. Physiological measurements and procedures. In: The effects of stress and pollution on marine animals; B.L. Bayne, D.A. Brown, K. Burns, D.R. Dixon, A. Ivanovici, D.R. Livingstone, D.M. Lowe, M.N. Moore, A.R.D. Stebbing & J. Widdows (eds). Praeger, New York, pp. 3-45 and 161-179

Widdows, J., 1993. Marine and estuarine toxicity tests. In: Handbook of ecotoxicology, Vol. 1; P. Calow (ed). Blackwell, Oxford, pp. 145-166

Widdows, J. & D. Johnson, 1988. Physiological energetics of *Mytilus edulis*: scope for growth. Mar. Ecol. Prog. Ser. 46: 113-121

Widdows, J. & P. Donkin, 1991. Role of physiological energetics in ecotoxicology. Comp. Biochem. Physiol. 100C: 69-75

Widdows, J. & P. Donkin, 1992. Mussels and environmental contaminants: bioaccumulation and physiological aspects. In: The mussel *Mytilus:* ecology, physiology, genetics and culture; E. Gosling (ed). Elsevier, Amsterdam, pp. 383-424

Widdows, J. & D.S. Page, 1993. Effects of tributyltin and dibutyltin on the physiological energetics of the mussel *Mytilus edulis*. Mar. Environ. Res. 35: 233-249

Widdows, J., D.K. Phelps & W. Galloway, 1981. Measurement of physiological condition of mussels transplanted along a pollution gradient in Narragansett Bay. Mar. Environ. Res. 4: 181-194

Widdows, J., T. Bakke, B.L. Bayne, P. Donkin, D.R. Livingstone, D.M. Lowe, M.N. Moore, S.V. Evans & S.L. Moore, 1982. Responses of *Mytilus edulis* on exposure to the water-accommodated fraction of North Sea oil. Mar. Biol. 67: 15-31

Widdows, J., P. Donkin & S.V. Evans, 1985. Recovery of *Mytilus edulis* L. from chronic oil exposure. Mar. Environ. Res. 17: 250-253

Widdows, J., P. Donkin & S.V. Evans, 1987a. Physiological responses of *Mytilus edulis* during chronic oil exposure and recovery. Mar. Environ. Res. 23: 15-32

Widdows, J., P. Donkin, P.N. Salkeld & S.V. Evans, 1987b. Measurements of scope for growth and tissue hydrocarbon concentrations of mussels (*Mytilus edulis*) at sites in the vicinity of the Sullom Voe oil terminal - a case study. In: Fate and effects of oil in marine ecosystems; J. Kuiper & W.J. van den Brink (eds). Nijhoff, Dordrecht, pp. 269-277

Widdows, J., K.A. Burns, N.R. Menon, D.S. Page & S. Soria, 1990. Measurement of physiological energetics (scope for growth) and chemical contaminants in mussels (*Arca zebra*) transplanted along a contaminant gradient in Bermuda. J. Exp. Mar. Biol. Ecol. 138: 99-117

Winberg, G.C., 1960. Rate of metabolism and food requirements of fishes. Transl. Ser. Fish. Res. Bd. Can. 194: 1-202

Appendix 11

General Experimental Design of Scope for Growth Measurements

In this Appendix the procedures for the preferred technique are presented, which include:

- the use of native mussels or transplanted mussels from a reference stock;
- measurements of individuals under standard conditions in the laboratory;
- the use of flow-through grazing chambers for clearance rate and absorption efficiency, and closed chambers for respiration (and excretion) rates.

This Appendix is largely based on Widdows (1985). The apparatus is schematized in Figure 4.

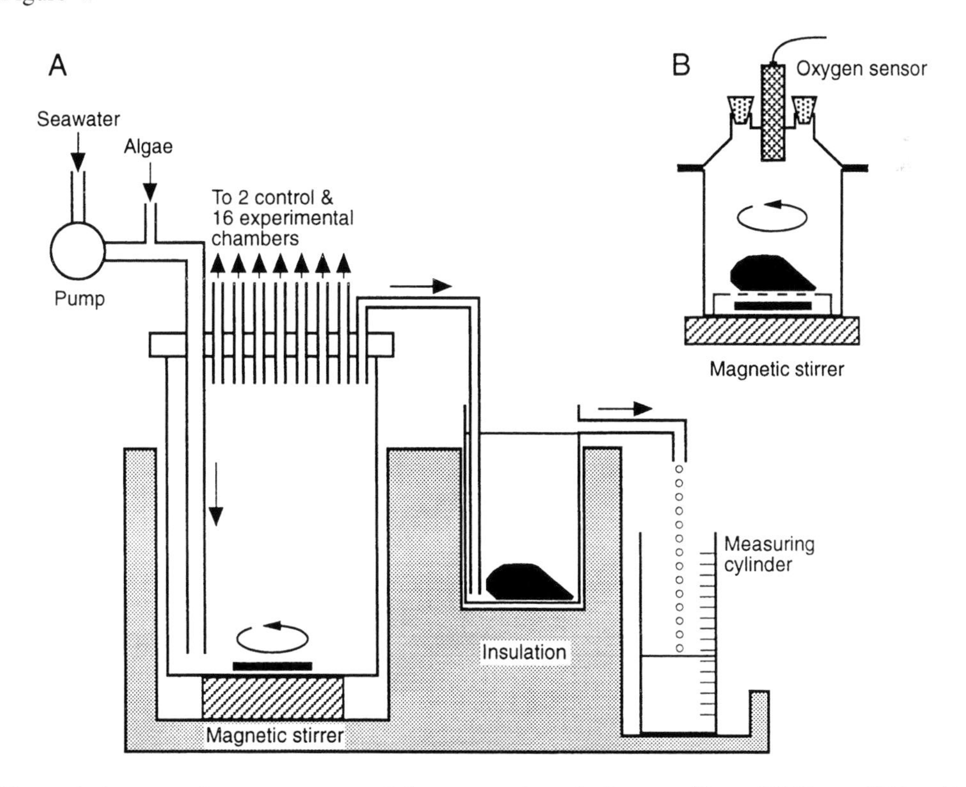

Figure 4. Apparatus for measurement of clearance and respiration rate (From: Widdows, 1993; with permission)

TREATMENT OF THE MUSSELS

Mussels (*e.g. Mytilus edulis*) used for transplantation are collected from a clean reference site and cleaned of epibionts. Prior to the monitoring programme the allometric coefficients of this base line population should be estimated, following the description of Widdows (1985) for expressing clearance, respiration and excretion rates on the basis of a standard body size, according to the equation:

$$Y = aW^b \quad (A1)$$

where Y is the physiological rate and W is the dry flesh weight in grams; a and b are intercept and slope respectively when expressed on a logarithmic basis:

$$\log Y = \log a + b \log W \quad (A2)$$

Once the b coefficient has been estimated, it can be used to correct measured rates to a standard body size during the course of the monitoring programme.

Prior to transplantation, mussels of the same size class are selected.

In the Dutch field programmes mussels are put into cages, acclimatized at a quiet and clean site to allow byssal attachment of the mussels to the cages, and then transported by ship to various exposure sites where the cages are connected to buoys. The exposure period is six weeks, taking care not to expose the mussels during their reproductive period. Every two weeks the buoys are checked and possible fouling of the cages is removed. After six weeks the mussels are collected and transported to the laboratory with a supply of high quality uncontaminated seawater.

When native mussels are used, sampling, transportation and handling is done in the same way as with transplanted mussels; for these mussels the allometric coefficients have to be estimated as well.

Transportation is performed under standard conditions in a cool box at 5 °C within 24 hours, but mussels are kept in the cool box for a standard period of exactly 24 hours. The mussels are then recovered in high quality (*i.e.* from an uncontaminated source) running seawater for 6 hours and acclimatized in the grazing chambers overnight (16 hours) with the diet used for the clearance rate measurements, prior to the SFG measurement. All measurements are performed at ambient or standard temperature and full salinity.

SCOPE FOR GROWTH MEASUREMENT

Clearance rate

The clearance rate is measured in 16-20 flow-through grazing chambers with individual animals, supplied with clean natural or filtered seawater and algal cells added. Some chambers have no animals and serve as a control. The particle concentration should be below the pseudofaeces threshold of 3-5 $mg.l^{-1}$ (depending on the size of the animals), to avoid variability due to different selection efficiencies in pseudofaeces formation, as well as mucus formation. There should be enough material to produce sufficient faeces for absorption estimates. The particle concentrations are measured with an electronic particle counter of the Coulter or Particle Data type, counting particles larger than 2-3 μm diameter. The flow rate is adjusted so that the outflow particle concentration is not more than 30 % below the inflow concentration, in order to use the inflow concentration as an estimate of the concentration in the chamber (A3). If this cannot be achieved, thorough mixing in a larger chamber is required and then the internal concentration is assumed to be presented by the outflow concentration (A4). The clearance rate is calculated as:

$$CR = \{(C_{in}-C_{out})/C_{in}\} \times Q \quad (A3)$$

or

$$CR = \{(C_{in}-C_{out})/C_{out}\} \times Q \quad (A4)$$

where CR = clearance rate ($l.h^{-1}$), C_{in} = inflow particle concentration, C_{out} = outflow particle concentration, Q = flow rate ($l.h^{-1}$). CR is corrected for a standard body size of 1 g DW by: $CR' = CR.W^{-b}$

The clearance rate is multiplied by the concentration of organic matter of the inflow, to estimate the consumption rate.

The clearance rate measurement starts after the 16 hour (overnight) acclimatization to the diet of natural or filtered seawater with algal cells (*e.g. Isochrysis galbana* or *Phaeodactylum tricornutum*) in a concentration range of 10,000-15,000 particles.ml^{-1}. The grazing chambers are cleaned of faeces produced during the preceding 16 h period to ensure that faeces reflect the standard laboratory diet.

The flow rates are adjusted prior to the measurement. They normally range from 5 to > 10 $l.h^{-1}$, depending of the activity *i.e.* the size of the animals. The mussels are left undisturbed for at least 1 hour before the clearance rates are determined. The flow rates and the outflow concentrations of the control and experimental chambers are measured at regular intervals over a period of two hours. These are performed at least three times for each chamber, unless large variation occurs; then measurements are repeated. A minimum of three clearance rate estimates are registered, and out of these the maximum value is generally used as the clearance rate for the calculations of SFG. There is a good agreement between the average and maximum values, but the maximum value is preferred for calculation of the scope for growth, because this reflects the maximum scope for growth and does not include brief periods of inactivity or partial valve closure.

Absorption efficiency

The Conover method (Conover, 1966) requires collection of suspended particulate matter (SPM) and faeces in sufficient quantities to analyze dry weight and ash-free dry weight. This is carried out directly after the clearance rate measurements by sampling the faecal production with a pipette. The faeces produced during the oxygen uptake and ammonia excretion measurement can be collected as well. Faeces of 3 individuals are pooled and treated as one sample. Thus 15 mussels provide 5 replicates. The concentrations of suspended particulate matter (SPM) and particulate organic matter (POM) are required to estimate food consumption and absorption efficiency. A minimum of 10 mg of sample (food or faeces) is required. The procedures for sampling and analysis of food and faeces are similar. Each sample, either a known volume of control outflow (1-4 l), or a faecal sample, is filtered onto a washed, ashed and pre-weighed 45 mm glass fibre filter (Whatman GF/C) with a mesh size of ca 1 μm. Salts are washed out with 50 ml distilled water. The filters are oven dried at 90 °C for 24 hours and weighed before and after ashing at 450 °C for >2 hours. Blank filters are used as control, and filters are stored in desiccators during cooling and transfer. Weighing is done with an analytical balance to 10^{-5} g sensitivity, to obtain the dry weight and ash weight of food and faecal samples. The ash-free dry weight is the difference between the dry and ash weight.

The absorption efficiency (ae) is calculated as:

$$ae = (F-E)/(1-E)F \quad (A5)$$

where F = ratio ash-free dry weight/dry weight of food, and E = ratio ash-free dry weight/dry weight of faeces.

Respiration rate

The respiration rate is measured as the oxygen uptake rate during incubation in closed chambers with seawater. The temperature is controlled by placing the chambers in a water bath or in running seawater. The animals are left undisturbed for one hour before the water flow through the chambers is stopped. The decline of the oxygen tension in the chamber is monitored during 1 hour with a polarographic oxygen electrode connected to an oxygen meter and chart recorder. Oxygen tension should be above 60 % air saturation to avoid an oxygen dependent respiration rate at lower oxygen tensions. Depending on the number of oxygen electrodes available, a number of individuals can be measured in parallel. A minimum of 15 mussels are measured.
Prior to measurements, the oxygen sensors must be calibrated to zero and full air saturation at the appropriate temperature and salinity, according to the specific procedure of the type of probe (see Widdows, 1985).
The oxygen consumption rate (OR) is calculated for a period of linear decline in oxygen tension and converted into ml, mg or preferably µmol O_2 l^{-1}, based on oxygen solubility tables (Gnaiger & Forstner, 1983):

$$OR = ([O_2]_{t=0} - [O_2]_{t=1}) \times (V_{chamber} - V_{animal}) \times 60/(t_1 - t_0) \quad (A6)$$

where OR is the rate of oxygen consumption (µmoles O_2 h^{-1}), $[O_2]_{t=0}$ and $[O_2]_{t=1}$ are oxygen concentrations (in µmoles.l^{-1}) at the start and at the end of the linear decline respectively, V = volume in litres of chamber and mussel, and t=time expressed in minutes.
As an alternative, the oxygen concentration can be measured by (micro) Winkler titration of a water sample before and after the incubation, under the assumption that there is a continuous rate of oxygen consumption during the incubation period. The Winkler titration is done on a subsample of an exact volume according to the special Winkler bottles available, normally 115 ml, and based on the fixation of oxygen by $MnCl_2$, NaOH and KJ, and titration with $Na_2S_4O_6$.
The oxygen consumption rate per mussel is corrected for body size according to the allometric b coefficient for respiration measured on the base line population prior to the monitoring programme.

Excretion rate

Measurement of ammonia excretion rate is optional for scope for growth estimation, because the energy loss through excretion is in many cases negligible. The method according to Widdows (1985), consists of 2 hours incubation of the animal in 200 ml filtered seawater, at ambient water temperature. Another 200 ml without animal acts as a control. Ammonia samples after incubation of experimental and control beakers are analyzed according to the standard method of Solorzano (1969), or with an auto-analyzer. The excretion rate per mussel is

$$U = ([Test] - [Control]) \times V/t \quad (A7)$$

where U = rate of ammonia excretion in µmoles NH_4-N h^{-1}, [Test] = µM NH_4-N in experimental chamber and [Control] = µM NH_4-N in control chamber, V = volume of incubation seawater (in l) and t = incubation time (h). Correction to standard body size by using the previously estimated b coefficient.

SCOPE FOR GROWTH CALCULATION

The scope for growth is expressed in $J.h^{-1}.g^{-1}$ DW according to:

$$SFG = ae \times C - (R + U) \tag{A8}$$

The conversion of the physiological components to $J.h^{-1}.g^{-1}$ DW is as follows:

$$C = CR.W^{-b} \times POM \times 23.5 \quad [1 \text{ mg POM} = 23.50 \text{ J}] \tag{A9}$$

$$R = OR.W^{-b'} \times 0.46 \quad [1\ \mu\text{mole } O_2\ h^{-1} = 0.465\ J.h^{-1}] \tag{A10}$$

$$U = ER.W^{-b''} \times 0.349 \quad [1\ \mu M\ NH_4\text{-}N\ h^{-1} = 0.349\ J.h^{-1}] \tag{A11}$$

where:
CR = clearance rate per mussel ($l.h^{-1}$);
b = allometric coefficient for clearance;
OR = oxygen consumption rate per mussel (μmoles O_2 h^{-1});
b' = allometric coefficient for respiration;
ER = excretion rate per mussel (μM NH_4-N h^{-1});
b'' = allometric coefficient for excretion.

Chapter 12

Survival of the Fittest: Resistance of Mussels to Aerial Exposure

Richard H.M. Eertman and **Ab de Zwaan**

Netherlands Institute of Ecology, Centre for Estuarine and Coastal Ecology, Vierstraat 28, 4401 EA Yerseke, The Netherlands

ABSTRACT

Mussels possess the natural ability to tolerate anoxic or hypoxic conditions. When suddenly exposed to air, subtidally acclimatized mussels close their valves and oxygen becomes virtually unavailable to them. During prolonged aerial exposure mussels suppress their metabolic rate and their metabolic processes become predominantly anaerobic. Mussels have adopted alternative glycolytic pathways, thereby enhancing anaerobic ATP yield and improving resistance to anoxia. Field and laboratory experiments have demonstrated that accumulated contaminants reduce the tolerance of mussels to aerial exposure, presumably caused by a failure to reduce their scope for metabolic depression and an inhibited ATP generation during anaerobiosis. Significant correlations can be demonstrated between tissue contaminant levels and tolerance to aerial exposure. In addition to its sensitivity to contaminants, the determination of the 'survival in air' response is a simple and low-budget method which has worldwide applicability in coastal monitoring programmes.

1. INTRODUCTION

In his impressive monograph entitled "Anaerobiosis in invertebrates", von Brand (1946) has written that many bivalves are found in surroundings which are very poor in oxygen (hypoxia) and that they are able to withstand experimental deprivation of oxygen for long periods. He also points to the fact that bivalves in the tidal zone, in contrast to gastropods, lead a predominantly anaerobic life during low tide. More recent studies have shown that the latter statement may be too general. Brinkhoff *et al.* (1983) collected one gastropod and three intertidal bivalves from their habitat during low tide in order to assess whether these species showed an anaerobic metabolism.

Biomonitoring of Coastal Waters and Estuaries. Edited by Kees J.M. Kramer.

From succinate accumulation and from decreased energy states it was established that the mussel *Mytilus edulis* and the clam *Scrobicularia plana* switch to (partially) anaerobic metabolism during emersion, but that aerial respiration is sufficient to maintain a fully aerobic metabolism in the limpet *Patella vulgata* and the cockle *Cerastoderma (Cardium) edule*. The relationship of anaerobic metabolism and respiration to total energy metabolism was quantified for *M. edulis* (and *C. edule*) under ecologically realistic conditions with tidal cycles with aerial exposure (Widdows & Shick, 1985; Shick *et al.*, 1988). Animals were acclimated to either intertidal or subtidal regimes before measurement of the rates of heat dissipation and oxygen uptake. Mussels that were continuously fed under subtidal conditions had no measurable oxygen consumption when subjected to acute exposure to air, whereas intertidal mussels had a rate of oxygen consumption which represented 40% of the total heat dissipation in air. However, in the latter group a non-feeding period of four days reduced aerial oxygen uptake to zero. Some adaptation and conditioning to an intertidal regime was observed, including the enhancement of the ability to continue the aerobic processing of food ingested during the immersion period.

In summary, mussels acclimated to subtidal conditions are unlike some other bivalves, unable to consume significant amounts of oxygen when suddenly exposed to air. Therefore, aerial exposure of mussels, like exposure to environmental anoxia, will progressively result in the tissues becoming anoxic and in metabolic processes switching from aerobic to anaerobic. Alternatively, in species that do consume oxygen during aerial exposure, anoxia tolerance can be determined by exposing animals to anoxic seawater (tolerance to aerial exposure is not exclusively related to anaerobic processes).

In this chapter, the adaptive mechanisms used by bivalves to tolerate anoxia will be reviewed and the effects of environmental variables (both natural and anthropogenic) on the tolerance of mussels to aerial exposure will be evaluated. The measurement of many biological effect parameters as an integral part of routine monitoring programmes is often time consuming, expensive and requires a great deal of skill. The determination of the 'survival in air' response of mussels has few of the above mentioned restrictions and can be performed routinely in most laboratories.

2. REVIEW OF THE FIELD

2.1. Resistance to anoxia

2.1.1. Anoxia tolerance in adult species

A comparative survey of the resistance to anoxia among 14 different macrofaunal invertebrate species from various biotopes of the North Sea and the Baltic by Theede *et al.* (1969) illustrated the generally impressive anoxic resistance of the bivalves. Four of the five bivalves included in the survey, survived longer in anoxic seawater than species belonging to other taxonomic groups (Gastropods, Polychaetes, Echinoderms, Crustaceans) which were studied. Anoxic exposure times resulting in 50 % mortality (LT_{50}) at 10 °C, these four species ranged from 55 to 21 days. *M. edulis,* with a value of 35 days was also included among them. The fifth bivalve, *C. edule*, differed distinctly from this group with a relatively low LT_{50} of 4 days.

As the disappearance of oxygen in the marine environment is often correlated with the formation of hydrogen sulphide (H_2S), the anoxia resistance of species was also studied where this compound was present. Hydrogen sulphide is known to inhibit the aerobic metabolism by blocking cytochrome aa_3. In general, the addition of sulphide reduced the anoxia tolerance. This has been confirmed for several polychaetes (Groenendaal, 1980; Bestwick *et al.*, 1989) and bivalves (Shumway *et al.*, 1983; de Zwaan *et al.*, 1991a). The experiments of

Theede *et al.* (1969) have played a crucial role in the development of anoxic survival time as a stress indicator. They showed that perturbing effects on energy metabolism by components dissolved in seawater may be expressed by a decreased tolerance to anoxia. In areas in which access to oxygen strongly fluctuates (*e.g.* intertidal and strongly eutrophicated zones), the success or failure of populations may (partly) rely on their ability to survive anoxic conditions. This is shown in the northern Adriatic Sea where eutrophication has caused a decline in the size of the benthic populations of a variety of species, but has simultaneously stimulated the establishment of the blood clam *Scapharca inaequivalvis.* This clam, originating from the Indo-Pacific region, was introduced into the Adriatic Sea by man only some 20 years ago (Ghisotti & Rinaldy, 1976), but is currently rapidly outnumbering native and economically important species such as *Venus gallina* to become the dominant bivalve species. It has been established that *S. inaequivalvis* has a superior tolerance for anoxia as compared to *V. gallina*; LT_{50} values at 18 °C were 17 and 4 d respectively (Brooks *et al.*, 1991).
It is, therefore, likely that natural and/or anthropogenic factors which reduce anoxia tolerance will lead to a higher mortality or reduced fitness of the population. When this is the case, such factors can be considered as stress indicators according to the definition of stress formulated by Bayne (1980). This author states that, from a physiological point of view, stress is the result of the alteration of processes (physiological or other) in such a way as to render the individual less fit for survival. By applying anoxic survival as a stress index, the damage (caused by accumulated contaminants) is expressed as a reduced capability to resist environmental change in the form of a blocked supply of oxygen. The occurrence of stress is thus evaluated from data concerning the mortality rate. This is a convenient and a low-budget experimental approach. The observations can be made over a short period of time, which is limited to the survival time of anoxia.

2.1.2. Anoxia tolerance in juveniles

Tolerance to anoxia changes during the lifetime of bivalves. During the early stages of larval development in mussels the median mortality time increases from 15 hours for early prodissoconch larvae (106 µm) to 39 hours for veliconch larvae (224 µm) (Wang & Widdows, 1991). In contrast to mussel larvae, settlement of oyster larvae appears to be more sensitive to moderate hypoxia. Oxygen concentrations down to 1.3 mg O_2 l^{-1} (3.16 kPa pO_2, 15 % saturation at 25 °C) do not seem to influence settlement of mussel larvae, whereas a similar oxygen concentration strongly inhibits the settlement of oyster larvae (Baker & Mann, 1992). In oxygen limiting environments, energy costly activities, such as settlement and feeding, will be avoided. Oyster larvae only settle once and remain sessile for the rest of their lives, whereas post-larval mussels migrate repeatedly before arriving at a final settlement site (Lane *et al.*, 1985). Therefore, mussel larvae do not need to be as discriminating as oyster larvae when selecting a suitable settlement habitat.
The establishment of substantial glycogen reserves during larval development and an increasing capacity to suppress metabolic activity during anoxia are thought to be the main factors improving the anoxia tolerance of mussel and oyster larvae. Rhodes & Manzi (1989) studied the survival rates of larval and juvenile bivalve molluscs (the clam *Mercenaria mercenaria* and the bay scallop *Argopecten irradians*) following transportation under standard conditions. The greatest mortality occurred in the smallest sized bivalves shipped over the longest period of time, and the highest survival rates were recorded in the largest sized bivalves shipped over the shortest period.
Anoxia tolerance further improves during the growth of young mussels. Mussels from the 23 - 37 mm length class have a significantly higher mean anoxic survival time in comparison to mussels from the 12 - 21 mm length class (Borsa *et al.*, 1992). When using the 'survival in air' response in monitoring programmes, age related differences in anoxia tolerance should be avoided by using adult mussels (or other bivalves) of the same age category. Selecting

mussels from one size class (*e.g.* 20-25 or 40-50 mm) minimizes the possibility of age dependent variation in anoxia tolerance.

2.1.3. Genetic adaptations

Bivalves that are periodically exposed to natural anoxic conditions may have the ability to adapt genetically in order to improve tolerance to this stressful condition. The venerid bivalve *Ruditapes decussatus* inhabits the shores of the Balaruc peninsula in southern France, where natural anoxic stress, the so called 'malaïgue' which is generated by eutrophication, occurs during hot summers.
Borsa *et al.* (1992) have demonstrated that animals that survived the period of anoxia had a significantly higher proportion of heterozygosity in comparison to animals that had not been exposed to natural anoxia. Koehn & Bayne (1989) postulated that heterozygotes perform better in stress conditions as a consequence of their reduced energy expenditure for maintenance metabolism. Of the seven enzyme loci studied by Borsa *et al.* (1992), heterozygosity at the locus transcribing the enzyme directly involved in energy production processes (phosphoglucomutase), showed the strongest correlation with survival. Although some extensive studies have failed to detect consistent relationships between the occurrence of heterozygosity and fitness related parameters, more often than not these relationships were positive (Gaffney, 1990).
As genetic variability between individuals may also affect the anoxia tolerance of mussels, it is preferable that mussels for various experiments be selected from within one single reference population.

2.2. Scope for metabolic arrest

Analysis of the underlying biochemical mechanisms determining the anoxic survival time may provide us with information about the physiological/biochemical processes that most likely undergo alteration during periods of oxygen unavailability. For a comprehensive account of our present knowledge about the mussel we refer to a recent review by de Zwaan & Mathieu (1992). Here we present only an overview.
Von Brand (1946) summarized literature which illustrates that in bivalves the length of the survival time under anaerobic conditions is inversely related to temperature. Metabolism depends on enzymes and most biological reactions possess a $Q_{10^\circ C}$ of 2 to 3, where $Q_{10^\circ C}$ is defined as:

$$Q_{10^\circ C} = (\text{rate at } a + 10\ ^\circ C) / (\text{rate at } a\ ^\circ C) \qquad (1)$$

In ectothermic animals like the mussel, where the body temperature fluctuates with the environment, temperature will have a great impact on metabolic rate. For the mussel, a $Q_{10^\circ C}$ value of 3 was established for the rate of mortality during anoxia (Veldhuizen-Tsoerkan *et al.,* 1991). Weber *et al.* (1992) investigated the effects of sublethal and lethal Cu concentrations on mortality under various combinations of anoxia, salinity and temperature. In all combinations of conditions, the $Q_{10^\circ C}$ of the mortality rate values for the blue mussel (*Mytilus edulis*) were in the range of 2 - 2.5.
These results point to a strong relation between metabolic rate and anoxic survival time. The lower the metabolic rate, the higher the anoxic survival time. Metabolic rate depression has the greatest quantitative impact of all biochemical adaptations known that contribute to anaerobiosis (Storey & Storey, 1990). Good anoxia resistant species, therefore, dramatically reduce their metabolic rate during anoxia (so-called metabolic arrest). A relationship between metabolic rate and anoxia tolerance is also shown in interspecific comparisons with adult bivalves. In the above cited comparative study by Theede *et al.* (1969), *Artica*

(Cyprina) islandica survived longest in oxygen-depleted and sulphide-containing water with 50 % mortality after 55 days. Calorimetric measurements showed that after prolonged anoxia, energy release decreases to less than 1 % of aerobic rates, the lowest rate found in marine invertebrates so far (Oescher, 1990). De Zwaan & Wijsman (1976) calculated from oxygen uptake and biochemical studies that the anoxic energy demand in *M. edulis* was only 5.4 % of the aerobic value. A value of about 4.5 % has recently been established for metabolic rate depression in *Mytilus galloprovincialis* (de Zwaan *et al.*, 1991b) and *S. inaequivalvis* (Brooks *et al.*, 1991). These species have well-developed and similar tolerances of anoxia.
The inability to strongly arrest ATP turnover appears to explain the relatively low anoxia tolerance of *C. edule* and *V. gallina*. The former species maintains a high metabolic rate by "air-gaping" during air exposure (50 - 75 % of aquatic rate; Widdows & Shick, 1985), whereas immersion in anoxic seawater results in the accumulation of high amounts of lactate which is the main fermentative product in this species (Meinardus & Gäde, 1981; Kluytmans & Zandee, 1983). Lactate accumulation reflects a high rate of energy expenditure relative to succinate fermentation (see below). The lower anoxia survival time displayed by *V. gallina*, as compared to *S. inaequivalvis* and *M. galloprovincialis,* is consistent with a much greater accumulation of anaerobic end products due to a lesser ability to reduce ATP turnover rate during anoxia. Calculated ATP turnover rate was 12.7 % of the aerobic rate in this species during anoxic exposure. Accordingly, it was shown that *S. inaequivalvis* and *M. galloprovincialis* respond to anoxia with changes to the properties of phosphofructokinase and glycogen phosphorylase which indicate that the enzymes are converted by covalent modifications to less active enzyme forms during anaerobiosis (Brooks *et al.*, 1991, de Zwaan *et al.*, 1991b). *Venus gallina* appears to lack the equivalent mechanisms for reducing glycolytic rate in anoxia (Brooks *et al.*, 1991). Other mechanisms that might be involved in the control of glycolytic rate depression are summarized by Storey & Storey (1990) and de Zwaan & Mathieu (1992). The clam *Mulina lateralis* is also relatively intolerant of prolonged anoxia. Again it was observed that this bivalve has an anoxic rate of heat dissipation which is slightly less (97 %) than that of normoxia (Shumway *et al.*, 1983).

2.3. Anaerobiosis

Anaerobiosis caused by incomplete external oxygen supply is termed environmental or organismal anaerobiosis. It is also known as the low-power output mode of anaerobic metabolism, because of its connection to energy conservation. The low cellular energy expenditure in Cnidarians, Molluscs, Annelids and Sipunculids is initially met by coupling additional pathways for energy extraction to the glycolytic pathway (the aspartate-succinate pathway). They respond to prolonged anaerobiosis by diverting the main carbon flow from glycogen at the level of phosphoenolpyruvate towards succinate and then to propionate. Compared to classical glycolysis (glycogen-lactate), the glycogen-propionate pathway enhances anaerobic ATP yield by a factor of 2.14 (see: De Zwaan, 1983). Therefore, modification of glycolysis is a biochemical adaptation which enhances anoxic survival resistance and also causes also metabolic rate depression. Another advantage is that propionate can be easily excreted, which helps to avoid self-pollution by the accumulation of toxic waste products. The glycogen-propionate pathway is a high efficiency (in terms of ATP yield per glycosyl unit), low rate of energy production (in terms of ATP output per unit time) pathway. Anaerobic energy metabolism also augments respiration in order to meet cellular energy demand when this is beyond the aerobic scope. This may be due to exercise or restoration of normal physiological function. Among all multicellular organisms the high-power output mode of anaerobiosis is linked to classical glycolysis. In the mussel it is involved in recovery anaerobiosis (de Zwaan & Zurburg, 1981) and repeated phasic contractions (Zange *et al.*,

1989). Classical glycolysis in marine invertebrates - except Arthropods and Echinoderms - mainly terminates in so-called opines with relatively small amounts of, or no, lactate.
Opines are formed by reductive condensation of pyruvate with an amino acid. Strombine is the most important opine in the adductor muscle of the mussel, but octopine and lactate are also formed (de Zwaan & Zurburg, 1981). The opine dehydrogenases are functionally analogous to lactate dehydrogenase. Both catalyse reactions in which an equimolar amount of NADH is oxidized to the product formed; this compensates the oxidative step catalysed by glyceraldehyde phosphate dehydrogenase in glycolysis. Replacement of lactate by opine formation, therefore, does not result in a higher ATP output per glycosyl unit. Classical glycolysis can be characterized as a low efficiency, high rate of energy production pathway. The opine pathway, therefore, operates preferentially when there is an increased glycolytic flux. Livingstone (1982) and de Zwaan & Putzer (1985) have calculated the actual rates of energy production by the opine and the succinate (propionate) pathway in different molluscs. For bivalves, the rate of energy production (μmoles ATP equivalents $g^{-1}.min^{-1}$) of the former pathway was in the order of 0.1 - 1.0, compared to a range of 0.005 - 0.010 in the case of the succinate route.
The implications of the fact that the two types of anaerobic pathways have a different intrinsic value for ATP power output is that the anaerobic end product spectrum accumulating during anoxia depends upon the degree of energy expenditure. When the ratio of: succinate + propionate / opines (+ lactate) decreases, this indicates that energy utilization probably has increased, and a decreased anoxia resistance is, therefore, to be expected. De Zwaan & de Kock (1988) proposed that the end product ratio could be developed into a useful stress index or pollution indicator on the basis of these considerations.
A first field exercise for the purpose of evaluating this index was carried out in 1985 in combination with active biomonitoring programmes (ABM, see De Kock & Kramer,this Volume). Mussels sampled from a clean area were transplanted to relatively clean sites and a transect through an area with a pollution gradient for cadmium and some of the higher polychlorinated biphenyl (PCB) congeners (River Scheldt estuary). The mussels were recovered and subjected to anoxia stress. The above-mentioned ratio index could clearly distinguish between the clean areas versus the polluted sampling sites but was not directly correlated to the pollution gradient (de Zwaan & de Kock, 1988). However, propionate appeared to be a good pollution indicator which was also applicable for a "fine" discrimination of sites within the contamination gradient.

2.4. Energy expenditure during stress

When mussels have to cope with stress, operational adaptive mechanisms and/or detoxification systems will increase their energy demand. This may reduce their scope for metabolic depression during anoxia survival. Various experiments support this assumption. De Zwaan *et al.* (1983) stressed mussels by keeping the shell valves tightly closed with a rubber band. This resulted in a much higher energy expenditure. Strombine and, to a lesser extent, octopine accumulated in much higher amounts than in untreated anaerobic control mussels. Recently it has been established that this treatment reduced anoxic survival time to 73 % of that of the controls (De Zwaan, 1989).
De Zwaan *et al.* (1991a) studied the survival and metabolic adjustment of the blood clam *S. inaequivalvis* to environmental anoxia and tissue anoxia induced by sulphide. This clam belongs to the primitive Arcidae and possesses erythrocytes with haemoglobin. The LT_{50} for clams kept in nitrogen flushed seawater at 18°C was 15.2 days. The addition of sulphide reduced this value to 9.5 days. Anaerobic metabolism was studied in live animals and in red blood cells incubated *in vitro*. In both experiments, sulphide caused greater changes in the levels of aspartate and the pyruvate derivatives alanine and strombine and/or alanopine,

compared to environmental anoxia. In addition, H_2S induced a marked accumulation of octopine in the adductor muscle, a compound which is absent in red blood cells.
Wang *et al.* (1992) observed that anoxic rates of heat dissipation by mussels were enhanced after exposure to pentachlorophenol (PCP) and tributyltin (TBT), both uncouplers of oxidative respiration. Biochemical measurements showed that there was a shift from the succinate to the lactate anaerobic pathways (the more important pyruvate derivative strombine was not measured), consistent with a reduction in anoxic survival time.

3. SURVIVAL IN AIR: EXTERNAL INFLUENCES

3.1. Temperature

The survival time in air response of mussels can be determined accurately in any laboratory with constant temperature room facilities. Veldhuizen-Tsoerkan *et al.* (1991) demonstrated that the aerial survival time of mussels is influenced directly by the ambient air temperature (Figure 1). LT_{50} values decrease with increasing air temperature, due to accelerated metabolic processes at higher temperatures. When introducing the 'survival in air' test in a laboratory, an exposure temperature may be selected from within a relatively large range. It is important, however, that the selected temperature remains constant (± 0.5 °C), not only during one single experiment but also over a longer period of time when more experiments are going to be performed. One should realise that if a relatively low air temperature is selected (*e.g.* 12 °C), the aerial survival time of the mussels will be relatively long ($LT_{50} \approx 18$ days). At this low temperature the LT_{50} is three times higher in comparison to an exposure temperature of 22 °C ($Q_{10°C} = 3$). For mussel species of temperate zones, temperatures above 22 °C should be avoided as all mussels would die within a few days and it would be difficult to discriminate statistically between groups. For practical purposes it is recommended that an exposure temperature within the 15 to 22 °C temperature range be selected. Under (sub)-tropical conditions higher temperatures could be established for which the defined range should be determined experimentally.

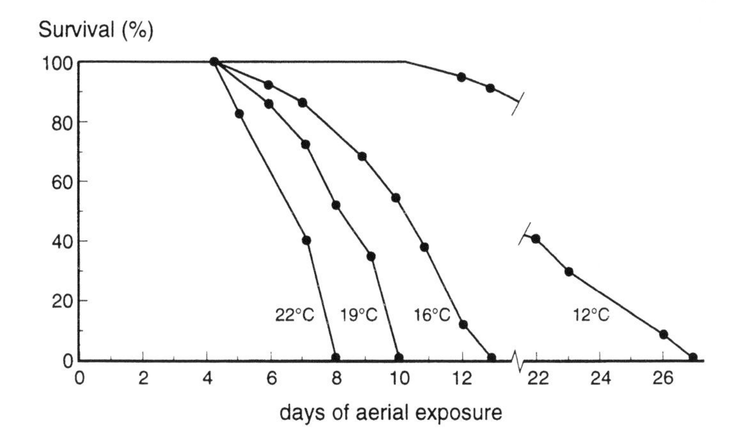

Figure 1: The effect of air temperature on the aerial survival time of mussels (From: Veldhuizen-Tsoerkan *et al.*, 1991; with permission)

3.2. Salinity

Field studies in the coastal environment often include estuaries where an increasing pollution gradient may coincide with a decreasing salinity gradient. To be able to assess the possible effects of lowered salinity on the 'survival in air' response, laboratory experiments had to be conducted where mussels were exposed to unpolluted seawater with lowered salinity. Eertman *et al.* (1993a) demonstrated that 15 days exposure to salinities ranging from 23 to 35×10^{-3} did not have a significantly adverse effect on the aerial survival time of mussels. Short-term acclimation (1 - 4 days) may affect the survival in air response negatively. Acute exposure of mussels to anoxic seawater of 14×10^{-3} salinity had a negative effect on the anoxia tolerance (Weber *et al.*, 1992). From these experiments it may be concluded that, following sufficient acclimation to seawater media with salinities as low as approximately 23×10^{-3}, tolerance to aerial exposure of mussels is independent of seawater salinity.

3.3. Seasonal variability

The 'survival in air' response of mussels shows seasonal variation. The LT_{50} values decline in winter, reaching lowest values in March (in the northern hemisphere) and gradually increase during summer, with highest values towards the end of the year (Eertman *et al.*, 1993a). Figure 2 shows the seasonal variation in LT_{50} values in 1991, determined in mussels from a reference population in the Oosterschelde in the southwestern part of the Netherlands. Anaerobic metabolism relies predominantly on endogenous glycogen stores, but during the transition phase aspartate conversion to succinate is coupled to its degradation. Maintenance of high reserves of glycogen is, therefore, an additional biochemical adaptation that contributes to anoxic survival. Depending on the organ and the season, mussel tissue glycogen accounts from 5 to 40 % of dry weight (de Zwaan & Zandee, 1972; Zandee *et al.*, 1980). Although the seasonal pattern in 'survival in air' response resembles the glycogen cycle, the aerial survival time does not seem to be influenced directly by the glycogen content of mussels as the total glycogen content of mussels is much larger than the amount of glycogen fermented during anoxia. It was estimated that mussels ferment approximately 5 to 20 % of the body glycogen reserves during anoxia (Eertman *et al.*, 1993a).

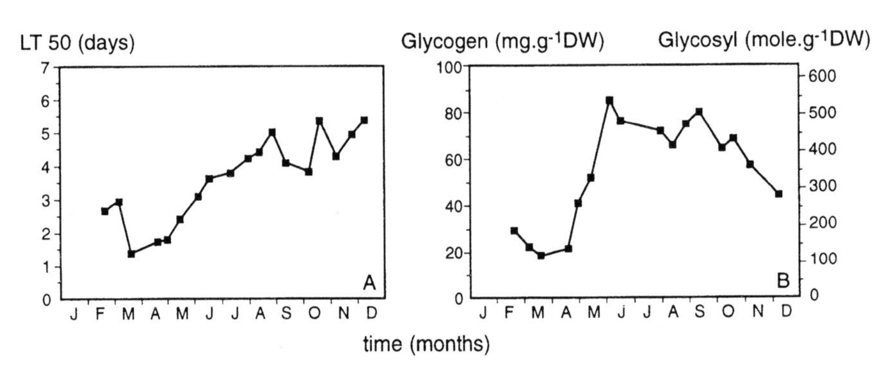

Figure 2: Seasonal variation during 1991 in (A) survival in air response and (B) glycogen content in mussels from the relatively unpolluted Oosterschelde (From: Eertman *et al.*, 1993a; with permission)

The reproductive process of mussels, which itself is activated by the seawater temperature (Seed, 1975), seems to be the main seasonal factor influencing the survival in air response. During the final stages of gonadal development in spring (March - April in the temperate zones of the northern hemisphere) up to 94 % of metabolic energy production is required for the reproductive process (Bayne, 1976), causing potential energy deficiencies for other metabolic processes.

Additionally, the 'survival in air' response may be influenced negatively by a temperature shock. Following transfer from field locations to the laboratory, the mussels are allowed to recover in running seawater at ambient temperature. Subsequently, the mussels are exposed to air in a constant temperature room of *e.g.* 18 °C, introducing a temperature shock which may range from approximately +14 °C in winter to -2 °C in summer. Due to the presence of a seasonal cycle, LT_{50} values measured in different seasons may not be compared directly. Variations in LT_{50} values between experimental groups that were exposed simultaneously within one season may be tested statistically.

4. 'SURVIVAL IN AIR': EXPERIMENTAL EVALUATION

The tolerance of mussels to aerial exposure, the 'survival in air', can be used as a biological effect parameter which may be indicative for the overall stress experienced by the animals under natural conditions. The applicability of this method as a biomonitoring technique was tested both in laboratory and field situations.

4.1. Laboratory exposures to toxicants

Veldhuizen-Tsoerkan *et al.* (1991) demonstrated toxic effects of short-term (2 - 4 weeks) and long-term (3 - 10 months) laboratory exposure of mussels under semi-field conditions to 50 $\mu g.l^{-1}$ Cd on the 'survival in air' response. Six months exposure to 1 $\mu g.l^{-1}$ PCBs (as Clophen A50) resulted in a significantly reduced tolerance to aerial exposure. Under similar conditions, Eertman *et al.* (1993a) found a significantly reduced aerial survival time after 3 weeks exposure to 1 $\mu g.l^{-1}$ ΣPCBs (Figure 3). At the end of the 3 weeks exposure period the mussels had accumulated 494 μg ΣPCBs kg^{-1} DW, corresponding with the highest values found at the eastern, most contaminated, exposure site in the River Scheldt estuary.

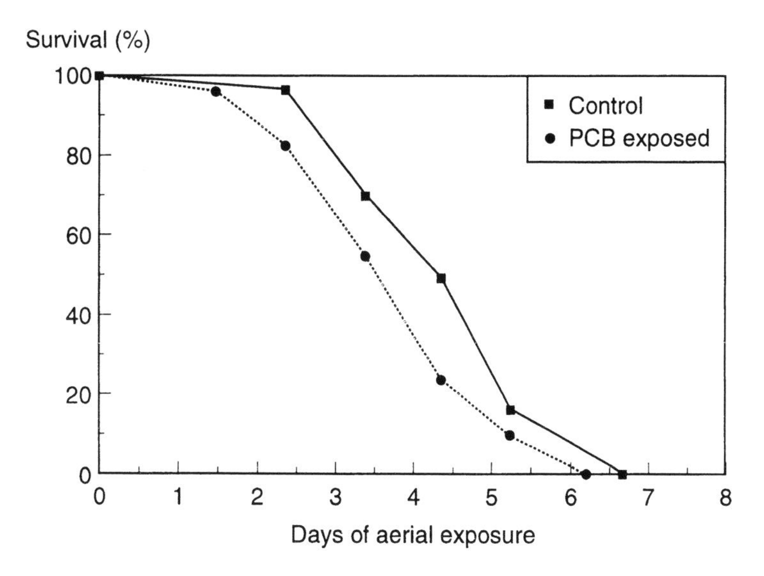

Figure 3: Survival curves of control and PCB exposed mussels. See text for details (From: Eertman *et al.*, 1993a; with permission)

However, Weber *et al.* (1992) observed that chronic exposure to sublethal Cu concentrations (20 $\mu g.l^{-1}$), and subsequent subjection to temperature and salinity stress had no significant effect on the survival of mussels under aerial exposure at either 6 or 15 °C. These authors also applied anoxia mortality in deoxygenated seawater in order to discern synergy mechanisms between natural and pollution related environmental stress factors. High copper concentrations (200 $\mu g.l^{-1}$) drastically decreased anoxic survival, which further deteriorated when combined with temperature increase and salinity decrease. In addition to the above-mentioned effects of PCP and TBT, Wang *et al.* (1992) also observed a reduced anoxia tolerance in mussels following short-term exposure (2 days) to the polycyclic aromatic hydrocarbon naphthalene, although this compound had no significant effect on the anoxic rate of heat dissipation. Exposure of mussels to the PAHs benzo[a]pyrene and fluoranthene gave an inconsistent response. At relatively low tissue concentrations of both aromatic compounds tolerance to aerial exposure improved in comparison to control animals, while a reduced tolerance to aerial exposure was observed at higher tissue concentrations (Eertman *et al.*, 1993b). Stimulatory effects of subinhibitory levels of toxicants, a phenomenon known as hormesis, is well known in toxicology and has been ascribed to transient overcorrections by control mechanisms to inhibitory challenges well within the animal's capacity to counteract (Stebbing, 1979). More research is required in order to discern the mechanisms of action of various contaminants that result in an altered anoxia tolerance.

4.2. Field applications

In coastal monitoring programmes a variety of biological effect or stress parameters (sometimes called biomarkers) at all levels of biological organisation have been used to determine the adverse effects of contaminants. In The Netherlands the 'survival in air' response of mussels has been included in a coastal monitoring programme in which mussels are exposed at locations with a varying degree of pollution on a regular basis (Figure 4) (Smaal *et al.*, 1991; Eertman *et al.*, 1993a).

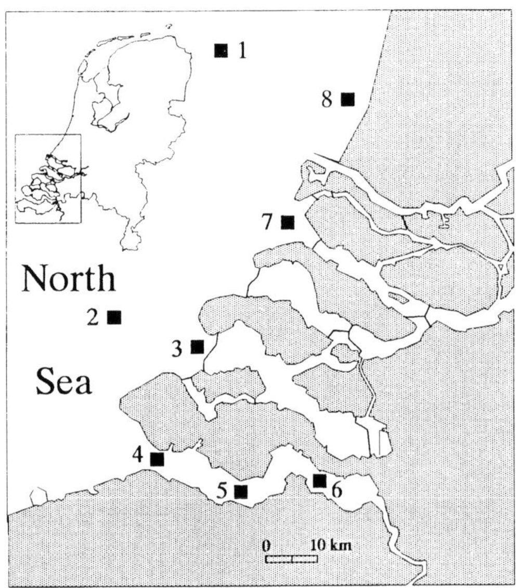

Figure 4: Topographic map of study area in the south-western part of The Netherlands. Exposure sites: [1] and [2]: North Sea (NS-1 and NS-2); [3]: Oosterschelde (OS); [4], [5] and [6]: River Scheldt estuary (WS-west, centre and east); [7]: entrance to the Haringvliet (HV), a basin receiving water from the rivers Rhine and Meuse; [8]: a station in the northerly coastal current (NAM)

Tolerance to aerial exposure showed a negative correlation with increasing tissue contaminant levels at relatively low levels of contamination. Figure 5 shows an example of the 'survival in air' response of mussels following six weeks of exposure in Dutch coastal waters in the autumn of 1990. Significant correlations could be demonstrated between tissue contaminant levels, especially for PCBs and PAHs, and LT_{50} values. Mussels exposed at relatively unpolluted North Sea locations possessed the best tolerance to aerial exposure.

Table 1: Range of tissue contaminant concentrations found in mussel species around the world. Concentration of ΣPCBs and ΣPAHs in μg.kg^{-1} DW, Cd in mg.kg^{-1} DW

Country	ΣPCBs	ΣPAHs	Cd	reference
Netherlands	25 - 430	30 - 1,300	0.3 - 2.5	1
Finland		440 - 1,200*)		2
USA-east	low - 930	low - 3,600	1.0 - 6.0	3
USA-west	low - 4,250			3
Norway	76 - 276	2,200 - 15,400	1.5 - 2.7	4
France #)	250 - 1,350	4,000 - 20,000	1,0 - 4.0	5
Hong Kong	23 - 4,480*)			6

Tissue concentrations of contaminants were measured in *M. edulis* unless stated otherwise. Data taken from: 1) Eertman *et al.*, 1993a; 2) Rainio *et al.*, 1986; 3) Farrington *et al.*, 1983 [USA-west: *M. edulis* and *M. californianus*]; 4) Widdows & Johnson, 1988; 5) Claisse & Simon, 1991; 6) Tanabe *et al.*, 1987 [*Perna viridis*]

*) Concentrations were transformed from WW to DW, assuming a WW:DW ratio of 8

#) read: lower than - higher than.

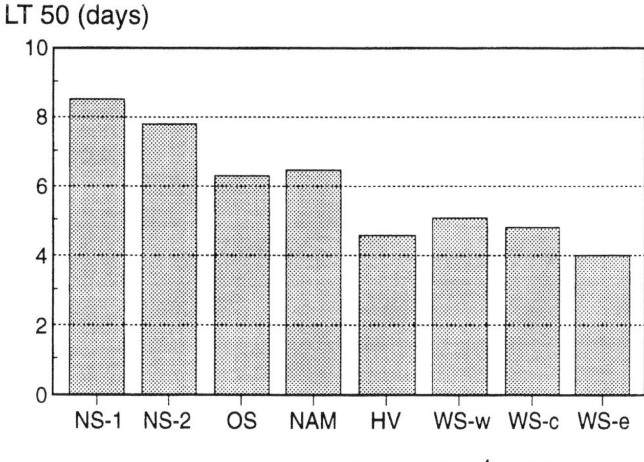

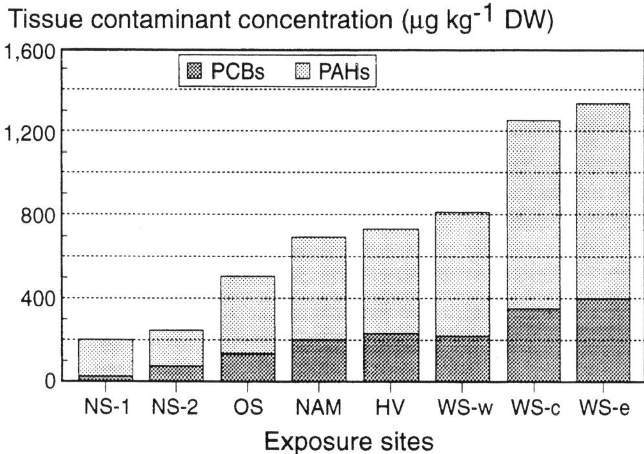

Figure 5: Tolerance to aerial exposure and tissue PCB and PAH concentrations in mussels following 6 weeks exposure in Dutch coastal waters. Abbreviations: see Figure 3

LT_{50} values declined at the exposure site near the entrance of the Oosterschelde and the stations under the influence of the Haringvliet basin (HV and NAM in Figure 5), whereas mussels exposed at the three stations in the polluted River Scheldt estuary (WS), with increasing contaminant levels from west to east, had the lowest LT_{50} values.
In comparison to field surveys in other parts of Europe or the (industrialized) world where mussels were used as a bio-indicator species, the tissue contaminant levels in The Netherlands belong to the lower ranges of values measured (Table 1). Although data from other field applications of the 'survival in air' response parameter are not available, it is expected that the 'survival in air' response of mussels has world wide applicability.

CONCLUSIONS

Biochemical adaptation mechanisms contributing to anoxic survival are possible targets of stress. Mechanisms involved include:

- control of glycolytic rate depression in order to largely arrest metabolism;
- the rerouting of the carbon flow from the glycogen degradation pathway to enhance anaerobic ATP yield;
- the formation and maintenance of high reserves of fermentable fuel (glycogen and to some extend aspartate).

Experimental evidence is presented in order to illustrate that the 'survival in air' response of mussels is a simple and low-budget method for the investigation of whether environmental variables (single or in combination) might have perturbing effects on cellular processes which lead to reduced fitness of the individual.

The first author's work was performed under contract from the BEON*EFFECT project of the Tidal Waters Division, Ministry of Transport and Public Works. This is publication no. 678 of NIE-CECE.

REFERENCES

Baker, S.M. & R. Mann, 1992. Effects of hypoxia and anoxia on larval settlement, juvenile growth, and juvenile survival of the oyster *Crassostrea virginica.* Biol. Bull. 182: 265-269

Bayne, B.L., 1976. Aspects of reproduction in bivalve molluscs. In: Estuarine Processes 1; M. Wiley (ed). Academic Press, London, pp. 432-448

Bayne, B.L., 1980. Physiological measurements of stress. Rapp. P.-v. Réun. Cons. int. Explor. Mer, 179: 56-61

Bestwick, B.W., I.J. Robbins & L.M. Warren, 1989. Metabolic adaptations of the intertidal polychaete *Cirriformia tentaculata* to life in an oxygen-sink environment. J. Exp. Mar. Biol. Ecol. 125: 193-202

Borsa, P., Y. Jousselin & B. Delay, 1992. Relationships between allozymic heterozygosity, body size, and survival to natural anoxic stress in the palourde *Ruditapes decussatus* L. (Bivalvia: Veneridae). J. Exp. Mar. Biol. Ecol. 155: 169-181

Brinkhoff, W., K. Stöckmann & M. Grieshaber, 1983. Natural occurrence of anaerobiosis in molluscs from intertidal habitats. Oecologia, 57: 151-155

Brooks, S.P.J., A. de Zwaan, G. van den Thillart, O. Cattani & K.B. Storey, 1991. Differential survival of *Venus gallina* and *Scapharca inaequivalvis* during anoxic stress: Covalent modification of phosphofructokinase and glycogen phosphorylase during anoxia. J. Comp. Physiol. 161(B): 207-212

Claisse, D. & S. Simon, 1991. Le 'Mussel Watch' français: résultats acquis sur les moules et les huîtres du littoral de la France. Exploitation de ces résultats dans le cas particulier de l'estuaire de la Seine. In: Estuaries and coasts: Spatial and temporal intercomparisons; M. Elliot & J.-P. Ducrotoy (eds). ECSA19 Symposium, University of Caen, France, pp. 341-347

de Zwaan, A., 1983. Carbohydrate catabolism in bivalves. In: The Mollusca 1; P.W. Hochachca (ed). Academic Press, New York, pp. 137-175 de Zwaan, A., 1989. Physiological and biochemical approaches to the assessment of pollution of European estuarine and coastal systems. Progress report of EC Research Contract EV4V.0122/NL(GDF), pp. 43

de Zwaan, A. & W.C. de Kock, 1988. The development of a general biochemical stress index. Mar. Environ. Res. 24: 254-255

de Zwaan, A. & M. Mathieu, 1992. Cellular biochemistry and endocrinology. In: The mussel *Mytilus*: ecology, physiology, genetics and culture. Developments in aquaculture and fisheries science 25; E. Gosling (ed). Elsevier, Amsterdam, pp. 223-307

de Zwaan, A. & V. Putzer, 1985. Metabolic adaptations of intertidal invertebrates to environmental hypoxia (a comparison of environmental anoxia to exercise anoxia). In: Physiological adaptations of marine animals 39, Symposia of the Society of Experimental Biology; M.S. Lavarack (ed). Company of biologists Ltd., University of Cambridge, pp. 33-62

de Zwaan, A. & T.C.M. Wijsman, 1976. Anaerobic metabolism in Bivalvia (Mollusca). Characteristics of anaerobic metabolism. Comp. Biochem. Physiol. 54(B): 313-324

de Zwaan, A. & D.I. Zandee, 1972. Body distribution and seasonal changes in the glycogen content of the common mussel *Mytilus edulis*. Comp. Biochem. Physiol. 43(A): 53-58

de Zwaan, A. & W. Zurburg, 1981. The formation of strombine in the adductor muscle of the sea mussel *Mytilus edulis*. Mar. Biol. Lett. 2: 179-192

de Zwaan, A., A.M.T. de Bont & J. Hemelraad, 1983. The role of phosphoenolpyruvate carboxykinase in the anaerobic metabolism of the sea mussel *Mytilus edulis* L. J. Comp. Physiol. 153: 267-274

de Zwaan, A., O. Cattani, G. Isani & P. Cortesi, 1991a. Survival and anaerobic metabolism of the arcid clam *Scapharca inaequivalvis* under anoxia and oxia in the presence of sulphide and cyanide. In: Cellular defence strategies to hypoxia; P.L. Lutz (ed). Second IUB Satellite Symposium, Noordwijkerhout, Netherlands, 1991. Abst., p. 17.

de Zwaan, A., P. Cortesi, G. van den Thillart, J. Roos & K.B. Storey, 1991b. Differential sensitivities to hypoxia by two anoxia-tolerant marine molluscs: a biochemical analysis. Mar. Biol. 11: 343-351

Eertman, R.H.M., A.J. Wagenvoort, H. Hummel & A.C. Smaal, 1993a. 'Survival in air' of the blue mussel *Mytilus edulis* L. as a sensitive response to pollution induced environmental stress. J. Exp. Mar. Biol. Ecol., 170: 179-195

Eertman, R.H.M., M.L.F.M.G. Groenink, B. Sandee, H. Hummel and A.C. Smaal, 1993b. Response of the blue mussel *Mytilus edulis* L. following exposure to PAH's or contaminated sediment. *Mar. Environ. Res.* (in press)

Farrington, J.W., E.D. Goldberg, R.W. Risebrough, J.H. Martin & V.T. Bowen, 1983. U.S. "Mussel Watch" 1976-1978: An overview of the trace-metal, DDE, PCB, hydrocarbon, and artificial radionuclide data. Environ. Sci. Technol. 17: 490-496

Gaffney, P.M., 1990. Enzyme heterozygosity, growth rate, and viability in *Mytilus edulis*: another look. Evolution, 44: 204-210

Ghisotti, F. & E. Rinaldi, 1976. Osservazione sulla populazione di *Scapharca*, insediatasi in questi ultimi anni su un tratto del litorale Romagnolo. Conchiglie, 12: 183-195

Groenendaal, M., 1980. Tolerance of the lugworm (*Arenicola marina*) to sulphide. Neth. J. Sea Res. 14: 200-207

Hamilton, M.A., R.C. Russo & R.V. Thurston, 1977. Trimmed Spearman-Karber method for estimating median lethal concentrations in toxicity bioassays. Environ. Sci. Technol. 11: 714-719

Kaplan, E.L. & P. Meier, 1958. Nonparametric estimation from incomplete observations. J. Amer. Statist. Assoc. 53: 457-481

Kluytmans, J.H. & D.I. Zandee, 1983. Comparative study of the formation and excretion of anaerobic fermentation products in bivalves and gastropods. Comp. Biochem. Physiol. 75(B): 729-732

Koehn, R.K. & B.L. Bayne, 1989. Towards a physiological and genetical understanding of the energetics of the stress response. Biol. J. Linn. Soc. 37: 157-171

Lane, D.J.W., A.R. Beaumont & J.R. Hunter, 1985. Byssus drifting and the drifting threads of the young post-larval mussel *Mytilus edulis*. Mar. Biol. 84: 301-308

Livingstone, D.R., 1982. Energy production in the muscle tissue of different kinds of molluscs. In: Exogenous and endogenous influences on metabolic and neural control 1; A.D.F. Addink & N. Spronk (eds). Pergamon Press, Oxford, pp. 257-274

Meinardus, G. & G. Gäde, 1981. Anaerobic metabolism of the cockle, *Cardium edule*. IV. Time dependent changes of metabolites in the foot and gill tissue induced by anoxia and electrical stimulation. Comp. biochem. Physiol. 70(B): 271-277

Oescher, R., 1990. Long-term anaerobiosis in sublittoral marine invertebrates from the Western Baltic Sea: *Halicryptus spinulosus* (Priapulida), *Astarte borealis* and *Arctica islandica* (Bivalvia). Mar. Ecol. Prog. Ser. 59: 133-143

Rainio, K., R.R. Rainio & L. Ruotsila, 1986. Polycyclic aromatic hydrocarbons in mussel and fish from the Finnish Archipelago Sea. Bull. Environ. Contam. Toxicol. 37: 337-343

Rhodes, E.W. & J.J. Manzi, 1989. Interstate shipment of larval and juvenile bivalves: effects of shipping duration and method on survival. J. Shellfish Res. 7: 130-131

Seed, R., 1975. Reproduction in *Mytilus* (Mollusca; Bivalvia) in European waters. Pubbl. Staz. Zool. Napoli, 39(Suppl.): 317-334

Shick, J.M., J. Widdows & E. Gnaiger, 1988. Calorimetric studies of behaviour, metabolism and energetics of sessile intertidal animals. Amer. Zool. 28: 161-181

Shumway, S.E., T.M. Scott & M.J. Shick, 1983. The effects of anoxia and hydrogen sulphide on survival, activity and metabolic rate in the coot clam, *Mulina lateralis* (Say). J. Exp. Mar. Biol. Ecol. 71: 135-146

Smaal, A.C., A. Wagenvoort, J. Hemelraad & I. Akkerman, 1991. Response to stress of mussels (*Mytilus edulis*) exposed in Dutch coastal waters. Comp. Biochem. Physiol. 100(C): 197-200

Stebbing, A.R.D., 1979. An experimental approach to the determinants of biological water quality. *Phil. Trans. R. Soc. London Ser. B*, Vol. 286, 465-481.

Storey, K.B. & J.M. Storey, 1990. Metabolic rate depression and biochemical adaptation in anaerobiosis, hibernation and estivation. Quart. Rev. Biol. 65: 145-174

Tanabe, S., R. Tatsukawa & D.J.H. Phillips, 1987. Mussels as bioindicators of PCB pollution: a case study on uptake and release of PCB isomers and congeners in green-lipped mussels (*Perna viridis*) in Hong Kong waters. Environ. Pollut. 47: 41-62

Theede, H., A. Ponat, K. Hiroki & C. Schlieper, 1969. Studies on the resistance of marine bottom invertebrates to oxygen-deficiency and hydrogen sulphide. Mar. Biol. 2: 325-337

Veldhuizen-Tsoerkan, M.B., D.A. Holwerda & D.I. Zandee, 1991. Anoxic survival time and metabolic parameters as stress indices in sea mussels exposed to cadmium or polychlorinated biphenyls. Arch. Environ. Contam. Toxicol. 20: 259-265

von Brand, T., 1946. Anaerobiosis in invertebrates. No. 4 of the 'Biodynamica monographs'. B.J. Luyet (ed). Published by Biodynamica, Normandy 21, Missouri, pp. 328.

Wang, W.X. & J. Widdows, 1991. Physiological responses of mussel larvae *Mytilus edulis* to environmental hypoxia and anoxia. Mar. Ecol. Prog. Ser. 70: 223-236

Wang, W.X., J. Widdows & D.S. Page, 1992. Effects of organic toxicants on the anoxic energy metabolism of the mussel *Mytilus edulis*. Mar. Environ. Res. 34: 327-331

Weber, R.E., A. de Zwaan & A. Bang, 1992. Interactive effects of ambient copper and anoxic, temperature and salinity stress on survival and haemolymph and muscle tissue osmotic effectors in *Mytilus edulis*. J. Exp. Mar. Biol. Ecol. 159: 135-156

Widdows, J. & J.M. Shick, 1985. Physiological responses of *Mytilus edulis* and *Cardium edule* to aerial exposure. Mar. Biol. 85: 217-232

Widdows, J. & D. Johnson, 1988. Physiological energetics of *Mytilus edulis*: Scope for growth. Mar. Ecol. Prog. Ser. 46: 113-121

Zandee, D.I., J.H. Kluytmans, W. Zurburg & H. Pieters, 1980. Seasonal variations in biochemical composition of *Mytilus edulis* with reference to energy metabolism and gametogenesis. Neth. J. Sea Res. 14: 1-29

Zange, J., H.O. Pörtner & M.K. Grieshaber, 1989. The anaerobic energy metabolism in the anterior byssus retractor muscle of *Mytilus edulis* during contraction and catch. J. Comp. Physiol. 159(B): 349-358

Appendix 12

'Survival in Air': a Practical Approach

SELECTION OF ORGANISMS

In biological monitoring programmes, transplanting mussels (*e.g. Mytilus* spp.) from a single reference population to contaminated locations is preferred above using resident mussel populations. Mussels from a single reference population have lower genetic variability between animals in comparison to animals from various field populations.

The 'survival in air' test should preferably be performed using subtidally acclimatized mussels. When acutely exposed to air these animals close their valves and become largely anoxic as a result of which tolerance to aerial exposure is predominantly related to anaerobic processes (Shick *et al.*, 1988). Intertidal mussels are less suitable for the determination of the 'survival in air' response. As intertidal mussels are capable of oxygen consumption during aerial exposure, accounting for up to 40 % of total heat dissipation in air (Shick *et al.*, 1988), the aerial survival time will be influenced by both aerobic and anaerobic processes. Additionally, mussels that are periodically exposed to air may have adapted genetically to improve tolerance to this stressful condition.

In order to exclude age dependent variation in tolerance to aerial exposure, about 50 adult mussels from one single length class (*e.g.* 40 to 50 mm) should be selected for each monitoring site. Subtidally acclimatized mussels may be obtained from a commercial mussel farm or, as *e.g.* in The Netherlands, from the Department of Fisheries.

PRE-TREATMENT OF MUSSELS

The aerial survival time may be affected significantly by variations in transportation time. In coastal monitoring programmes the various collection and/or exposure sites are often distributed over a large area. Mussels from locations in the vicinity of the laboratory may be transported to the laboratory within a few hours, while it may take much longer (*e.g.* 20 hours) for mussels from distant locations to reach the laboratory. It is important that mussels from all locations are treated equally before the start of the 'survival in air' test. Therefore, following collection or retrieval of mussels in the field, it is imperative that the transportation time back to the laboratory and conditions during transportation be standardized.

Immediately following collection, the mussels from each field location should be wrapped in paper tissues (*e.g.* disposable nappies) and stored and cooled between freezepacks in a polystyrene box. The paper tissue prevents the mussels from freezing. When stored in this way, the temperature of the mussel compartment will vary between 4 and 6 °C. At this temperature the organisms will have low metabolic activity.

Mussels from all field locations should arrive at the laboratory within 24 hours after collection. In case transportation from a distant location takes longer than 24 hours, the mussels from this location should be excluded from the 'survival in air' test. In order to standardise for the various transportation times, the boxes are left untouched until the standard 24 hour transport/storage period has elapsed. At the completion of the storage period the boxes can be opened and the mussels are transferred to tanks with running seawater, where they are allowed to recover for a period of 24 hours.

AERIAL EXPOSURE

At the start of each test, 30 mussels are collected at random from each experimental group and placed on a tray. The mussels are damp dried with paper tissue and the tray is placed in a constant temperature room, incubator or suchlike. The selected temperature should be recorded continuously, or at least checked daily when continuous recording facilities are not available. Temperature fluctuations of more than 0.5 °C should be avoided. It is recommended that the air temperature be recorded continuously. Preferably, humidity should be controlled as well to eliminate excessive dehydrating effects.
Daily observations are made of the number of animals alive. Animals are considered dead when shell-gape occurs and an external stimulus (*e.g.* prodding tissues with a probe or squeezing the shells) does not generate any closure response. Occasionally, dead animals do not show shell-gape. Therefore, it is necessary to observe whether any animals produce a specific smell indicating decay. As it is difficult to be certain about the death of a bivalve, the animals considered dead should be placed in a tank with running seawater for 24 hours in order to observe whether any animals recover. Organisms that do recover were obviously not yet dead and should be excluded from the experiment. Accurate recordings should be made of the time at which the animals are inspected each day. This information is necessary for the exact determination of the LT_{50}.
It should be remembered that instead of 'survival in air', survival under anoxia in nitrogen purged seawater under flow through conditions may be used as an alternative. Temperature regulation may be easier following this method. However, it is anticipated that different results, to those arrived at in the 'survival in air' method will be obtained.

STATISTICS

As the survival time of animals does not conform to a normal distribution range of data, parametric tests such as Student's 't-test' or ANOVAs may not be applied to test the data statistically. In medical research the nonparametric Kaplan-Meier test (Kaplan & Meier, 1958) has proven to be an accurate method for estimating survival curves. A confidence limit of 95 % is used to test the significance of differences between groups. Figure 3 shows an example of survival curves of mussels following 4 weeks laboratory exposure to 1 $\mu g.l^{-1}$ ΣPCBs. LT_{50} values can be estimated using the trimmed Spearman-Karber method (Hamilton *et al.*, 1977). To prevent the LT_{50} estimation from being influenced disproportionally by irregular observations the a % extreme values are trimmed from each tail of the tolerance distribution (usually a = 10 %). The trimmed Spearman-Karber test has maintained the good properties of the conventional Spearman-Karber estimator, but is less sensitive to unusual responses of individual mussels.

Chapter 13

Monitoring of Hard Substrate Communities

Rong-Quen Jan[1], **Chang-Feng Dai**[2] and **Kun-Hsiung Chang**[1]

[1] Institute of Zoology, Academia Sinica,
Taipei, Taiwan, Republic of China
[2] Institute of Oceanography, National Taiwan University,
Taipei, Taiwan, Republic of China

ABSTRACT

Marine organisms living on hard substrates generally lack mobility, therefore they are less able to avoid harmful conditions. Also, they are relatively sensitive to pollution and usually accumulate contaminants. Such properties taken as a whole mean that they can function as suitable indicators for detection of environmental stress. In practice, information on changes in species composition, species diversity and the relative abundance of those species can be useful when it comes to measuring the nature and extent of environmental impacts on a community. Similarly, some groups of organisms, or single species, which play a central role in the structuring process of a community, can also function as good indicators.
The idea of using sessile communities as stress indicators has been widely applied in field studies in order to demonstrate that man-made stress accounts for a variety of observed changes.
A successful monitoring programme on a sessile community normally involves the use of a wide spectrum of appropriate techniques. Experience gained with such techniques are presented here, while case studies in several coastal environments which show the effects of ecological fluctuation, sewage, oil pollution and thermal stress demonstrate their application. In order to assist in the development of a practical monitoring programme, technical details on sampling are summarized at the end of the chapter.

1. INTRODUCTION

Succession, which may be principally defined as the non-seasonal, directional and continuous pattern of colonization and extinction on a site by species populations (Connell & Slayter, 1977; Begon *et al.*, 1986), is often used to describe the community patterns in time.

Biomonitoring of Coastal Waters and Estuaries. Edited by Kees J.M. Kramer.

Succession is mainly governed by underlying mechanisms involving life history characteristics such as recruitment and individual growth of the organism, as well as various environmental factors (Connell & Slayter, 1977; Dayton, 1984; Witman, 1985). When unfavourable environmental conditions arise, they are likely to cause stress to a species, which leads to changes in the succession of its population. In cases, species will not be able to survive if appropriate conditions are not available or resources do not exist in that environment (Begon *et al.*, 1986).

Many ideas on the biomonitoring of the effects of man-made disturbances to coastal waters and estuaries can be traced back to the concept of ecological succession, particularly to that of communities of species (Lewis, 1978; Pearson & Rosenberg, 1978; Nichols, 1979; Westman, 1985; Jan & Chang, 1991). Communities of benthic and/or sessile organisms, such as sea grass, algae, corals, polychaetes, ascidians, etc., are good stress indicators (Jones *et al.*, 1980; Brown, 1986; Burd *et al.*, 1990; García & Salzwedel, 1991). Since the majority of these organisms lack mobility, they are less able to avoid potentially harmful conditions than are the non-sessile organisms such as fish. Thus, these organisms can function as indicators for detection of long-term environmental effects.

Inshore benthic communities are closely associated with the pelagic food web. There they constitute a link for the transport of contaminants to higher trophic levels. This also suggests that organisms at such a trophic position would normally accumulate contaminants earlier than those at higher levels. In addition, benthic/sessile organisms are highly sensitive to some man-made environmental disturbances, *e.g.* sedimentation and sewage run off, that have a profound influence on the quality of the substrata which is an essential for larval settlement (Rogers, 1990). Therefore assemblages of benthic organisms form bio-indicators which are particularly suitable for monitoring such environmental changes.

Of all the coastal substrates, those made up of rocks or coral reefs are the most densely inhabited by macro-organisms (Dayton, 1984; Sebens, 1985; Buss, 1986). Organisms commonly observed on a hard substrate comprise algae, sponges, cnidarians, polychaetes, crustaceans, molluscs, echinoderms, ascidians, etc. (Sebens, 1985). Communities on hard substrates form a striking contrast to those on soft bottoms (Parker, 1975). Most animals live below the sediment surface in a soft substrate and are uncovered only when some appropriate sampling method is used or a disturbance of the sediment takes place (Parker, 1975; Burd *et al.*, 1990). When and where hard substrates occur their communities may be considered as suitable stress indicators, due to the fact that these areas are relatively easy to access, and the organisms are readily identifiable.

In order to assess environmental effects, baseline data must be available for relevant local communities. Ecological processes underlying the community structure are complex. These primarily involve slow processes, rare events (episodic phenomena) and high annual variability. To detect such phenomena, long-term data sets are often required (Jackson *et al.*, 1989; Likens, 1989; Southward, 1991).

This is particularly true for the encrusting communities of long-lived bryozoans, compound ascidians, colonial cnidarians and sponges (Dayton, 1984). Long-term data sets are also essential for the demonstration of the relative contribution of different factors. Without such knowledge it will be very difficult to separate pattern from noise. In the case of drastic environmental changes, which cause instant damage to organisms, short-term data might be useful for demonstrating these acute effects. On the other hand, it may take a decade to establish a trend in a rocky shore environment and several decades to detect a change in a coral reef. Data collected over a relatively short period, for example two or three years, may not be revealing in terms of detecting change or in detecting trends. Sometimes such data could even be misleading (Southward, 1991).

Most hard substrata in both intertidal and sub-tidal zones, are made up of boulders, coral reefs and rocky extensions of land masses. Artificial hard substrates can be found *e.g.* along dykes, quays and piers. These habitats vary greatly in size, and are separated from other habitats like mud or sand. Hard substrates are used by a variety of plants and sessile animals. Some general characteristics can be observed in marine sessile communities. Many organisms living in the intertidal zone are solitary. In addition to the solitary form, the colonial form is observed for many organisms in the sub-tidal zone in tropical and sub-tropical environments (Sebens, 1985). In contrast, solitary organisms are abundant in the temperate regions. Sessile colonial organisms pervade open substratum by vegetative growth of existing colonies or settlement of propagules. Many of these animals and marine algae release large numbers of free-swimming propagules into the sea which, after dispersal, settle on rocks or on the surfaces of other animals and plants. The length of the planktonic stage varies depending on, for the most part, the specificity of the species itself. Ecological conditions also contribute to modification of the length of the planktonic stage. This planktonic stage, although particularly susceptible to predation and environmental stresses, is assumed to add chance elements to the recruitment and to enhance species dispersal (Scheltema, 1986).

In the marine environment biological succession can vary due to physical and physiological stress (wave exposure, nutrient availability, light, tidal height) and to biological factors (recruitment, predation, competition, mutualism), and their integrated effect. Space available for attachment and/or resources associated with open substrates are often assumed to be the prime limiting factors in both the intertidal and the sub-tidal areas. A variety of mechanisms may apply in the competition for substrate and such competition is one of the major forces structuring sessile communities (Buss, 1986; Dai, 1990).

Many hypotheses have been proposed in order to try to account for the community structure of marine organisms. Most have emerged in the contexts of chance elements, resource partitioning, disturbance and density-dependant life history parameters. Parallel hypotheses based heavily on the assumption of either resource partitioning (Mapstone & Fowler, 1988; Jan, 1989; *cf.* Clarke, 1988), or recruitment-limitation (Victor, 1986; Mapstone & Fowler, 1988) have been proposed on the structuring processes of a fish community. Assumptions for hypotheses on sessile community structure have centred on either natural disturbance (Connell, 1978; Connell & Keough, 1985; Sousa, 1984), food web regulation (Paine, 1980), competition for space (Paine, 1984), or recruitment limitation (Gaines & Roughgarden, 1985; Sutherland, 1987) and a combination of these factors.

Although any of the factors proposed in these hypotheses may be critical to the co-existence of sessile organisms, it does not necessarily mean that each forms a sufficient reason on its own, to explain the complexity of the community structure. Different factors may operate at different times or places, or simultaneously at the same place and time (Matthews *et al.*, 1982).

However, information on changes in community species composition, species diversity and the relative abundance of those species can be useful in measuring the nature and extent of impacts on that community (Herricks & Cairns, 1982; Westman, 1985; Smith *et al.*, 1988). For example, in a community exposed to stress, species diversity is expected to decrease, coupled with an increase of species dominance. Moreover, both the size and life span of organisms might also decrease (Freedman, 1989). Some animals and plants may give an indication of potential effects by their absence. Some are good indicators because they have narrow tolerances to physiochemical factors, or have a restricted distribution because of narrow habitat requirements (Spellerberg, 1981; Herricks & Cairns, 1982; Stephens *et al.*, 1988). The use of 'keystone' species has been proposed to identify the linkage of species in food webs. A keystone species not only indicates community characteristics (as an indicator species would), but also plays a role in controlling community structure (Paine, 1984). When an indicator species is selected, the examination of individual or colony size frequencies is one approach which can provide information on the response of the population to favourable or unfavourable conditions (Hughes, 1985; Hughes & Connell, 1987). When evidence of a decline appears or if there is a complete absence of any recolonization, it may be indicative of a substantial environmental change.

Some symbiotic relationships may also be considered as a bio-indicator. For example, most reef building corals harbour symbiotic dinoflagellates, the zooxanthella. Zooxanthellae contain chlorophylls, which facilitate photosynthesis but also add various colorations to host corals. When corals are exposed to *e.g.* sublethal hyperthermal or hypothermal environments, the symbiotic zooxanthellae will be expelled. Such bleaching effects have been widely recognized as an indicator for both the seriousness caused by, and the range of thermal pollution (Coles *et al.*, 1976; Glynn, 1983).

With the perception gained from bio-monitoring, this review is arranged into three parts. Firstly, techniques which cover a spectrum from fieldwork to data interpretation are described for monitoring marine sessile communities. Secondly, various anthropogenic stresses and their impacts on such communities are reviewed along with discussions on the processes likely to account for the observed pattern changes. Thirdly, case studies on biological fluctuation, sewage, oil and thermal stress are described in order to give examples of monitoring of hard substrate communities.

The emphasis in this chapter is placed upon coral reef communities. However, many of the ideas supporting it may also apply to coastal and estuarine ecosystems where hard substrates occur.

2. TECHNIQUES FOR MONITORING SESSILE COMMUNITIES

In the planning phase two sets of questions usually need to be answered. One concerns the proposed methods of data collection (sampling), and the other is concerned with the (statistical) analysis and interpretation of these data (Spellerberg, 1991). Clear answers to these questions would certainly help in the initiation of the programme.

2.1. Data collection

2.1.1. General considerations

Where changes in a sessile community are concerned, there may be a need for long-term data collection. Such a monitoring programme is bound to be a laborious occupation and it is worth summarizing the reasoning involved in setting it up. In designing the sampling programmes for hard substrate communities, the following principles need to be considered:

- methods of sampling should be tested, and adopted if they give a good quantitative representation of the flora and the fauna (Westman, 1985). When species abundance is considered as the prime parameter, the number of species sampled should be as large as possible and should include all sedentary species;
- the sampled unit should be the species level and the recorded data should retain this information rather than contracting it into an index, or other aggregated form to validate data reexamination (Shulman, 1983);
- sampling methods should be objective, repeatable and standardized so that different workers can produce similar results for the same test quadrates;
- destructive sampling should be avoided;
- it is suggested that the sampling programme be planned if possible 12 months in advance. Maintaining a standardized programme, that is carried out regularly, is considered be very important and the investigators would have to give this work priority. In practice, this would keep a full-time staff occupied.

2.1.2. Selection of study sites

The selection of the sites should be based on the need to cover a wide range of micro-habitats. Particular attention should be paid to sites close to areas of human activity and places of special scientific interest.

Characteristics of the local community should be taken into account when it comes to site selection. Intertidal or sub-tidal sites, or both, could be considered. Manual collection of samples is usually essential both to be sure of having an adequate record of the organisms, and to prevent causing as much damage as possible. For intertidal areas, low tide periods offer opportunities for observation, but in sub-tidal areas divers are necessary. Studying the rocky intertidal habitats has some disadvantages, *e.g.* sometimes hazardous working conditions and periodic inaccessibility occur. In contrast, however, sub-tidal studies face more constraints. These include viewing limitations (underwater visibility, size overestimation); mobility and stability problems (currents, bottom surge); communication problems (recording, cooperation of divers with one another) and time restrictions (air supply, decompression, water temperature) (John *et al.*, 1980). Though these constraints will not limit the scope of sub-tidal monitoring, they obviously are among the primary factors affecting the selection of a study site as well as the design of a field study.

Changes in the locations of sampling sites should be avoided and the continuity of sampling over as long a period as possible should be ensured.

2.1.3. Sampling frequency

Samples collected in a survey undertaken in one summer do not form a well-founded base for the evaluation of the effect, for example, of an oil spillage on the site in mid-winter.

Cyclical changes over periods of years can be demonstrated only by sampling over a considerably longer period than the duration of one cycle. On the other hand, long-period sampling raises many problems, including the continuity of control, interest and, in many cases, funds. Monthly sampling is most useful in the assessment of seasonal and other short term changes while quarterly and half-yearly sampling is less valuable and adds little to the information obtainable by an annual visit. A higher sampling frequency is also useful in that it allows the trend of change to be detected and minimizes the effects of occasional errors in sampling.

An annual review is suggested in order to evaluate the frequency of sampling, and to decide whether a change will improve data quality.

2.1.4. Survey methods for sampling at the community level

The sampling system chosen would have to allow rapid, objective quantitative sampling and, where different organisms require different methods, the variables recorded have to be commensurable. It is recommended that if species lists are prepared before and after any environmental impact, that they be compiled by the same worker. Comments should be given wherever possible on the basic characteristics of the species, *e.g.* the habitat type, location (or zone) of occurrence, the abundance or density, the reproductive status and the health of the organism. Assessing change necessarily implies that communities are sampled regularly by standard, repeatable methods. Many standard ecological field methods used in terrestrial vegetation studies are applicable. With slight modification, they may be used in underwater work (John *et al.*, 1980). Methods widely used in surveys of marine sessile organisms include collection along fixed transects or in fixed quadrates, photography and/or video recording and remote sensing. Quantitative data can be collected by various methods, *e.g.* random sampling, stratified random sampling, or systematic sampling (John *et al.*, 1980; Krebs, 1989). The survey methods and some supporting techniques are summarized here and described in more detail in the Appendix.

Fixed belt transects

The transect method involves the random placing of a line or tape along the substratum and recording everything it covers, either at a single point, at a particular distance, or the part of the line that covers a particulate organism (Figure 1). When supplemented with a pole, the transect can also be transformed into a survey zone (Figure 2) (Loya, 1978; John *et al.*, 1980; Jan & Chang, 1991).

Figure 1. The length of a coral colony was measured *in situ* in an underwater survey in Taiwan

Figure 2. A pole was used to highlight the zone along the transect in an underwater survey in Taiwan. Note that the pole was scaled to improve accuracy

Fixed quadrates at selected areas
Quadrates are squares placed on the substratum at several locations. Coverage of each species in each quadrate, in its simplest form in percentages, provides valuable information on biological succession (Sebens, 1986; Dai, 1991). The accuracy of quadrate data collection can be greatly improved with the help of a point frame (Figure 3).

Figure 3. A point frame method was demonstrated by dr. Peter Hogarth (middle right) in an intertidal survey at Millport, Scotland

Colonization studies
For many years, panels or blocks of cement, slate, earthenware, glass or various plastics have been suspended from rafts or fixed to solid surfaces in order to collect algae and sessile animals at the settling stages for use as an indication of recolonization (Chang *et al.*, 1977; Woodhead & Jacobson, 1985).

Photography and video recording
Photographs and video recordings of permanent transects or quadrates provide valuable information for the detection of the changes in a community (Holme & Barrett, 1977; George, 1980; Holme, 1984; Potts *et al.*, 1987).

Remote sensing
Remote sensing techniques range from satellite imagery to aerial photography (Hopley, 1978; Coulson *et al.*, 1980). Photo-interpretation often needs to be combined with information collected from fieldwork (Budd, 1991). It is not always necessary to use airplanes, which can be very expensive, for carrying out aerial photography. Zeppelins or even kites have proven successful.

2.1.5. Species identification
The basic parameters in almost all benthic ecological investigations, as in other community studies, include species lists and the abundance and biomass of each species. Environmental changes in either space or time often disturb community stability and give rise to changes in these parameters. One major problem of species listing is the taxonomic status of the groups.

Many marine sessile invertebrates are rare or are poorly described. The capability of species identification will hopefully increase with experience. The resulting 'lumping' or 'splitting' of old species names will result in changes of the listing. This will make it difficult to draw conclusions based on an increase or decrease in the number of species (Harding, 1991). Under such a circumstance, retrospective updating of the species list is suggested in order to systematize bias. In addition, further taxonomic checking may be necessary when apparently anomalous records appear in the data.

2.1.6. Abundance, frequency, cover, density

Abundance can be transformed into frequency, cover or density for comparisons (Jones *et al.*, 1980). By definition, frequency is a non-absolute measure. A change in frequency is very difficult to interpret because it is influenced by the size and shape of the sampling quadrate and, more importantly, by species density and distribution patterns. Cover is a parameter widely used in vegetation surveys and is defined as the proportion of ground occupied by the perpendicular projection on to it of the species under consideration. In marine sessile community studies, cover is as widely used to describe colonial plants and animals as density is for solitary ones.

2.1.7. Supporting (co)variables

The structuring process of a community is generally governed by both biotic and abiotic factors, so caution is needed in the selection of the environmental factors to be included in the sampling programme. Natural abiotic factors in the marine environment such as water temperature, salinity, light, turbidity, current speed, depth, substrate type, etc. all have a profound influence on the local sessile community. Moreover, interactions between factors, *e.g.* water temperature and depth, also add complexity to the interpretation.

2.1.8. Data storage and processing

In general, the amount of data collected by this type of monitoring programme is very large. Therefore, computers have been widely used in the storage, processing and retrieval of such data. The basic hardware/software requirement generally includes a rapid and easily up-dated storage/processing system, which has the ability to search data for inspection, and to yield data in a format compatible with popular statistical and other analytical programs.

A monitoring scheme generally produces the following types of data:

- species performance values (abundance or importance value);
- collection method: quadrates or transects;
- location and time.

Such a data set constitutes elements of performance value in a four dimensional matrix. For advanced comparisons, some summarization would be adopted due to the sheer quantity of material involved. Pooling of results is also necessary in order to indicate large-scale trends. However, pooling of data with high variance around the mean (abundance, biomass or other) should be avoided. Unfortunately, such data are frequently found, particularly when collection has been carried out at intertidal areas where zonation of the biota is distinct. Under such circumstances it is suggested that data collected at each zone (or patch) be treated separately in order to avoid loss of detail.

Removal of outliers may sometimes help smooth the data. However, this must be checked by referring back to the original data on species performance. One common assumption in parametric analyses is the normality of data distribution and homogeneity of variance. However, the truth often departs if data are collected from aggregated or clumped assemblages where most species do not follow a normal distribution.

In order to make this assumption conform, transformation of the data is widely practised (Burd *et al.*, 1990). Transformation of the data may also be used in order to avoid the risk of overemphasizing the dominant species (Ludwig & Reynolds, 1988). Methods commonly used include square root, cube root and log transformations. For zero entries in the data matrix, adding a small value to each entry (usually 0.1-1) will solve the undefined log 0 problem in the log transformation (Burd *et al.*, 1990).

2.2. Data analysis and interpretation

Analysis of changes in benthic communities which occur in response to environmental variations usually involves two steps. Firstly, the biological data are analyzed in order to determine community patterns and secondly, the environmental data are correlated with the observed community patterns.

Intuitive summation of biological information has been widely practised when indicator species or groups are used diagnostically for particular sets of conditions. Where variables for the dynamics of both control and experimental populations are available, goodness of fit tests can be used to test the differences between them. Various mathematical techniques can be used to summarize community data, including diversity indices, similarity indices, equatability indices and multivariate analysis (Digby & Kempton, 1987; Ludwig & Reynolds 1988; Smith *et al.*, 1988; Burd *et al.*, 1990; Warwick & Clarke, 1991). Many factors must be dealt with where community interpretation is concerned.

With the limited space available here, descriptions on analytical processes can only be described briefly below. For details of the analytical processes the reader is referred to *e.g.* Pielou (1984), Westman (1985), Digby & Kempton (1987), Ludwig & Reynolds (1988), Smith *et al.* (1988) and Burd *et al.* (1990).

2.2.1. Classification of communities

Communities can be compared using similarity indices or distance coefficients. Similarity indices are based solely on the presence or absence of data. It varies from a minimum of 0 (when two samples are completely different) to 1 (when two samples are identical). The distance coefficients are the opposite - they range from 0 (when two samples are identical) to a maximum value (in some cases infinity). Three groups of distance indices are commonly used in community studies: the Euclidean distance coefficient, the Bray-Curtis dissimilarity index and the relative Euclidean distance coefficient. The use of the relative Euclidean distance coefficient, which is expressed on standardized or relative scales, is recommended by Ludwig & Reynolds (1988) for its good performance over a diverse set of data.

The similarity or distance matrix alone holds limited value in determining community patterns (Williamson, 1987). Through further application in multivariate analysis, the similarity index can be transformed into dissimilarity, by a simple function: 1 - similarity.

Multivariate analyses such as cluster analysis and ordination techniques have been widely used in community pattern delineation. In application, both methods use the dissimilarity matrix as the starting point. Cluster analysis delimits groups of biologically similar samples (Pielou, 1984). It falls into four categories: agglomerative, divisive, constructive and direct optimization. Agglomerative, hierarchical cluster analysis is broadly used in benthic ecological studies. This method involves successive pairing of the most similar samples, or groups of samples, until all samples are in one group (Smith *et al.*, 1988). The results summarized by this process can then be conveniently plotted in a tree-like presentation known as dendrogram. Because this classification method lacks an objective criterion for determining the number of legitimate clusters, the number of groups is often determined by a level which has been subjectively selected along the similarity axis (Burd *et al.*, 1990).

Ordination techniques are used to display the biological patterns in a multidimensional space (Pielou, 1984; Smith *et al.*, 1988). The distance between any two points (representing two samples) in the space should be proportional to their dissimilarity. The dimensions are axes and the projections of the points onto the axes are scores. These methods attempt to display the maximum amount of the biological variation in the data in a minimum number of ordination axes. The axes are ordered according to the degree of variability among scores, with the first axis having the greatest variation, and the last axis the least (Westman, 1985; Smith *et al.*, 1988).

2.2.2. Environmental interpretation/correlation

In a monitoring study, when variation is found in community patterns, the next step is to provide an environmental interpretation for this variation. The structuring process of a community is generaly governed by both biotic and abiotic factors, so caution is needed in the selection of the environmental factors to be included in the analysis. Natural abiotic factors in the marine environment all have a profound influence on the local sessile community. Interactions between factors, *e.g.* water temperature and depth, also add complexity to the interpretation. Results from experimental community studies of well-defined environmental factors are relatively easy to interpret. These may include monitoring studies which have pollution singled out as the experimental factor.

There are several effective statistical approaches suitable for examining environmental effects. Analysis of variance (ANOVA) and multiple discriminant analysis are two commonly used methods; others include the t-test, the sign test, the Wilcoxon rank test, principal components analysis (PCA), canonical correlation etc. (Ludwig & Reynolds, 1988; Burd *et al.*, 1990; García & Salzwedel, 1991).

The One-Way ANOVA can be used to test the means of an environmental variable, *e.g.* concentration of a heavy metal across clusters. Rejection of the null hypothesis that the means are from the same population may lead to the implication that the environmental factors in question have contributed to the differences between clusters. The parametric ANOVA is strongly based on the assumptions of normality, and homogeneity of variance. Where these assumptions are not met, transformation of data before analysis will be necessary in order to avoid misinterpretation and error. Alternatively, non-parametric methods such as the Kruskal-Wallis test can be used. This test treats the numeric vector containing the environmental variable with ranks, and therefore it does not follow the assumption of data normality. Discriminate analysis is a multivariate analysis used in testing for significant differences in abiotic characteristics among communities delimited by classification. The test is conducted by computing a multivariate distance statistic and an F ratio. Under a number of simultaneous environmental factors, it has the ability to determine the relative importance of each in the separation of clusters. In application, careful selection of the environmental factors is crucial. Using a large number of variables introduces the risk that some may be highly correlated.

Ludwig & Reynolds (1988) recommend in the discriminate analysis the use of those environmental factors that are of high ecological rather than purely statistical importance. Finally, time series data are very useful in terms of understanding the long-term trend of a population or a community. In reality, however, apart from plankton and fishery statistics (Williamson, 1987; Cushing, 1990), little of such data are available in marine ecological studies.

3. ANTHROPOGENIC IMPACTS ON HARD SUBSTRATE COMMUNITIES

An evaluation of the living communities on a periodic basis may provide valuable information at the stage when a change in the marine regime has taken place.

Many references describe damage sustained by reef communities due to natural events, such as storms (Lassig, 1983; Connell & Keough, 1985) or man-made disturbances (Brown, 1986; Kuhlmann, 1988). Further discussion on the effects caused by natural disturbances is beyond the scope of the present review. Effects of trace metals upon coral reefs have been discussed by *e.g.* Howard & Brown (1984). In this section our area of interest is mainly anthropogenic stress and its effect upon coastal sessile communities caused by increased sedimentation, sewage, oil pollution and thermal stress. The emphasis in this section is placed on coral reef communities. However, many of the ideas supporting it may also apply to other coastal ecosystems where hard substrates are involved.

3.1. Sedimentation

Increased sedimentation is probably the most common and serious anthropogenic problem for organisms in coastal waters. Sediment stress frequently arises when activities such as dredging for maintainance and harbour construction take place.

Heavy sedimentation may also be caused by run-off from urban development along shorelines causing erosion (Salm & Clark, 1982; Sakai & Nishihira, 1991).

Suspended particulate matter in the water column will increase turbidity, cutting down on the light available for photosynthesis, and, by sedimentation, eventually increasing the sediment load on the organisms. Excessive sedimentation is likely to bury and smother bottom dwelling organisms, and alter both the physical and biological processes. As a result, it adversely affects the structure and function of the benthic ecosystem and causes degradation of sessile communities (Rogers, 1990).

Sedimentation can kill major reef building corals, leading to collapse of the reef framework. Many papers report a correlation between heavy sedimentation and fewer coral species, less live cover, lower coral growth rates, greater abundance of branching forms of coral, reduced coral recruitment, and decreased net productivity of corals *(e.g.* Loya, 1976; Birkeland, 1977; Cortes & Risk, 1985; Tomascik & Sander, 1987a,b). Consequently, the decline in the amount of shelter the reef can provide, causes a reduction in both the number of individuals and in the number of species that inhabit the reef. In addition, sedimentation constitutes a major controlling factor affecting the distribution and abundance of sessile organisms (Pearson & Rosenberg, 1978; Hubbard, 1986; Rogers, 1990) because different species often have different capabilities of clearing themselves of sediments or surviving low light levels (Loya, 1976; Hubbard, 1986). Sedimentation may reduce reproduction and recruitment of sessile organisms on hard substrates. Tomascik & Sander (1987a) suggested that lower light levels may inhibit the development of coral larvae by reducing the amount of energy available to maturing ova and embryos. The pelagic larvae of corals settle on hard substrates where they barely protrude above the surface. Consequently, they are more likely to be smothered by sediment deposition than are the adults. Wittenberg & Hunte (1992) showed that settlement rates and juvenile abundance were lower on high sediment reefs than on low sediment reefs. Just as differences appeared between species in their susceptibility to suspended sediments and the reproduction and recruitment hindrance it causes, excessive sedimentation may also result in a significant change to the species composition of a sessile community (Loya, 1976; Tomascik & Sander, 1987a).

Following a severe damage, sessile communities find it difficult to recover. Increased sedimentation is often accompanied by other stress, e.g. by pollutants present in the particulates. These stresses may delay or inhibit recovery (Rogers, 1990). For example, Maragos (1972) estimated that 80% of the coral communities in the lagoon in Kaneohe Bay, Hawaii, died because of a combination of dredging, increased sedimentation, and sewage discharge. Moreover, these stresses also delayed or prevented recolonization of coral communities (Maragos *et al.*, 1985).

Nevertheless, a number of studies exist which show that sediment stress does not always result in serious damage (Brown & Howard, 1985). For example, limited effects were detected after a major spill of 2200 tons of kaolin clay on a reef off the island of Hawaii (Dollar &

Grigg, 1981), during dredging activities in Diego Garcia Lagoon (Sheppard, 1980) and also while drilling through mud in the Palawan Islands, The Philippines (Hudson *et al.*, 1982). However, Brown & Howard (1985) argue that a more detailed and longer survey may reveal subtle changes in the community structure at these sites.

3.2. Sewage and agricultural drainage pollution

The discharge of sewage and agricultural drainage into coastal waters is experienced all over the world. Sewage, which brings large amounts of chemicals and small particles to the water, may or may not have had some treatment before being discharged.

The contents of sewage can be grouped roughly into three categories, *i.e.*, nutrient subsidy, toxic materials or toxic byproducts (from pesticide, herbicides, chlorine, or heavy metals, etc.) and sediments (Pastorok & Bilyard, 1985).

Impacts caused by sewage on sessile communities vary in accordance with its contents. However, it should be noted that negative effects of the three components will also involve a degree of synergism.

Nutrient enrichment of benthic communities produces various direct and indirect effects. On coral reefs, low levels of nutrient input usually causes an indirect stress, affecting the structure of coral communities through increased production at other trophic levels (Tomascik & Sander, 1985). When the nutrient levels are elevated, however, high abundance of benthic macro-algae and filamentous algae occur, together with phytoplankton bloom in the water column (Tomascik & Sander, 1987a). Increased primary production in the water column favours benthic filter feeders which, along with the abundant growth of benthic algae, may out-compete corals, coralline algae, and other reef building organisms.

Toxic effects on benthic organisms could result from one or more of the chemicals commonly found in sewage effluent, *e.g.* free chlorine, heavy metals, pesticides, PCBs, PAHs and petroleum hydrocarbons. Toxic substances may induce metabolic changes in benthic organisms, decrease rates of growth and reproduction or reduce their viability (Peakall, 1992). In addition, toxic substances accumulated in bottom sediments near sewage outfalls may also affect sessile communities. For instance, resuspension of sediments during storms may transfer toxic materials to the water column, increasing the likelihood of sessile biota being affected (Pastorok & Bilyard, 1985).

Suspended solids in sewage discharges originate from particles contained in effluents, particulate organic matter produced by nutrient enrichment and natural seston. The relative importance of these sources depends on the level of waste water treatment. Common effects of sediment stress on sessile communities is given in the previous section. It is difficult to separate the negative effects of one category of sediment from the other since they often happen simultaneously. After sewage diversion, recolonization of species and recovery of sessile communities are likely to occur. However, this is again governed by many factors, including the species composition of the community, location of the damaged habitat, intensity and frequency of the disturbance, availability of a reproductive population and the availability of a hard substrate for larval settlement. Because the relative weights of these and other related environmental factors vary between sites, the time taken for recolonization and recovery may also vary (Maragos *et al.*, 1985; Grigg & Dollar, 1990).

3.3. Oil pollution

Oil pollution of the sea mostly comes from sources such as tanker operations, tanker accidents, other shipping operations, offshore oil production, coastal oil refineries etc. (Dawson, 1980). In a few instances it has occurred due to (inter)national wars. It is estimated that some 5 million tonnes of petroleum hydrocarbons reach the world's seas and oceans each year (Clark, 1986). In the Amoco Cadiz supertanker accident, which occurred off the coast of

Brittany (France) in 1978, about one third of the 233,000 metric tonnes of oil spilled is estimated to have contaminated beaches, rocky shores and other parts of the marine environment (Ganning *et al.*, 1984).

Many studies of the effects of oil on benthic organisms have been conducted in the laboratory. With corals taken as an example, their response has been reported to include a decrease in growth rates, reproduction and colonization capacity (Loya & Rinkevich, 1979; Peters *et al.*, 1981), damage to tactile stimuli and normal feeding mechanisms (Reimer, 1975) and excessive mucus secretion (Peters *et al.*, 1981).

In contrast, some other studies suggest persistent sublethal effects of oil on corals (Bak & Elgershuizen, 1976; Elgershuizen & de Kruijf, 1976). Moreover, it is also likely that in the case of a major spill, benthic organisms could be threatened more by the clean-up with chemical detergents than by the oil itself (Lewis, 1971; Elgershuizen & de Kruijf, 1976). This was especially true for the old type of dispersants.

Results of field studies suggest harmful effects and lasting damage by oil to coral populations, as measured by abundance, mortality, reproduction and recruitment (Lewis, 1971; Loya, 1976; Jackson *et al.*, 1989; Guzman *et al.*, 1991). However, where reef communities are concerned, single-event episodes of oil exposure have rarely been documented to have detrimental effects. In a summary of effects of 16 oil spills near coral reefs, there is no indication of specific reports of damage to corals (National Research Council, 1985). One important reason for these conflicting results may lie in the problem of the temporal scale used in monitoring (Guzman *et al.*, 1991). In addition, the nature and the extent of biological consequences of oil spillage are likely to be determined by a complex interaction of factors, which include the type of oil, oil dosage, physical environmental factors, prevailing weather conditions, nature of biota, seasonal factors, prior exposure of the area to oil, presence of other pollutants and the type of remedial action taken, etc. (Loya & Rinkevich, 1980).

In the Red Sea, Loya (1976) reported on the recovery and recolonization patterns of an oil-stressed reef and a control reef following an unpredicted and catastrophic low tide. Ten years after the event, the control reef was fully recolonized with a community structure similar to the original community, while no recolonization was evident on the chronically oil-polluted reef. It has been speculated that the chronic oil spills may have damaged the reproductive system of corals, preventing normal settlement and development of coral larva (Rinkevich & Loya, 1977).

3.4. Thermal stress

Thermal stress in the marine environment mainly comes from heated effluent used to cool generators in power plants (Salm & Clark, 1982). The potential threat of elevated temperatures to marine organisms has been reviewed by Johannes (1975) and Grigg & Dollar (1990). The sensitivity of benthic organisms to thermal stress often depends on the temperature range they experience.

Drastic damage to benthic organisms has often resulted from thermal discharge into shallow, inshore waters in the tropics and subtropics, where ambient water temperatures are closer to upper thermal limits of resident organisms (Coles *et al.*, 1976). By contrast, in the temperate regions the effects of thermal discharge seem less critical (Jokiel & Coles, 1974).

Living corals are very sensitive to hyperthermal effects on water temperatures. In tropical waters an increase of 4 °C above the ambient temperature could kill the local fauna. For example, Jokiel & Coles (1974) evaluated the effects of heated effluents at Kahe Point, Oahu, Hawaii, before and after the expansion of Kahe Power Plant. The abundance of dead and damaged corals correlated strongly with their proximity to discharge from the plant and therefore with the level of thermal enrichment. In areas characterized by a temperature increase from 2 to 4 °C, corals lost zooxanthellar pigments and suffered high mortality. Other sublethal temperature effects on corals include depressed feeding responses, reduced reproductive rates, increased zooxanthellae and mucus extrusions and a decrease in the P/R ratio. Nearly all corals were killed when water temperature increased 4 to 5 °C above the ambient temperature. Damage to corals was most severe in late summer, and coincided with annual ambient temperature maxima. After prolonged exposure to temperatures of 30 to 31 °C, corals at Kahe Point, appeared to be bleached. Eventually, most of the coral species died.

4. CASE STUDIES

4.1. Monitoring the changes in a rocky shore community

Distinct zonations of the biota are common on a rocky shore. In an intertidal habitat, occurrences of such zonations could mostly be attributed to physical environmental factors, such as tidal fluctuation and wave exposure (Menge & Farrell, 1989). Nevertheless, in a monitoring study biological interactions within the habitat should not be overlooked because such factors may also contribute much to the community changes, as in an example given by Hawkins & Hartnoll (1983), who monitored the community on a moderately exposed rocky shore on The Isle of Man.

They established three permanent quadrates along the seashore; the 'high' quadrate at the boundary between the brown algae *Fucus spiralis* and *F. vesiculosus* zones, the 'mid' quadrate in the middle of the barnacle *Semibalanus balanoides/Fucus vesiculosus* zone and the 'low' quadrate at the lower limit of the *Semibalanus balanoides* zone. Each quadrate measured 2 m along the shore by 1 m perpendicular to the coast line. These quadrates were visited at 6-8 weekly intervals for 30 months from January 1977 to June 1979. For sampling, each quadrate was divided into eight 0.5 m sided subquadrates which were marked by holes drilled in the rock. In each subquadrate the percentage cover of algae, barnacles and bare rock was estimated by the point frame method. Moreover, water temperature and meteorological data for the study period were analyzed and manipulative field experiments were conducted in order to determine the possible causative factors for observed changes.

Their survey showed that in the 'high' quadrate the balance between the algae *Fucus spiralis* and *F. vesiculosus* changed and an outburst of ephemeral algae occurred following a decline in overall canopy cover. In the 'mid' quadrate, *F. vesiculosus* decreased considerably from its initial cover and disappeared by the end of the study in 1979. In the 'low' quadrate *F. vesiculosus* increased from being absent to comprising 20% of the canopy, while *Semibalanus balanoides* and understory algae decreased in cover. The beadlet anemone *Actinia equina,* the dog whelk *Nucella lapillus* and the common limpet *Patella vulgata* showed marked changes in their numbers at all levels where they occurred.

Results from manipulative experiments showed that most of the above community changes were biologically controlled. For example, the blooms of ephemeral algae and the decline of *Patella* in the high shore quadrate, and the decline of *Actinia* in the mid-shore quadrate, were due to decreases in the fucoid canopy. These canopy changes were, in turn, the result of biological cycles which are characteristic of these communities.

4.2. Monitoring impacts caused by sewage and eutrophication

Some of the most thorough studies of sewage stress on coral reefs were caried out in Kaneohe Bay on the island of Oahu, Hawaii. These studies were very substantial since they were intended to provide information on the environmental effects resulting from sewage discharge and the responses of reef communities following the elimination of the sewage stress.

Kaneohe Bay, stretching 13 km along northeast coast of Oahu, is Hawaii's largest bay. From 1963 to 1977, point source discharges of secondary sewage entered the bay from three outfalls, with a total peak flow rate of 20,000 $m^3.day^{-1}$. In 1978, the sewage was diverted to a new offshore outfall, eliminating all but a small volume of discharge in the northwest lagoon. Monitoring studies had revealed some impacts caused by discharged sewages on the community structure of local sessile organisms. The fixed belt transect sampling method was applied in these studies. In 1971, 15 transect surveys were undertaken (Maragos, 1972). These transects were resurveyed in 1983 (Maragos *et al.*, 1985). At each underwater station, a 25 m long transect line was laid out by scuba divers, from the top to the bottom of the reef slope and perpendicular to the depth contour. A 1 x 1 m quadrate frame, divided with wires into 100 equal grids, was centred over the transect line. Coral species, coralline algae, green algae (*Dictyosphaeria cavernosa)* and other algae, and other bottom characteristics (sand, mud, rubble, dead coral) under each square were identified and their coverages were measured. Results showed that during peak sewage discharge in 1971, essentially no live corals were found in the southern bay. This was possibly a result of the release of hydrogen sulphide caused by anaerobic conditions. In the central and northern sectors of the bay, sewage nutrients supported dense growths of *D. cavernosa,* which, as a consequence, smothered many reef corals and the associated fauna (Maragos, 1972).

In 1983, five years after diversion of the sewage, total live coral coverage had increased dramatically, almost doubling in abundance (Figure 4). Many small coral colonies were present in the southern bay. Notable coral recovery on old dredged surfaces was observed, but deteriorated reef surfaces covered with sediment were still devoid of coral colonization. On the other hand, by 1983 *Dictyosphaeria* had decreased by 75% overall, with the greatest reduction occurring in the areas where the alga had previously exhibited peak abundance. By contrast, dense algal growth remained at the station near the small sewage outfall where sewage continued to be discharged (Maragos *et al*, 1985). Data collected from these studies are so convincing that later they formed the major element in a simulation of the impacts of sewage disposal on coral reefs by Berwick & Faeth (1988).

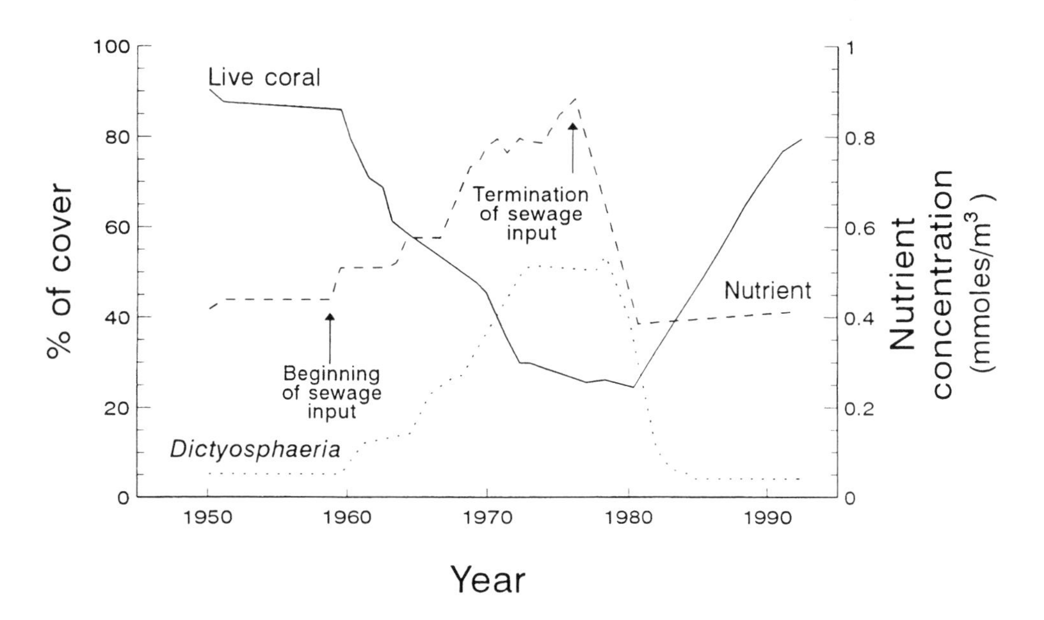

Figure 4. Percent covers of live coral, green algae *Dictyosphaeria,* and nutrient concentration as a function of sewage-induced nutrient loading in Kaneohe Bay, Hawaii (After: Berwick & Faeth, 1988)

4.3. The effects of oil spill on coastal marine communities

Oil spill often causes adverse effects on many local marine organisms. Jackson *et al.* (1989) and Guzman *et al.* (1991) studied such a case on the Panama coast. On 27 April, 1986, more than 8 million litres of crude oil was spilled into a complex region of mangroves, seagrass and coral reefs just east of the Caribbean entrance to the Panama Canal. In their studies, the types and extent of damage to coastal populations and communities in the first 1.5 years after the spill were compared with findings of earlier works.

Impacts of the spill were assessed differently, depending on the types of data available before the oil spill. For example, surveys of reefs for coral cover, abundance, size and diversity were performed using 1 x 1 m quadrates. Six reefs within the study area, including four non-polluted, one moderately polluted and one heavily affected, were surveyed before (in July and October, 1985) and after the spill (in July and August, 1986, and in May and June, 1988). At each reef, four or five transects were extended perpendicular to the shore from the shoreward edge to the bottom of the reef (Guzman *et al.*, 1991). Along each transect, quadrates were placed contiguously or spaced at regular intervals up to 3 m apart, depending on the length of the transect and the profile of the reef. Quadrates were divided by string into one hundred 10 x 10 cm^2 grids. The corals in each grid were identified and their size estimated visually.

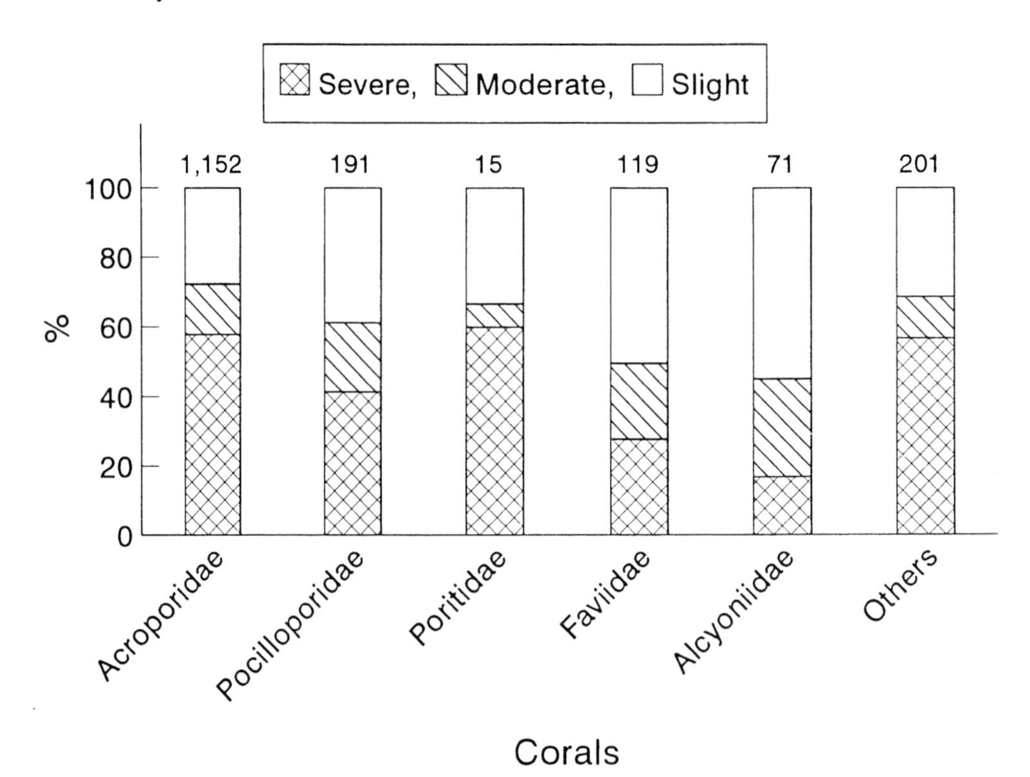

Figure 5. Percentage composition of three levels of coral bleaching caused by thermal discharges from a nuclear power plant, based on data collected at Nanwan Bay, Taiwan on August 17-19, 1988. Five major local coral families, and some others, are included. Where Severe: more than 80% of the colony showing signs of bleaching, Moderate: 20-80% of the colony showing signs of bleaching.
Slight: lower than 20% of the colony showing signs of bleaching; number above each stacked bar denotes sample size (After: Hung, 1989)

Soon after the spill, intertidal mangroves, sea-grasses, algae and associated invertebrates were covered by oil and died. The intertidal sea-grass beds were killed entirely on some heavily oil polluted reef flats but sub-tidal *Thalassia* survived. At the seaside edge of the reef flat, the most common sessile animals before the spill were zoanthids *(Zoanthus sociatus* and *Palythoa* spp.), hydrocorals *(Millepora* spp.) and scleractinian corals *(Porites* spp.). Populations of all these animals were severely reduced and only *Zoanthus* had returned to a typical abundance after 18 months. There was also extensive mortality of shallow, sub-tidal reef corals observed (Guzman *et al.*, 1991). Sublethal effects on corals were substantial. Most of the scleractinian corals which were still alive in depths of less than 3 m showed signs of recent stress including bleaching or swelling of tissues and conspicuous production of mucus. Overall, the spill harmed organisms in all intertidal and sub-tidal environments, infauna and epifauna, and members of all trophic levels including primary producers, herbivores, carnivores and detrivores. Not all organisms had recovered in areas exposed to the open sea.

These studies are a good example of an ecological monitoring programme for two reasons. Firstly, detailed baseline information is available as populations of many organisms in both oiled and unoiled sites had been studied since a previously oil spill in 1968. Thus, the time-series data on the physical environment and biota before and after the oil spill could be compared. Secondly, documentation of the spread of oil and its biological effects started immediately the oil reached the shore. This helped the data interpretation considerably.

4.4. Monitoring of a benthic community for thermal pollution

When a power plant uses seawater as its cooling medium, it is likely to cause an ecological impact on the marine environment. A case of thermal pollution has been studied in southern Taiwan, where commercial operation of a nuclear power plant was commenced in 1984. To assess the ecological impact of the thermal discharge on the marine environment, a succession of local marine macro-organisms has been monitored since 1979, five years before the plant's operation began (Jan & Chang, 1991). The monitoring was done in two ways. Firstly, the living status of coral colonies was monitored within fifty metres around the outfall of the water discharge channel. Secondly, sessile macro-organisms along fixed transects were surveyed seasonally at sub-tidal stations more than fifty metres outside the outfall. In data collection, the macro-organisms were divided into four categories, *i.e.* algae, soft corals, hard corals and 'other invertebrates'.

In the summers of 1987 and 1988 respectively, bleached corals were widely observed around the water outfall (Figure 5) (Hung, 1989). Most corals at depths of 0-3 m were found to be dead after bleaching occurred. However, in the transect study, no consistent change was found in the coverage by each category of organism between data collected before and after the plant coming into operation. Currently, this long-term study is still being carried on and the latest findings, based on time series transect data, have shown that an inconsistency in trends between transects persists (Figure 6). The overall results demonstrate that drastic negative effects, caused by thermal discharge on marine organisms, were evident. However, the observed damage was limited to the area near the cooling water outfall.

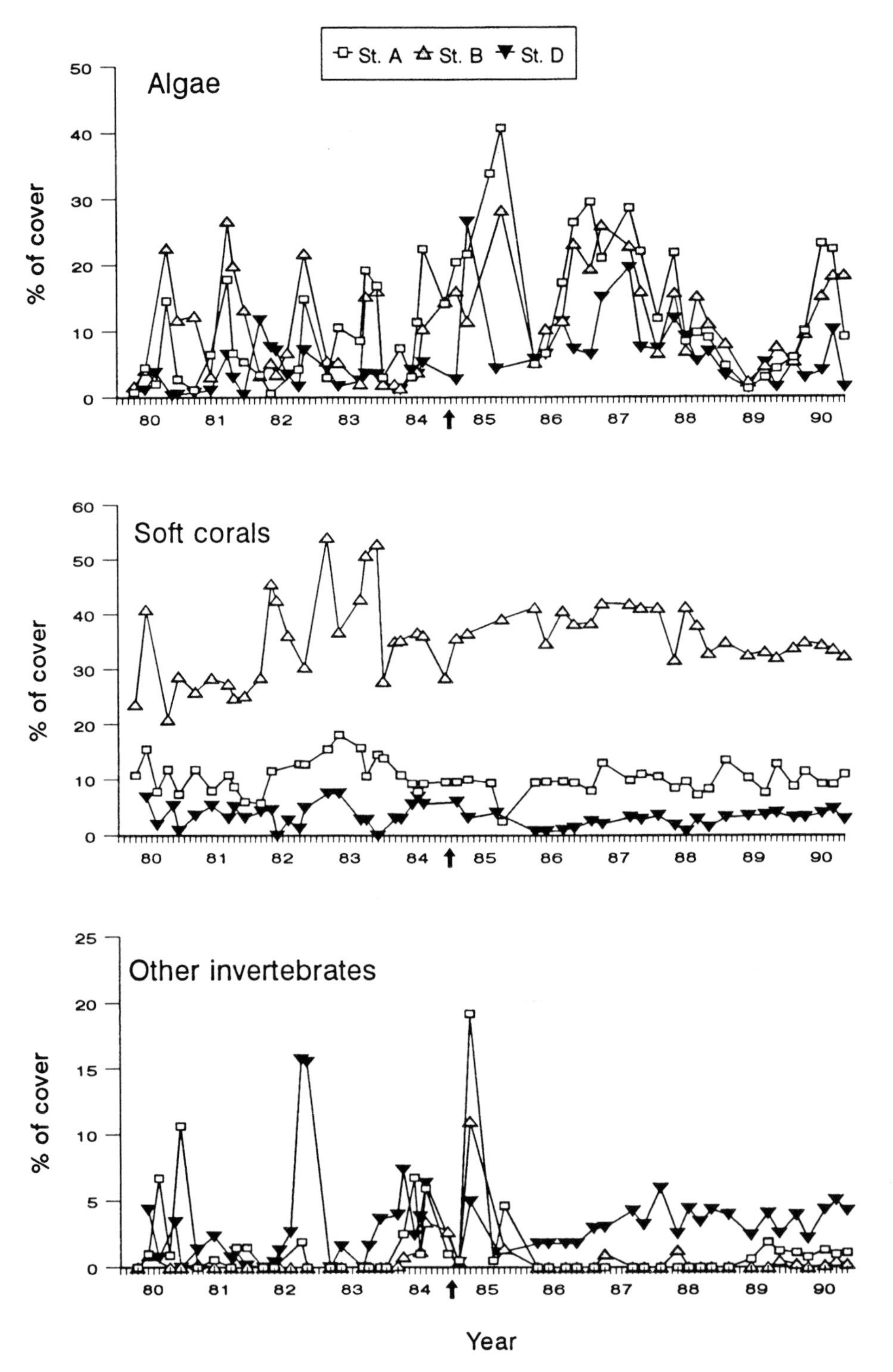

Figure 6. Successions of percentage coverage of some macro-organisms surveyed at three subtidal stations in southern Taiwan during July 1979 - May 1990. Arrow indicates the time the commercial operation of the nuclear power plant was commenced. Station A was assigned to an angular reef located outside the surge zone. The length of the transect was 10.1 m. Station B was located on a wide non-limestone terrace at the reef slope. The length of the transect was 20 m. Station D was composed of six adjacent rocks at depths between 10 and 12 m. The length of the transect was 13.4 m

CONCLUSION

From the various biomonitoring techniques developed for environmental quality assessment, the use of hard substrate community monitoring is recommended in respect of its ability to form an outline of the structure of a natural assemblage of organisms. Community monitoring may function in two ways. With regard to a community study, it is advisable to search for patterns in nature, in order to provide information for further identification of ecological generalization. Such information may, in turn, be used as a baseline in the monitoring of environmental changes. Gross effects caused by the occurrence of high, man-made, environmental stress could be readily examined through changes in community characteristics like species diversity, species dominance, etc. Moreover, time series data may reveal long term effects, if any, of a prolonged stress.

Satisfaction with the application of community monitoring depends mainly on a thorough plan design, careful data collection, proper data analysis and adequate interpretation of results. In reality, the time scale of community monitoring is relatively large when compared of that of toxicological studies, using a single species, in the laboratory. Survey methods are simple, yet their application could be time-consuming and therefore, labour intensive.

Besides, there are limitations. Few variables can be controlled in the field leading to difficulties in isolating the effects of the subject for study from other factors such as site, predation, etc. Thus the identification of causal mechanisms will require data collected from many sites or over a long period at one site. Furthermore, monitoring hard substrate communities is not very useful when it comes to providing information on an early warning of environmental deterioration. An indicator species is better for this purpose. Therefore, it is not suggested that community monitoring be used alone for the purpose of environmental quality assessment.

REFERENCES

Bak, R.P.M. & S.H.B.W. Elgershuizen, 1976. Patterns of oil-sediment rejection in corals. Mar. Biol. 37: 105-113

Begon, M., J.L. Harper & C.R. Townsend, 1986. Ecology, individuals, populations and communities. Blackwell, Oxford, pp. 876

Birkeland, C., 1977. The importance of rate of biomass accumulation in early successional stages of benthic communities to the survival of coral recruits. Proc. 3rd Int. Coral Reef Symp. Vol. 1: 15-21

Berwick, N.L. & P.E. Faeth, 1988. Simulating the impacts of sewage disposal on coral reefs. Proc. 6th Int. Coral Reef Symp. 2: 353-361

Brown, B.E., 1986. Human induced damage to coral reefs. Unesco reports in marine science, no. 40. UNESCO, Paris, pp. 180

Brown, B.E. & L.S. Howard, 1985. Assessing the effects of 'stress' on reef corals. Adv. Mar. Biol. 22: 1-63

Budd, J.T.C., 1991. Remote sensing techniques for monitoring land-cover. In: Monitoring for conservation and ecology; F.B. Goldsmith (ed). Chapman & Hall, London, pp. 33-59

Burd, B.J., A. Nemec & R.O. Brinkhurst, 1990. The development and application of analytical methods in benthic marine infaunal studies. In: Advances in marine biology, Vol. 26; J.H.S. Blaxter & A.J. Southward (eds). Academic Press, London, pp. 169-245

Buss, L.W., 1986. Competition and community organization on hard surface in the sea. In: Community ecology; J. Diamond & T.J. Case (eds). Harper & Row, New York, pp. 517-536

Chang, K.H., C.P. Chen, H.L. Hsieh & K.T. Shao, 1977. An experiment on the evaluation of artificial reefs with invertebrate community. Bull. Inst. Zool., Academia Sinica, 16: 37-48

Clark, R.B., 1986. Marine pollution. Oxford Univ. Press, Oxford, pp. 215

Clarke, R.D., 1988. Chance and order in determining fish-species composition on small coral patches. J. exp. mar. Biol. Ecol. 115: 197-213

Coles, S.L., P.L. Jokiel & C.R. Lewis, 1976. Thermal tolerance in tropical versus subtropical Pacific reef corals. Pacif. Sci. 30: 159-166

Connell, J.H., 1978. Diversity in tropical rain forests and coral reefs. Science, 199: 1302-1310
Connell, J.H. & M.J. Keough, 1985. Disturbance and patch dynamics of subtidal marine animals on hard substrata. In: The ecology of natural disturbance and patch dynamics; S.T.A. Pickett & P.S. White (eds). Academic Press, San Diego, pp. 125-151
Connell, J.H. & R.O. Slayter, 1977. Mechanisms of succession in natural communities and their role in community stability and organization. Am. Nat. 111: 1119-1144
Cortes, J. & M.J. Risk, 1985. A reef under siltation stress: Cahuita, Costa Rica. Bull. Mar. Sci. 36: 339-356
Coulson, M.G., J.T.C. Budd, R.G. Withers & D.N. Nicholls, 1980. Remote sensing and field sampling of mudflat organisms in Langstone and Chichester Harbours, Southern England. In: The shore environment, Vol. 1: Methods; J.H. Price, D.E.G. Irvine & W.F. Farnham (eds). Academic Press, London, pp. 242-263
Cushing, D.H., 1990. Plankton production and year-class strength in fish populations: an update of the match/mismatch hypothesis. In: Advances in marine biology, Vol. 26; J.H.S. Blaxter & A.J. Southward (eds). Academic Press, London, pp. 249-293
Dai, C.F., 1990. Interspecific competition of Taiwanese corals with special reference to interactions between alcyonaceans and scleractinians. Mar. Ecol. Prog. Ser. 60: 291-297
Dai, C.F., 1991. Distribution and adaptive strategies of alcyonacean corals in Nanwan Bay, Taiwan. Hydrobiol. 216: 241-246
Dawson, J., 1980. Superspill, the future of ocean pollution. Jane's, London, pp. 126
Dayton, P.K., 1984. Processes structuring some marine communities: are they general? In: Ecological communities: conceptual issues and the evidence; D.R. Strong, D. Simberloff, L.G. Abele & A.B. Thistle (eds). Princeton Univ. Press, Princeton, New Jersey, pp. 181-197
Digby, P.G.N. & R.A. Kempton, 1987. Multivariate analysis of ecological communities. Chapman & Hall, London, pp. 206
Dollar, S.J. & R.W. Grigg, 1981. Impact of a kaolin clay spill on a coral reef in Hawaii. Mar. Biol. 65: 269-276
Elgershuizen, J.H. & H.A.M. de Kruijf, 1976. Toxicity of crude oils and dispersant to the stony coral *Madracis mirabilis*. Mar. Pollut. Bull. 7: 22-25
Freedman, B., 1989. Environmental ecology. Academic Press, San Diego, pp. 424
Gaines, S.D. & J. Roughgarden, 1985. Larval settlement rate: a leading determinant of structure in an ecological community of marine intertidal zone. Proc. nat. Acad. Sci. U.S.A. 82: 3707-3711
Ganning, B., D.J. Reish & D. Straughan, 1984. Recovery and restoration of rocky shores, sandy beaches, tidal flats, and shallow subtidal bottoms impacted by oil spills. In: Restoration of habitats impacted by oil spills; J.L. Cairns & A.L. Buikema (eds). Butterworth, Boston, pp. 7-36
García, C.B. & H. Salzwedel, 1991. Structure of soft-bottom macrobenthos in shallow areas off the Caribbean coast of Colombia: introducing a new analysis strategy. In: Estuaries and coasts: Spatial and temporal intercomparisons; M. Elliott & J.P. Ducrotoy (eds). Olsen & Olsen, Fredensborg pp. 239-249
George, J.D., 1980. Photography as a marine biological research tool. In: The shore environment, Vol. 1: Methods; J.H. Price, D.E.G. Irvine & W.F. Farnham (eds), Academic Press, London, pp. 45-115
Glynn, P.W., 1983. Extensive 'bleaching' and death of reef corals on the Pacific coast of Panama. Environ. Conserv. 10: 149-154
Grigg, R.W. & S.J. Dollar, 1990. Natural and anthropogenic disturbance on coral reefs. In: Ecosystems of the world, Vol. 25. Coral reefs; Z. Dubinsky (ed). Elsevier, Amsterdam, pp. 439-452
Guzman, H.M., J.B.C. Jackson & E. Weil, 1991. Short-term ecological consequences of a major oil spill on Panamanian subtidal reef corals. Coral Reefs, 10: 1-12
Harding, P. T., 1991. National species distribution surveys. In: Monitoring for conservation and ecology; F.B. Goldsmith (ed). Chapman & Hall, London, pp. 111-188
Hawkins, S.J. & R.G. Hartnoll, 1983. Changes in a rocky shore community: an evaluation of monitoring. Mar. Environ. Res. 9: 131-181
Herricks, E.E. & J. Cairns, 1982. Biological monitoring, Part III. Receiving system methodology based on community structure. In: Biological monitoring in water pollution; J. Cairns (ed). Pergamon, Oxford, pp. 141-153
Holme, N.A., 1984. Photography and television. In: Methods for the study of marine benthos; N.A. Holme & A.D. McIntyre (eds). Blackwell, Oxford, pp. 66-98
Holme, N.A. & R.L. Barret, 1977. A sledge with television and photographic cameras for quantitative investigation of epifauna on the continental shelf. J. mar. biol. Ass. U.K. 57: 391-403
Hopley, D., 1978. Aerial photography and other remote sensing techniques. In: Coral reefs: research methods; D.R. Stoddart & R.E. Johannes (eds). UNESCO, Paris, pp. 23-44

Hubbard, D.K., 1986. Sedimentation as a control of reef development: St. Croix, U.S. Virgin Islands Coral Reefs, 5: 117-125
Hudson, J.H., E.A. Shinn & D.M. Robin, 1982. Effects of offshore drilling on Philippine reef corals. Bull. Mar. Sci. 32: 890-908
Hughes, T.P., 1985. Life history and population dynamics of early successional corals. Proc. 5th Int. Coral Reef Congr. 4: 101-106
Hughes, T.P. & J.H. Connell, 1987. Population dynamics based on size or age? A reef coral analysis. Am. Nat. 129: 818-829
Hung, T.C., 1989. The preliminary report for the assessment of ecological impact of the operation of the third nuclear power plant on the Nanwan Bay. National Scientific Committee on Problems of the Environment, Academia Sinica, Taipei, pp. 34
Jackson, J., J. Cubit, B. Keller, V. Batista, K. Burns, H. Caffey, R. Caldwell, S. Garrity, C. Getter, C. Gonzalez, H. Guzman, H. Kaufmann, A. Knap, S. Levings, M. Marshall, R. Steger, R. Thompson & E. Weil, 1989. Ecological effects of a major oil spill on Panamanian coastal marine communities. Science, 243: 37-44
Jan, R.Q., 1989. Aspects of reproduction ecology of damselfishes (Pomacentridae, Teleostei), with emphasis on substrate utilisation. PhD. thesis, York Univ., York, pp. 244
Jan, R.Q. & K.H. Chang, 1991. A monitoring study of the succession of marine sessile macro-organisms five years before and after the operation of nuclear power plant. In: Bioindicators and environmental management; D.W. Jeffrey & B. Madden (eds). Academic Press, London, pp. 21-35
Johannes, R.E., 1975. Pollution and degradation of coral reef communities. In: Tropical marine pollution; E.J. Ferguson-Wood & R.E. Johannes (eds). Elsevier, Amsterdam, pp. 13-51
John, D.M., D. Lieberman, M. Lieberman & M.D. Swaine, 1980. Strategies of data collection and analysis of subtidal vegetation. In: The shore environment, Vol. 1: Methods; J.H. Price, D.E.G Irvine & W.F. Farnham (eds). Academic Press, London, pp. 265-284
Jokiel, P.L. & S.L. Coles, 1974. Effects of heated effluent on hermatypic corals at Kahe Point, Oahu. Pacif. Sci. 28: 1-18
Jones, W.W., S. Bennell, C. Beveridge, B. McConnell, S. Mack-Smith, J. Mitchell & A. Fletcher, 1980. Methods of data collection and processing in rocky intertidal monitoring. In: The shore environment, Vol. 1: Methods; J.H. Price, D.E.G. Irvine & W.F. Farnham (eds). Academic Press, London, pp. 137-170
Krebs, C.J., 1989. Ecological methodology. Harper & Row, New York, pp. 654
Kuhlmann, D.H.H., 1988. The sensitivity of coral reefs to environmental pollution. Ambio, 17: 13-21
Lassig, B.R., 1983. The effects of a cyclonic storm on coral reef fish assemblages. Env. Biol. Fish. 9: 55-63
Lewis, J.B., 1971. Effects of crude oil and an oil-spill dispersant on reef corals. Mar. Pollut. Bull. 2: 59-62
Lewis, J.R., 1978. The implications of community structure for benthic monitoring studies. Mar. Pollut. Bull. 9: 64-67
Likens, G. E., 1989. Long-term studies in ecology: approaches and alternatives. Springer, Berlin, pp. 214
Loya, Y., 1976. Recolonization of Red Sea corals affected by natural catastrophes and man-made perturbations. Ecology, 57: 278-289
Loya, Y., 1978. Plotless and transect methods. In: Coral reef: research methods; D.R. Stoddart & R.E. Johannes (eds). UNESCO, Paris, pp. 197-218
Loya, Y. & R. Rinkevich, 1979. Abortion effects in coral induced by oil pollution. Mar. Ecol. Prog. Ser. 1: 77-80
Loya, Y. & R. Rinkevich, 1980. Effects of oil pollution on coral reef communities. Mar. Ecol. Prog. Ser. 3: 167-180
Ludwig, J.A & J.F. Reynolds, 1988. Statistical ecology. Wiley, New York, pp. 337
Mapstone, G.M. & A.J. Fowler, 1988. Recruitment and structure of assemblages of fish on coral reefs. TREE, 3: 72-77
Maragos, J.E., 1972. A study of the ecology of Hawaiian reef corals. PhD. thesis, Univ. of Hawaii, Honolulu, pp. 292
Maragos, J.E., C. Evans & P.J. Holthus, 1985. Reef corals in Kaneohe Bay six years before and after termination of sewage discharge. Proc. 5th Int. Coral Reef Congr. 4: 189-194
Matthews, B.A., A.L. Buikema, J. Cairns & J.H. Rodgers, 1982. Biological monitoring, Part IIA - Receiving system functional methods, relationships and indices. In: Biological monitoring in water pollution; J. Cairns (ed). Pergamon, Oxford, pp. 129-140

Menge, B.A. & T.M. Farrell, 1989. Community structure and interaction webs in shallow marine hard-bottom communities: test of an environmental stress model. Adv. Ecol. Res. 19: 189-262
National Research Council, 1985. Oil in the sea, inputs, fates and effects. National Academy Press, Washington, pp. 601
Nichols, D., 1979. Monitoring the marine environment. Symposia of the Institute of Biology, No. 24, Institute of Biology, London, pp. 205
Paine, R.T., 1980. Food webs: linkage, interaction strength and community infrastructure. J. Anim. Ecol. 49: 667-685
Paine, R.T., 1984. Ecological determinism in the competition for space. Ecology, 55: 1339-1357
Parker, R.H., 1975. The study of benthic communities. Elsevier, Amsterdam, pp. 279
Pastorok, R.A. & G.R. Bilyard, 1985. Effects of sewage pollution on coral-reef communities. Mar. Ecol. Prog. Ser. 21: 175-189
Peakall, D., 1992. Animal biomarkers as pollution indicators. Chapman & Hall, London, pp. 291
Pearson, T.H. & R. Rosenberg, 1978. Macrobenthic succession in relation to organic enrichment and pollution of the marine environment. Oceanogr. Mar. Biol. Ann. Rev. 16: 229-311
Peters, E.C., P.A. Meyers, P.P. Yevich & N.J. Blake, 1981. Bioaccumulation and histopathological effects of oil on a stony coral. Mar. Pollut. Bull. 12: 333-339
Pielou, E.C., 1984. The interpretation of ecological data. Wiley, New York, pp. 263
Potts, G.W., J.W. Wood & J.M. Edwards, 1987. Scuba diver-operated low-light-level video system for use in underwater research and survey. J. mar. biol. Ass. U.K. 67: 299-306
Reimer, A.A., 1975. Effects of crude oil on corals. Mar. Pollut. Bull. 6: 39-43
Rinkevich, R. & Y. Loya, 1977. Harmful effects of chronic oil pollution on a Red Sea scleractinian coral population. Proc. 3rd Int. Coral Reef Symp., Miami, 2: 285-291
Rogers, C.S., 1990. Responses of coral reefs and reef organisms to sedimentation. Mar. Ecol. Prog. Ser. 62: 185-202
Sakai, K. & M. Nishihira, 1991. Immediate effect of terrestrial runoff on a coral community near a river mouth in Okinawa. Galaxea, 10: 125-134
Salm, R.V. & J.R. Clark, 1982. Marine and coastal protected areas: A guide for planners and managers. IUCN, Gland, pp. 302
Scheltema, R.S., 1986. On dispersal and planktonic larvae of benthic invertebrates: an eclectic overview and summary of problems. Bull. Mar. Sci. 39: 290-322
Sebens, K.P., 1985. The ecology of the rocky subtidal zone. Amer. Sci. 73: 548-557
Sebens, K.P., 1986. Spatial relationships among encrusting marine organisms in the New England subtidal zone. Ecol. Monogr. 56: 73- 96
Sheppard, C.R.C., 1980. Coral fauna of Diego Garcia lagoon following harbor construction. Mar. Pollut. Bull. 11: 227-230
Shulman, M.J., 1983. Species richness and community predictability in coral reef fish faunas. Ecology, 64: 1308-1311
Smith, R.W., B.B. Bernstein & R.L. Cimberg, 1988. Community-environmental relationships in the benthos: applications of multivariate analytical techniques. In: Marine organisms as indicators; D.F. Soule & G.S. Kleppel (eds). Springer, Berlin, pp. 247-326
Sousa, W.P., 1984. The role of disturbance in natural communities. Ann. Rev. Ecol. Syst. 15: 353-391
Southward, A.J., 1991. Forty years of changes in species composition and population density of barnacles on a rocky shore near Plymouth. J. mar. biol. Ass. U.K. 71: 495-513
Spellerberg, I.F., 1981. Ecological evaluation for conservation. Edward Arnorld, London, pp. 59
Spellerberg, I.F., 1991. Monitoring ecological change, Cambridge Univ. Press, Cambridge, pp. 334
Stephens, J.S., J. Ellen Hose & M.S. Love, 1988. Fish assemblages as indicators of environmental change in nearshore environments. In: Marine organisms as indicators; D.F. Soule & G.S. Kleppel (eds). Springer, Berlin, pp. 92-105
Sutherland, J.P., 1987. Recruitment limitation in a tropical intertidal barnacle: *Tetraclita panamensis* (Pilsbury) on the Pacific coast of Costa Rica. J. exp. mar. Biol. Ecol. 113: 267-282
Tomascik, T. & F. Sander, 1985. Effects of eutrophication on reef corals. I. Growth rate of the reef-building coral *Montastrea annularis*. Mar. Biol. 87: 143-155
Tomascik, T. & F. Sander, 1987a. Effects of eutrophication on reef corals. II. Structure of scleractinian coral communities on fringing reefs, Barbados, West Indies. Mar. Biol. 94: 53-75
Tomascik, T. & F. Sander, 1987b. Effects of eutrophication on reef corals. III. Reproduction of the reef-building coral Porites porites. Mar. Biol. 94: 77-94
Victor, B.C., 1986. Larval settlement and juvenile mortality in a recruitment-limited coral reef population. Ecol. Monogr. 56: 145-160
Warwick, R.M. & K.R. Clarke, 1991. A comparison of some methods for analyzing changes in benthic community structure. J. mar. biol. Ass. U.K. 71: 225-244

Westman, W.E., 1985. Ecology, impact assessment and environmental planning. Wiley, New York, pp. 532

Williamson, M., 1987. Are communities ever stable. In: Colonization, succession and stability; A.J. Gray, M.J. Crawley & P.J. Edwards (eds). Blackwell, Oxford, pp. 353-371

Witman, J.D., 1985. Refuges, biological disturbance, and rocky subtidal community structure in New England. Ecol. Monogr. 55: 421-445

Wittenberg, M. & W. Hunte, 1992. Effects of eutrophication and sedimentation on juvenile corals. I. Abundance, mortality and community structure. Mar. Biol. 112: 131-138

Woodhead, P.M. & M.E. Jacobson, 1985. Biological colonization of a coal-waste artificial reef. In: Wastes in the ocean, Vol. 4: Energy wastes in the ocean; I.W. Duedall, D.R. Kester, P.K. Park & B.H. Ketchum (eds). Wiley, New York, pp. 597-612

Appendix 13

Benthic Sampling Surveys for Hard Substrates

FIXED BELT TRANSECTS

Transect methods have been widely adopted in monitoring programmes of hard substrates (Loya, 1978; John *et al.*, 1980; Jan & Chang, 1991). The simplest transect method involves placing a line randomly along the substratum and recording everything it covers either at a single point, at a particular distance, or the part of the line that covers a particular organism (Figure 1). When supplemented with a pole, the transect can also be transformed into a survey zone (Figure 2). Data collection can also be supplemented by photography. Transects fixed permanently to the substrate provide some advantages over the simple transects mentioned above in terms of data collection. It allows for effortless relocation of the sampling area for subsequent visits and, with a constant sampling area, species performance within it could be consistently assessed. Variation associated with site difference can therefore be avoided.

Such transects are often made by fixing numbered tags or marks to substrates by means of stainless steel nails to which nylon strings can be attached. Alternatively, a semi-fixed transect can be used to alleviate the effects of a fixed string on the biota. This is made by attaching the string (or tape) to the tags when, and only when, data collection is required. In this method, organisms fouling the marks need to be removed and missing marks must be replaced during each survey in order to ensure successful relocation of the transect in the next survey (Jan & Chang, 1991).

FIXED QUADRATES AT SELECTED AREAS - POINT FRAME METHOD

For sessile organisms such as algae, sponges and corals, their coverage in each quadrate provides valuable information on biological succession (Sebens, 1986; Dai, 1991). In its simplest form, percentage coverages is estimated merely by subjective inspection. The data may be acceptable when collected by experienced surveyors; otherwise they can vary alarmingly.

The accuracy of quadrate data collection can be greatly improved by means of a point frame (Figure 3). A point frame is a grid made up of cross-lines. Percentage coverage of an organism can be calculated from the co-incidence of the organism and the intersection of cross-lines. For example, when a 100-point frame is used, one co-incidence represents 1% cover. For convenience, with regard to repetitive surveys, permanent quadrates can be made by marking plot corners. The quadrates can be surveyed repetitively and the study can be supplemented by photography.

Sessile organisms often occur, particularly in the intertidal area, in patches or zones. One may never assume that sessile organisms are randomly placed. Under such circumstances it may be desirable to apply the method of stratified sampling. This method has improved accuracy because each patch or zone has a separate quadrate. If the heterogeneity in density is in question, it can be detected by dividing the site into patches and calculating the coefficient of the variation between them.

COLONIZATION STUDIES

For many years, panels or blocks of cement, slate, earthenware or various plastics have been suspended from rafts or fixed to solid surfaces in order to collect the settling stages of algae and sessile animals as an indication of recolonization (Chang *et al.*, 1977; Woodhead & Jacobson, 1985). Panels of a selected texture can not only be placed at any desired orientation and tidal level, but can also be set out during any season and left on site for a chosen period of time. After recovery, panels may be sampled *in situ* and brought to the laboratory with minimal disturbance caused to the attached biota.

PHOTOGRAPHY AND VIDEO RECORDING

Photography has been applied as a means of data collection. Photographs taken of permanent transects or quadrates provide valuable information for detecting the changes in a community (Holme & Barrett, 1977; George, 1980; Holme, 1984). It can be used to aid site and habitat descriptions, to record the spatial relationship of plants and animals, to measure population numbers and the size of individual organisms, and to record temporal changes in community structure.

Recently, the use of video has also been introduced in underwater survey, particularly benthic surveys (Potts *et al.*, 1987). A transcript can be made during the playback of the tape. Any image of particular interest can be grabbed by graphic software/hardware packages for further analysis by computers.

REMOTE SENSING

Remote sensing techniques range from satellite imagery to aerial photography (Hopley, 1978; Coulson *et al.*, 1980). With remote viewing we can acquire images for selected times of the day or stages in the development of hostile environments. Photo interpretation often needs to be combined with information collected from fieldwork (Budd, 1991). Application of remote sensing can be extended to intertidal studies. For example, by retrospective checking of colonial organisms at a mudflat, Coulson *et al.* (1980) were able to identify fourteen species groups including algae such as *Enteromorpha, Ulva, Fucus,* etc. on infrared, false colour films taken from the air.

Chapter 14

Benthic Biological Pollution Indices in Estuaries

James G. Wilson[1] and **David W. Jeffrey**[2]

[1] Environmental Science Unit, Trinity College, Dublin 2, Ireland
[2] School of Botany, Trinity College, Dublin 2, Ireland

ABSTRACT

The principal qualities of an index are in communication, particularly between scientist and non-specialist, in integration and simplification of a mass of heterogeneous data and in interpretation whereby it is possible to set quality and management goals.

Pollution indices are in widespread use in air pollution and in freshwater, but no one index has been universally accepted for use in estuaries. This situation is partly a reflection of the perceived priorities in pollution research, and partly a consequence of the highly dynamic nature of the estuarine environment itself. However, there have been a number of such indices proposed and they can be divided into two main types: community and system indices and indices of biological community structure.

Of the community and system indices, the National Water Council (NWC) system is based primarily on water column data, the Irish Estuarine Research Programme (IERP) scheme is based on both biological and chemical analyses of the sediment, while the sediment quality triad adds a sediment toxicity assay to biological and chemical sediment analyses. The first two (NWC and IERP) alone provide a numerical system index, while the triad presents the assessment on a site-by-site basis. The latter can also be interpreted visually in a manner akin to the AMOEBA approach, into which arbitrary management goals for any desired parameter can be entered.

Indices of biological community structure are likewise set on a site-by-site basis. Analyses of community structure can be divided roughly into univariate methods, which result in true index values, graphical methods, from which numerical indices can be extracted if desired, and multivariate methods which have excellent powers of integration and to a lesser extent interpretation but not as yet communication. Of these methods, the Shannon-Weiner, or the more recent variations on it, is probably the most useful if only because it is commonly used as the standard to which all others are compared.

Several examples are given to illustrate these methods.

Biomonitoring of Coastal Waters and Estuaries. Edited by Kees J.M. Kramer.

1. INTRODUCTION

In freshwater there are numerous biological pollution indices which receive widespread scientific support and are in widespread use as management tools. In proposing indices for use in estuaries, it is first necessary to explain their properties:

- communication,
- integration, and
- interpretation.

Scientists may mistrust indices, in that the clarity of original data is blurred by the indexing procedure. On the other hand, administrators who require information on estuarine quality cannot be expected to understand data. The first property of an index is that it serves as a mode of communication. The usual communications net includes the following: scientist - engineer - administrator - politician - member of the general public.

Data on estuarine ecology and estuarine environments will inevitably include material from many sources and in many forms. There is thus a scientific need for integration to permit the interfacing of, say, algal records and infauna records. Similarly, comparisons in time and space, for example trends for the whole of Europe and over decades, require integration to allow trends to be seen.

In order to make optimum use of data and information for management, interpretation is essential. It is possible to embody critical settings within indices that permit interpretation. The index itself may serve as a benchmark or target for quality management. These three qualities of communication, integration and interpretation compensate for any sacrifice of scientific data.

The indices described here are generally 'system state' quality yardsticks. They need to be complemented with more dedicated bioindicators for particular conditions as diagnosed through the indexing procedure *e.g.* bioindicators and sensitive species. These latter have been described as 'specific local indicators' and 'general local indicators' (Jeffrey, 1988). Furthermore, it may be desirable to formulate some form of environmental index to complement the interpretative overview of estuarine quality.

We thus assert that the use of biological indexing is part of a hierarchy of management tools, of use because of the robustness and quality of application.

2. THE ESTUARINE ENVIRONMENT

Estuaries are commonly defined with reference to physical and/or chemical characteristics, such as that of Cameron & Pritchard (1963) that:

'an estuary is a semi-enclosed body of water which has a free connection with the open sea and within which sea water is measurably diluted with fresh water derived from land drainage.'

This definition is one on which most field workers would base their activities.

However, it is difficult to find a biological definition of an estuary that would gain such wide acceptance. There have been several systems proposed that classified brackish waters on the basis of critical salinities for particular groups of species, but no one scheme has gained universal acceptance. The Venice system (1959) is probably the most widely used. Carriker (1967) and more recently McLusky (1989) have tried to combine the physical, chemical and biological elements of the system into a coherent whole, and their classification schemes are summarised in Table 1.

The classic work of Remane (1971) showed the paucity of species in brackish water as opposed to both fresh- and seawater. The situation in the Baltic where the mean salinity declines with distance from the North Sea, mirrors what is found in the estuary. The major difference is that the salinity at a given point in the Baltic varies much less, or on a much larger time-scale than in an estuary, where the major variability is around the semi-diurnal tidal cycle. In estuaries, it is the variation itself in salinity (and other physical conditions such as temperature) that can be the major source of stress to the organism. This means that the effects of contaminants must always be viewed against this background of environmental variation.

In the context of potential indices, two important things are evident from Table 1. Firstly, there do seem to be certain types of organisms 'typical' of estuarine conditions. Secondly, these species are relatively few in number, and this in particular has important implications for many of the community-based indices.

There are some species groups, or more accurately species clines whose distribution reflects the estuarine zonation summarised in Table 1. Two such clines are listed in Table 2.

Can we use some of these species as a basis for an index or as bioindicators?

At its most basic, a bioindicator is an organism whose presence can be used, as the name suggests, to indicate certain more-or-less well-defined environmental conditions. It has also taken on the meaning (see *e.g.* Phillips, 1980) of an integrator, over time, of contaminant

Table 1. Physical, chemical (salinity) and biological zones in an estuary (After Remane, 1971; Carriker, 1967; McLusky, 1989)

Estuary division	Substrate	Salinity range (10^{-3})	Zone	Organism type	No. of Species
river	gravels	<0.05	limnetic	freshwater	>50
head	becoming finer	0.5 - 5.0	oligohaline	oligohaline, freshwater migrants	20
upper reaches	mud, currents minimal	5.0 - 18.0	mesohaline	true estuarine, limit of non-transient migrants	12
middle reaches	mud, some sand	18.0 - 25.0	polyhaline	estuarine, euryhaline	20
lower reaches	sand/mud depending on tidal currents	25.0 - 30.0	polyhaline	estuarine, euryhaline, marine migrants	40
mouth	clean sand/rock	30.0 - 35.0	euhaline	stenohaline, all marine	>100

Table 2. Salinity ranges of estuarine Hydrobiidae and Cardiidae. Data from Muus (1963) and Russel & Petersen (1973)

Taxon	Species	Salinity range (10^{-3})
Hydrobiidae	*Potamopyrgus jenkinsi*	0 - 15
	Hydrobia ventrosa	6 - 20
	Hydrobia neglecta	10 - 24
	Hydrobia ulvae	10 - 33
Cardiidae	*Cardium hauniense*	6 - 12
	Cardium glaucum	3 - 25
	Cerastoderma (Cardium) edule	15 - 35
	Cardium exiguum	25 - 35
	Acanthocardia aculeata	?30 - 35

loads on a system. The lack of a precise definition means that the term often is used loosely in any biological monitoring context, and most frequently - from a quick survey of current literature anyway - in articles on pollution. The various interpretations of the term 'bioindicator' and their use in estuaries have been explored by Wilson & Elkaim (1991b) and Wilson (1993), and these will be returned to later.

The key to the meaning is in a bioindicator's capacity for integration. The greater the precision with which we can specify the relationship between the organism and its environment, the better a bioindicator it will be.

There are two questions in particular which are of great importance if we are recommending estuarine organisms as bioindicators.

The first of these is taxonomy. Much confusion has resulted from phenotypic variation. For example Brock (1987) has recently shown that *C. glaucum* and *C. lamarcki*, considered by Russel & Petersen (1973) as conspecific, should be considered as distinct forms within the species complex, albeit as forms between which interbreeding and hybridisation are possible.

The second is the match between organism and environment, and in particular a better understanding of the long-term effects of environmental variables on estuarine species and on their processes of recruitment and distribution. While there is data available on the relationship between salinity and distribution, there is, with the exception of *H. ulvae* and *C. edule*, an almost total lack of information on the effects of pollution on the species listed in Table 2. Consequently they have not figured in any way as bioindicators of system health or in indices of estuarine pollution.

There have been some species such as *Capitella capitata* recommended, at least initially, as pollution indicators. However, it rapidly became evident that they were by no means confined to polluted situations (see *e.g.* Eagle & Rees, 1973) and the species once proposed as indicators have now been incorporated into a much longer list variously described as 'transgressive', 'opportunist' or 'r-strategist' (Pearson & Rosenberg, 1978; Gray, 1979). Pearson & Rosenberg (1978) list some 90 species as belonging to this list, but only a few of these actually belong to a specific organic enrichment category and the great majority respond to any disturbance. Wilson & Elkaim (1991b) have analyzed this list and have shown that estuarine species seem to be quite capable of acting as pollution opportunists should the chance arise, but that the converse (*i.e.* that opportunist species might be found in estuaries) does not hold (Table 3).

Table 3. Number of times selected species have been reported from situations of estuarine and fully marine pollution (From: Pearson & Rosenberg, 1978)

Species	Type	Estuarine pollution	Marine pollution
Corbula gibba	Stenohaline bivalve	1	8
Thyasira flexuosa	Stenohaline bivalve	0	4
Mya arenaria	Euryhaline bivalve	5	7
Macoma balthica	Euryhaline bivalve	7	4
Mytilus edulis	Euryhaline bivalve	3	4
Nereis diversicolor	True estuarine annelid	7	5
Corophium volutator	True estuarine crustacean	3	2

These transgressive species have variously been labelled 'opportunist' or 'pollution indicator' species, although the initial enthusiasm with which they were greeted was quickly dampened by the realisation that they were by no means confined to polluted situations (*e.g.* Eagle & Rees, 1973). Nevertheless, the search for species to act as pollution indicators continues, and a useful contribution to the debate has been made by Ducrotoy *et al.* (1989). They suggested that estuarine organisms could be classified as 'key' species such as *C. edule*, whose populations respond quickly to any environmental perturbation and 'target' species such as *M. balthica* whose populations appear to be much more stable in the long-term and can thus be used for large scale (spatial and temporal) comparisons (Desprez *et al.*, 1986, 1991). While this definition does seem to align 'key' species with opportunists, the concept is broadened to include population structure and inter-annual fluctuations to allow greater discrimination, and it does suggest a way round the difficulty of the comparative paucity of estuarine species in general.

However, although it is not possible to rely solely on the presence/absence of certain species (except perhaps in the broadest possible sense), one can use the distinguishing feature of this group of opportunist species - that is the ability of r-strategists to quickly take advantage of vacant niches and to prosper if only temporarily - in changing community structure as the basis for possible pollution indices.

In addition to the change in community structure induced by opportunist species, there is also a change induced by the different reactions of the indigenous species to change in environmental quality. When challenged by a change in environmental conditions, including a change in pollution status, an organism will attempt to adapt to it. If the adaptation required is beyond its capacity the organism dies (or moves away). As a result, we can assess the health of the system, by some measure of the change in community structure caused by death, lack of recruitment or emigration of individuals.

3. COMMUNITY AND SYSTEM INDICES

A major barrier to the use of conventional pollution indices at the community level is that estuaries are naturally places of few species often present at great densities. Wilson (1988 p 120 *et seq.*) has taken a number of these commonly-used indices and demonstrated that they indicate a 'stressed' community even in an unpolluted situation.

Currently, there are only two indices which have been used in estuaries and one which has been used in near-shore embayments to obtain an overall index of quality at system level. These are the:

- National Water Council U.K. (NWC) index,
- Irish Estuarine Research Programme (IERP) indices and
- the sediment quality triad in the USA.

3.1. National Water Council U.K. index (NWC)

The first of these was developed by the NWC in 1981 in the UK, but it has also been used in Ireland, where similar monitoring approaches have been employed (Portmann & Wood, 1985; WPAC, 1983). The NWC scheme allocated points on the basis of a number of criteria (Table 4), which can then be summed to give an overall score. The scores are then used to allocate the estuary, or a defined stretch of an estuary, to one of four quality classes. Scores of 30 - 24 are indicative of good quality, 23 - 16 of fair quality, 15 - 9 of poor quality and 8 - 0 of bad quality (Portmann & Wood, 1985). The overall status is expressed as the length of each stretch assigned to the different categories, although a fairer picture might be gained by using the volume, since many of the heavily polluted sections are relatively short until they reach the sea.

The NWC index was devised along the lines of freshwater indices and to use data that was being routinely collected. The results to date suggest that it is reasonably accurate, and agrees with other assessment methods in the majority of cases (Table 5). However, it relies heavily on oxygen as a measure of quality, and it does produce a slightly optimistic quality evaluation in cases where only chemical pollution is involved (Wilson & Jeffrey, 1987).

Table 4. Allocation of points under the NWC quality evaluation scheme. (Portmann & Wood, 1985)

Category 1:	**Chemical quality**		**Points**
	dissolved oxygen:	> 60%	10
		40 - 60%	6
		30 - 40%	5
		20 - 30%	4
		10 - 20%	3
		< 10%	0
Category 2:	**Biological quality**		
a)	allows passage of migratory fish		2
b)	resident fish population		2
c)	resident benthos		2
d)	no toxic/tainting substances in biota		4
	(summed total maximum =		10)
Category 3:	**Aesthetic quality**		
a)	no pollutant inputs/no detectable effects		10
b)	some pollution, no serious impairment		6
c)	inputs affect usage		3
d)	inputs cause widespread nuisance		0

3.2. Irish Estuarine Research Programme indices (IERP)

In the IERP system two indices are designed to be used in a complementary manner. To evaluate contaminant load, Jeffrey *et al.* (1985) recommended that a minimum of six mandatory substances (organic matter, N, P, Cd, Cr, Zn) be assessed to cover the three most common categories of pollution (organic matter, nutrients, heavy metals). Other substances were to be added to the list if indicated by preliminary assessment of the catchment. The estuarine *pollution load index* (PLI) was then calculated as the geometric mean of individual site PLI values which in turn were derived as the geometric mean of the contaminant scores at that particular site. To standardise the contaminant scores, the concentrations of each were fitted into a scale based on 'Baseline' (the concentration expected from an unpolluted situation) and 'Threshold' (the concentration at which signs of damage or impairment of the system could be detected) values. Thus a score of 10 (Baseline) indicated no contamination by that substance, a score of 1.0 (Threshold) indicated that concentrations were at a level at which damage might be expected, and scores considerably below 1.0 were associated with damaged or sub-optimal systems (Wilson & Jeffrey, 1987). This system provided a built-in mode of interpreting contaminant values.

The PLI was designed to be used in conjunction with a *biological quality index* (BQI). The estuary is divided into its constituent biosedimentary zones, the number depending on the size and heterogeneity of the system, which are then assigned to one of three categories A (abiotic); B (opportunistic) and C (stable). An abiotic zone is defined as one in which there is no macrobiotic life, an opportunistic zone is one which is dominated by opportunist (*i.e.* r-strategist) species, and a stable zone is one in which there is a normal range of both species and/or year classes - shown diagrammatically in Figure 1 with the SAB (species/abundance/biomass) curves of Pearson & Rosenberg (1978) for comparison. Plants and animals are

evaluated in this classification. Some examples of opportunist animal species have already been given in Table 3, while on the plant side, species such as *Enteromorpha* and *Ulva*, when present in large quantities, would be considered opportunistic.
From this, the proportion of the estuarine area assigned to each quality category can be measured, and the overall estuarine index calculated from the biological quality index formula

$$BQI = \text{antilog}_{10}\,(C - A) \quad \text{or} \quad BQI = 10^{(C - A)} \qquad (1)$$

where A = proportion abiotic and C = proportion stable and (A + B + C) = 1.0. An unpolluted estuary (*i.e.* where C = 1.0) would score 10, and an estuary totally dominated by opportunists (*i.e.* where B = 1.0 and both A and C = 0.0) would score 1.0. To date, only one estuary has scored less than 1.0 (Wilson & Elkaim, 1991a). Figure 2 and section 5.1 show an example of the BQI and PLI based on the Tolka Estuary in Dublin.
There has been generally good agreement between the two elements of the scheme, which has now been used to assess a variety of estuaries in Ireland and in France. These are shown in section 5.2.
A similar approach to the BQI was adopted independently by Leppakowski (1977) to quantify the extent of pollution in sites in the Baltic, in which the community type could be assigned to various categories (see Figure 1c) from which a *benthic pollution index* (BPI) could be calculated. The formula for the BPI was:

$$BPI = 10 - \{5x + 4y + 2(z - a/2)\}/a \qquad (2)$$

where a = whole area, x = area of secondary minimum, y = area of secondary minimum and z = area of the primary maximum (Figure 1).

3.3. Sediment quality triad

The final illustration in Figure 1 is based on the criteria of Long & Chapman (1985) for the sediment infauna component of their sediment quality triad. The criteria for allocation into 'U' (urban, contaminated) locations and 'R' (reference, uncontaminated) categories were based firstly on the dominance of polychaetes and bivalves and secondly on the presence or absence of phoxocephalid amphipods. Those sites at which bivalves and polychaetes represented over 50% of individuals and from which phoxocephalids were absent were classified as 'urban'. Unfortunately, their data set contained relatively few uncontaminated sites, and no intermediate situations, so the full range was not available (Figure 1). Their applications were also confined to almost fully marine embayments, so that the wider extension of the technique into estuaries may not be appropriate - for example the use of phoxocephalids as indicator species (see *e.g.* Table 1). Finally, although this is presented as a system index, the evaluations are still on the basis of individual sites, and an overall numerical value is not generated.
The other elements of their sediment quality triad were a chemical assessment based on the sums of selected heavy metal concentrations and of polychlorinated biphenyls (PCBs) and selected polycyclic aromatic hydrocarbons (PAHs), and a range of sediment toxicity tests. While there was generally good agreement between all three elements of the triad, this predictability could not be maintained on a station-by station basis.
For management purposes therefore, any first approximation or evaluation of pollution status should include both a biological and a chemical component. Some element should also be introduced of a reference state for the system, either by including a toxicity test (Long & Chapman, 1985) or concepts such as 'baseline' and 'threshold' (Jeffrey *et al.*, 1985).
An interesting extension of this is the AMOEBA approach developed by the Rijkswaterstaat in The Netherlands, in which departure from the target or uncontaminated condition for any

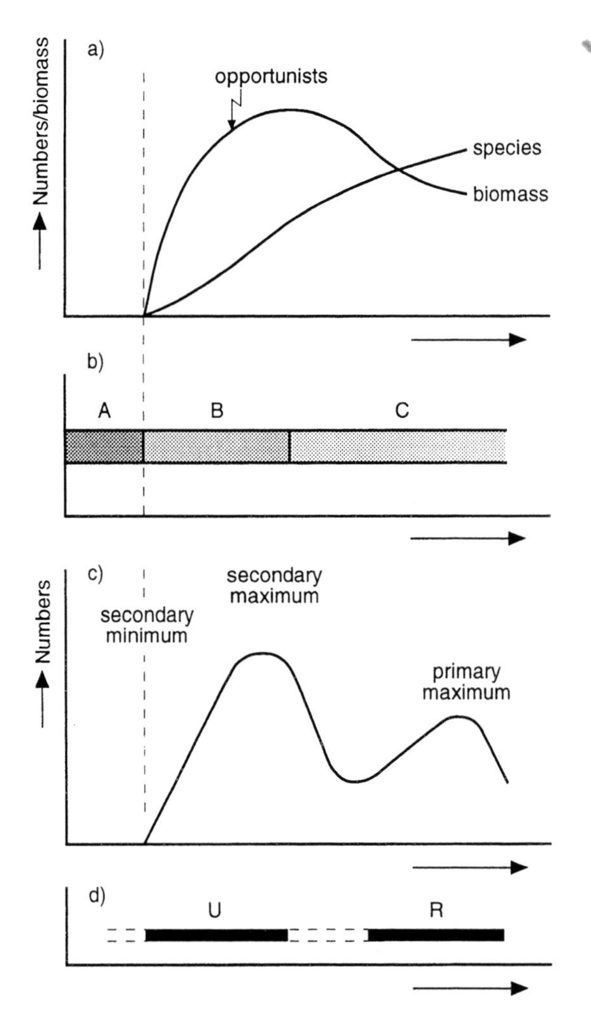

Figure 1. Community response with distance from a pollution source: after a) Pearson & Rosenberg (1978); b) Jeffrey *et al.* (1985); c) Leppakowski (1977); d) Long & Chapman (1985). See also text for explanation of lettering

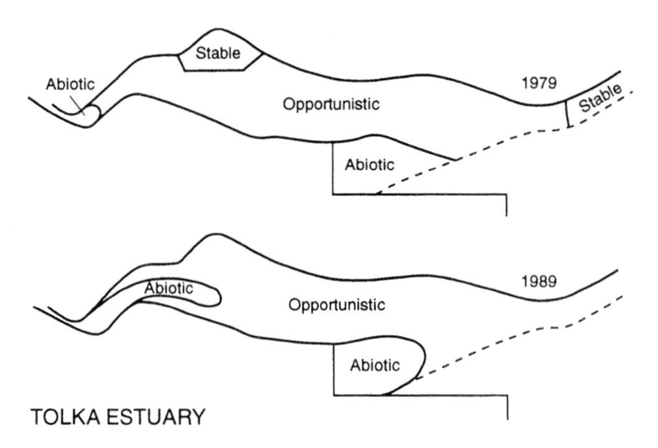

Figure 2. Change in the Tolka Estuary, Dublin, from 1979 to 1989. The designations of the zones A (abiotic), B (opportunistic) and C (stable) are marked along with the site PLI scores at each station. See also text for explanation

number of parameters is presented visually as an aid to effective communication to non-specialists (ten Brink *et al.,* 1991). The AMOEBA approach in fact mirrors a similar sort of presentation of the sediment quality triad (Chapman *et al.*, 1991) in which the three elements of the triad are plotted in triaxial plots as RTR (ratio to reference) values. The RTR values are calculated by dividing the value of the parameter (*e.g.* species richness or total abundance or % polychaeta *etc.* for the benthic infauna) by the average of the parameter at the control stations. The plot thus reflects not only the magnitude of departure from reference, but also the type of perturbation in the system. In the location (Gulf of Mexico) studied by Chapman *et al.* (1991) certain locations demonstrated a chemical enrichment which was not manifest in either benthic composition or toxic effects, reinforcing both the shortcomings of concentration factors as pollution indices (Wilson, 1993) and the need for both biological and chemical elements in a comprehensive system.

4. CHANGES IN COMMUNITY STRUCTURE

Under polluted conditions, the number of individuals of a few species increase dramatically (Figure 1) and the organisation of the community changes. The degree of change can be quantified or assessed though a variety of techniques and a number of these approaches have been reviewed by Gray & Pearson (1982) and more recently by Warwick & Clarke (1991) who divided the techniques into three categories:

- univariate methods, in which the relative abundances of the individuals of the different species are reduced to a single index;
- graphical methods, in which a visual interpretation of the changes can be made;
- multivariate methods of classification and ordination.

4.1. Univariate methods

The most commonly used of the univariate techniques is the Shannon-Weiner Index (H'), for which sufficient experience exists in the freshwater situation that pollution status can be encapsulated in a single Shannon-Weiner value.

In the marine environment, the Shannon-Weiner has also been used with some success to detect changes in pollution status either in space (*e.g.* Gray, 1976) or in time (*e.g.* Elkaim, 1981). However its uncritical use in estuarine situations has been deemed inappropriate due to firstly, the low number of species naturally found in estuaries (since $H_{max} = \log_2 S$ where S = number of species), and secondly the response of the index, which to any environmental stress (including salinity and shore height) mimics the response to pollution (Wilson, 1983).

The problem of low species numbers can be partially circumvented by the calculation of evenness ($J = H'/H_{max}$), as recommended by Gray (1981).

This idea has been further refined by McManus & Pauly (1990) who have adapted the Shannon-Weiner evenness index to Warwick's (1986) abundance/biomass concept (of which more later) to give the *Shannon-Weiner evenness proportion* (SEP). In essence it is a comparison of the evenness of biomass over evenness of numbers with the formula:

$$SEP = H' \text{ biomass}/H' \text{ numbers} \quad (3)$$

This index has been calibrated using Pearson's (1975) data which gave values of around 0.5 from unstressed sites to around 2.5 from stressed sites. Again the concept is based around stress, and does not seem to distinguish between pollution stress and, say, salinity stress. Examples are given in section 5.3.

However, even in marine situations where sufficient numbers of species may be (potentially) found, it seems that while the Shannon-Weiner index or its derivatives may detect changes or

trends in the data, it is not particularly sensitive (Gray & Pearson, 1982) and depends heavily for its interpretation on the reference data to which it is being compared.
The same criticisms can be levelled at any of the diversity- or dominance-based indices, and since the main purpose of an index is to provide an instantly interpretable summary of the data, much of the justification for their use is lost if they have to be hedged around with caveats.

4.2. Graphical methods

Accordingly, several workers have recommended graphical methods which achieve the same goal of rapid communication through visual interpretation.

4.2.1. RSA-plots

Warwick (Warwick, 1986; Warwick & Clarke, 1991; Warwick *et al.,* 1987) has demonstrated the use of k-dominance or ABC (abundance biomass comparison) curves derived from *rank species abundance* (RSA) curves (Shaw *et al.,* 1983), in which cumulative percentage abundance and biomass are plotted against species rank in a number of situations, including estuarine data sets. The degree of disturbance of the community type can be gauged from the degree of curvature of the plot. As a refinement, it appears that in polluted situations, the curve for biomass and the curve for abundance intersect, while in unpolluted situations, the curve for abundance is always above that for biomass.
While these type of graphical methods retained more of the information that a single index figure, they were not necessarily any more sensitive in detecting community disturbance (Warwick & Clarke, 1991). In addition, Wilson (1988, p. 124) has shown that, as with the univariate indices, community response as judged by graphical methods such as RSA curves can behave to any environmental stress, such as that imposed by salinity in the estuary, as they would to pollution stress. See section 5.4 for examples.
Wilson (1988) did conclude that for non-specialists, RSA curves did offer a significant advantage in that the degree of taxonomic expertise, or time spent in identification could be reduced, as only the first few ranked species need be identified and plotted to see the shape of the curves. This is in contrast to the Shannon-Weiner, in which not only does the value depend on the number of species, but also the splitting or lumping of one of the few abundant species - see for example the comments on the taxonomy of *Hydrobia* or *Cardium* above - could drastically change the final H' and J values.
Strictly speaking, the RSA curves are not indices, but numerical values can easily be extracted. Since the curve becomes more pronounced as the degree of stress increases (Shaw *et al.,* 1983), then the degree of curving approximates to an index of pollution. By fitting the curves into a quadratic equation such that:

$$y = b_o + b_1 x + b_2 x^2 \quad (4)$$

where y = percentage and x = rank (see Figure 3), the quadratic coefficient b_2 gives a measure of the departure from linearity (*cf.* Mangum & van Winkle, 1973). In the example above in Figure 3, b_2 values for the Tolka Estuary and Bull Island were identical (1.33), while that for Bannow Bay, the uncontaminated site, was 0.14.

4.2.2. S(N-B)-plots

Other workers have attempted to extract summary values from Warwick's (1986) curves. Beukema (1988) summarised the differences between the numbers curve and the biomass curve to give S(N-B) (sum of numbers minus biomass) values and McManus & Pauly (1990) used a microcomputer programme to calculate essentially the same difference from discrete segment ogives. The latter index, known as difference in area by percent (DAP), gives a range of values from -1 in unstressed conditions to +1 in stressed conditions. Values from -0.3 (unstressed) to +0.2 (stressed) were found using Pearson's 1975 data from Loch Eil and Loch Linne (McManus & Pauly, 1990).

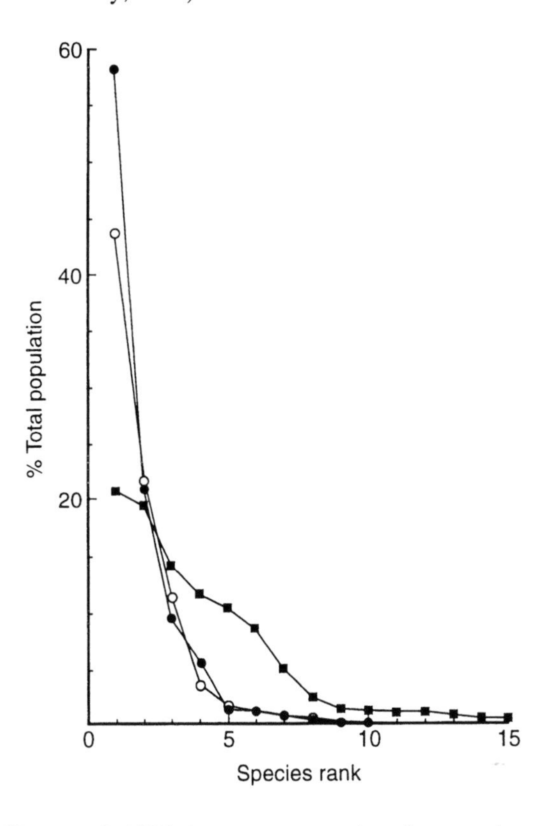

Figure 3. RSA curves (Shaw *et al.,* 1983) from an uncontaminated estuary (squares), a slightly contaminated estuary (filled circles) and a contaminated estuary (open circles). See also text for locations. Data from Magennis (1987) and Wilson (unpublished data)

However, Beukema (1988) has tested Warwick's (1986) curves on an extensive data set from the Wadden Sea intertidal, with not altogether satisfactory results. In particular, Beukema noted that temporal and spatial variability in the S(N-B) were marked to the extent that not only did the magnitude of the difference change, but also the sign (*i.e.* positive or negative). Much of the variability seemed to centre around small mobile species such as *Hydrobia ulvae*, leading Beukema (1988) to conclude that the method appeared useless for assessing pollution status of a benthic community. As *H. ulvae* is one of the more prominent members of estuarine benthic communities, extreme caution may be necessary in employing these methods in estuaries.

4.2.3. Log-normal plots

There are also a number of theoretical models of the distribution of individuals in the community among the constituent species, and a useful summary, with illustrations, of these is given in Magurran (1989). Of these models, the one which has been most keenly advocated for use in detecting pollution in the marine environment is the log-normal (Gray, 1979; Gray & Mirza, 1979; Gray & Pearson, 1982). For these plots, cumulative percentage of species, transformed to probits, is plotted against geometric classes of abundance. Unpolluted situations result in a straight line, under polluted conditions, the plot assumes a 'broken stick' configuration and under gross pollution, the line becomes straight again albeit with a reduced slope (Figure 4).

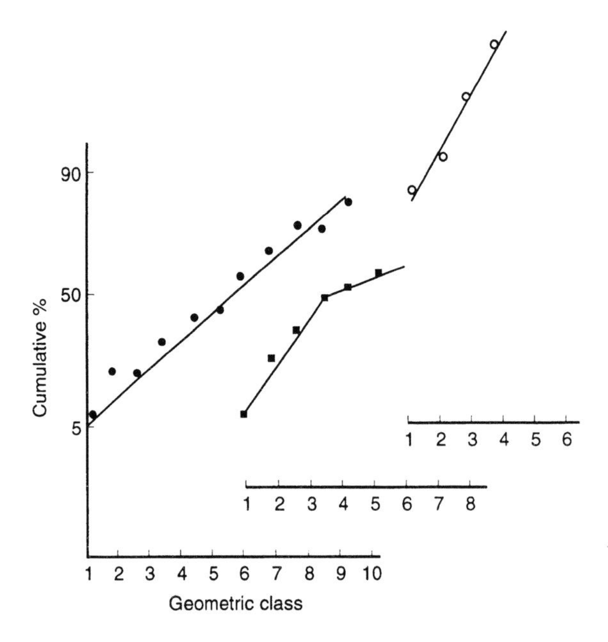

Figure 4. Log-normal plots of an unpolluted site (squares), a slightly polluted site (filled circles) and a polluted site (open circles). See also Figure 3 and text.

There have been occasions where the log-normal has been successfully employed to assess change in pollution status of an estuary (*e.g.* Andrews, 1984), so its use is not automatically precluded. However, it may be wise to test first to which theoretical model (Magurran, 1989) the data conforms before trying to fit it into the plots. Other workers have criticised, on theoretical grounds, the assumption of the log-normal distribution for natural populations and have suggested instead, models such as the neutral model, which makes no underlying assumptions about the data (Platt & Lambshead, 1985). These models are largely untested, and while the neutral model, like all these techniques can detect change or disturbance, neither the degree nor indeed the direction of change are predictable from the model. Accordingly, until compelling evidence is presented to the contrary, we conclude that they offer no advantage over an index such as the Shannon-Weiner, which has at least the merit of being widely-applied and understood.

4.3. Multivariate methods

As with the other approaches we have considered, there are a considerable number of multivariate approaches which have been used to discriminate between samples either in space (for a review and comparison of some current techniques, see Shillabeer, 1991) or in time (*e.g.* Souprayen *et al.,* 1991). In their comparison of univariate, graphical and multivariate analyses, Warwick & Clarke (1991) concluded that the multivariate technique, which in this case was *multi-dimensional scaling* (MDS), offered several advantages over the other two kinds. Firstly, MDS discriminated between groups of sites which neither the Shannon-Weiner nor ABC curves could separate; secondly, the same pattern of response was obtained from different groups - in other words, the classification was independent of what kind of data was used; and thirdly, environmental (or pollution) variables could be encoded and aligned alongside biological variables. In this way it may be possible to refine selection of indicator species, such that not only could the response of selected species be correlated with degree of contamination but also different species might be distinguished to differentiate between environmental and pollution stress or even the different types of pollution.

By their nature, multivariate analyses lend themselves to statistical techniques, such that levels of significance for differences may be easily calculated. If a satisfactory unpolluted or baseline data set can be included, then significantly polluted sites, defined as those statistically different from the control, can be identified. Presented graphically, in the same manner as ABC or RSA curves as above, departure of conditions from baseline or uncontaminated can be relatively easily visualised, but presentation of the same departure as a single eigen vector value would mean little without a frame of reference. Thus, multivariate analyses do not yet provide indices of pollution status.

As we have seen, many field indices both lack discrimination and have difficulties separating natural perturbations or environmental stresses from pollution. For that reason, many workers have concentrated on laboratory-based or derived methods which of necessity, work at the level of the individual or below. A number of these have been compared amongst themselves and also with field techniques such as the Shannon-Weiner for the Bull Island, Dublin Bay (Wilson & Elkaim, 1991b). Examples of some of these techniques can be found elsewhere in this volume, and a useful comparative exercise has also been carried out in Oslofjord (Bayne *et al.*, 1988).

5. CASE STUDIES

The following case studies are taken from investigations based mostly in and around the Tolka Estuary in Dublin Bay, Ireland, which was chosen partly because it was an area with which the authors were familiar and partly because it offered a true estuarine situation in which a pollution gradient ran alongside the estuarine gradient. For more detailed information on this location the reader is invited to follow up the references cited, and for full details on the procedures or indices, the primary sources cited in the preceding sections should be consulted.

5.1. PLI and BQI based on the Tolka Estuary

Figure 2 shows an example of the BQI based on the Tolka Estuary in Dublin, and that of a repeat survey some ten years later. PLI values ranged from 4.3×10^{-3} at a site at the head of the estuary (biological classification: abiotic) to 1.69 at a site near the mouth of the estuary (biological classification: stable). The overall estuary PLI was 0.12.

The repeat survey showed an increase in the proportion of the areas designated abiotic (from 0.14 to 0.32) and a decline in the proportions designated opportunistic (from 0.76 to 0.68) and stable (from 0.10 to 0.0). Consequently the BQI declined from 0.93 to 0.47.

5.2. Comparison of the PLI and BQI indices and the NWC designation

A comparison of the BQI and the PLI for a number of Irish and French estuaries and the corresponding NWC classifications for the Irish estuaries is shown in Table 5.

The trend in all three indices presented in Table 5 is broadly similar, with discrepancies between the NWC index and the BQI and PLI such as that noted already for the Avoca Estuary, although on direct comparison, the PLI is consistently less optimistic than the BQI. See also text for discussion.

Table 5. PLI and BQI indices for Irish and French estuaries, along with NWC designation for the Irish estuaries. See also text

Estuary	PLI	BQI	NWC
Irish:			
Boyne	2.33	8.81	fair
Rogerstown	1.67	3.38	n.a.
Tolka	0.12	1.83[3)]	poor
Liffey	0.01	0.68	poor[1)]/good[2)]
Avoca	0.00	0.01	poor
Slaney	1.74	6.88	good
Lee	1.23	8.71	n.a.
Shannon	2.56	9.42	fair[1)]/good[2)]
French:			
Somme	3.37	5.29	n.a.
Seine	0.84	3.87	n.a.
Baie de Veys	7.37	7.65	n.a.
Loire	0.31	4.66	n.a.
Gironde	0.61	6.44	n.a.

[1)] inner estuary; [2)] outer estuary; [3)] but see also Figure 2 and text for inner estuary survey; n.a.: not analyzed

5.3. Comparison of three Irish estuaries by univariate methods

Three estuarine situations in Ireland - one polluted, one slightly polluted and one unpolluted - have been compared using the Shannon-Weiner indices and the SEP (Table 6). The polluted site is from the Tolka Estuary in Dublin Bay (see the BQI and PLI example above), the slightly polluted site is from Bull Island, Dublin Bay (Magennis, 1987; Brennan, 1988), while the unpolluted site is from Bannow Bay on the south-east coast of Ireland (Magennis, 1987).

Table 6. Number of species, Shannon-Weiner indices (H') for numbers and biomass, and evenness proportion (SEP) for sites in the Tolka Estuary, Bull Island and Bannow Estuary. See also text for explanation. Data from Wilson (1983) and Magennis (1987)

Estuary	No. spp.	H' biomass	H' numbers	SEP
Tolka	7	1.80	2.31	0.78
Bull Island	26	2.26	1.33	1.70
Bannow	25	0.94	2.86	0.33

The SEP for the unpolluted site (Bannow) is remarkably low. This was due to the presence of *Scrobicularia plana* which accounted for just under 12% of the individuals but almost 73% of the biomass. This dominance is reflected in the very low H' biomass value, while the H' numbers (of individuals) indicated a much more even spread among the species. While reasonable numbers of species were found both at Bull Island and in Bannow Bay, few were present in the Tolka Estuary. In fact the chosen site (number 175, Wilson (1983)) was one of the most diverse in the location, and only one species (*Nereis diversicolor*) was reported from other sites within a few hundred metres of that shown in Table 6.

5.4. Use of RSA-plots from three estuaries

Figure 3 shows an example of rank species abundance (RSA) plots from a polluted estuarine location a slightly polluted estuarine location and an unpolluted estuarine location, using the same three locations as in Table 6 above. The curve from the unpolluted site, (Bannow Bay) while clearly distinct from either of the contaminated sites, still displays a profile more characteristic of 'stressed' than 'unstressed' communities (Shaw *et al.,* 1983). It is also impossible to distinguish between the two contaminated sites (Bull Island and Tolka Estuary), although these two areas have been clearly separated by other techniques (Jeffrey *et al.,* 1978; Wilson, 1983; Jeffrey *et al.,* 1991).

5.5. Use of a log-normal plot for three estuaries

Figure 4 plots the same three sites as Figure 3 and Table 6, and it can be seen that, unlike the RSA plot, the three locations can be clearly distinguished. However, the shape of the plots do not conform to the expected. The unpolluted site (Bannow Bay) has the 'broken stick' configuration, while it is the polluted site (Tolka Estuary) that assumes the straight line with steeper slope. It has been shown, for example, that the community reaction to metal contamination, and the subsequent log-normal plots, differs from that to organic pollution (Rygg, 1986). Under metal contamination, all species are more or less equally affected, while under organic pollution, some species may greatly increase in numbers (see Figure 1). In Figure 4 we may be seeing a situation in which the effect of the salinity and of the different kinds of contamination present at the two polluted sites are interacting.

CONCLUSION

Evaluation of system health is dependent on either a spatial or temporal comparison with an identical situation either remote from or before contamination. The value of such an assessment is lessened as the two situations diverge. For this reason, many current indices are laboratory-based where conditions can be controlled, while field approaches lack discriminatory power. Field biomonitoring should concentrate on species such as *M. balthica*, which some workers have designated as a 'target' species, whose natural long-term variation is slight, allowing a high signal/noise ratio for the detection of pollution (Desprez *et al.,* 1986).

Any of the indices discussed in this chapter can be used to detect temporal changes in community structure. However, the implications of such changes in terms of changes in biological 'quality' or pollution status are by no means fully understood. Extreme caution should be employed when making spatial comparisons, and for this reason, a broader approach such as the system index (BQI) of Jeffrey *et al.* (1985) is recommended.

REFERENCES

Andrews, M.J. 1984. Thames estuary: pollution and recovery. In: Effects of pollution at the ecosystem level. SCOPE 22; D.R. Miller, G.C. Butler & P. Bourdeau (eds). Wiley, Chichester, pp. 195-227

Bayne, B.L., K.R. Clarke & J.S. Gray, 1988. Biological Effects of pollutants. Results of a practical workshop. Mar. Ecol. Progr. Ser. 6: 1-278

Beukema, J.J., 1988. An evaluation of the ABC-method (abundance/biomass comparison) as applied to macrozoobenthic communities living on tidal flats in the Dutch Wadden Sea. Mar. Biol. 99: 425-433

Brennan, B.M., 1988. The geochemistry and bioavailability of heavy metals in coastal inshore sediments. PhD. thesis, University of Dublin, pp. 378

Brock, V. 1987 Genetic relations between the bivalves *Cardium (Cerastoderma) edule, C. lamarcki* and *C. glaucum* studied by means of crossed electrophoresis. Mar. Biol. 93: 493-498

Bryan, G.W., W.J. Langston & L.G. Hummerstone, 1980. The use of biological indicators of heavy metal contamination of estuaries. Mar. biol. Ass. U.K. Occas. Publ. 1: 1-73

Bryan, G.W., W.J. Langston & L.G. Hummerstone, 1985. A guide to the assessment of heavy metal contamination in estuaries using biological indicators. Mar. biol. Ass. U.K. Occas. Publ. 4: 1-92

Cameron, W.M. & D.W. Pritchard, 1963. Estuaries. In: The Sea, Vol. 2; M.N. Hill (ed). Wiley, New York, pp. 306-324

Carriker, M.R., 1967. Ecology of estuarine benthic invertebrates: a perspective. In: Estuaries; G.H. Lauff (ed). Am. Ass. Adv. Sci. 83: 442-487

Chapman, P.M., Power, E.A., Dexter, R.N. & Andersen, H.B., 1991. Evaluation of effects associated with an oil platform, using the sediment quality triad. Environ. Toxicol. Chem. 10: 407-424

Desprez, M., G. Bachelet, J.J. Beukema, J.-P. Ducrotoy, K. Essink, J. Marchand, H. Michaelis, B. Robineau & J.G. Wilson, 1991. Dynamique des populations de *Macoma balthica* dans les estuaires du Nord-Ouest de l'Europe. In: Estuaries and coasts: spatial and temporal intercomparisons; J.-P. Ducrotoy & M. Elliott (eds). Olssen and Olssen, Fredensborg, Denmark, pp. 159-165

Desprez, M., J.-P. Ducrotoy & B. Sylvand, 1986. Fluctuations naturelles et evolution artificielle des biocoenoses macrozoobenthiques intertidales des trois estuaires des côtes françaises de la Manche. Hydrobiol. 142: 249-270

Ducrotoy, J.-P., M. Desprez, B. Sylvand & B. Elkaim, 1989. General methods of study of macrotidal estuaries: the biosedimentary approach. In: Developments in estuarine and coastal study techniques; J. McManus & M. Elliott (eds). Olssen and Olssen, Fredensborg, Denmark, pp. 41-52

Eagle, R.A. & E.I.S. Rees, 1973. Indicator species - a case for caution. Mar. Pollut. Bull. 4: 25

Elkaim, B., 1981. Effets de la marée noire de l'Amoco Cadiz sur le peuplement subtidal de l'estuaire de la Penzée. Actes du Colloque Internationale CNEXO, novembre 1979. Publications Scientifiques et Techniques CNEXO, Brest, pp. 527-539

Gray, J.S., 1976. The fauna of the polluted River Tees estuary. Estuar. Coast. Mar. Sci. 4: 653-676

Gray, J.S., 1979. Pollution-induced changes in populations. Phil. Trans. Roy. Soc. (London), 286: 545-561

Gray, J.S., 1981. The ecology of marine sediments. Cambridge Univ. Press, Cambridge, pp. 185

Gray, J.S. & F.B. Mirza, 1979. A possible method for detecting pollution-induced-disturbance on marine benthic communities. Mar. Pollut. Bull. 10: 142-146

Gray, J.S. & T. Pearson, 1982. Objective selection of sensitive species indicative of pollution-induced change in benthic communities. 1. Comparative methodology. Mar. Ecol. Progr. Ser. 9: 111-119

Jeffrey, D.W., 1988. Biomonitoring of freshwaters, estuaries and shallow seas: a commentary on requirements for environmental quality control. In: International Symposium on Biomonitoring the State of the Environment; M. Yasuno & B. Whitton (eds). Tokai Univ. Press, Tokyo, pp. 75-90

Jeffrey, D.W., J.G. Wilson, C.R. Harris & D.L. Tomlinson, 1985. The application of two simple indices to Irish estuary pollution status. In: Estuarine management and quality assessment; J.G. Wilson & W. Halcrow (eds). Plenum Press, London, pp. 147-165

Jeffrey, D.W., B. Madden, B. Rafferty, R. Dwyer & J.G. Wilson, 1991. Indicator organisms as a guide to estuarine management. In: Bioindicators and environmental management; D.W. Jeffrey & B.M. Madden (eds). Academic Press, London, pp. 55-64

Jeffrey, D.W., P. Pitkin & B. West, 1978. Intertidal environment of northern Dublin Bay. Est. Coast. Mar. Sci. 7: 163-171

Leppakowski, E., 1977. Monitoring the benthic environment of organically polluted river mouths. In: Biological monitoring of inland fisheries; J.S. Alabaster (ed). Applied Science Publ., Barking, Surrey, pp. 125-132

Long, E.R. & P.M. Chapman, 1985. A sediment quality triad: measures of sediment contamination, toxicity and infaunal community composition in Puget Sound. Mar. Pollut. Bull. 16: 405-415

McLusky, D.S., 1989. The estuarine ecosystem, 2nd Ed. Chapman and Hall, London, pp. 215

McManus, J.W. & D. Pauly, 1990. Measuring ecological stress: variations on a theme by R.M. Warwick. Mar. Biol. 106: 305-308

Magennis, B.M., 1987. The assessment of pollution at the Bull Island, Dublin. PhD. Thesis, University of Dublin, Dublin, pp. 252

Magurran, A.E., 1989. Ecological diversity and its measurement. Croom Helm, London, pp. 185

Mangum, C.P. & W. van Winkle, 1973. Responses of aquatic invertebrates to declining oxygen conditions. Am. Zool. 13: 529-541

Muus, B.J., 1963. Some Danish Hydrobiidae with the description of a new species. Proc. Malacol. Soc. London, 35: 131-138

Pearson, T.H., 1975. The benthic ecology of Loch Linnhe and Loch Eil a sea loch system on the west coast of Scotland. IV. Changes in the benthic fauna attributable to organic enrichment. J. exp. mar. Biol. Ecol. 20: 1-41

Pearson, T.H. & R. Rosenberg, 1978. Macrobenthic succession in relation to organic enrichment and pollution of the marine environment. Oceanogr. Mar. Biol. Ann. Rev. 4: 481-520

Phillips, D.J.H., 1980. Quantitative aquatic biological indicators. Applied Science Publ. Barking, Essex, pp. 488

Platt, H.M. & P.J.D. Lambshead, 1985. Neutral model analysis of patterns of marine benthic species diversity. Mar. Ecol. Prog. Ser. 24: 75-81

Portmann, J.E. & P.C. Wood, 1985. The UK national estuarine classification scheme and its application. In: Estuarine management and quality assessment; J.G. Wilson & W. Halcrow (eds). Plenum Press, London, pp. 173-186

Remane, A., 1971. Ecology of brackish water. In: Die Binnengewässer, Band 25; A. Remane & C. Schlieper (eds). Wiley Interscience, New York, pp. 348

Russel, P.J. & G.H. Petersen, 1973. The use of ecological data in the elucidation of some shallow water European *Cardium* species. Malacologia, 14: 223-232

Rygg, B., 1986. Heavy-metal pollution and log-normal distribution of individuals among species in benthic communities. Mar. Pollut. Bull. 17: 31-36

Shaw, K.M., P.J.D. Lambshead & H.M. Platt, 1983. Detection of pollution-induced disturbance in marine benthic assemblages with special reference to nematodes. Mar. Ecol. Prog. Ser. 11: 195-202

Shillabeer, N., 1991. Analysis of a standard set of benthic fauna data. A comparison of the computer based techniques used by different laboratories. In: Space and time series data analysis in coastal benthic ecology; B.F. Keegan (ed). CEC, Brussels, pp. 501-517

Souprayen, J., K. Essink, F. Ibanez, J.J. Beukema, D. Michaelis, J.-P. Ducrotoy, M. Desprez & D.S. McLusky, 1991. Numerical analysis of long-term trends of west-European intertidal sedimentary macrozoobenthic communities. In: Space and time series data analysis in coastal benthic ecology. B.F. Keegan (ed). CEC, Brussels, pp. 65-235.

ten Brink, B.J.E., S.H. Hosper & F. Colijn, 1991. A quantitative method for the description and assessment of ecosystems: the AMOEBA approach. Mar. Pollut. Bull. 23: 265-270

Venice System, 1959. Symposium on the classification of brackish waters, Venice, April 8-14, 1958. Arch. Limnol. Oceanogr. (suppl.), 11: 1-248

Warwick, R.M., 1986. A new method for detecting pollution effects on marine benthic communities. Mar. Biol. 92: 557-562

Warwick, R.M. & K.R. Clarke, 1991. A comparison of some methods for analysing changes in benthic community structure. J. mar. biol. Ass. U.K. 71: 225-244

Warwick, R.M., T.H. Pearson & T. Ruswahyuni, 1987. Detection of pollution effects on marine macrobenthos: further evaluation of the species abundance/biomass method. Mar. Biol. 95: 193-200

WPAC, 1983. A review of water pollution in Ireland. Water Pollution Advisory Council, Department of the Environment, Dublin, pp. 243

Wilson, J.G., 1983. The littoral fauna of Dublin Bay. Ir. Fish. Invest. Ser. B 26: 1-20

Wilson, J.G., 1988. Biology of estuarine management. Croom Helm, London, pp. 197

Wilson, J.G., 1993. The role of bioindicators in estuarine management. Estuaries (in press).

Wilson, J.G. & B. Elkaim, 1991a. A comparison of the pollution status of twelve Irish and French estuaries. In: Estuaries and coasts: spatial and temporal intercomparisons. J.-P. Ducrotoy & M. Elliott (eds). Olssen and Olssen, Fredensborg, Denmark, pp. 317-322

Wilson, J.G. & B. Elkaim, 1991b. The toxicity of freshwater: estuarine bioindicators. In: Bioindicators and environmental management; D.W. Jeffrey & B. Madden (eds); Academic Press, London, pp. 311-322

Wilson, J.G. & D.W. Jeffrey, 1987. Europe wide indices for monitoring estuarine pollution. In: Biological indicators of pollution; D.H.S. Richardson (ed). Royal Irish Academy, Dublin, pp. 225-242

Species Index

Subject Index